AF531088

GENERAL HISTORIES

Aviation in the U.S. Army, 1919–1939

Maurer Maurer
United States Air Force Historical Research Center

OFFICE OF AIR FORCE HISTORY
UNITED STATES AIR FORCE
WASHINGTON, D.C., 1987

Library of Congress Cataloging-in-Publication Data

Maurer, Maurer.
Aviation in the U. S. Army, 1919-1939.

(General histories)
Bibliography: p. 587.
Includes index.
1. Aeronautics, Military–United States–History. 2. United States. Army–Aviation–History. I. Title. II. Title: Aviation in the U. S. Army, 1919-1939. III. Series.
UG633.M323 1987 358.4'00973 87-12257
ISBN 0-912799-38-2

For sale by the Superintendent of Documents, U.S. Government Printing Office
Washington, D.C. 20402

United States Air Force
Historical Advisory Committee

(As of December 31, 1986)

Ms. Kathleen A. Buck
The General Counsel, USAF

Mr. DeWitt S. Copp
United States Information Agency

Dr. Norman A. Graebner
University of Virginia

Dr. Warren W. Hassler, Jr.
Pennsylvania State University

Brig. Gen. Harris B. Hull,
USAF, Retired

Dr. Haskell M. Monroe, Jr.
(*Chairman*)
University of Texas at El Paso

Dr. John H. Morrow, Jr.
University of Tennessee at
Knoxville

Gen. Thomas M. Ryan, Jr.,
USAF, Retired

Lt. Gen. Winfield W. Scott, Jr.,
USAF
Superintendent, USAF Academy

Lt. Gen. Truman C. Spangrud,
USAF
Commander, Air University

The Author

MAURER MAURER received his B.S. from Miami University, Ohio, and his M.A. and Ph.D. from The Ohio State University. Before joining the U.S. Air Force Historical Program in 1950, he taught in Ohio public schools and served as a musician in the U.S. Navy. While employed in the historical office of the Air Materiel Command from 1950 to 1955, he taught during his off-duty hours at Wittenberg College in Ohio. From 1955 until his retirement in 1983, Dr. Maurer worked in the United States Air Force Historical Research Center, Maxwell Air Force Base, Alabama. There he produced numerous monographs, special studies, and articles on military and aviation history. Among his books are the four-volume *The U.S. Air Service in World War I*; *Air Force Combat Units of World War II*; and *Combat Squadrons of the Air Force, World War II*. Dr. Maurer's published works also include a number of papers on music in eighteenth-century Colonial America.

Foreword

Historians generally agree that the birth of American air power occurred in the two decades between the world wars, when airmen in the U.S. Army and Navy forged the aircraft, the organization, the cadre of leadership, and the doctrines that formed a foundation for the country to win the air war in World War II. Nearly every scholarly study of this era focuses on these developments, or upon the aircraft of the period; very few works describe precisely what the flyers were doing and how they overcame the difficulties they faced in creating air forces. In this detailed, comprehensive volume, Dr. Maurer Maurer, retired senior historian of the United States Air Force Historical Research Center, fills this void for land-based aviation.

As Dr. Maurer explains in his personal note, this book grew out of his previous editing of the documents of the American Air Service in World War I. He decided to write a descriptive rather than an analytical book, taking the vantage point of the Army flyers themselves. While policy, organization, and doctrine form the background, they are not addressed or explained explicitly. Instead, Dr. Maurer focuses on men and planes, describing in the process how the Army Air Corps came to possess a supporting structure and the nationwide network of airfields. He exposes the difficulties encountered in training and organizing tactical units. However, Dr. Maurer does not write solely about problems and setbacks. In his capable narrative hands, readers cross the country and the continents on the many dramatic record flights with the flyers of the Army Air Corps.

The value of this book is twofold: the wealth of detail Dr. Maurer provides about the scope, structure, and activities of interwar Army aviation; and the comprehensive portrait that emerges of a military service struggling with limited resources to develop a new weapon of tremendous destructive potential. As such, the book fills a gap in the literature and contributes to knowledge about the history of the Army air arm.

The Office of Air Force History wishes to express its special appreciation to General Bryce Poe II, USAF (Ret.), Major General Haywood S. Hansell, Jr., USAF (Ret.), Dr. Robert Coakley, retired Deputy Chief Historian of the U.S. Army, and Dr. Edgar Raines, historian at the U.S. Army Center of Military History, all of whom served, along with historians from this office, as members of the final panel on the manuscript. The panel met at Bolling Air Force Base in November 1981, reviewed the volume, and recommended its publication.

RICHARD H. KOHN
Chief, Office of Air Force History

A Personal Note

This book, a product of the U.S. Air Force historical program, has grown out of a long-time personal interest in the early history of the United States Air Force. After editing four volumes of documents on *The U.S. Air Service in World War I* (Washington: Office of Air Force History, 1978–1979), I set out to learn more about Army aviation between World Wars I and II. Books provided broad treatment of the subject; monographs and articles covered many topics. But they did not tell all I wanted to know about organization, manning, equipment, training, and operations, and especially about what the airmen were doing and what Army flying was like in the twenties and thirties. Many writings, cast in the same mold as the histories one finds of armies without soldiers, told of air forces that never got off the ground. I resolved to keep an eye on the sky.

Historical records and published works in the USAF Historical Research Center and the Air University Library at Maxwell Air Force Base, Alabama, enabled me to relive the past and fly with the Army. Once more the bark of a motor overhead sent me (and every other boy in town) running out to see the flying machine. Begging some money, I paid to go up and fly once around the cow pasture. I soloed at Carlstrom Field, flew to Alaska with "Bill" Streett, crossed the continent with Kelly and Macready, went around the world with Lowell Smith, rode with "Elmie" Elmendorf to test an O–2, and made the night mail run from Cleveland to Chicago. I saw "Billy" Mitchell sink the *Ostfriesland* and heard him threaten coal miners with tear gas. I followed the battle over organization, sat through "Benny" Foulois' efforts to defend himself before a hostile subcommittee, took in all the Pulitzer races, watched Three Men on a Flying Trapeze put on their act, sailed through a thunderstorm in a balloon with "Bill" Kepner, went out on the range with the 19th Bombardment Group, and froze on an arctic patrol with the 1st Pursuit Group. I learned much about the state of military aviation in the early 1920s by watching a pilot whittle his rain-chewed propeller, and from seeing a sergeant slide out on the tail of a DH–4 to weight it for a fast landing. To find out what mobility meant to a bombardment group in the mid–1930s, I not only flew with the bombers to an advance base but made a trip with the trucks bearing the group's impedimenta—all without leaving Building 1405 at Maxwell Air Force Base.

Curious about everything, I wanted to know where the Air Service got its pilots, and how "Jimmy" Doolittle flew "blind." How did "Bert" Dargue and company prepare for a flight through South America? What went wrong with the five-year program? How did George Brett miscalculate requirements for depot reserves? How did "Barney" Oldfield land a formation of B–10s through clouds? What went into survival kits for flights over water or jungle? Why did pilots of A–12s run out of gas and crash with a full tank? While looking into all of these things and more, I took time to accept an invitation to Billy Mitchell's *despedida* for the USS *Alabama*, watch Clarence Irvine crash a plane for the movie *Wings*, attend one of "Hap" Arnold's wingdings, drop in on one of Mrs. Lackland's teas, play polo at Kelly Field, and dance to the sounds of the Tantalizing Hounds of Syncopated Jazz at Langley. Along the way, I met hundreds of enlisted men, but few with names.

I put this into the draft of a book, which I planned as a descriptive rather than an analytical work, detailed and specific rather than generalized, telling of events rather than interpreting them, and dealing more with flying than with national policy and airpower doctrine. Review of the manuscript in the Office of Air Force History led to substantial changes. Revision deleted much detail, summarized some topics, dropped others, omitted many examples, and added some explanation and interpretation. So the work became something more and something less, and something much different, than originally intended.

Maurer Maurer

Acknowledgments

Maj. Gen. John W. Huston, former Chief of the Office of Air Force History, approved the original project; his successor, Dr. Richard H. Kohn, directed its completion. Lloyd H. Cornett, Jr., Director, USAF Historical Research Center, provided strong support, good advice, and excellent criticism. Dr. Stanley Falk and Mr. Max Rosenberg, both formerly of the Office of Air Force History, reviewed the early chapters; Mr. Herman S. Wolk of the Office of Air Force History, and Col. John F. Shiner, then Acting Head of the Department of History at the U.S. Air Force Academy, critiqued the entire manuscript. Dr. Kohn, Gen. Bryce Poe II (USAF Ret.), Maj. Gen. Haywood S. Hansell, Jr. (USAF Ret.), Col. John Schlight, Dr. Robert W. Coakley, Dr. Edgar F. Raines, Mr. Warren A. Trest, Mr. Cornett, and Mr. Wolk took part in a very profitable seminar on the revised draft. Drs. Coakley and Raines submitted written notes and comments of great value.

Lois Wagner of the Historical Research Center handled typing and editorial work on the draft and its several revisions. The author is indebted also to Robert Lane, Director, and the staff, of Air University Library for their help over many years, and especially to Bill Mardis and the crew at the Circulation Desk, Helen Hopewell in the Periodical Room, Ruth Griffin in the Interlibrary Loan Office, and Gurvis Lawson of the Cartographic Division. Special thanks are owed to Ray Del Villar, Publishing Division, Directorate of Administration, USAF, for his excellent and painstaking work on the maps and charts.

Illustrations chosen for use in this volume are from the large body of Signal Corps and Air Corps photography collected over the years in official U.S. Army, Army Air Corps, and U.S. Air Force repositories. This material, dating from the first decade of this century to the year 1954, has now been consolidated in the holdings of the National Air and Space Museum, Smithsonian Institution, Washington, D.C. Special thanks are due to Mr. Dana Bell, Assistant Chief of Records Management at the National Air and Space Museum, for his assistance in providing copies of rare aviation photographs and identifying obscure variants of early military aircraft. Sources for those few photographs procured from collections outside official

Air Force holdings are specified in captions to provide appropriate credit to individuals or institutions authorizing their use.

In the Editorial Branch of the Office of Air Force History, a number of people made important contributions to preparing the manuscript for publication. Mr. Eugene P. Sagstetter performed the editing of this manuscript with grace and dedication. Also deserving much credit are Sgt. Rosalyn L. Culbertson for typing the entire manuscript, Ms. Ann Caudle for proofing the work, Ms. Bobbi Levien for copyediting the appendices, and Mrs. Joyce Truett for applying the final corrections. Ms. Laura Hutchinson selected the illustrations and designed the book. Mr. Jacob Neufeld, Acting Chief of the Editorial Branch, closely reviewed the entire manuscript for publication.

Contents

Page

Part Three: GHQ Air Force 1933–1939

Photographs

Charts and Maps

Introduction

When the United States declared war on Germany on April 6, 1917, Army aviation consisted of fewer than 1,200 men, some 250 airplanes, and 5 balloons under Signal Corps control to provide observation and courier service for infantry, cavalry, and artillery. The war brought rapid growth, greater operational capabilities, and status as a separate combatant arm. At the Armistice on November 11, 1918, the Army had more than 190,000 men on aviation duty, 40 percent in Europe as part of the Air Service of the American Expeditionary Force (AEF), commanded by Gen. John J. Pershing. Except for a few men assigned to garrisons in Hawaii, the Philippines, and the Panama Canal Zone, the remainder were at home training pilots and mechanics and turning out airplanes and other aeronautical equipment for the AEF.

Army aviation during the war employed both heavier and lighter-than-air craft, the former for bombardment and pursuit operations, and both for observation. In Europe, General Pershing removed aviation from the Signal Corps and established an Air Service, of which Brig. Gen. Mason M. Patrick eventually became chief. Air Service units at the front were attached to divisions, army corps, and armies for operations under the control of the division, corps, or army commander. At each level the senior aviation officer served in a dual capacity as a staff officer and a commander. Thus, assigned to Headquarters First Army during the Saint-Mihiel and Meuse-Argonne offensives, Brig. Gen. William Mitchell was Chief of Air Service, First Army, and commanded aviation units attached directly to the First Army. However, he had no command authority over units attached to the First Army's corps and divisions.

In the Air Service, AEF, observation aviation was attached to divisions, corps, and armies to work with ground forces. Aerial observers in captive balloons floating aloft behind the lines looked out over no-man's-land to report enemy positions and activities. Other observers flew in airplanes along the lines and over enemy-held territory to obtain similar information by visual and photographic means. So the Air Service performed duties historically belonging to the Cavalry. When friendly troops came out of the trenches to attack the enemy, aviators flew over the battle area to observe and report movements and positions of ground forces to command posts in the rear. Aerial observers also monitored friendly artillery fire to enable batteries to lay their guns on the target. At times observation planes assigned to army headquarters flew reconnaissance missions well into enemy territory, but for the most part observation aviation operated in the battlezone in direct support of ground forces. As for results, aerial observation and reconnaissance provided much valuable information to ground commanders, while aerial observers in both heavier- and lighter-than-air craft proved helpful in directing artillery fire. Infantry support missions, labeled "contact" and "liaison," generally were not very successful. The absence of good air-ground radio communications made visual signals and dropped messages the chief means of communication between aerial observers and ground forces. Then, too, there was insufficient joint training for such operations.

Pursuit aviation, assigned to armies, shared with antiaircraft batteries the protection of friendly forces from observation and attack by enemy aviation. Pursuit planes and pilots on alert at airdromes rose to intercept, engage, and destroy enemy planes reported in the vicinity. At other times, singly or in formation, they patrolled assigned areas against hostile aircraft. Often they escorted observation or bombing planes to fight off enemy aircraft threatening the mission. And they attacked and destroyed enemy observation balloons. Pursuit operations made heroes of such men as Capt. Edward V. Rickenbacker, the top ace of the Air Service, AEF, and 2d Lt. Frank Luke, Jr., who won fame as a "balloon buster." Pursuit work also included strafing and bombing ground targets in the battle area, a business that became increasingly frequent in the last days of the war, and led to plans to form attack squadrons for such work.

Bombardment units, assigned to armies, attacked such objectives as troop concentrations, supply areas, munitions dumps, roads, trains, and railway centers, the range of the planes in use restricting operations to targets no more than about seventy-five miles away. In comparison with the work of other branches of aviation, the bombing effort was relatively small. Of the forty-five aero squadrons at the front on Armistice Day, only seven were bombardment, as compared with eighteen observation and twenty pursuit. AEF plans called for a powerful air force for long-range bombardment of strategic targets, but the fighting ended before units became available for such missions. As conceived, the air force would be directly under a general headquarters that in turn would report to the AEF Commander. Because of their limited scope and character, AEF bombardment operations before the Armistice gave little or no indication of what might be achieved by strategic air warfare.

Hence the U.S. Army's experience in aerial operations during World War I was with units controlled by division, corps, and army commanders and used to support ground forces. Having found aviation a valuable auxiliary, ground commanders wanted to keep it as part of their forces and under their own control.

At home Army aviation focused on training men, producing airplanes and equipment, and forming units to send to the AEF. In April 1917 the United States had no significant capacity for airplane production. Consequently, the Army turned to America's Allies for the AEF's aircraft until the United States could take over. In the United States, businessmen and engineers, many newly commissioned in the Army, took charge of designing planes and motors, creating factories and converting existing industrial facilities to make aeronautical materiel, and managing production programs under Signal Corps direction.

The Signal Corps attempted to solve production problems by reorganization, but production lagged. Eventually, President Woodrow Wilson used wartime powers granted by Congress to remove Army aviation from the Signal Corps. At the same time, on May 20, 1918, Wilson created two agencies under Secretary of War Newton D. Baker to manage Army aviation at home: The Bureau of Aircraft Production, headed by Mr. John D. Ryan, former president of the Anaconda Copper Company; and the Division of Military Aeronautics, headed by Maj. Gen. William L. Kenly, an artilleryman who had recently returned from France and had taken charge of training and operations. President Wilson then authorized Secretary Baker to form the two agencies into an Air Service with one man responsible for coordinating the work of the two. It was not until August 27, 1918, however, that Secretary Baker announced the appointment of Mr. Ryan as Second Assistant Secretary of War and Director of Air Service.

By the Armistice the Army had received about 11,000 planes of the 27,000 ordered. Of these, some 7,800 were trainers, the majority the famous JN–4D (Jenny), of which the Army acquired more than 5,000. The only combat plane produced in quantity in the United States for the Army was the DH–4. Built from British plans, the DH–4 used the Liberty engine designed by the Signal Corps' equipment division after the United States entered the war. The DH–4 program moved slowly, with only 15 planes produced in April 1918. By November, however, more than 3,000 had been completed. The first DHs went to France in May 1918, but a shortage of shipping delayed movement of large numbers overseas. Although some 1,200 arrived in France, only 196 reached the front before the Armistice. In battle U.S. Army aviators flew mostly foreign-built planes, including Nieuports, SPADs, Breguets, and Salmsons.

The U.S. Army entered the war with three airfields in the United States. The principal one was the aviation school, founded by the Signal Corps in December 1912, on North Island, San Diego, California, the facility which became Rockwell Field. Another school, opened in 1916 at Mineola, Long Island, New York, later Hazelhurst Field, trained candidates for the National

Guard and Reserve. The third, a temporary training field at Essignton, near Philadelphia, Pennsylvania, had been set up just five days before America entered the war. During the war the Army acquired forty-five more flying fields, as well as nineteen depots and numerous other facilities scattered all over the country.

A young man who enlisted in the Army as a flying cadet began training with two months in ground school at one of eight universities. He next went to an Army flying school for an eight-week course. If successfully completed, he received a pilot's rating of Reserve military aviator (RMA) and a commission as a second lieutenant. During the war eighty-six hundred cadets graduated from primary schools in the United States. Others attended schools in France and Italy, bringing the total number of pilots trained during the war to ten thousand. New pilots moved on to other fields at home or overseas for advanced training in gunnery and in observation, pursuit, or bombardment. Thousands of other men were schooled as balloonists, airplane mechanics, engine mechanics, armorers, storekeepers, radio operators, and photographers. In addition, many mechanics underwent training overseas in England or in France.

To move officers and men to Europe for training and operations, the Division of Military Aeronautics formed them into units. Included were service (observation, pursuit, or bombardment) squadrons, and balloon companies. Some units contained officers and men, but many consisted chiefly of men with one or two officers in charge. For example, service squadrons received their pilots and observers after arriving in France. Altogether, some 5,700 Air Service officers and 74,000 enlisted men joined the AEF. Many worked at training centers, at depots, or in administrative or staff positions. Plans called for 202 service squadrons at the front by June 30, 1919, but, as noted earlier, only 45 arrived before the Armistice. Of the many trained flyers then in the Army, only 767 pilots, 481 observers, and 23 aerial gunners were on duty with those squadrons on November 11, 1918. Thus the war ended before the full weight of America's production and training programs came to bear upon the enemy.

Americans greeted the Armistice with wild celebration. They had fought and won a war to end war, a war to make the world safe for democracy. President Wilson went to Europe to arrange a just and lasting peace based upon his famous Fourteen Points. But there were Americans who, out of conviction or partisanship, rejected the President's program. People who had sacrificed much for a just cause saw their dreams shattered at the conference table, where European statesmen quarreled over the spoils of war, and the President of the United States compromised his ideals to gain a League of Nations. The Senate defeated the treaty Wilson brought home from Europe and thus rejected American membership in the league. Finally, in 1921, the United States concluded a separate treaty with Germany.

The United States meanwhile demobilized the civilian army it had created during the war, abolished wartime agencies, and liquidated activities undertaken as emergency measures during the conflict. Veterans went back to work. Industry turned from production of war materiel to the usual peacetime business. The transition was not easy, being attended by labor and racial unrest, with strikes of steelworkers and coal miners. Wartime hatred of German sympathizers shifted to Bolshevists and persons suspected of Bolshevik sympathies. Conservatism in the interest of private enterprise replaced the progressive liberalism of Wilson's New Freedom. So America turned to what a new President, Warren G. Harding, called "normalcy."

The inauguration of Harding in 1921 ushered in a period of general prosperity. Industry grew, national income rose, and the prices of stocks soared on Wall Street. Businessmen, tradesmen, professional people, and white-collar workers (but not farmers and laborers) did well financially. Economy under Republican Presidents Harding, Calvin Coolidge, and Herbert Hoover permitted the government not only to lower taxes but at the same time to reduce the national debt, which as a result of the war had grown to $24 billion or about twenty times the prewar amount.

The period which has been labeled the Roaring Twenties was one of "wonderful nonsense" and "tremendous trifles," of flappers, lounge lizards, torch singers, jazz, and the Charleston, of speakeasies and bathtub gin, bootlegging and gang warfare, flagpole sitters and daredevil aviators, and newspaper ballyhoo which reached new heights with Charles A. Lindbergh's flight across the Atlantic in 1927. The spirit of the age found expression in popular songs like "Making Whoopee," "Yes, We Have No Bananas," and "Show Me the Way to Go Home." It also was reflected in movie titles like *Flaming Youth*, but producers also turned out *Ben-Hur, The Gold Rush,* and *The Three Musketeers*. Mary Pickford, Douglas Fairbanks, Clara Bow, and Charlie Chaplin appeared among the galaxy of stars. New theaters opened and attendance grew at a phenomenal rate, further stimulated when movies with sound started to replace silent movies in 1927. The first of the "talkies" featured Al Jolson in *The Jazz Singer*. Sports held a big place in American life. A nation of spectators made heroes of Babe Ruth (the Sultan of Swat), golf champion Bobby Jones, football greats Knute Rockne and Red Grange, tennis stars Helen Wills and Bill Tilden, swimmer Gertrude Ederle, prize fighters Jack Dempsey and Gene Tunney, and a thoroughbred named Man o' War.

Radio became a novel form of entertainment as well as a source of information for the American people. Station KDKA in Pittsburgh, Pennsylvania, began broadcasting just in time to bring the Harding-Cox election returns to the few people who owned radios in 1920. By 1929, more than six hundred broadcasting stations operated, and one family in three owned a receiver. Establishment of the National and the Columbia networks afforded nationwide audiences for singer Rudy Vallee, commentator Hans V. Kaltenborn, and advertisers extolling the merits of Lucky Strike cigarettes, Campbell's soup, and Ipana toothpaste.

The automobile became popular during the twenties. Henry Ford led the industry with his mass-produced, low-priced, utilitarian, tin lizzie, which he designated Model T. The T's chief competitors were the General Motors Corporation's Chevrolet, and the Chrysler Corporation's Plymouth, but both companies also produced higher priced, better looking, more stylish cars. The willingness of people to pay more to get something better forced Ford to produce a car of better appearance and more style. His introduction of the Model A, one of the great events of 1927, kept him in the lead and sent sales to new heights. By the end of the decade, more than twenty-five million cars and trucks were in operation in the United States.

The aviation industry amounted to little until the mid-1920s, when Congress authorized contracts with private organizations to carry airmail, a business previously handled by the Post Office Department's airmail service. The profits accruing from such contracts gave aviation a tremendous boost, helped along by public interest aroused by a series of spectacular feats topped by Lindbergh's trans-atlantic hop in 1927. Meantime, in 1926, the government created a bureau in the Commerce Department to aid, encourage, supervise, and regulate civil aviation.

Millions of dollars went into the production of airplanes and aviation facilities for transportation of mail, passengers, and express. The Commerce Department developed a network of federal airways with emergency landing fields, radio and teletype communications, lighting for night operations, and weather service. Closed cabins replaced open cockpits as considerations for passenger comfort and safety altered aircraft design. By the end of 1929, forty-five airlines, including Delta, Eastern, National, and Pan American, offered scheduled services. That year alone, scheduled airlines averaged 68,881 miles a day (up from 11,830 in 1926) on routes criss-crossing the United States from coast to coast and border to border. In four years the weight of airmail carried by contract increased nearly ten times, and the number of passengers more than two hundred times, while airline revenues rose from $765,000 to $20.7 million. Air transport accordingly became big business and a part of the American way of life.

The scene changed with the stock market crash of October 1929 and the ensuing economic depression. Some had warned that all was not well with the nation's economy, but Americans clung to their belief that prosperity was here to stay. They could not see the disaster that would befall them as the result of wild speculation, over-extension of credit, inadequate controls over banking and the stock market, tariff barriers, and a decline of agricultural prices. Breadlines lengthened as unemployment mounted. Some six million people were without work late in November 1930, and the number reached sixteen million in March 1933. Creditors foreclosed mortgages. Banks, utility companies, and businesses of all kinds failed. Panicking, people ran to withdraw their savings from banks. The Hoover administration's ineffectual efforts to furnish relief and reverse the course of the depression led to a steep rise in federal expenditures. This and a sharp drop in receipts, yielded a large deficit in 1932, the first since the close of the war. In the national election of 1932, the American people turned to Franklin D. Roosevelt, who promised a "New Deal."

Roosevelt came to the presidency on March 4, 1933, resolving to act quickly. He commenced by declaring a bank holiday until the sound institutions could be identified. Then on March 12 he took to the radio, and in his first "fireside chat" reassured the American people and urged them to redeposit their money. Reopening of banks soon afterwards ended the immediate crisis.

The President meanwhile sought ways to reduce government expenditures. Viewing an anticipated deficit of $5 million by 1934 as a threat to U.S. credit, he quickly won congressional support for cutting government salaries and veterans' compensations. There followed in quick succession a series of emergency acts. These insured bank deposits, curbed speculation, protected investors, and assisted farmers and mortgage holders. The acts fostered industrial recovery by eliminating overproduction and destructive competition. They enhanced employment by setting maximum hour and minimum wage limits, and by undertaking a gigantic public works program. The President's methods slowly pulled the nation from the depths of depression. Progress toward recovery quickened in the late 1930s as international developments prompted the United States to strengthen its defenses.

The buildup of American land, sea, and air forces in the late 1930s stemmed from a growing threat to national security rather than from any fundamental change in national defense policy. Having no aggressive design on any nation, and seeing no major power as an immediate threat, the United States in 1920 had adopted the policy of maintaining active armed forces at minimum strength in peacetime, which suited most Americans. The policy pleased those who did not want millions of the taxpayers' money spent on the Army and Navy. It found favor among those who believed that Europe's affairs were none of America's business and that security could be found in political isolation. Internationalists accepted the policy, urging cooperation with other nations in working for world stability and peace. Support came from pacifists who thought war wrong but believed some defense necessary in a less than ideal world. There were antimilitarists, however, who would have abolished all armed forces. Some of them advocated—and were willing to accept the consequences of—unilateral action if international agreement could not be reached. Then there were those who embraced the defense policy but regarded existing preparations inadequate.

Such classification does not imply well-defined groupings of the American people. Wide differences of opinion prevailed within each camp, and some people embraced more than one philosophy. There arose a proliferation of groups bearing such names as American Peace Society, League to Enforce Peace, Fellowship of Reconciliation, Interorganizational Council on Disarmament, Committee on Militarism in Education, and War Resisters League. Some groups exerted little influence but others played significant roles in American life. Those with widely divergent views often joined forces to support some cause or attack some threat to peace. Many Army men believed pacifism prevented them from obtaining the appropriations they thought necessary for preparedness. The National Council for the Prevention of War led a campaign which forced the War Department to abandon the annual mobilization day it instituted in 1924. Peace organizations compelled President Coolidge in 1928 to abandon an ambitious naval construction program. Although pacifists and irate taxpayers failed to stop U.S. Army Air Corps maneuvers over eastern cities in 1931, they succeeded in embarrassing government officials. Other examples included the work of the American branch of the Women's International League for Peace and Freedom, whose executive secretary and lobbyist, Dorothy Detzer, claimed responsibility for the Nye Committee's investigation of the munitions industry. Attacks like these provoked responses from government officials, including military officers, as well as from special interest groups like the Reserve Officers Association, Navy League, and Aircraft Manufacturers Association.

Philosophically, most Americans were pacifists in that they loved peace and hated war. Pacifist groups drew their leadership chiefly from the nation's clergy, liberals, intellectuals, and women, and found their opponents among advocates of preparedness. Aroused, pacifists condemned military men and the "one hundred-percent Americans" of patriotic organizations as "militarists." The latter retaliated with "communist," the vilest political epithet of the time.

The peace movement in America and in Europe brought forth various schemes for outlawing war by international agreement. Suggestions for enforcement included economic sanctions, moral suasion, and, paradoxically, use of military power. In 1927 French Foreign Minister Aristide Briand proposed that France and the United States agree to renounce war as an instrument of national policy. U.S. Secretary of State Frank B. Kellogg suggested that other powers be permitted to subscribe to it. Fifteen nations signed the Kellogg-Briand Pact at Paris in August 1928, and ultimately the number rose to sixty-three. But the pact lacked enforcement provisions, and some nations made reservations in signing. Secretary Kellogg himself believed the treaty reserved the right to self-defense.

Isolationism and internationalism went hand in hand, intertwined with pacifism. The isolationism that kept America out of the League of Nations and the World Court aimed at preventing America from being caught up in conflicts abroad. While refusing to take any responsibility for preserving peace and world order, the United States worked with other nations on arms limitation, disarmament, outlawing of war, and on numerous social and humanitarian projects. As early as 1922, this country commenced sending "unofficial" observers to League of Nations committees concerned with such subjects as narcotics traffic and white slavery. Afterwards, U.S. representatives attended many other nonpolitical conferences. Before long, they were engaging in political matters. In 1931, for instance, America took part in league discussions on the Manchurian crisis, and the next year participated in the World Disarmament Conference called by the league.

The United States itself sponsored an arms limitation conference in Washington in November 1921 to stop a naval race among the great powers. A bold plan offered by Secretary of State Charles Evans Hughes became the basis of a treaty, signed by Great Britain, the United States, Japan, France, and Italy in February 1922. In this treaty the five powers undertook to scrap some old capital ships, agreed to stop building new ones for ten years, and set forth ratios for naval forces based on existing strength. A conference at Geneva, Switzerland, in 1927 to deal with land as well as other naval armaments, failed in both. At London in 1930, Great Britain, the United States, and Japan extended the holiday on capital ship construction to 1936 and limited most classes of vessels not covered in the 1922 treaty. After grabbing Manchuria, Japan announced in 1933 it would withdraw from the London treaty on expiration of the agreement in 1936. Further efforts at arms reductions and naval limitations proved fruitless.

The Japanese invasion of Manchuria and China and the bellicose attitudes assumed by Benito Mussolini and Adolf Hitler in the 1930s raised in the American mind the specter of a second world war and United States involvement. The prospect made the nation all the more isolationist. Mussolini's designs on Ethiopia produced the Neutrality Act of 1935, whereby America sought to avoid some of the situations that led her to war in 1917. This act, not as strong as isolationists desired, embargoed shipments of arms and munitions to belligerents when the President recognized the existence of a state of war between two or more foreign nations. It did not, however, prohibit trade in oil, scrap iron, and other things convertible to war use. Nor did it prohibit Americans from traveling on ships owned by belligerents, but they did so at their own risk. After the outbreak of the civil war in Spain in 1936, Congress extended the Neutrality Act of 1935 to encompass civil war.

With the threat of war in Europe and Asia growing, Congress enacted a new neutrality act in 1937. It prescribed an arms embargo when the President proclaimed the existence of international or civil war. The act prohibited loans to belligerents, use of American ships to carry war materiel to belligerents, travel by Americans on ships owned by belligerents, and arming of American merchant ships. A "cash and carry" clause allowed the President to identify oil, scrap iron, and similar items that could be sold if the belligerent paid cash and carried the goods away in his own ships.

The deteriorating international situation and the expansion of military power abroad in the late 1930s not only caused the United States to take measures to avoid being drawn into a foreign war, but caused her to look to her own defenses. Finding the existing level of preparedness inadequate, the United States undertook to modernize and strengthen land, sea, and air forces.

America was blessed with good natural defenses in the form of two great oceans. The points most vulnerable were territories beyond the continental United States—the Philippine Islands, Guam, Hawaii, and the Panama Canal Zone. A defeated Germany and an exhausted Europe seemed, at the end of World War I, to leave the United States free from any serious threat from that direction for a long time to come. The situation was different on the Pacific side, for America's relations with the Japanese were not good. Thus the most serious threat to U.S. security lay in the area of greatest vulnerability. Whatever the threat, however, from either the west or east, the Navy constituted the nation's first line of defense. But naval treaties and the climate of opinion in America kept the Navy at relatively low strength during the 1920s and early 1930s.

One of the most significant developments in the Navy between the world wars came in the area of amphibious warfare. Assuming that in war the Japanese would capture Guam and the Philippines, the Navy prepared to win them back. Accordingly, the Navy developed landing craft and the Marine Corps devised techniques and practiced amphibious operations, all proving to be valuable preparation for the Pacific campaigns of World War II.

Another meaningful naval development was the rise of naval aviation and its integration with the fleet. Immediately following the Armistice of 1918, the U.S. Navy resumed the work, interrupted by the war, of adapting aviation to fleet operations. It met opposition, however, from propagandists, including the Army's Brig. Gen. William Mitchell, who held that aviation had made surface fleets obsolete. Mitchell unwittingly gave naval aviation a large boost by sinking a former German battleship, the *Ostfriesland*, by aerial bombardment during ordnance tests in 1921. The Navy quickly converted a collier, the USS *Jupiter*, to an aircraft carrier renamed the USS *Langley*, launched in March 1922. Then came two more carriers, the USS *Lexington* and USS *Saratoga*, converted from unfinished battle cruisers laid down after the war and scheduled to be scrapped under the Naval Treaty of Washington, February 6, 1922. In 1938 the Navy laid keels for four more carriers: USS *Ranger*, USS *Yorktown*, USS *Enterprise*, and USS *Wasp*.

In the early 1920s Navy officers generally expected to use carrier aviation for scouting and for fleet defense against carrier or shore-based airplanes. Exercises in the twenties demonstrated the use of carrier aviation against shore bases. Later exercises employed carriers with fast cruisers to screen the fleet and strike enemy forces at sea or ashore. Therefore, the concept of a fast task force of carriers, battleships, cruisers, and destroyers emerged in the mid-1930s, an idea destined for full development during World War II. The interval between the wars also witnessed great improvements of planes, as well as the development of dive bombing of moving targets, and antiaircraft guns and gunnery for fleet defense. The period was further attended by intense rivalry between Army and Navy aviation, and by protracted conflict over the role of each in national defense.

Unlike the Navy, the Army never played a key role in international affairs through the 1920s and 1930s. Isolated politically and geographically, the United States found little use for it except for civil projects and the defense of outlying territories. The end of the fighting in Europe had made nearly all of the Army's 188,000 officers and 3,000,000 enlisted men eligible for release from military service. Its goal was to discharge the troops quickly and equitably without unduly disrupting the nation's economy and while preserving forces to serve on occupation duty in Europe. Sufficient soldiers would be needed to guard the southern border as dictated by revolutionary disorders in Mexico, and to suppress domestic disturbances arising from labor disputes and racial unrest. Others had to be on hand to occupy overseas possessions and to carry on normal peacetime duties.

Holding to traditional methods, the Army demobilized by units with many exceptions for individuals and some occupations. Concurrently, it recruited to replace some of the men discharged. Within a year the Army shrunk to 19,000 officers and 205,000 enlisted men, rendering it once again a volunteer force. During the same time, the government disposed of facilities and materiel excess to its current and future requirements.

With materiel needs met from stocks held over from the war, the Army's chief concern for a number of years was obtaining the personnel for its peacetime functions. The United States began withdrawing troops from the occupation force after the Paris Peace Conference ended in 1919, but the last of the U.S. contingent did not depart Europe until January 1923. The demand for forces on the Mexican border soon eased. Reorganization of the National Guard in 1921 let states deal with their own domestic disturbances. So the Army settled down to creating and maintaining a military establishment that could meet a future emergency. In addition it would perform various nonmilitary duties, for example, those associated with rivers and harbors and with inland and coastal waterways.

The War Department urged approval of a permanent Army of 600,000 men with three-month universal training to permit its rapid expansion in an emergency. Regarding the Navy as the first line of defense, and seeing no great power as an immediate threat, Congress and the public generally rejected universal training and a large standing army. They would support in peacetime only the minimum military force necessary to defend the United States and its territories, keep military arts alive, and train volunteers for active duty in an emergency.

Congress set peacetime military policy for the nation with a 1920 amendment to the National Defense Act of 1916. The new law established the Army of the United States, made up of a professional Regular Army and two civilian components, the National Guard and the Organized Reserve, the last consisting of the Officers' Reserve Corps and the Enlisted Reserve Corps. It authorized the Regular Army a maximum of 17,726 officers and 280,000 enlisted men, and put all officers except doctors and chaplains on a single promotion list. The act continued all of the arms and services established before 1917, gave legislative sanction to the Air Service and Chemical Warfare Service which had come into being by executive order during the war, created a Finance Department, and charged the War Department with mobilization planning.

In the reorganization after approval of the National Defense Act of 1920, the War Department retained the three existing overseas departments for command and administration of field forces in Hawaii, the Philippines, and the Panama Canal Zone. At home it eliminated the nine geographical departments that had previously administered field activities, and replaced them with six corps areas to which field forces were assigned. The division supplanted the regiment as the basic unit for mobilization, with each corps area assigned 6 divisions—1 Regular Army, 2 National Guard, and 3 Reserve. Thus 54 divisions, formed into 18 army corps to make up 6 field armies totaling 2,000,000 men, constituted the initial mobilization of the Army of the United States. The War Department Chief of Staff, assisted by the General Staff, supervised field forces as well as the War Department's military branches. The General Staff, its position and authority strengthened during the war, consisted at war's end of four divisions: Military Intelligence; War Plans; Operations; and Purchase, Storage, and Traffic. When General of the Armies John J. Pershing became Chief of Staff in 1921, he remodeled the staff along the lines of General Headquarters American Expeditionary Force, which he had commanded during the war. This reorganization gave the staff five divisions: G–1 (Personnel); G–2 (Intelligence); G–3 (Operations and Training); G–4 (Supply); and War Plans.

The Regular Army in the 1920s and early 1930s never reached the strength authorized by law. Approval of the National Defense Act of 1920 found the Army about a third below the authorized strength. Its future strength depended upon the amount of money Congress provided in annual appropriations. Budgetary restrictions took the enlisted authorization down to, first 175,000, and then to 150,000 in 1921, with a further decline to 125,000 and then to 118,000 in 1922. The officer authorization fell from 17,726 to 12,000. There the authorizations stabilized for a number of years. Military expenditures also stabilized at around $300 million a year until the mid-1930s.

Except for aviation, Army appropriations in the 1920s and early 1930s permitted almost no new equipment and provided little for development of new weapons. Units of all arms existed at skeletal strength, some solely on paper. Protesting that appropriations did not meet the minimum required for defense, the War Department repeatedly attempted to obtain more money. Failure year after year so conditioned some Army leaders that instead of asking for what they believed necessary, they tended to seek only such lesser amounts as they thought Congress might appropriate.

General Douglas MacArthur, serving as Chief of Staff from 1931 to 1935, drew up mobilization plans based on four field armies. In 1932 he set up headquarters for field armies to furnish higher organization of field forces, something hitherto lacking. The next year saw the organization of headquarters for General Headquarters (GHQ) Air Force, the aerial component of Army field forces, followed a year and a half later by creation of the GHQ Air Force itself in March 1935. But General MacArthur made little progress in mechanizing and motorizing Regular Army combat units and filling them with troops to make an effective fighting machine to meet an emergency.

International developments led to an increase in the enlisted authorization of the Regular Army to 165,000 in 1935 and to considerably larger appropriations for the following years. Expenditures for equipment and facilities grew. The Army adopted the Garand M–1 .30-caliber semiautomatic rifle as replacement for the 1903 Springfield, built light and medium tanks of greatly improved speed, perfected the 105-mm mobile howitzer, and replaced horses with motors. So the U.S. Army expanded and became stronger during the late 1930s.

The U.S. Army air arm evolved during the 1920s and 1930s from experience gained during World War I and within the political, social, economic, and military environment of the time. Development of Army aviation in the areas of organization, manning, equipment, training, and operations is the subject of this work, consisting of three parts: the Air Service, the Air Corps, and the GHQ Air Force.

The air arm's first task following the Armistice in 1918 became demobilization (Chapter I). It involved discharging war-service officers and enlisted men and disposing of surplus facilities, equipment, and supplies. While demobilization was under way, the Air Service tried to stimulate interest in and advance aviation with flying circuses, cross-country flights, and aerial contests (Chapter II). The National Defense Act of 1920 gave Army aviation statutory existence as a separate arm and prescribed its peacetime organization, but low appropriations and personnel ceilings prevented the Air Service from attaining the strength authorized by law (Chapter III). In its postwar reorganization, the Air Service established primary and advanced flying, balloon and airship, technical, tactical, medical, and engineering schools to prepare individual Army officers and enlisted men for duty with the Air Service (Chapter IV). Unit training took place mainly in the tactical units, the organization of which became a controversial matter concerning not only the Air Service but also the War Department, Congress, and the American people (Chapter V). Creation of Reserve components of the Air Service, (i.e., the National Guard and Air Reserve) provided for expansion of the Regular Army Air Service in a national emergency (Chapter VI). With responsibilities for defense of both land and sea frontiers, the Air Service patrolled the Mexican border and made plans and preparations for coastal defense (Chapter VII). Bombing tests not only failed to answer the question of what effect the advent of aviation had on sea power, but aggravated controversy between the Army's air arm and the Navy (Chapter VIII). Besides carrying out many civic operations (Chapter IX), the Air Service undertook projects for advancing civil as well as military aviation in the United States (Chapter X). It also stirred interest in aviation and hastened development through numerous record-making altitude, speed, distance, and endurance flights, including the pioneering flight around the world (Chapter XI).

The Air Corps, which replaced the Air Service in 1926, benefited greatly from a five-year expansion begun in 1927 even though the program failed to reach all goals within the allotted period (Chapter XII). Better equipment and training (Chapter XIII), as well as annual maneuvers (Chapter XIV), brought some improvements in tactical capabilities and performance. In the six years following its establishment, the Air Corps continued to make aviation history in a series of flights. Included were the maiden flight from California to Hawaii, setting an endurance record, and the first ones where pilots relied entirely on instruments to take off, fly, and land without sight of ground (Chapter XV).

The establishment of GHQ Air Force headquarters in 1933 (Chapter XVI) and GHQ Air Force itself in 1935 (Chapter XVIII) were highly significant advances in U.S. Army aviation. Between these two events came the airmail emergency of 1934, disclosing deficiencies in equipment and training and giving impetus to their correction (Chapter XVII). New airplanes, hard-surface runways, more bombing and gunnery ranges, a good bombsight, better training, and numerous training exercises raised the combat effectiveness of the Air Force in the late 1930s (Chapters XIX to XXI). A big expansion program begun early in 1939 proved to be the beginning of the U.S. Army air arm's mobilization for World War II (Chapter XXII).

All told, these two decades marked the real birth of American air power.

They witnessed the transition of the Air Service that came out of World War I to a far different, more modern, and flexible Air Corps that stood on the threshold of the greatest cataclysm of the twentieth century.

Part One

The Air Service 1919–1926

Chapter I

Demobilization

At the Armistice on November 11, 1918, the United States Army had more than 3.6 million men in uniform. Some 2 million were in Europe with General Pershing's American Expeditionary Force (AEF). The newest arm, the Air Service, had grown from fewer than 1,200 officers and men in April 1917 to more than 190,000. Seven million men and women were in the war industries. A nation thus geared to war now confronted the gigantic task of demobilizing. The goal was to get back to normal as quickly as possible without wrecking the nation's economy.[1]

With the signing of the Armistice, Secretary of War Newton D. Baker cut war production, halted the induction of enlisted men into the Army, ended the commissioning of officers, and stopped the shipment of men and materiel to Europe. The Air Service began demobilizing the same day. In the weeks ahead it found itself discharging troops, recruiting for peacetime service, and disposing of surplus facilities and materiel.

The Armistice brought Air Service activities in the United States to a standstill. Maj. Gen. William L. Kenly, Director of Military Aeronautics, responsible for training, could not foresee whether men already enrolled would be allowed to complete their courses. The Armistice, he said, "has created the greatest uncertainty here as to the immediate future." How was demobilization to be accomplished? What was to be left for the peacetime Air Service? How was the transition to be made?[2]

Combat groups and wings of the Air Service, AEF, disbanded overseas. Aero squadrons and balloon companies returned to the United States where most disbanded. Some remained active: the 27th, 94th, 95th, and 147th Pursuit Squadrons that had comprised the 1st Pursuit Group in France; the 11th, 20th, 96th and 166th Bombardment Squadrons that had constituted the AEF's 1st Day Bombardment Group; a number of observation squadrons (including the 1st, the oldest unit of the Air Service); and some balloon companies. The pursuit and bombardment squadrons and four of the observation squadrons (8th, 12th, 90th, and 104th) were formed into a wing of three groups (pursuit, bombardment, and surveillance) in Texas during the summer of 1919 for service on the Mexican border.[3]

Officers of the AEF, entitled to first-class passage, returned as casuals on one of the transatlantic liners. Enlisted men, under the command of one or two junior officers, returned by troopship.[4] Troops returning from France assembled at Brest, Bordeaux, or Saint-Nazaire, where they went through "the mill" to be deloused and cleaned up before boarding ship. Most returnees landed at Hoboken, New Jersey, or Boston, Massachusetts, and some at other ports.[5]

Units and companies arriving at east coast debarkation camps were broken up and new companies formed for shipment to demobilization centers throughout the country. Each man went to the center nearest his home. There he was given a physical examination, his records were completed, and he was paid. With money in his pocket and discharge in hand he was ready to go. But first he had to talk to a member of the Red Cross or Young Men's Christian Association. The government was concerned about what might befall him when he walked out the gate with his back pay, the sixty-dollar bonus voted by Congress, and transportation money at five cents a mile from the demobilization center to his home. He was informed that the railroads offered a reduced rate of two cents a mile if he bought his ticket within twenty-four hours, was told of dangers lurking in the city, and advised to go straight home.[6]

Often the veteran could not find a job. General Kenly wanted an all-out effort to employ ex-members of the Air Service. He asked officers to canvas their men and divide them into two classes: those who had been assured employment, and those who desired assistance by the U.S. Employment Service of the Department of Labor. The latter were given cards which, when filled in, were collected, segregated by geographical regions, and sent to the federal employment office nearest the place where the man was to be discharged.[7] The War Department not only worked with the U.S. Employment Service, but set up a special committee to deal with unemployment among ex-servicemen during the transition from war to peace. A man with no prospect for a job could be retained for a reasonable time to look for work. On the other hand, if work awaited him, he could be released even though not in line for immediate discharge.[8]

All officers, except those of the Regular Army, were to be separated as

Secretary of War
Newton D. Baker

soon as their services could be spared. Release of officers commissioned during the emergency meant termination of commissions. As a general policy, however, officers were offered new Reserve appointments. Those desiring full and immediate separation were discharged first. Those seeking prompt separation with subsequent appointment in the Officers' Reserve Corps (ORC) came next. Those wishing commissions in the Regular Army, if opportunity permitted, were last. About half of the Air Service officers said they preferred to enter the Reserves, about three-eighths wanted to join the Regular Army, and just one-eighth requested complete separation.[9]

Cadets in training at ground or flying schools could choose immediate discharge or complete their training. Those who elected to continue were discharged and sent home after finishing primary and advanced training. In due course each received a commission as second lieutenant in the Officers' Reserve Corps.[10]

Under pressure from a people crying, "Bring the boys home," the Army discharged more than 1 million men by February 1, 1919, and over 2.7 million by June 30. The Air Service dropped from 190,000 (112,000 in the United States and 78,000 overseas) at the time of the Armistice to 81,000 at the end of January 1919, and 27,000 (5,500 officers and 21,500 enlisted) at the end of June 1919.[11] During this period Air Service planning rested on a peacetime Army of 500,000, the Air Service portion being 1,200 officers and 22,000 enlisted men.[12] Congress, however, had not yet enacted legislation for the peacetime military establishment.

Many months were to pass before Congress gave a definite answer.

Director of Military Aeronautics
Maj. Gen. William L. Kenly

Meanwhile, in September 1919 it authorized the Army to retain some emergency officers (those commissioned during the war) until June 30, 1920, but the total number of officers was not to exceed 18,000 after October 31, 1919. Many Regular officers had received temporary promotions in the Regular Army and even higher rank in the Reserve. The act let them keep their temporary rank for the time being. The Air Service authorization of emergency officers was 1,200, of whom at least 85 percent were to be qualified flyers. On September 26, 1919, Maj. Gen. Charles T. Menoher, Director of Air Service, distributed a list of 1,200 temporary officers to be retained on active duty. Temporary officers not listed were to be discharged by October 31, 1919.[13]

Recruiting

The Adjutant General, War Department, ordered that all men who had been drafted or had enlisted for the emergency be discharged by September 30, 1919.[14] With a temporary authorization for fifteen thousand enlisted men, the Air Service had begun in the spring of 1919 to replace the losses during demobilization. "Men Wanted for Air Service" was the heading of a news release. "Have you a good job today? If not, what can you find better than the Air Service? . . . There will always be a large demand for skilled aeroplane pilots and mechanics. Now is the time to learn at Government expense."

Director of Air Service
Maj. Gen. Charles T. Menoher

A man could take his discharge, collect the sixty dollars bonus and transportation money, reenlist (usually at the same grade), and get a month's furlough. The Air Service offered training for airplane mechanics, motor mechanics, propeller workers, fabric workers, magneto repairmen, instrument repairmen, radio electricians, carpenters, machinists, welders, vulcanizers, photographers, draftsmen, and, among others, chauffeurs. The future of commercial aviation looked bright. The time would come when there would be a big demand for skilled mechanics and other aviation experts.

The pay was good. A sergeant's base pay was $360 a year. Add room and board ($480), clothing ($170), and savings for retirement ($500), and a sergeant got $1,510 a year. If he made sergeant first class (SFC), the total went up to $1,690. In addition he received such things as free medical attention, free amusement, a chance to travel, and commissary privileges for married men. Compare this, the Air Service said, with the $885 the average civilian of 20–25 years of age made in 1 year.

Chances for advancement were excellent. "A bright, energetic, young man with initiative, should not remain a private long." The enlisted man with a high school education who applied himself, had a good knowledge of planes and motors, and could pass the physical examination could learn to fly. He then would have a good opportunity for a commission. Here, Air Service recruiting literature said, was "an opportunity of a lifetime."[15]

Seventeen flying fields, one repair depot, and five balloon stations took part in an Air Service recruiting campaign beginning in April 1919. Capt. Henry E. Reece and Lt. Nathan P. Oakes of Bolling Field flew a Curtiss training plane on a two-week recruiting trip to Rhode Island. Langley Field, Virginia, sent a recruiting expedition into Pennsylvania. Ellington Field near Houston formed a recruiting squadron with four DH–4s to work an area from Denver to Cheyenne, Wyoming, then to Lincoln, Nebraska, and Kansas City, Missouri. The 12th Balloon Company put a recruiting party into the same general area. Lt. Col. John D. Carmody visited flying fields in search of men who would volunteer to transfer to fill eight hundred vacancies then existing in balloon and airship companies. The largest recruiting expedition in the spring and summer of 1919, however, apparently was one led by Lt. Col. Henry B. Clagett. The seven DH–4s under his command flew from Dallas to Boston and back, the meandering 3,400-mile tour through the central and northern parts of the United States taking more than four months. These and other efforts produced more than nine thousand enlistments by mid-October 1919.[16]

To encourage enlistment the Air Service offered to fly the applicant from his home to the nearest field where his enlistment could be completed. Commanding officers at Air Service flying fields spread the word. If a person wanted to enlist, he notified the commander by telephone, letter, or postcard, and an airplane went to get him. The only requirements were that the applicant be sincere in purpose and willing to sign a waiver relieving the government of responsibility in case of accident.

A year wrought great change. Of the men who had become aces in France, the Air Service could find only nine on active duty a year after the Armistice.[17] The current allotment of 1,340 officers and 11,000 men constituted "a woefully inadequate force," in General Menoher's opinion. He recommended a force of about 42,000 men and 4,500 officers as the minimum to meet "reasonable requirements of the Air Service in time of peace." And that allowed nothing for an expeditionary force after providing for the training establishment and fixed fortifications in the United States, Hawaii, and Philippines, and the Panama Canal Zone. He asked for at least 600 more officers and 8,500 more enlisted men to put the Air Service into a position where in an emergency it could expand at a rate more in step with the remainder of the Army.[18] No increase in the personnel authorization was immediately forthcoming. In fact worse times loomed ahead.

Expiration of one-year enlistments taken in 1919 meant another recruiting campaign in the spring of 1920. The School of Aerial Photography at Langley Field, Virginia, offered a learn-while-you-earn plan in a well-paying profession. An Air Service recruiting party from Aberdeen Proving Ground in Maryland returned at the end of the week with thirty-four recruits, all of whom appeared to be excellent material. Carlstrom Field outside Arcadia, Florida, like many other Air Service stations, found the practice of bringing in recruits by plane to be very popular. In some weeks most of the work at Selfridge Field (then a storage depot), Mount Clemens,

Michigan, was given over to recruiting in the local area. The Aviation Repair Depot at the Speedway in Indianapolis, Indiana, reported in June that recruiting was slow, with only twenty-six enlistments in a month's time. The number might have been greater if recruiters had not been so careful to accept only the very best for the Air Service.

In the west, March Field, outside Riverside, California, billed 1st Lt. Harold H. George as a "famous overseas flyer" when he carried the recruiting message to the Rotary Club of San Bernardino. Two hundred boys soon to graduate from high school at Riverside were guests at March Field, where they were shown every department and entertained with an aerial exhibition. Many seemed interested in the cadet course that would soon begin. The 9th Aero Squadron of Rockwell Field, not far from San Diego, had little success during a recruiting trip in the San Joaquin valley, where wages were high.

The Observation School at Post Field, Fort Sill, Oklahoma, was one of many Air Service activities with severe personnel shortages in the spring of 1919. Capt. Follett Bradley, director of the school, had trouble getting the school organized for its opening on June 1. He was short both officers and enlisted men. He sent out recruiting parties but without much success. The natives were friendly and interested in aviation, but they showed no desire to enlist. The area around Post Field was devoted to farming. Farm wages were high, but there nevertheless was a shortage of farm help. The oil fields paid more.

Units seemed interested mainly in filling their own ranks. The Air Service prepared no master plan and provided no coordination between units. Waco, Texas, became "one of the most thoroughly recruited districts known to man." The city had permanent recruiting parties from the general recruiting services of the Army, Navy, and Marine Corps, as well as from Rich Field, the Air Service station at Waco. In addition recruiting parties from Brooks Field and Kelly Field near San Antonio visited Waco. Having little success, recruiters from Rich Field doubted if the practice of sending parties into districts covered by local Air Service posts would be continued if the expense of such trips was checked against the results obtained.[19]

Recruiting in the spring of 1920 brought in 1,209 enlistments by June 30. At that time enlisted strength stood at 8,428. The Air Service then had only 155 Regular officers, the remaining 1,013 on active duty still having temporary appointments. Thus, the Air Service was considerably below the 1,516 officers and 16,000 enlisted men authorized by Congress in the National Defense Act of 1920.[20]

Surplus Property

During the war the government had established flying fields, balloon stations, supply and repair depots, acceptance parks, schools, a large experimental station, and other installations for the Air Service. In most cases the government leased these facilities or constructed them on leased land. The Air Service abandoned some installations soon after the Armistice. It no longer needed acceptance parks, for instance, after completion of deliveries on airplane contracts. Primary flying training, previously given at fifteen flying fields, was consolidated at Carlstrom and March Fields. The Air Service could not immediately dispose of fields no longer needed for training. It needed time to figure out what facilities would be required for peacetime, and that awaited congressional action on peacetime military policy. It also needed time to dispose of property, including hangars, airplanes, engines, and other equipment, at the various fields. Surplus flying fields became temporary storage depots.[21]

Liquidation of the United States Spruce Production Corporation took a long time—twenty-eight years. Authorized by Congress, the corporation produced wood for American and Allied aircraft production during the war. At the Armistice its properties encompassed thousands of acres of timber, five large sawmills, nearly thirty million feet of lumber, four railroads with locomotives and other equipment, automobiles and trucks, and even a hotel. The government disposed of the bulk of the property rather quickly, but the contract for the sale of one of the railroads contained a payment schedule that ran to December 1946.[22]

Nearly the entire aviation industry in being at the Armistice had been created during the war, either anew or by conversion of other industry. Some production capacity was of necessity carried over into peacetime to furnish a base for mobilization. The same situation existed with regard to ordnance material, so the Army Ordnance Department retained plants on standby. The Air Service hoped development of civil aviation would create a large demand for aircraft and thus support an industry available for expansion in any future emergency.[23]

When hostilities ceased, hundreds of factories in America were turning out aviation equipment for the United States and her Allies. Factories in Europe were also making equipment for the U.S. Air Service. A War Department claims board, headed by Assistant Secretary of War Benedict Crowell, was responsible for terminating War Department contracts and settling claims with American industry. The bureaus themselves, however, settled most claims.[24]

In closing contracts with American industry, the Air Service sought to ensure economical use of materials and parts on hand and to keep factories going until manufacturers could shift from military to commercial production. It asked airplane companies to stop producing when they used up subassemblies already on hand. At the Armistice, firms had delivered less then half of the planes contracted for. The Air Service reduced the balance due, some 15,000 planes, to 2,000. Of the more than 68,000 motors on order, nearly 50,000 remained to be delivered; it cut this number to under 30,000. By June 30, 1919, the Air Service claims board liquidated 91.5 percent of the 5,000 orders outstanding on November 11, 1918. The value of production terminated amounted to about $300 million.[25]

A War Department board went to Europe to settle the business of the Allies in the United States, including large contracts for Liberty motors. Another closed out American contracts and disposed of surplus military property in Europe.[26] Much of the aviation material held by the AEF was sold to France and other European countries. Disposing of the remainder, the Air Service, AEF, sent home some 2,000 airplanes and about 1,000 engines. Among the planes were about 600 American-built de Havillands, 1,100 of various types purchased from the French and British, and 300 captured from the Germans. These constituted the AEF's entire supply of good planes except for the few retained in Europe for the U.S. Army of Occupation. The War Department ordered sold or destroyed nearly 2,300 other planes not worth sending home. There being no market for them in Europe, the task of destroying them fell to Maj. Gen. Mason M. Patrick, Chief of Air Service, AEF.

General Patrick, a veteran of more than thirty years of commissioned service, took the precaution of having each plane surveyed by several boards. He saved everything usable (engines, propellers, machineguns, bomb racks, radios, instruments, wheels, tires, and sometimes whole wings). Patrick sent metal parts to the quartermaster to sell as junk, and burned what remained (wooden framework, doped fabric, and little more). As he anticipated, the Air Service had to explain why equipment costing the American people millions of dollars was destroyed: The planes were in very bad condition; it was cheaper to destroy old planes in Europe and buy new, up-to-date models later. But this did not convince everyone.[27]

With tons of equipment excess to current needs, the Air Service could sell some but had to consider the effect on the economy. It transferred items to satisfy the wants of other government departments, but kept enough to meet future requirements and to build up a reserve for tapping in an emergency. Still some items seemed of no value to anyone.

Right after the Armistice the Air Service set up an organization to dispose of more than $100 million in airplanes, engines, parts, materials, tools, equipment, and other items.[28] The plan was to sell several hundred training planes and engines on the open market, which should do much to popularize flying in the United States. South America and other countries were also looked to as possible markets for aircraft.[29]

Maj. Gen. Mason M. Patrick, Chief of Air Service

In the spring of 1919 the Air Service announced that anyone wanting to buy a plane or motor should send his name to the Salvage and Sales Branch, 6th and B Streets, Washington, D.C. The branch would notify him when the opportunity arrived to make a purchase. The offer was later withdrawn because the branch could not handle the many transactions involved in selling to individuals. Moreover, the Air Service lacked people and shop facilities to put the equipment in safe operating condition before sale, and did not want to be responsible for placing unsafe equipment in the hands of the American people. The Air Service and War Department were likewise concerned about the ill effects of competing with private business. They decided to sell to the industries that produced the materials.[30]

The disposal program moved slowly at first but was helped along by a few large sales. The Nebraska Aircraft Corporation of Lincoln, Nebraska, bought 280 Standard J–1 training planes and 280 Hispano-Suiza motors. The Curtiss Aeroplane and Motor Company, Inc., Garden City, N.J., purchased 4,608 Curtiss OX–5 motors, 1,616 JN–4 airplanes without motors, and 1,100 Standard planes, also without motors, all for $2.72 million.[31]

Many of the JNs sold to Curtiss came from Love Field near Dallas, which also served as a storage site for surplus DHs. The men at Love took great pride in the ingenious and efficient system they devised for de

Havillands. Having removed the wings, they dovetailed the fuselages on the floor of a hangar and placed the wings on racks overhead. That way they got sixty-eight planes in a single hangar.[32]

Excess equipment was scattered among many facilities across the country. The Air Service often had to move items from one place to another as it closed temporary facilities or withdrew material to be reworked for use. On one occasion, for example, it sent forty-eight Hispano-Suiza H motors from Little Rock, Arkansas, where the Air Service had some thirteen thousand motors of various types in storage, to the repair depot at Fairfield (near Dayton), Ohio. In time, accidents and normal wear and tear on aircraft and equipment consumed much of the material left over from the war. The 12th Aero Squadron at Nogales, Arizona, was delighted to exchange old planes for new ones, the new ones being DH–4s recently reassembled after many months in storage.[33]

Over several years the Air Service withdrew many DH–4s from storage for conversion to DH–4Bs. One project in 1923 took 260 planes from storage at San Antonio and Fairfield for conversion at several different airplane factories. About the same time all of the DH–4s, along with Liberty motors and spare parts, previously sent to the Philippines to be stored as a reserve, came back to the United States for conversion to DH–4Bs. When workmen at Rockwell Field outside San Diego opened one of the crates, they found a motor with a remarkable history. Built in Detroit, it went to France, back to the United States, then to the Philippines, and now to Rockwell—without ever being used.[34]

The Air Service endeavored to concentrate certain classes of material at one or two points. Little Rock, for instance, became the principal but by no means the only storage facility for engines. Furthermore, the Air Service took great pains in preparing material for storage when long-time preservation was desired. To retard deterioration of airplanes, workmen stripped off the fabric, applied an extra coat of varnish to glued joints, painted the metal with red or white lead, and placed wings in racks to keep them from warping. Motors were thoroughly cleaned, covered inside and out with a rust-inhibiting grease, and put in a cool, dry place. Thousands of propellers were stored at Middletown, Pennsylvania, under controlled humidity and temperature.[35]

The Air Service revised the storage and disposal programs from time to time, shifting material from one program to another as policies and requirements changed. By June 1920 items costing $94.2 million had been reported for disposal. Sales, transfers to other agencies, and other transactions that included withdrawals for retention by the Air Service, accounted for another $79.1 million. This left about $15.1 million worth for disposal. Unserviceable airplanes, motors, and other equipment made up more than two-thirds of this residue.[36]

The closing of a number of stations in 1921 and 1922 made large quantities of surplus, obsolete, unserviceable, and nonstandard equipment available for disposition. Regulations requiring sales by sealed bid or auction were changed to authorize negotiated sales for small lots costing not more than $2,000. General Patrick, who succeeded General Menoher as Chief of Air Service, reported in 1922 on the substantial progress in cataloging excess materials which, except for small lots, were selling at auction. Items awaiting disposition on June 30, 1922, cost almost $29 million. The disposal program, Patrick said, was retarding the development of an efficient, peacetime service.[37]

The closure of more facilities in Fiscal Year 1923 added nearly $33 million worth of material to the disposal program. There seemed to be a good market for cheap planes among ex-flyers, would-be flyers, speculators, and Reserve officers who wanted to keep in flying trim. Dorr Field, east of Arcadia, Florida, offered surplus Jennies with OX–5 motors for $400. Having sold more than 150 Jennies within a few months, Rockwell Field began removing nearly 200 S–4 Thomas-Morse Scouts from storage to sell to the public.[38] During Fiscal Year 1923 the Air Service held fifteen auctions, fifty sales by sealed bids, and numerous fixed-price sales and transfers. General Patrick reported that all surplus property had been disposed of by June 20, 1923.[39]

Such in brief was the course of demobilization. The Air Service stopped production of equipment, terminated contracts, ended training programs, disposed of excess facilities and equipment, and disbanded the wartime force. Within a year, all troops enlisted or drafted during the war were discharged. Coincidentally, a smaller number were recruited and reenlisted for peacetime duty. During the same period, the Air Service discharged about ninety-five percent of the officers commissioned during the emergency. There being so few officers with permanent Regular Army commissions, some temporary officers were retained until Congress enacted legislation for the peacetime officer corps.

Disposition of facilities and equipment took longer than to discharge personnel. Soon after the Armistice, the Air Service began to abandon air parks, depots, flying fields, and other facilities as they became excess to existing or foreseeable needs. For a time, however, it kept some extra training fields for temporary storage of materiel. After preserving items for future use, the service disposed of surplus stocks by transferring some to other government agencies, selling others, and destroying those no longer of value.

At the completion of demobilization, the Air Service found its personnel strength far below the level its leaders deemed necessary for the nation's defense. As for materiel, the service retained enough from the war to last several years.

DEMOBILIZATION

Demobilization proved a large undertaking, placing Air Service officials under tremendous pressures. These came from a public wanting troops discharged immediately, from persons and organizations who saw profits from dealing in surplus property, and from a citizenry and their elected representatives ever watchful for mismanagement and fraud. The absence for twenty months of an approved peacetime national military policy rendered the task more difficult. Though of the first order of importance, demobilization was but part of the Air Service story in the years after the Armistice.

Chapter II

The Flying Game

During the war, most Army flying in the United States took place in training pilots, observers, bombardiers, and gunners, or in developing and testing equipment. Seldom did it involve going across country from place to place, and then not very far. But this changed with the Armistice. With plenty of men, time, money, and planes (plenty at least for the moment), the Air Service flew off to arouse interest in aviation, win public support, and collect information for the further development of military and civil aeronautics. Aerial performances for the entertainment and enlightenment of the public, pioneering cross-country flights, and aerial contests during the year following the Armistice reveal something of the character of the U.S. Army aviation and the nature of flying in that day.

The Flying Circus

The U.S. Army Air Service found that a county fair, a horse show, a patriotic meeting—any large gathering of people—could be grasped as the occasion for an air show. With no such event scheduled, the sound of planes and the prospect of some stunts seemed reason enough for a crowd to gather. The Air Service encouraged units and stations to put on such performances and report them along with their cross-country flights and other newsworthy activities to Washington for a newsletter to be distributed in the Air Service and released to the press.

Rockwell Field outside San Diego put on a big show for the public in February 1919. Earlier, to celebrate the Armistice, the Commanding Officer, Lt. Col. Harvey B. S. Burwell, led 212 airplanes (described as the greatest number ever in the air at one time in the United States) in a flight over the city. Now San Diegans would have an opportunity to inspect the field. They came by train and motor car to see the hangars, machine shops, schools, and airplanes, listen to the bands, watch athletic events, eat barbeque, dance at the cabaret, see broncobusting by troopers from a nearby Army remount station, and take in the sideshow. A sign at the shooting gallery read: "Get 5 planes and become an ace." The entertainment included a balloon ascension and parachute drop, airplanes flying in formation, combat between two aces just back from France, some trick flying, and a bombing raid. The profit from the circus ($4,742) went to the Rockwell Field athletic fund. The climax was a ceremony in which Col. Henry H. Arnold, Supervisor of the Western District, Department of Military Aeronautics, presented decorations for action in the war in Europe. A French Croix de Guerre went to Maj. Kenneth Marr, who had fought with the Lafayette Escadrille and commanded another famous squadron, the U.S. 94th; there was a Distinguished Service Cross for Maj. Carl Spatz for heroism while flying in combat with the 13th Aero Squadron.[1] (The latter officer did not become Spaatz until 1938, when he changed his name in an attempt, only partially successful, to get people to call him "Spots" instead of "Spats.")

Kelly Field Flyers that are part of the "Flying Circus," with Maj. George E. Stratemeyer (center, standing) in command.

Each field had its own idea of how to run a circus. A show at Kelly Field on the evening of April 12, 1919, depicted "War Swept France" and "the horrors of war, as carried out by the Huns." The scene was a city attacked by the Germans. After artillery shelling and bombing from airplanes, the Germans entered, looting and burning homes, capturing and killing people. Suddenly, a squadron of American planes swept across the sky and engaged the German planes. As American flyers won the battle in the air, American infantry went over the top and drove out the Huns. Here, the ads said, was "a rare opportunity for the general public to see a wonderful show."[2]

The Air Service grasped every chance for display and publicity. When Theodore Roosevelt died in January 1919, Lt. Col. Millard F. Harmon, Jr., Commander at Hazelhurst Field on Long Island, sent planes to patrol and drop wreaths over the ex-President's home.[3] When President Woodrow Wilson returned to Washington from the Paris Peace Conference at the end of February, Capt. Roy Francis flew his new Martin bomber low over the parade route, taking motion pictures as he went. On that occasion, Bolling Field contributed a variety of other planes to the "air parade," which had become "an essential part of any self-respecting celebration." The 29th Balloon Company came from Aberdeen Proving Ground with a balloon to tow in the parade, but the balloon floated off with 1st Lt. G. H. McMillan and a cameraman in the basket.[4]

An aeronautical exposition sponsored by the Manufacturers Aircraft Association in New York during the first half of March 1919 afforded good exposure to the Air Service as well as to the aviation industry. The Air Service displayed planes and equipment. Aerial traffic to New York was "fairly heavy" on Friday, March 14, Air Service Day. Among those flying from Washington for the event were Assistant Secretary of War Benedict Crowell, Maj. Gen. Charles T. Menoher, Maj. Gen. William L. Kenly, Brig. Gen. William Mitchell, and Senator Key Pittman of Nevada. Captain Francis kept busy taking newsmen up in his Martin bomber.[5]

The newsmen were fortunate. Many people would have liked a ride with an Air Service pilot, but joyrides for civilians were prohibited. Pilots could take any member or employee of the Air Service, or any officer or enlisted man of the Army, Navy, or Marine Corps as a passenger. If the President, the Chief Justice, or a member of Congress wanted to go, a pilot could take him. Anyone else needed permission of the Secretary of War. "No exceptions," General Kenly said.[6]

With businessmen, actors, movie stars, and other celebrities contributing their services to the U.S. Government's Victory Loan campaign in 1919, the Air Service helped by putting on what it called the "greatest flying program the United States has yet witnessed."[7] It organized a Victory Loan Flying Circus of three flights under the command of Maj. Ora M. Baldinger. Each flight consisted of about 22 officers, 50 enlisted men, 18 airplanes (5 Fokkers captured from the Germans, 4 SPADs, 4 SE–5s, and 5 Curtiss-Hs), and a train of 9 baggage cars, 3 sleepers, and a diner. Maj. Henry J. F. Miller commanded the eastern flight, based at Hazelhurst Field; Maj. George E. Stratemeyer the middle western, at Ellington Field; and Major Spatz the far western, at Rockwell Field.

On the 30-day tour commencing April 10, 1919, the 3 circuses covered more than 19,000 miles in one-night stands, playing in 88 cities in 45 states. They also put on sideshows at nearby cities off the scheduled route. The performance, which ordinarily began at about 1330, featured exhibition flying, acrobatics, and a sham battle in which Fokkers attacked Curtiss planes which were rescued by SPADs and SE–5s. There were Victory Loan speeches, and the city was attacked with pasteboard bombs containing Victory Loan leaflets. The rule against joyrides being relaxed, persons at each place who subscribed the largest amounts to the loan got free rides in Air Service planes.

The Air Service used the Victory Loan campaign in April 1919 to recruit and to collect information on landing fields, the nature of the terrain, and other things "of aid to military, postal, commercial, and sporting aeronautics." When Major Spatz flew a Fokker across the mountains from Helena, Montana, on April 26, 1919, to put on a sideshow at Great Falls, he carefully recorded his flying altitudes. Like other pilots of the day, Spatz constantly watched for places where he might land in an emergency. He reported seeing on this trip only one stretch, about fifteen miles, without a landing place.

Using unfamiliar and makeshift fields presented many hazards to the men of the flying circuses, as Major Spatz could testify from what happened to him when he arrived at Great Falls. Clearing one telegraph line and gliding under another, he landed safely. But when he started to taxi to the side of the field he ran into a ditch, breaking the tail skid and damaging the rudder.

Master Electrician Victor V. Allison of the middle western Flying Circus averted a wreck in Milwaukee by heroic action. Trees and houses surrounded the field, Washington Park, which was enclosed by a fence and a trench on 3 sides. Curtiss planes, arriving first, landed safely. Then came 1st Lt. Franklin O. Carroll in a SPAD. Members of the Flying Circus and 25,000 spectators watched him attempt to land 3 times and saw him pull up each time to go around again. On the fourth try he made a good landing about 200 feet from the end of the field. There was no way he could avoid crashing into the fence. Allison ran out and, with the SPAD rolling about 40 miles an hour, grasped the left wing and turned the plane aside. Although dragged about 75 feet, he sustained only minor bruises, and immediately returned to his duties as crewchief of another plane.[8]

The general policy of the Air Service immediately following the Armistice was to accept invitations to put on exhibitions and demonstrations. Any reasonable request to the commander of a flying field received prompt approval. The President of the National Association of Air Pilots complained that Air Service demonstrations deprived civilian pilots of a living. General Menoher thought such demonstrations a good thing because they enabled people in different parts of the country to become acquainted with the character of the people in the Air Service and with the use and performance of Army aircraft.[9] The Air Service, however, was involved in so many things that Menoher decided it should not take part in a big aeronautical exhibition at Atlantic City in May 1919.[10]

Cross-Country Flying

Flying across country, pilots kept a constant lookout for landing places, as Spatz did between Helena and Great Falls. It was nice to have a place in sight if motor trouble—a rather frequent occurrence—made landing necessary. Some flights were for the specific purpose of locating landing fields and charting air routes. On one of the first, Maj. Albert D. Smith, director of training at Rockwell Field, took off on December 4, 1918, with five JN–4s to map a route to El Paso. Receiving permission to extend the flight, he continued to Jacksonville then to Washington and New York City, returning to San Diego on St. Valentine's Day, 1919.

One crew came back much earlier. The very first day Lt. Bruce Johnson became separated from the flight in a fog and landed at El Centro, California. Wrecking his plane while trying to take off, he shipped it home by rail. The other four planes (piloted by Major Smith and Lts. H. D. McLean, Robert S. Worthington, and Albert F. Pyle; with Maj. James H. McKee, flight surgeon; Lt. John W. Evans, photographer; and Sgts. William G. Lew and Robert P. Blanton, mechanics) made it safely to El Paso by December 7, 1918. They reached Houston on the 11th, New Orleans, the 14th, Montgomery, the 16th, Americus, Georgia, the 17th, and Jacksonville, the 18th. Having made the U.S. Air Service's first transcontinental flight,[11] Smith and his men flew south to Dorr Field to spend Christmas before heading northward to Washington. (*Map 1*)

Since leaving California the men had met with no sizable problems or serious difficulties. Soon after leaving Savannah, however, one of the planes landed because of motor trouble. So as not to separate, the others also came down. Repairs did not take long, but the field was too small for takeoff. The men disassembled the planes, hauled them down the road to a bigger field, and reassembled them—costing a day. They lost more time at Raleigh due to bad weather, and did not make Washington until January 6, 1919. One plane needed its engine overhauled, so they borrowed a plane for the flight to New York City, touching down there on January 7.

Major Smith's flu delayed the return trip until January 26. The men left New York in the same four planes they had flown from San Diego early in December. The aircraft were overhauled at Mineola, New York, but the only major changes on the way east had been replacement of four propellers and one engine.

Rough weather, high winds, and soft fields plagued the flight westward. On February 7 Lieutenant Worthington hit a tree while trying to take off in the mud at Vernon, Alabama, some seventy-five miles west of Birmingham. Abandoning the plane, Worthington and McKee took the train back to California.

The other three planes reached El Paso on February 11, and the men planned to leave early the next morning. But during the night a sixty-mile-an-hour wind badly damaged two of the planes (which were later sent to Rockwell Field by train). After being delayed by strong winds, Smith and Blanton arrived at Rockwell on February 14, 1919. In 73 days they had covered some 6,600 miles in a flying time of about 110 hours.[12]

The story of Smith's flight reveals some of the difficulties and hazards airmen contended with in those days. Forced landings were a common, accepted fact of flying. They often resulted in claims against the government for damages. A farmer in Texas told 2d Lt. Harold W. Beaton the wheels did not do much harm to his cotton but the "stinger" on the tail "tore up quite a bit."[13] Another farmer thought $25 would cover the damage done when "U.S. Flying Ship No. 4217 landed in my cotton patch . . . and it nock [sic] and blew out about 400 lbs of cotton and thrashed out about a bushel of peas . . . "[14] Spectators sometimes paid little heed to a flyer's warning that one could get hurt by a turning propeller. The flyer needed to keep his eyes open to see that spectators did not carry off souvenirs and that farm animals did not damage his plane. Cattle seemed to have a taste for airplane dope. After making a forced landing near Fort Worth, 2d Lt. Robert W. C. Wimsatt notified the officer in charge of flying at Brooks Field: "Delayed due to cow eating wing. Home tomorrow." When he returned, his plane bore patches as proof that "some unprincipled bovine with a low sense of humor and depraved appetite had eaten large hunks out of the lower wing panels and stabilizer." Such hazards were part of the "flying game."[15]

Although a forced landing often destroyed the plane, flyers frequently walked away from their wrecks. Close scrapes made news. Accidents were not bad publicity unless the fatality rate went too high.

Accidents rose right after the Armistice. Pilots learned acrobatics as part of their tactical training during the war, but for use only in combat. Stunting for the thrill of it, to show off, to entertain someone, was forbidden (though not entirely suppressed). The barriers came down when pilots took to the air to celebrate the Armistice. Never had the American people witnessed such a display.

The accident rate went up at once owing to what the Air Service called "indiscriminate flying and failure to observe even the common rules of flying." General Kenly did not, however, want to prohibit all stunting. On December 12, 1918, he directed commanders to stop low flying and acrobatics in the vicinity of cities, towns, and buildings, and to see that all acrobatics finished at an altitude of at least 1,500 feet. Kenly told them to stringently carry out the order. Ground any pilot, he said, who failed to comply. If the offending officer was in the Regular Army, he should be court-martialed; if a Reservist, discharged.[16]

By March 1919 the rate for fatal accidents still stood well above what it had been in the United States during the war,[17] and it stayed high. Unreliable equipment, improper maintenance, bad gasoline, inadequate facilities, poor instruments, lack of navigational aids, and perilous operating conditions caused accidents for even the most experienced and careful flyers. The 8th Aero Squadron on border patrol in Texas "washed out" twenty-two planes in one year. The Mechanics School at Kelly Field reported three wrecks in one week.[18] And Bolling Field, to cite but one more example, recorded two wrecks on Monday, April 19, 1920, and three more the following Thursday.[19]

The "indiscriminate" kind of flying that drew Kenly's attention brought complaints from sportsmen and others interested in wild life. Pilots disturbed migratory ducks by buzzing their feeding places. And that was not all. They were shooting the birds with their machineguns. Kenly declared such hunting "absolutely forbidden." He ordered flying to be conducted so as to interfere as little as possible with wild fowl. Offenders in the future would be brought to trial.[20]

It took more than orders and regulations, however, to keep some men in line—2d Lt. James H. Doolittle, for example. His uninhibited style of flying and frequent wrecks kept him in trouble much of the time with his commander at Rockwell Field. Colonel Burwell nevertheless gave Doolittle and two companions, Lts. Walter Smith and Charles Haynes, three Jennies for a coast-to-coast flight. This venture ended suddenly when Smith and Haynes wrecked their planes before they got out of California.[21]

Another who tried the coast-to-coast flight early in 1919 was Maj. Theodore C. Macauley, Commander of Taliaferro Field, near Fort Worth. He and his mechanic, Private Staley, flew westward from Fort Worth to begin the transcontinental flight at Rockwell Field, on January 21, 1919. They used a DH-4 with tanks for an additional fifty-seven gallons of gasoline and ten gallons of oil. On the way east, the failure of first the generator then the battery forced them down near Hot Wells, Texas. After changing connections to the lighting battery, Macauley and Staley flew on to Pecos, Texas, where they obtained a new battery. Landing at Baton Rouge on the 23d, the plane went up on its nose, but without damage. Turning southward at Americus,

Georgia, Macauley and Staley reached Arcadia, Florida, on January 24. Thus far they experienced no major difficulties or long delays.

En route to Jacksonville, motor trouble on January 25 caused Major Macauley to land in tall weeds at the north end of Lake Okeechobee. Getting help, he and Staley pulled the damaged plane to the New River Canal and took it by barge to Fort Lauderdale, and from there to Miami to be restored to flying condition at the Marine base.

Major Macauley and Private Staley resumed their trip on February 1. Shortly after they left Americus, the water temperature went up and the motor quit. In landing, Macauley hit a wire fence and the plane turned over, damaging the propeller, the radiator, and three wing sections. Men from Souther Field picked up the wreck and sent it to the repair depot at Montgomery, Alabama. The commander at Souther gave Macauley another de Havilland.

Between Montgomery and West Point, Mississippi, rain damaged the propeller of the second plane. Payne Field outside West Point supplied a new propeller, but the field was too soft for Macauley to take off. Leaving the plane, he and Staley boarded the train to Fort Worth. Two weeks later on February 16, a JN–4H carried Macauley to West Point. He flew the DH–4 to Taliaferro Field the following day, thus completing what the Air Service called the "first round trip" across the United States.[22]

Major Macauley made another transcontinental flight in April 1919. Starting from Fort Worth on Saturday, April 12, he flew to Tucson, then to San Diego and back to Tucson on Sunday, and back to Fort Worth on Monday. Delayed by bad weather farther eastward, he did not reach Jacksonville until Thursday. His flying time from the west coast to the east (2,366 miles) was 19 hours and 15 minutes. He started back at once and, after being forced down by motor trouble in Mississippi, arrived at Fort Worth the next day. He had completed the round trip in 7 days with a flying time of 44 hours and 15 minutes.[23]

Major Macauley was only one of the many flyers with propellers damaged by rain. DHs used props of wood covered at the tip with doped fabric. Raindrops cut the fabric and chewed the wood to bits, throwing the prop out of balance and causing the motor to vibrate with all sorts of dire consequences. The Engineering Division at McCook Field, Dayton, Ohio, was trying to develop better propellers, but meantime pilots were told to avoid heavy rain. When they could not, they might reduce the impact of the drops by throttling down the motor.

A pilot who took a colonel from Bolling Field to Asheville, North Carolina, in a DH–4B flew in heavy rain for two or three minutes near Bedford, Virginia. From the vibration that started he knew his prop had been badly chewed. At Asheville he shaved the propeller smooth and got the radiator soldered, but as soon as he started the motor the cracks opened and the radiator began leaking again. On the way back to Washington they ran into more rain. The prop began to thump because of air pockets in the fabric tips. The generator burned out. The radiator still leaked. The gas supply ran low. Developing what he called a "pardonable curiosity" about where he was, he landed in a field beside a creek and found he was at McDowell, west of Staunton, Virginia. He "once more whittled the prop smooth with a jacknife," got gas from a general store, filled the radiator from the creek, checked the battery, and took off again. He and the colonel arrived at Bolling Field "with the still leaking radiator streaming water, the propeller . . . screaming like a squad of fire engines doing double time, the battery singing its 'Swan-Song' . . . , the motor reciting in spits and coughs . . . , and with struts and wires putting the most vigorous of 'shimmy dancers' to shame, on account of woodwork warped from the rain." "But what cared we," the pilot said, "we were home."[24]

While Major Macauley was crossing the American continent, other airmen were attacking the Atlantic Ocean, seeing it as the biggest challenge of the day. Who would take the $50,000 the London *Daily Mail* had offered in 1913 for the first nonstop crossing? A U.S. Navy NC–4 flying boat, with Lt. Comdr. Albert C. Read in command, made the first crossing in May 1919 but stopped at the Azores on the way from Newfoundland to Portugal. The British team of Harry Hawker and Lt. Comdr. Kenneth F. MacKenzie-Grieve attempted a nonstop flight the same month but were forced down at sea.

The U.S. Air Service planned for Capt. Roy Francis to fly the Atlantic in a Martin bomber, but later decided the Martins needed more testing before venturing on a long flight to sea. Francis would make a one-stop transcontinental flight on June 10, 1919. But the Martin factory at Cleveland did not get the bomber ready until June 14 to fly to McCook Field for testing. The following day Capt. John Alcock and Lt. Arthur Whitten-Brown of Great Britain won the *Daily Mail's* $50,000 by flying nonstop from Newfoundland to Ireland. Work on the Martin bomber at McCook further delayed Captain Francis. On July 25 he flew the plane to New York where plans for the transcontinental trip came to a sudden end. Wind wrecked the Martin bomber as it sat in a hangar at Hazelhurst Field.[25]

On July 24, 1919, while Captain Francis was still at McCook, Lt. Col. Rutherford S. Hartz took off from Bolling Field in another Martin bomber on a flight around the rim of the United States. His crew included 2d Lt. Ernest E. Harmon, assistant pilot, and two master signal electricians, Jack Harding and Jerry Dobias, to serve as mechanics. Their main purpose was to test the endurance of the Martin bomber and the Liberty engine, but they also expected to arouse interest in aviation, encourage enlistments in the Air Service, chart routes, and locate landing fields.

Colonel Hartz and crew experienced no major difficulties on their way

Lt. Comdr. Albert C. Read completes first crossing of Atlantic by air, May 1919.

to Hazelhurst Field on Thursday or on the flight to Augusta, Maine, on Friday. However, en route to Cleveland on Saturday, heavy rain east of Lake Champlain, New York, and trouble with the compass forced them to land. A strut broke, the right wing caught, and the plane went up on its nose, hurling Harmon out over Hartz's head. Hartz reached out—"instinctively," he said—and grabbed Harmon's collar, turning him over in the air so he landed on his feet instead of his head. Everyone else was still safely seated. A month passed before the plane could be flown from the field near Jay, New York, where it came down.

Colonel Hartz made six more forced landings during the expedition, one due to failure of gasoline gauges, another owing to forest fires in Montana, and four because of weather. The men stayed in Cleveland two days to dope and varnish their plane, 1 day in Detroit to repair a radiator, a week in Milwaukee awaiting replacements for four blown-out tires, 10 days at San Diego for overhaul, and a day at Los Angeles for rest. Other stops included 1 day for a speech at a Rotary Club in St. Paul, another for a speech at the Aviation Club in Seattle, and yet another for a Chamber of Commerce lunch at Portland. Thus, Colonel Hartz and his men did not get back to Bolling Field until November 9, 1919. In 108 days they had flown 114 hours and 25 minutes covering 9,823 miles around the rim of the United States.[26] (*Map 1*)

Meanwhile, Maj. Ora M. Baldinger led a flight of nine Curtiss planes on a pathfinding expedition across Pennsylvania, Ohio, Indiana, and Wisconsin to Minneapolis. This group collected data on aerial routes and landing fields, encouraged cities and towns to build flying fields to government specifica-

MAP 1

Above. Martin bomber used on Round–the–Rim flight; center: Lt. Ernest E. Harmon poses with family after the flight; left: Lt. Col. Rutherford Hartz stands in front of plane.

tions, inspected proposed sites, and made recommendations for improving and maintaining them. During the two-month survey, the "All-American Pathfinders," as the group was called, found keen interest and enthusiasm everywhere.[27]

Aerial Contests

The general tendency during 1919 was toward centralizing control over Air Service flying. The object was to prevent the indiscriminate flying referred to previously, reduce the accident rate, avoid dissipation of energies and resources on ill-conceived and poorly executed ventures, and in general regain the discipline that had been lost after the Armistice. The policy of allowing local commanders to approve cross-country and exhibition flights changed in September. Only routine training flights could be authorized locally. All others required approval by the Director of Air Service. Exhibition flights were discouraged except for recruiting. Any request for an exhibition went to Washington with the commander's comments on his ability to take on the extra work.[28] Afterwards the Air Service tended to take part in fewer but larger, more significant events.

One of the first big competitive events in which the U.S. Air Service participated was an airplane race and handicap contest between New York City and Toronto, in the summer of 1919. A few months earlier, General Menoher ruled that Air Service pilots could not compete with civilian pilots in events offering rewards. He thought it "improper to use government time and property in competing for prizes and trophies."[29] Later he relaxed the rule to permit Air Service personnel to join in such events if they did not accept prizes. Thus civilians who entered the New York-Toronto contest, sponsored by the American Flying Club of New York, could compete for prizes offered by New York City's Hotel Commodore, but Air Service pilots could not.

Regarding the contest as an opportunity to gain much valuable information about the relative merits of various machines and motors, the Air Service entered thirty-two planes of various kinds. They came from Hazelhurst, Bolling, Selfridge, Langley, and McCook Fields, the depot at Middletown, Pennsylvania, and the Office of the Director of Air Service. Those from McCook and Selfridge started at Toronto, the others at Roosevelt Field, Long Island, New York.

The rules gave contestants two days (daylight to dark) to complete the round trip. At control stops at Albany, Syracuse, and Buffalo, each pilot circled the field once, landed, refueled, remained thirty minutes and, after receiving clearance from the control officer, took off again. Delays caused by bad weather could be added to extend the two days. The Army Coast Artillery put searchlights at landing fields to aid pilots who arrived after dark.

As indicated by the name of the event, there actually were two contests—a speed test to see who could complete the round trip in the shortest time, and a reliability test pitting each plane against its theoretical performance as calculated by aeronautical engineers. Further, in each of these tests there were two contests, one among civilian pilots, the other among Air Service flyers. Although started by a civilian organization and supported by several flying clubs, the meet came to be more or less dominated by the Air Service. Of the forty-nine entries starting on August 25, 1919, about two-thirds were military.

Bad weather delayed the start from Roosevelt Field for five hours. Thirty of those who started completed the round trip of 1,040 miles. First Lieutenant Belvin W. Maynard, the "Flying Parson," won the speed contest for the Air Service by averaging 133.8 miles per hour in his DH-4. Maj. Rudolph W. Schroeder, flying a Vought VE-7, won the handicap contest. Roland Rohlfs, chief test pilot for the Curtiss Aeroplane and Motor Company, Inc., flew a Curtiss Oriole to win both the speed and handicap contest for civilians.[30]

The following month, September 1919, the Air Service competed with the U.S. Navy in a balloon race organized by the Missouri Aeronautical Society. Each service entered three balloons of about fifty thousand cubic feet, using coal gas, and carrying a crew of two (pilot and aide). The race began at St. Louis on September 26. One Navy balloon leaked gas and did not start. One Air Service balloon landed in the heart of St. Louis after a twenty-minute flight. Capt. Eldon P. Phillips and 1st Lt. Byron T. Burt of the Air Service landed in Lake Michigan off Wisconsin's Door Peninsula and were rescued by a passing boat. Lt. G. R. Emerson and Ensign F. L. Sloman of the Navy landed near Stittsville, Michigan. It was uncertain which of the two had traveled the greater distance until the Coast and Geodetic Survey determined that Phillips went 491.8 miles, Emerson, 486.4. The Navy's other balloon also landed in Michigan (479 miles), the other Air Service balloon in Wisconsin (470 miles). Phillips being the winner, Maj. Albert B. Lambert, President of the Missouri Aeronautical Society, awarded a silver trophy to the Balloon Division of the Air Service.[31]

The biggest and most important aerial contest held in the United States during 1919 was the first transcontinental reliability and endurance test. Brig. Gen. William Mitchell and the staff of the Training and Operations Group, Office of the Director of Air Service, commenced planning at once after the New York-Toronto contest. General Menoher envisioned a peacetime maneuver to test Air Service personnel and equipment. Mitchell also regarded the test as a field maneuver, "but one calculated to yield a far greater profit to the Air Service and to the cause of aeronautics in general than any field maneuver ever did before." One aim would be to determine what could be done with an Air Service that had lost a large percentage of its most valued people and was cumbered with obsolete equipment scattered among deserted flying fields across the nation. Another would be to subject personnel, organization, administration, equipment, the supply system, and communications to close scrutiny. In addition the test would be concerned with landing fields, meteorology, navigation, and other matters pertaining to air travel. Further, Mitchell saw it awakening interest in aviation which "promises to equal in importance the land and sea forces."[32] For the American people, however, the reliability and endurance test became a thrilling race across the continent.

Lt. Col. Harold E. Hartney, who prepared the plan in the Operations Division, laid out a course from New York to San Francisco by way of Buffalo, Cleveland, Chicago, Omaha, Cheyenne, Salt Lake City, and Reno, with twenty-nine control stations along the 2,701-mile route. To test the ability of the people and the organization to respond on short notice, General Menoher did not announce the race until September 20, 1919, eighteen days before the start. Any government plane and pilot could enter with the recommendation of the commander of a flying field or department, provided the plane was rated one hundred or more miles per hour and was standard in every respect. (*Map 1*)

The rules, similar to those for the New York-Toronto contest, called for planes to stop for at least thirty minutes and not more than forty-eight hours at each control point, counted time on the ground elsewhere as flying time, and prohibited flying at night and on Sunday. Planes would start at the same time at either end of the route, with three classes of competition each way: the shortest elapsed time, the shortest flying time, and the fastest flying time based upon the handicaps given the various kinds of machines.

Much work went into locating and establishing landing fields, arranging for spare parts, gasoline, oil, maintenance, weather forecasts, and communications, and taking care of the many other things necessary to carry out the test. Seventy-four planes entered: 46 DH–4s, 6 DH–4Bs, 1 DH–9, 7 SE–5s, 5 Fokkers, 3 Le Pere LUSAC–11s, 2 Martin bombers, 1 Ansaldo SVA–5, 1 Thomas-Morse MB–3, 1 SPAD, and 1 Bristol Fighter. Fifty-eight registered for the race from New York to San Francisco, 16 for the race from west to east. Two pilots, Col. Townsend F. Dodd and Maj. Patrick Frissell, died in wrecks before the race began. Several others did not start for one reason or another. Forty-six planes started from Roosevelt Field, Long Island, and 15 from the Presidio of San Francisco. All but two pilots belonged to the Air Service: Brig. Gen. Lionel E. O. Charlton of the Royal Air Force flew the Bristol, 1st Lt. G. B. Newman of the U.S. Marine Corps one of the DH–4s.

The starting time was 0900 in New York, 0600 in San Francisco, on Wednesday, October 8, 1919.

General Mitchell and Assistant Secretary of War Crowell were among the many people at Roosevelt Field for the start. Fuel problems delayed General Charlton, first in line for takeoff. Consequently, 2d Lt. Jerome B. Machle and his mechanic, SFC Jessie D. McClure got away first. Next came Colonel Hartney in a Fokker. To prevent congestion at Binghamton, New York, the first control stop, half of the planes were held at Roosevelt Field until afternoon.

Lieutenant Maynard, the ninth one off from Roosevelt Field, quickly captured the lead in the east-west race. Having driven his DH–4 810 miles the first day, he and his mechanic, Sgt. W. E. Kline, and his police dog, Trixie, spent the night at Chicago. Three planes were 160 miles behind at Bryan, Ohio, 4 at Cleveland, 18 at Buffalo, 9 at Rochester, and 2 at Binghamton. The SPAD and SVA–5 had not reached Binghamton. Other planes had made forced landings, and 4 had been wrecked. General Charlton hit a fence in making a forced landing in his Bristol Fighter near Ithaca, New York. Second Lieutenant George C. McDonald hit a ditch when motor trouble forced him to land at Plymouth, Pennsylvania. First Lieutenant D. B. Gish's DH–4 caught fire over western New York. Gish made an emergency landing, and neither he nor his passenger, Captain de la Vergne of the French air service, was injured, but the plane was destroyed. Gish, who had been in the hospital recovering from wartime injuries and had been let out long enough to take part in the test, was not about to give up. He went back to New York, obtained another plane, and started again. A forced landing resulted in the death of Sgt. W. H. Nevitt, who flew with Col. Gerald C. Brant. An oil line broke on their DH–4 and, as Brant tried to land, the Liberty motor cut out and he lost control. The plane plunged to the ground, killing Nevitt and injuring Brant.

Heavy fog Wednesday morning did not prevent flyers at the Presidio from winging eastward. Despite snowstorms causing hazardous flying in the mountains, eleven planes arrived at Salt Lake City before night. One DH–4 crashed in landing. Both the pilot, Maj. Dana H. Crissy, and his mechanic, SFC Virgil Thomas, died. Two planes reached Salduro, Utah. Motor trouble forced down two others.

The weather in the eastern part of the country was bad on Thursday, especially east of Chicago, which worked to Maynard's advantage. He got away early but the planes at Bryan, Ohio, could not take off until 1000. The weather was also bad in the west with snow, high wind, and low temperatures. West of Cheyenne a DH–4 crashed into the side of a mountain in a snowstorm, killing 1st Lt. Edwin V. Wales and badly injuring 2d Lt. William C. Goldsborough. Lieutenant Maynard, still leading in the east-west race, and Capt. Lowell H. Smith, in front in the west-east contest, met that afternoon at North Platte, Nebraska. Maynard spent Thursday night at Cheyenne, his nearest west-bound rivals being 573 miles behind at Des Moines.

Bad weather continued on Friday, October 10. A frozen radiator delayed Maynard's departure from Cheyenne. Consequently, that day he got only as far as Salduro, 518 miles from San Francisco. Three east-bound pilots—Major Spatz, Captain Smith, and 2d Lt. Emil C. Kiel—reached Bryan, 650 miles from New York. Another man died when Maj. Albert L. Sneed with his DH–4 nearly out of gas made a fast landing at Buffalo. His passenger, Sgt. Worth C. McClure, unfastened his safety belt and slid back on the fuselage to weight the tail for a quicker stop. The plane bounced and smashed nose first into the ground, and McClure was thrown off and killed.

When Lieutenant Maynard landed at the Presidio at 1312 on Saturday, October 11, a large crowd, including General Menoher, Colonel Hap Arnold, and the commanding general of the Western Department, Lt. Gen. Hunter Liggett, greeted him. Of the east-bound flyers, Captain Smith fell behind when he damaged his plane in landing near Cleveland to inquire the way to the control field. Major Spatz lost a few minutes when he made the mistake of landing at Hazelhurst Field. Thus Lieutenant Kiel landed first at Roosevelt Field, with Major Spatz close behind.

Lieutenant Maynard won the elapsed-time contest with 3 days, 6 hours, 47 minutes, and 11 seconds between Roosevelt Field and the Presidio; the speed contest with a flying time of 25 hours, 16 minutes, and 47 seconds; and the DH–4 handicap race from east to west. In the west-to-east race, Major Spatz won the elapsed-time contest with 3 days, 8 hours, 40 minutes, and 35 seconds; 2d Lt. Robert S. Worthington, in an SE–5, won the flying-time contest with 25 hours, 23 minutes, and 19 seconds.

As the test continued the second week, two more men lost their lives. Second Lieutenants French Kirby and Stanley C. Miller died in an emergency landing near the Wyoming-Utah border on Wednesday, October 15. Eventually, twenty-six of the forty-six planes leaving New York completed the flight to San Francisco; seven of the fifteen departing San Francisco reached New York.

Meantime, seventeen planes had begun the return flight, twelve headed eastward and five westward. The original plan for the test had been expanded from a one-way flight to a round trip for those who wanted to continue. No one who did not complete the first leg by October 18, however, was permitted to make the return flight. Six planes finished at New York and two at San Francisco within the twenty-day deadline. Five of the eight planes were DH–4s, two SE–5s, and the other was Colonel Hartney's Fokker. Second Lieutenant Alexander Pearson, flying a DH–4, won the speed contest for the round trip with a flying time of 48 hours, 14 minutes, and 8 seconds. But Lieutenant Maynard won the elapsed-time contest with 9 days, 4 hours, 25

Above: Planes "on the line" at Roosevelt Field on Wednesday morning, October 8, 1919, before the start of the transcontinental flight. Lt. Belvin W. Maynard (left) sets record in crossing continent in shortest time.

minutes, and 12 seconds for the round trip. "As an individual performance," General Mitchell said, "Maynard's record stands second to none in the annals of the air in time of peace. His judgment, ability, grit, and determination exhibit the quality shown by our pilots in the European war, and are typically American." The public agreed. The "Flying Parson" became the hero of the hour.

It had been a grueling ordeal for the men. Some pilots, like Colonel Hartney, turned their planes over to the mechanics at the control stops and tried to relax. They might get something to eat, watch Boy Scouts carry out five-gallon cans of gasoline to refuel the plane, or visit with the spectators who turned out to greet the airmen and watch what went on. If they stayed Friday night at Bryan, they might go to a high school football game; on Sunday at Rock Island they could listen to evangelist Billy Sunday. Other pilots, Capt. Lowell H. Smith for example, fussed over their planes and resented being hauled off to a dance when they wanted to work or needed to rest.

Crossing the continent the men contended with rain, snow, fog, ice, and cold, any and all making flying miserable and hazardous. They flew by compass or followed railroad tracks, depending on conditions and what risks they accepted. They skimmed low over treetops to keep going in fog and mist; they searched for altitudes with favorable winds. They watched continually for places to land in an emergency. Even in the Rockies they saw a surprising number of places which, as Colonel Hartney, said, would "afford a refuge for a scared pilot when his motor quit."

Bad weather could force a plane down and keep it on the ground for hours, even days. Pilots got lost; planes ran out of gas; motors stopped for any number of reasons. In an emergency a friendly farmer would help pull a plane out of a muddy field, haul gasoline to the plane, and give a stranded flyer supper and a place to sleep.

When asked about any remarkable experience during the trip, 2d Lt. Alexander Pearson, Jr., laughed, "Yes, I got there and back." Many others could not say the same. During the test 54 accidents wrecked or damaged planes. Twenty-nine resulted from motor trouble, 16 from bad landings, 5 from poor weather, 2 when pilots lost their way, 1 in takeoff, and 1 by fire. In 42 cases the accident meant the end of the race for the pilot. As noted before, seven fatalities occurred during the race, one in a DH–4B, the others in DH–4s.

Oil lines, pistons, gas pumps, and water connections broke and required repair. Landing gear and propellers smashed and needed replacement. Tires blew out and were fixed or changed. Water pumps froze. Radiators had to be soldered or cornmeal put in them to stop leaks. A new strut carved from a pine plank replaced one broken in landing. A wing was repaired with pieces from a wrecked plane. After a forced landing with a broken crankshaft, a pilot and his mechanic could continue their flight after changing their Liberty engine for one retrieved from another plane that had crashed. Lieutenant Maynard and Sergeant Kline did, with the help of men from the nearest control station. They took an engine from a Martin bomber which Capt. Roy Francis entered in the race only to fail in his second attempt to cross the continent. Many pilots in those days were good mechanics, few better than Maynard or Captain Smith.

Had the test been worthwhile? Some people thought not. The death toll had been too great. Air Service officials and other defenders of the test attributed the casualties to bad weather, pilot mistakes, violations of race rules, and the construction of the DH–4s. Colonel Hartney thought none of the deaths necessary. He saw "less danger in aeroplaning than in fast or careless automobiling," but the news of a death in an airplane was "broadcast throughout the land and creates the impression that flying is hazardous and that we have a long way to go before it will be a success commercially." This, Colonel Hartney said, tended to "stifle progress of this great and up-to-date science, which is developing despite its setbacks."[33] Glenn L. Martin, comparing the New York-Toronto and transcontinental contests with automobile races at Indianapolis, judged the airplane not only faster but safer and more reliable than the motor car.[34]

The test, which Colonel Hartney thought a success, confirmed many things already well known. Gas should be filtered through chamois to keep out dirt—sometimes it was not (as General Charlton could attest after surviving a crash caused by such neglect). Compasses ought to be properly oriented before a flight—pilots did not always take the trouble. Mechanics needed to be thoroughly trained—not all of them were (one put gasoline in the oil tank of 1st Lt. Russell L. Maughan's SPAD). A pilot was more than just an "aerial chauffeur"—knowledge and experience were a must (Major Crissy's inexperience had been the principal cause of the crash in which he and Sergeant Thomas died). Good organization and a proper supply system were important.

The test disclosed many weaknesses in the DH–4. For one thing the wheels sat too far back, giving the plane a tendency to go over on its nose in landing on soft ground. It had become rather common practice in such situations for the man in the backseat to slide out on the tail to hold it down. It was dangerous, of course, and Sergeant McClure had been killed that way at Buffalo. First Lieutenant Ralph B. Bagby, however, managed the maneuver successfully during the race while flying with Lt. Col. John N. Reynolds. Engineers at McCook Field claimed they never heard of such a thing. The landing gear obviously needed to be moved forward. Pilots should report unsatisfactory conditions to the Engineering Division so defects could be corrected.

The test showed the DH–4B to be better than the original plane. The principal change was relocation of the main gas tank from between the pilot and the observer to in front of the pilot's cockpit. This afforded a safer plane, improved communication between pilot and observer, and greater visibility for the pilot. More DH–4s would be turned into DH–4Bs.

The Liberty proved a good motor with a fairly long life if kept throttled down and handled carefully, but not if run full tilt as some pilots ran theirs during the race.

The Post Office maps used by the pilots showed the larger towns and railroads that carried mail but no data on altitude. Pilots who took along Rand McNally state maps found them helpful, but lacked special maps for cross-country flying.

The Air Service depended on commercial telephone and telegraph for communication between stations and with headquarters. Planes and control stations had no radios. Cross-country flying demanded radio communication between planes and landing fields and from each field to the one on either side, plus powerful stations at frequent intervals to relay messages to all parts of the country.

Better weather service was essential to rapidly collect and disseminate aviation data. Flying above the clouds needed development to let planes climb above the weather; this entailed frequent landing fields and ways for finding them. Searchlights aided flyers arriving at Binghamton, Rochester, and Buffalo after dark. Such lights should be part of the regular equipment of fields used for cross-country work.

General Mitchell did not belittle the importance of oiling thrust bearings and cleaning gasoline strainers every day, but for him the test meant much more. He saw aviation breaking down America's isolation from Europe and Asia, but the United States lagged behind other nations "in her actual development of air power—military and commercial." The nation should do more to build up aviation for both wartime and peacetime uses. He advocated "a consolidation of the air activities of the Government under a central direction and under personnel which is trained and expert in the knowledge and use of aircraft of all sorts."[35]

The hero of the transcontinental race went off to Georgia and Florida on a recruiting flight in his DH–4. In February 1920 the War Department attached Lieutenant Maynard to the General Recruiting Service for another drive in the south. Thousands of miles later, on his way back to New York in April 1920, he stopped at Middletown for gas and oil. His plane, still the same old one, was sad to behold. Someone at Middletown said it looked like "a trusted war horse after many hard knocks." Later, men at Mitchel Field examined the DH–4 more carefully. They discovered the radiator loose, the cowling split and hitting the propeller, five ribs in the lower right wing and four in the lower left cracked, all of the wires loose, and the horizontal stabilizer broken. The lower wings were scratched and written on by spectators during the tour. They believed the plane held a record for continuous flying and for time in the air. The motor was still the Liberty taken from Capt. Roy Francis' Martin bomber. The Air Service condemned the old DH–4 as unsafe.[36]

Flying circuses, cross-country flights, and aerial contests displayed the wonders of aviation to the American public. People had read of Eddie Rickenbacker and Frank Luke, and of U.S. Air Service victories and losses in the skies over France. Now they could see the Army's men and planes in action—a good show put on by brave, daring, and at times foolhardy young flyers. These performances kindled intense interest in aviation but failed to create public demand for a large, powerful, and expensive air force.

The Liberty Loan circuses, transcontinental flights, the tour along the rim of the United States, and other events convey some idea of how perilous and challenging flying was. There were no airways, improved landing grounds, or emergency fields. Absent were servicing facilities, communications, and lighting. The magnetic compass was the sole navigational instrument in an era where mechanical trouble and forced landings were routine to the game. Still, these Spartan conditions pointed the way and furnished incentives for improvements. The Air Service learned a lot about airplanes and engine performance, gaining technical information by solving the problems. And among other things it collected information on routes, landing places, terrains, and flying conditions. All this contributed in one way or another to progress in American aviation.

Chapter III

Reorganization

During the period when events of the two preceding chapters took place, Congress was considering a national military policy for peacetime. The National Defense Act of 1920 provided for a small professional Army and counted on civilian soldiers for large mobilizations. One of the principal peacetime jobs for Regular Army professionals was training the civilian components, the National Guard and the Organized Reserve. Together, the Regular Army, Guard, and Reserve made up the Army of the United States. The Act of 1920 authorized a maximum of 17,726 officers and 280,000 enlisted men for the Regular Army. Actual strength, however, depended upon the amount of money Congress voted in annual appropriations.

This chapter discusses provisions Congress made for aviation in the 1920 act and how the Air Service of the Regular Army fared with respect to money and people during 1920–26.

A National Policy

Working out national military policy for peacetime, Congress had to deal with the controversial problem of how to organize aviation. In Europe during the war, General Pershing removed aviation of the AEF from Signal Corps control and created an Air Service. At home, President Wilson used his wartime authority on May 20, 1918, to establish the Division of Military

Aeronautics (formerly part of the Signal Corps) and a newly created Bureau of Aircraft Production as separate agencies directly under the Secretary of War. The division, headed by General Kenly, trained officers and enlisted men for aviation duty. The bureau director, Mr. John D. Ryan, managed the production of aeronautical equipment. On August 28, 1918, Ryan became Second Assistant Secretary of War and Director of Air Service with supervision and control over both the Bureau of Aircraft Production and the Division of Military Aeronautics. Under this arrangement the Air Service enjoyed only a nominal existence until after the war.

Ryan's resignation soon after the Armistice led to General Menoher's appointment as Director of Air Service. A field artillery officer who had commanded the 42d (Rainbow) Infantry Division in France, Menoher assumed his new duties on January 2, 1919. In March he acquired responsibilities formerly assigned to the Directors of Aircraft Production and Military Aeronautics. At the same time, he reorganized the Office of the Director of Air Service. The principal officers on his new staff were an executive and four assistant executives who soon became chiefs of functional groups. Thus General Menoher's office came to include Col. Oscar Westover, Executive Officer; Brig. Gen. William Mitchell, Chief, Training and Operations Group; Col. William E. Gillmore, Chief, Supply Group; Maj. Horace M. Hickam, Chief, Information Group; and Lt. Col. William F. Pearson, Chief, Administrative Group.[1]

General Menoher flew frequently (he had his own flying suit) but always as a passenger; he held no aeronautical rating and was not a pilot. As Director of Air Service, he reported to the War Department General Staff. Menoher was responsible for training officers and enlisted men for aviation duty; developing, buying, and distributing aviation equipment and supplies; and supervising technical matters pertaining to aviation in the Army. He was also charged with organizing tactical units, but as each was formed it was assigned to one of the Army's territorial departments.[2] Each department commander controlled tactical training and operations, exercising his authority through the Air Service Officer on his staff. Thus, General Menoher commanded training fields, schools, supply and repair depots, and experimental facilities. But he did not command tactical aviation except when the War Department placed units under his control for exercises, maneuvers, or other special projects.[3]

This arrangement reflected the widely held view that aviation's chief function was to support troops and that ground force commanders should control their own aviation. That was the way it had been in France, where AEF aviation units were dispersed among the various divisions, corps, and armies for operational control. A board appointed by General Pershing in April 1919 considered organizational and tactical lessons learned during the war. Headed by Maj. Gen. Joseph T. Dickman, the board found Army aviation an auxiliary service that should stay under the control of ground force commanders.[4]

A committee appointed by Secretary of War Baker to investigate aviation in Europe after the Armistice recommended a major change in organization. Having visited France, Italy, and England, the committee, headed by Assistant Secretary Benedict Crowell, expressed grave concern about the effects of demobilization on the future of aviation in the United States. The government needed to safeguard the nation's interests, salvage something from the vast amount of money spent for aviation during the war, and prevent the aviation industry, ninety percent of which had already been liquidated, from disappearing entirely. Stressing the importance of developing civil as well as military aviation, the committee suggested creation of a single agency, headed by a civilian Secretary of Air appointed by the President, to direct all aviation activities in the United States. Secretary Baker did not agree. He thought the Army and Navy each needed complete control of its own personnel, training, equipment, and operations.[5]

On July 28, 1919, nine days after submission of the Crowell report but two weeks before Secretary Baker released it to the public, Representative Charles F. Curry of California laid before Congress a bill to create a department of aeronautics.[6] The new executive department, headed by a secretary of aeronautics, would deal with all military and commercial aviation matters, including those now handled by the Army, Navy, and Post Office Departments. The bill would create a United States Air Force consisting of a staff to prepare plans for national defense by air forces, and a combat force to operate independently or with land or sea forces. The President might detail air units to land and sea forces, at which time they would come under the control of such forces. Three days later, Harry S. New of Indiana offered a similar bill in the Senate.[7]

Secretary Baker convened a board comprising General Menoher and three other generals (all artillerymen) to study the New and Curry bills. The board's conclusions on October 27, 1919, said the Army and Navy should each have its own aviation, and no military air force should be created independent of Army or Navy control. The board did believe, however, that two separate government agencies should be formed. One would oversee development work common to Army, Navy, and commercial aviation. The other would procure all aircraft used by the government. Having approved the report, Baker sent a copy to the Chairman of the Senate Committee on Military Affairs.

After the Menoher Board filed its report, Senator New introduced another bill. Like the earlier one, it provided for a department of aeronautics, a United States Air Force, and the detail of air units to operate under the command of land or sea forces.[8] Several similar bills were set before Congress during 1919 and 1920, but only New's revised bill made it out of committee.

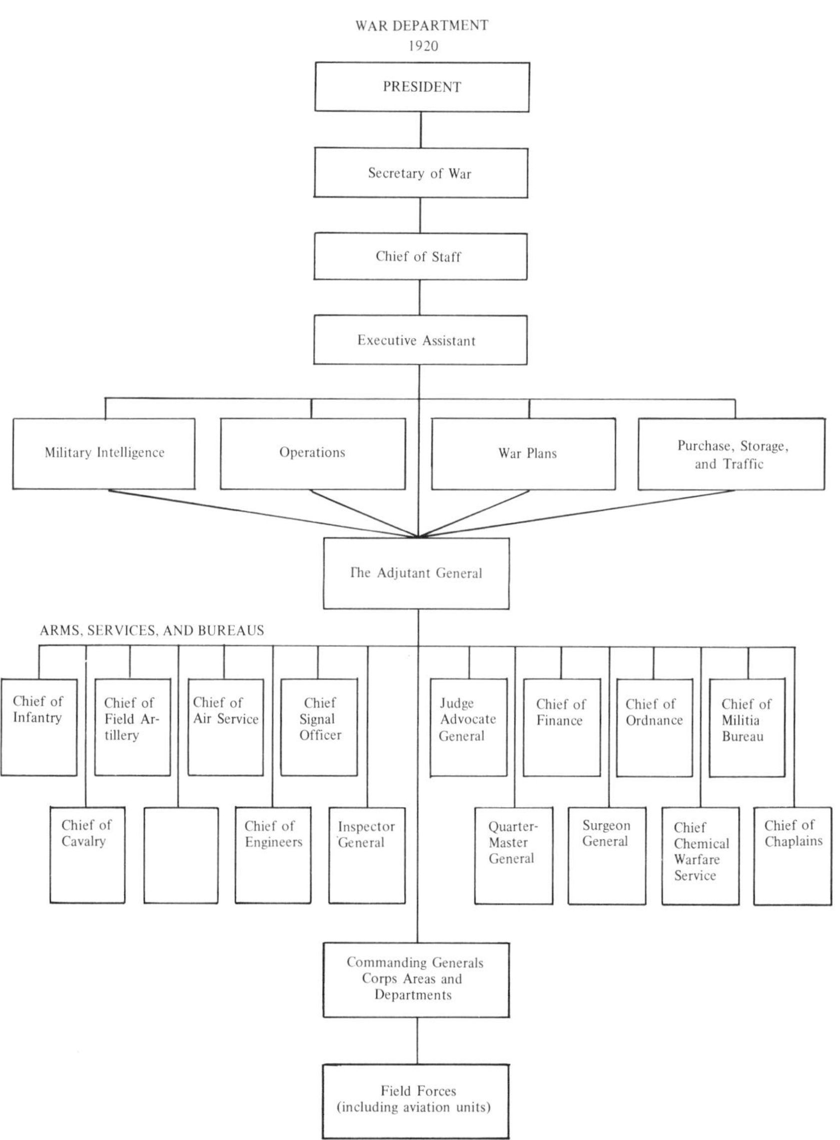
WAR DEPARTMENT
1920
PRESIDENT
Secretary of War
Chief of Staff
Executive Assistant
Military Intelligence
Operations
War Plans
Purchase, Storage, and Traffic
The Adjutant General
ARMS, SERVICES, AND BUREAUS
Chief of Infantry
Chief of Field Artillery
Chief of Air Service
Chief Signal Officer
Judge Advocate General
Chief of Finance
Chief of Ordnance
Chief of Militia Bureau
Chief of Cavalry
Chief of Engineers
Inspector General
Quarter-Master General
Surgeon General
Chief Chemical Warfare Service
Chief of Chaplains
Commanding Generals Corps Areas and Departments
Field Forces (including aviation units)

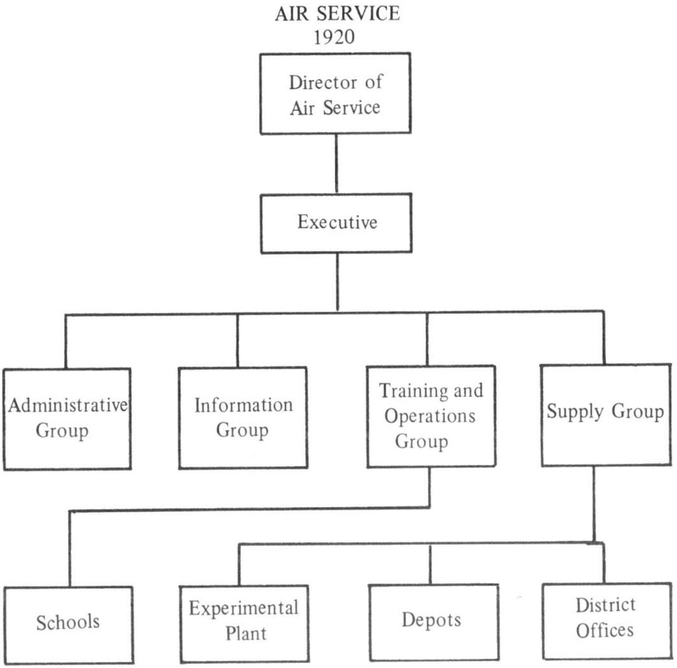

Floor discussion soon revealed the Senate's absence of understanding as to the bill's intent. On January 31, 1920, New therefore asked that the bill be recommitted to the Committee on Military Affairs for further consideration. That was the last heard of it during the session.

According to Representative Fiorello H. La Guardia of New York, a former AEF pilot who favored creation of a separate department of aeronautics, many Air Service officers of like mind were afraid to speak out in public. This is understandable, for their chief, General Menoher, opposed a separate department. So did his executive, Colonel Westover, and such top officials of the War Department and the Army as Secretary Baker, General Pershing, and Army Chief of Staff Gen. Peyton C. March. Nevertheless, some members of the Air Service spoke in behalf of an independent air force. Among them were Brig. Gen. William Mitchell, Col. Henry H. Arnold, Col. Charles de F. Chandler, and—taking the strongest stand of all—Maj. Benjamin D. Foulois.

Proponents of a separate department held it necessary strategically and tactically to give final authority in aviation matters to those persons most interested in aeronautics and best qualified to direct aerial operations, i.e., to aviators. Believing military aviation was no longer an auxiliary of ground forces but a powerful striking force in itself, they felt it should be organized separately for independent operations. They thought central control would eliminate the increased expense of duplication, and would encourage and stimulate the aviation industry. They cited England's success with an independent air service as reason for copying that model in the United States.

Opponents countered each argument. Aviation needed to be maintained as an integral part of the Army, or the Army would not be able to use it when and where needed. Cooperation with ground forces, aviation's principal role, would suffer if the Army did not control its air service. Centralization would not save money nor would it benefit industry. Further, the British experiment had proved a failure.[9]

After long debate, Congress eventually adopted the War Department's recommendations for the peacetime organization of the Army. Under the National Defense Act of 1920, the Air Service remained a combatant arm coordinate with the Infantry, Cavalry, and Artillery, with responsibility also for development and supply of its own technical equipment. The Air Service was already operating that way under a wartime executive order and temporary legislation. The law did not alter the status of aviation in the Army but made the existing scheme permanent.[10]

Money for Aviation

Coincident with passage of the National Defense Act, Congress appropriated money for the Air Service for the coming fiscal year. From wartime appropriations of some $460 million for the year beginning July 1, 1918 (Fiscal Year 1919), the amount appropriated had dropped to $25 million for Fiscal Year 1920. General Menoher said the total was too small, that all of the money allotted to the production and purchase of airplanes, balloons, airships, engines, and spare parts—more than $6.5 million—went for maintenance. Other major expenditures included $2.4 million for experimental work and research, $2.8 million for instruction and operations, $2.3 million for expenses connected with disposing of surplus material, and among other things $8 million for salaries and expenses of civilian employees.

The Air Service appropriation, however, contained but part of the money that went toward supporting Army aviation. Officers and enlisted men received their pay from a general appropriation for the pay of the Army. Other departments (Quartermaster, Ordnance, Engineer, Signal, and Medical) had their own appropriation. They furnished rations, machineguns, ammunition, bombs, flares, gas masks, motor cars and trucks, long-distance telephone service, as well as medical supplies and hospital services. Furthermore, the $25 million appropriated to the Air Service for 1920 did not include the more than $11 million authorized for aviation stations for seacoast defense of the United States and its insular possessions. Even so, more than a fourth of the $11 million could not be used because Congress prohibited the acquisition of the necessary lands.[11]

Army appropriations for Fiscal Year 1921 had $33 million for Air Service training, operations, procurement, and maintenance.[12] For the next year, Congress cut the amount to $19.2 million. Expressing sympathy with the government's efforts to reduce expenses, General Menoher warned that if reductions went too far the Air Service would not be able to do its job. He was particularly concerned about the curtailment in experimental work and in the procurement of new, up-to-date, aeronautical equipment. But his successor, General Patrick, proudly announced that in keeping with the policy of strict economy laid down by President Harding, the Air Service had reported a savings of $800,000 to the Bureau of the Budget and had ended the year with $450,000 unobligated.[13]

Looking forward to Fiscal Year 1923, Patrick requested $26.2 million, stressing the urgent need for new airplanes. Then, too, reserve supplies left over from the war were almost exhausted, equipment in use was inadequate and deteriorating rapidly, and no provisions had been made for replacement. His presentation caused the War Department to raise his tentative allotment from $13 million to $15 million. Congress gave him $12.9 million. At the end of the year, Patrick reported great progress in developing equipment, training personnel, and perfecting organization. He warned, however, that too little money left the service "entirely incapable of meeting its war requirments."[14]

For Fiscal Year 1924, Patrick wanted $19 million to buy sufficient equipment of up-to-date design. (Realizing the need for rigid economy, he pared the amount for new aircraft to $8.6 million.) His total estimate for the year came to $18.154 million, or nearly $6 million more than the War Department's tentative allotment. He finally received $12.6 million and spent all but $6,924.57.[15]

Changing his strategy, Patrick put in an estimate of $43.48 million for Fiscal Year 1925. He got $12.8 million. Discouraged, he urged that in the future Congress furnish enough money "to supply safe flying equipment for at least existing units and personnel."[16]

Due to the pinch in funding, most units fell short of the people and equipment authorized. Many were at less than half strength. Patrick figured it would take $40 million in Fiscal Year 1926 to give the units their full equipment. If that amount was not to be had in one lump sum, he wanted an annual appropriation of $25 million for three years to complete equipping the units by 1928. The War Department approved $10.664 million for procurement of aircraft, the total budget estimate being $22.435 million. The Bureau of the Budget shaved this amount to $14.7 million, plus a contract authorization of $2.15 million for new planes. The bill passed by Congress and signed by Coolidge gave the Air Service $14.9 million.[17]

People

The 1920 act authorized the Air Service a major general for Chief, a brigadier general as Assistant Chief, 1,514 officers below brigadier general, and 16,000 enlisted men, including 2,500 flying cadets.[18] This was not as many people as the Air Service wanted but more than it had.

Endorsing an earlier policy of the Air Service, Congress required assignment of flying officers to command flying units.[19] Before the war, airplane pilots had been rated military aviators (MAs) and junior military aviators (JMAs). Those rated during the war, when permanent appointments to the officer corps were not being made, were designated Reserve military aviators (RMAs). To advance to MA, an aviator had to serve three years as a JMA or distinguish himself in active service. Members of the lighter-than-air branch held similar ratings, the "A" of MA, JMA, and RMA standing for aeronaut instead of aviator. In addition there was an observer rating for each branch.

During the war, rated personnel wore embroidered badges sewn on their uniforms. Lieutenant Luke, Captain Rickenbacker, Maj. Raoul Lufbery, and other pilots with either JMA or RMA ratings wore silver wings and a shield with "US" in gold in the center. Capt. Elliott W. Springs, Maj. Reed G. Landis, and other MAs wore the same device with a star above the shield. The aeronaut's badge bore a spherical balloon in place of a shield. Observers wore an "O" on a single wing. After the war the Air Service made several changes in ratings and badges before settling on a system in 1921 that continued until the eve of World War II. Pilots and observers wore wings of oxidized silver (or an embroidered badge on the wool service coat) with a device in the center denoting the specific rating—a shield for airplane pilot; dirigible for airship pilot; "O" for airplane observer; and spherical balloon for balloon observer.[20]

The National Defense Act of 1920 brought significant changes in the rank and pay of Air Service officers. Under laws of 1916 and 1917 an aviation officer, required to fly frequently and regularly, was given an additional twenty-five percent of the pay of his grade and length of service. A captain or lieutenant qualifying for flight pay (the "aviation increase" as it was called) received temporarily the rank, pay, and allowances of the next higher grade, plus a fifty-percent raise if he held a junior rating and a seventy-five-percent one if a military aviator or aeronaut. The 1920 act did away with the temporary rank and the differential in flight pay except for officers awarded the military aviator rating for distinguished service. At that time, just three such distinguished men—Major Spatz and Majs. John N. Reynolds and Lewis H. Brereton—were on active duty. Other officers (and enlisted men) flying frequently and regularly got fifty-percent more pay.[21]

An exception to fifty percent for flight pay was made for flying cadets. During the war a cadet's pay was only $23 a month, but this went to $49.50 in November 1918. In mid-1919, on the recommendation of the Air Service, Congress set the cadet pay at $75 a month, with a ration allowance of $1 a day and other allowances as for a private first class in the Air Service. Congress carried this over into the 1920 act.[22]

Under this act, Regular Army officers earning temporary promotions in the war reverted to their permanent rank on July 1, 1920. Some officers dropped back more than one grade. Hap Arnold, for example, went from colonel to captain, but like many others he partially recovered by simultaneous promotion. Retaining his job as Air Officer of the Western Department, he kept on doing as a major what he had done as a colonel.

Officers who wanted to stay in the Air Service were transferred from the Signal Corps and other branches in which they had been commissioned. The way such changes were effected is illustrated nicely by the case of Lewis H. Brereton, a permanent captain in the Aviation Section, Signal Corps, and temporary lieutenant colonel. He was discharged as a lieutenant colonel on June 30, 1920, transferred from the Signal Corps to the Air Service as a captain on July 1, 1920, and promoted to major.[23]

All officers without Regular Army commissions had to be discharged by the end of December 1920. However, these men (1,013 of the 1,168 officers on duty with the Air Service on July 1, 1920) could apply for Regular commissions. Officers at Kelly Field, like those elsewhere in the Air Service, commenced cramming for exams. They attended classes every morning before working hours. The "shavetail's" head became a "seething riot of an algebraic, logarithmic and geometric equation," which according to reports from Kelly, "intermixed with geography and history."

Boards of officers, each with a surgeon, convened at various bases to examine applicants for permanent commissions. At Souther Field, for example, a board examined one group of officers from Georgia and Florida, as well as former officers from the southeast who wanted to return to service.[24] On September 17, 1920, the Air Service announced that thus far 9 men had been commissioned in the Regular Army as captains, 80 as first lieutenants, and 125 as second lieutenants. By the end of December 1920, the Air Service discharged all temporary officers except 15 in hospitals.[25]

The officer shortage seriously constrained aviation operations during the first half of the 1920s. Forest fire patrols were a case in point. After two successful seasons of patrol work in the west, the personnel pinch forced the Air Service to cut the size and scope of the operation drastically. General Patrick reported the officer situation "still acute" in 1926. Because of this, he said, school groups could furnish instructors for training no more than about sixty percent of the qualified applicants. The 1st Pursuit Group should have had 101 officers. To man 1 of the 2 flights in each of the group's 4 pursuit squadrons took at least 45 officers; the group (the only one of its kind in the U.S. Army) had 20. The 1st Bombardment and 3d Attack Groups fared little better. The garrisons in Hawaii, the Philippines, and the Canal Zone had around half of the officers authorized. Still, General Patrick said the number assigned was "more than ample in view of the material available for use."[34] The Air Service had fallen on hard times in more ways than one.

At the time of the National Defense Act of 1920, the Air Service had about half of the 16,000 enlisted men authorized. The Air Service immediately mounted a huge recruiting drive to fill more than 7,000 vacancies and replace men being discharged every day. Calling for recruits, General Menoher stressed the training opportunity. It was, he said, "the mechanics dream realized." The 18th Balloon Group at Aberdeen, Maryland, picked up 57 recruits during a two-month recruiting drive but other units did not do as well. High civilian wages made it hard to recruit in some areas, and worked against men staying in the service when their enlistments expired. The 1st Pursuit Group reported its enlistment rate near zero. Without a turn for the better the group soon would have sufficient men solely for kitchen police (KP). The big problem at Montgomery, Alabama, was not the area's wage scale but the illiteracy. Recruiters at the Montgomery repair depot were dismayed by the many young men who could barely read and write. March Field, California, discharged 18 men in 1 week; 7 men reenlisted and claimed their bonus; 3 recruits enlisted. A report from March Field told of a recruit flown from Red Bluff, California, for enlistment. On the way he got "rather shaken up" in an emergency landing. But being a "gritty fellow," he went through with the enlistment and turned out to be a loyal and interested member of the service.[35]

At the beginning of 1921, Air Service recruiters were doing better. The service's enlisted strength rose to 12,280 by early February then abruptly halted. As an economy measure, Congress lowered the Army's enlisted ceiling from 280,000 to 175,000 and stopped enlistments until the Army got down to the smaller number. When the War Department parceled out the cut among the various branches, the Air Service shrunk from 16,000 to 11,500.[36]

Expiration of enlistments and other attrition shaved the Air Service to around 11,000 men in June 1921. At this point, Congress directed Secretary of War John W. Weeks to grant discharges until the total number of enlisted men in the Army stood at 150,000. The Air Service's new ceiling was 10,300, 1,340 more than if the cut had been apportioned equitably among all Army branches.[37] Within 3 weeks, 247 of the 476 men at Mather Field, California, applied for discharge. During July the garrison at Fairfield Air Intermediate Depot in Ohio dropped from 440 to 177 men. At Carlstrom Field, Florida, 129 men left in 1 month. Chanute Field, Illinois, reported on August 12, 1921, that 203—nearly half—of its men had been discharged. The Air Service touched bottom in September 1921 with 7,160 enlisted men on its rolls. The situation of the 31st Balloon Company, Camp Knox, Kentucky, at the end of August was not unique among Air Service units of that time. Having lost 40 men, the company could still "function after a fashion," but it took every remaining man to get the balloon up and put it away. The morale of the company actually improved, because "those who are staying are contented with the Army and all are pulling together." Thus, the 31st Company could report the balloon up "bright and early" on September 1, 1921.[38]

Recruiting being resumed in September 1921, enlisted strength of the Air Service grew to more than 9,500 by April 1922. That number, however, included nearly 3,000 men reaching the end of their enlistments.[39] Some men were purchasing discharges, since for many years a law had permitted any enlisted man to buy a release in time of peace.[40] A private of the 12th Observation Squadron secured his discharge by purchase as did a corporal who gave $170 to the finance office at Kindley Field in the Philippines. A flying cadet who found a job as an airmail pilot paid to get out of the Air Service. Men who joined the service in 1920 and 1921 when jobs were hard to find wanted to get out when jobs became more plentiful in the fall of 1922. Mitchel Field, New York, was only one station reporting that it lost a lot of men that way.[41]

In June 1922 Congress once more cut the Army's enlisted authorization, this time to 125,000 with 8,500 going to aviation. The Air Service's allotment changed 3 times but never by more than 300 over the next four years, the number at the end of June 1926 being 8,342. During that time the Air Service succeeded in keeping its enlisted strength very near or even above the number allotted. In fact, at one time the excess came close to 1,000. The War Department did not object because other branches were understrength, enabling the Army to stay below the overall ceiling set by Congress.[42]

The enactment of a national military policy and program was attended by congressional rejection of a separate department of air and by the statutory establishment of military aviation in the role it had attained during the war as a combatant arm of the Army. The U.S. Army Air Service handled the development, procurement, and distribution of aeronautical equipment. It trained officers and enlisted men for aviation duty. And it organized units that the War Department then assigned to the Army's territorial departments for service with field forces. In short, the Chief of Air

Service acted as chief of a supply bureau rather than as a combat force commander.

In the National Defense Act of 1920, Congress authorized numbers of officers and enlisted men for the Regular Army as a whole and for various arms and services. During the next several years, however, Congress did not appropriate sufficient money for a force of the size it had authorized. Further, to save money it set ceilings on the numbers of officers and enlisted men on duty with the Regular Army. Consequently, the Air Service like other arms and services did not attain the strength contemplated by the 1920 act. The resultant weakness of the Air Service caused grave concern among officials of the service and the War Department.

Under the lower ceilings, the Air Service generally succeeded in recruiting and reenlisting men to fill enlisted vacancies. Its inability to fill all officer vacancies stemmed from several reasons—chiefly the high standards set for physical examinations and for pilot training, and the few promotion opportunities. This shortfall of officers seriously eroded the effectiveness of field forces.

Chapter IV

Training

The training of its officers and enlisted men was one of the chief functions of the U.S. Army Air Service, consuming a great deal of money, people, equipment, and time. Such training encompassed pilots, observers, mechanics, radio operators, other technicians and specialists, engineers, and doctors. After demobilization and the adjustments imposed by personnel ceilings and fund limitations, the Air Service opened a number of schools.

Flying Training

During the war, American colleges conducted ground schools for flying cadets commencing flying training. These schools were closed by the end of February 1919. Most fields where cadets received primary flying training soon became storage depots. For peacetime, the Air Service combined ground school and primary training in a new primary flying course offered at Carlstrom and March Fields. The first classes entered in January 1920.[1]

Congress authorized resumption of enlistment of flying cadets in July 1919 and limited the number on duty to thirteen hundred. To quickly eliminate the unfit, Congress let the service discharge a cadet at any time upon the recommendation of a board of three or more officers. The cadet who survived the rigorous course could serve out his enlistment or take a discharge and enter the Officers' Reserve Corps as a second lieutenant.[2]

The civilian wanting flying training had to be an unmarried, U.S. male citizen aged twenty to twenty-seven with a high school education, good character, sound physique, and excellent health. Those accepted were enlisted for three years. Regular Army enlisted men could also apply for training as flying cadets. Classes in Air Service flying schools likewise included, at various times and in varying numbers, officers of the Air Service and other branches of the Regular Army, members of the Officers' Reserve Corps and National Guard, Air Service noncommissioned officers, U.S. Navy and Marine Corps personnel, and students from foreign countries.[3]

In the National Defense Act of 1920, Congress increased the Air Service's authorization for flying cadets from 1,300 to 2,500. The allotment dropped to 500, however, when Congress stopped Army enlistments in February 1921. Fund shortages and ceilings on enlisted men made War Department allocations of flying cadet spaces even smaller—190 in 1923 and 196 in 1926. Even at these levels, the Air Service never trained the allotted number.[4]

There was no dearth of applicants, but many interested young men failed to meet the high qualifications. Further, the Air Service could not always accept all who qualified. Early in the new program (between July 1920 and February 1921), when the authorization was 2,500 and the average strength about 300, the Air Service turned down nearly one-third of the applicants. Of 1,288 who took the cadet examination in Fiscal Year 1925, just 362 passed. Of these, 254 could not be appointed because the training organization could handle no more. Since officers of the Regular Army took priority, how many cadets could be selected for any class depended on the vacancies remaining after the officers had been accommodated.[5]

Another factor affecting the numbers trained was the attrition rate. Though the students entering at Carlstrom and March Fields in January 1920 had been chosen with great care, many of them fell short of the schools' high standards. Instructors soon reported students unsuitable for further training. Boards composed of faculty officers, and with authority given by law, ruthlessly removed the unfit. At the end of 6 months, 61 of the 202 in the 2 classes had graduated from primary training; 43 were still under instruction; 98 had washed out. In the following class at Carlstrom, 51 percent graduated; in the next, 72 percent; and the one after that, 60 percent.[6]

Cadets of the 1920s feared the so-called Benzine Board that purged cadets who did not measure up. One of the first rumors Cadet Charles A. Lindbergh heard upon reporting to primary flying school in 1924 concerned the high washout rate. He recalled later that of his 103 fellow students in primary training, 17 graduated with him from the advanced school. Students brought before a board expected to be let out. They thought it "unprecedented and hard to account for" when a board called 7 men and weeded out 1.[7]

Charles A. Lindbergh

The bulk of the eliminations stemmed from unsatisfactory progress in learning to fly. The rest came from academic failure, physical defects not detected earlier by doctors, student requests to be dropped, and, rarely, breaches of discipline. Besides cadets, among those eliminated were Regular Army officers, Reservists, Guardsmen, and foreign students. General Patrick kept a big chart in his office to tell him how many persons were in each class, when the class started, the number relieved, the number remaining and when they would graduate. Over three years, the figures showed 1,235 starting primary training and 499 commencing advanced training.[8]

Young men attending flying training in the early 1920s came under strict military discipline. They endured the drilling, hiking, setting-up exercises, and other activities common to the enlisted regimen of the Army. When those coming from civilian life had been processed into the military, classes started. Ground school covered the theory of flying, aviation motors, navigation, radios, military administration, the manual of court-martial, and kindred subjects. After ground school was well under way, flying began, with each student eagerly awaiting his turn. Meantime he might find himself on KP or other work details to fill his time.

When the cadet's name came up on the flying list, he drew helmet, goggles, and leather coat and reported to the line. His instructor, usually a lieutenant, explained the controls and ran him through a complete inspection of the Curtiss JN in which he would learn to fly. Seated in the rear cockpit, the cadet was shown how to fasten his safety belt. On signal from the instructor, the crewchief swung the propeller. The engine coughed, sputtered, then took hold. Satisfied it was running properly, the instructor signaled the mechanics to pull the blocks from the wheels, opened the throttle, and lifted off to give the cadet a joyride.

After the first ride, the cadet was flown to a safe altitude, given the controls, and taught to fly by means of directions coming by way of a speaking tube from the front cockpit where the instructor was ever prepared to take control. A student at Carlstrom Field told what it was like to be in the backseat when the voice from the tube said:

> "She's all yours"—You take the stick and try to give her enough right rudder to overcome the torque and by that time the nose is either up or down or the wings aren't level, and by the time you correct those minor details she is drifting off sideways somewhere . . . and after you sweat blood for a while and go from rotten to rottener, the instructor grabs the stick and sets her level and directs you in the name of all that's holy, to keep her that way, and you don't and he makes a few choice remarks as to the amount of gray matter you are endowed with, and the horizon gets the St. Vitus dance, and the wings absolutely refuse to stay put, and the instructor tells you you are a goof, which is superfluous because you've known it for some time. Just about then your instructor's remarks are hot enough to scorch the tail surface and you wish you had a transmitter on the Gosport phone so you could talk back at him . . . then you get mad and pull yourself together and manage to hold her level for a minute or two and find it's not so bad after all. Just then, if your instructor nods his approbation, 'Oh, boy, ain't it a gran' and glorious feeling?[9]

So the training went, an hour a day, as the cadet learned to fly straight and level. After executing easy turns, he eventually learned to take off and—what was more difficult—to land. Then came a day when, after landing, the instructor climbed down, pinned a streamer on the Jenny's tail to warn others, and ordered: "Take her up. Go once around and land."

After his first solo, the cadet practiced taking off and landing one hour a day. Next he mastered a figure eight, a spiral, and other maneuvers. Thus flying continued in the morning with class and shopwork in the afternoon until time for graduation. Those who completed the four-month course (changed a little later to six months) appeared before an examining board that classified them for advanced training. In one graduating class of 24 at Carlstrom Field, the board designated 3 for training in pursuit, 6 in bombardment, and 15 in observation. The graduate's assignment depended first upon his qualifications and secondly upon his preference.[10]

Students destined for advanced training in observation went to the Observation School at Post Field, Fort Sill, Oklahoma. The Air Service lacked money, however, to maintain similar schools for pursuit and bombardment in the early 1920s. The 1st Pursuit and 1st Day Bombardment Groups provided advanced training in those branches of aviation at Kelly Field.[11]

Cadets arriving at the 1st Pursuit Group in the summer of 1920 had finished primary training, but had not yet graduated from Jennies. Their advanced instruction at Kelly started with formation flying, stunting in formation, combat with camera guns, observation, elements of bombing, and elementary gunnery in JN training planes. After flying in DH–4s the students progressed to SE–5s. They studied pursuit tactics; acrobatics; and fancy formation, night, and cross-country flying. They became proficient in landing on small fields, aerial gunnery, individual combat, battle maneuvers, and bomb dropping. Training exposed them to Army paperwork and the duties of operations, armament, radio, engineering, supply, and mess officers. Running about six months, the course aimed to do more than just qualify men for the silver wings of the airplane pilot. It sought to turn out first-rate pursuit pilots who were confident, accurate flyers and excellent shots, possessing quick keen judgment.[12]

Advanced training in bombardment and observation, like that in pursuit, entailed work in classrooms and hangars as well as in the air. Students assigned to the 1st Day Bombardment Group for advanced training transitioned to DH–4s and were schooled in flying, bombsights, camera obscura, gunnery, and, among other things, the history of the development of aviation. After the Army Surveillance Group became the 3d Attack Group in 1921, it accepted students for advanced training in attack aviation.[13]

At the Observation School, students transitioned to and learned DH airplanes. There were courses on formation and cross-country flying; visual and photographic reconnaissance; surveillance; intelligence; liaison with ground forces; observation and adjustment of artillery fire; mapreading; meteorology; maintenance and operation of radio, telephone, and telegraph; Liberty engines; and rigging.[14] Graduates of advanced training were rated airplane pilots. To be rated airplane observer, a student had to take an additional course at the Fort Sill artillery firing center.

Cadets completing advanced training were certified for appointment in the Officers' Reserve Corps. Most newly rated pilots who had enlisted as cadets took discharges and entered the Reserve Corps. Those who had been

Formation of Curtiss JN training planes.

enlisted men when they entered flying training reverted to their enlisted rank and remained on duty.[15]

Under Army and Air Service regulations, enlisted men who went through pilot training and were certified for or held Reserve commissions were permitted to pilot aircraft.[16] In fact, there were suggestions that the Air Service use more enlisted and fewer commissioned pilots, if for no other reason than to save money. Some people would go so far as to let enlisted men be pilots in tactical units, with commissioned officers serving as their flight leaders and commanders. The Air Service wanted all pilots commissioned. The mental, moral, and physical qualifications for pilots were extremely high. Consequently, a person having them could easily secure a commission in another branch of the Army if he could not get one in the Air Service. The pilot's degree of responsibility was that of a commissioned officer, not an enlisted man. Moreover, the Air Service contended, officers with permanent commissions could be expected to stay in service longer than men who enlisted for relatively short terms—a matter of great importance considering the cost of training.[17]

The Air Service successfully defended its stand that pilots should be commissioned officers. Nevertheless, it always had some enlisted pilots during the first half of the 1920s. Sergeants in the first three grades, who were twenty-two to thirty years of age and met cadet qualifications, could be detailed in grade to flying school and to the same course given cadets. Among the few noncommissioned officers selected for this training was SSgt. Alva L. Harvey, who had been with Maj. Frederick L. Martin on the world flight in 1924. Having completed primary training in 1925 and advanced training the following year, Sergeant Harvey received a Regular commission as second lieutenant.[18]

Sergeant Harvey took his primary instruction at Brooks Field near San Antonio, as did Cadet Charles A. Lindbergh and others who commenced training after mid-1922. A shortage of money and failure of the training program to become as large as planned, prompted the closing of the primary school at March Field in 1921. The financial pinch and need for better coordination brought on a major Air Service reorganization in 1922. San Antonio thereupon became the flying training center of the Air Service.[19]

The primary school at Carlstrom Field closed and a new one opened at Brooks Field. The Air Service revamped the course and extended it to nine months but later trimmed it to six with two classes a year. The first part consisted of ground school, stressing military training as well as classwork more directly related to the business of flying. After inspecting cadets and enlisted men at Brooks late in 1922, General Pershing, Army Chief of Staff, commended the Commandant, Maj. Ralph Royce, on their military bearing and aptitude for infantry drill. Pershing "was glad to see that soldiers as well as flyers were being made of the cadets."[20]

Flying training comprised the second half of the primary course. Starting with straight-and-level flying accompanied by an instructor in a dual trainer, the student advanced step by step until he could solo, execute a variety of maneuvers, and deal with many of the situations arising in flight. If the Benzine Board did not get him, he graduated with the rating of junior airplane pilot (JAP).[21]

One officer was awarded a JAP rating without going to the Primary Flying School. After becoming Chief of Air Service, General Patrick began taking flying lessons from Maj. Herbert A. Dargue. He felt he should learn to pilot to get a better idea of the skills required and a deeper appreciation of the dangers involved. Passing the examination given by a board of officers, the 59-year-old general received his wings during a luncheon at the Army and Navy Club in Washington on June 26, 1923. General Patrick skipped advanced pilot training.[22]

In the reorganization of 1922, the Air Service opened an Advanced Flying School at Kelly Field to take over training formerly performed by tactical units and by the Observation School at Post Field. The course at Kelly originally ran eight months but later was reduced to six with two classes a year. Each student chose one of the four branches of aviation. Sergeant Harvey, for example, took the attack course. Cadet Lindbergh specialized in pursuit; 2d Lt. Otto P. Weyland (of Lindbergh's class), observation; and 2d Lt. Elmer T. Rundquist (of the same class), bombardment. Graduates got the rating of airplane pilot.[23]

When primary training got under way at Brooks field, the dual trainer was the JN–6H. Brooks later accepted other planes, including VE–7s and TA–3s, for evaluation, but JNs were used until 1926. The beginning class in March of that year was the first without Jennies. Students now flew in the new Consolidated PT–1, with tandem seats and a Wright E engine. But the day of the Jennies was not over; they still served the Organized Reserve and National Guard.[24]

Balloon and Airship School

Demobilization in 1919 left the Air Service with balloon schools at Lee Hall, Virginia; Brooks Field; and Ross Field, California. Officers and cadets trained as observers at Lee Hall and Ross Field. Enlisted men at those two stations were trained there, while recruits went to Brooks Field for instruction. There being so few observers in school during 1920, the Air Service concentrated observer training at Ross Field.

Cadets studied such things as aerostatics, aerodynamics, topography, photography, meteorology, Army paperwork, and military justice. They did practical work in observation and artillery adjustment. The course covered both free and captive balloons. Ross Field received a Goodyear Pony Blimp in 1920 for observer training and for primary training in airships.[25]

The Air Service had no dirigibles during the war but wanted to get into the business after the Armistice. Col. Charles de F. Chandler, Lt. Col. Harold Geiger, and others concerned with lighter-than-air aviation saw a greater future in airships for regulating artillery fire, patrolling the border and coast, protecting harbors, performing reconnaissance, and perhaps transporting men and materiel. Early in 1919 the Air Service began building an airship station at Langley Field, Virginia, bought some small nonrigid airships, and planned larger, semirigid and rigid ships. Under a decision of the Joint Board, however, the Navy undertook development of rigid ships for U.S. forces.[26]

The first ships at Langley Field included two American-made blimps, C–2 and A–4, a British Mullion, and a French Zodiac. The collapse of a wooden hangar wrecked the Mullion. The Air Service used the other three ships for training and for work with coast defenses in the Chesapeake Bay area.[27]

The airships acquired by the Air Service in 1920 and 1921 included a large, semirigid ship, the *Roma*, purchased in Italy. Disassembled, shipped to the United States, and erected at Langley Field, the *Roma* made her first flight in America on November 15, 1921. Maj. John G. Thornell sailed her to Washington for dedication on December 21. Dissatisfied with the Ansaldo engines on the *Roma*, the Air Service replaced them with Liberties, which were ready for testing in flight on February 21, 1922.

With forty-five officers, enlisted men, and civilians aboard and Capt. Dale Mabry in command, the *Roma* headed out over Chesapeake Bay, passed Fort Monroe, and crossed Hampton Roads. She was sailing along smoothly at around 55 miles an hour, about 600 feet over the Army Supply Base at Norfolk, when suddenly the control box at the rear of the ship broke and the nose buckled. Plunging downward, the *Roma* hit high-voltage wires, exploded, and burned. Thirty-four persons aboard, including Mabry, died. Officers who investigated the accident could not determine the cause, but it was generally thought that the Liberty engines were too powerful for the *Roma*.[28]

The *Roma* disaster placed the airship program in jeopardy. Lighter-than-air activities were already being curtailed because of shortages of people and money. Perhaps airships should be eliminated. After reviewing the situation, the General Staff and Secretary of War permitted the Air Service to continue with airships. However, outside of the lighter-than-air branch of the service, there would be little enthusiasm or support.[29]

The loss of the *Roma* hastened the change from hydrogen to helium for Army and Navy airships. Helium cost more and did not have quite the lifting power, but when mixed with air it was not explosive like hydrogen. Helium

Roma airship (above) on first test flight at Langley Field, Virginia; and later (left) in flames after hitting high-voltage wires.

would not have saved the *Roma* from destruction, but the number of lives lost would undoubtedly have been smaller.[30]

The United States started producing helium for lighter-than-air aviation during the war, but the Navy's C–7 was the sole government airship using helium at the time of the *Roma* disaster. New ships were built with envelopes designed specifically for helium. But it would be some time before helium production would fully satisfy both Army and Navy needs.[31]

In 1922 the Air Service centralized lighter-than-air training at Scott Field, Illinois, where it set up a depot for balloon and airship supplies and equipment. At the same time it transferred experimental work in lighter-than-air aviation from Fort Omaha, Nebraska, to the Engineering Division at McCook Field, Ohio. The nine- (later ten-) month course of the new Balloon and Airship School at Scott offered free ballooning, operation of captive balloons, and piloting of airships, leading to ratings of airship pilot or balloon observer. Classes were small. The one commencing on September 15, 1924, had 26 members and 14 graduates (9 officers and 5 cadets).[32]

Above: Pony Blimp used for training at the Balloon and Airship School; below: the RS–1, which was the only American-built semi-rigid dirigible.

When the school opened in 1922, it used the Air Service's newest airship (the 180,000-cubic-foot AC–1) and Pony Blimps for training. A little later the school began getting ships of the TA and TC class. The former were 130,000-cubic-foot ships controlled by one man. The student pilot mastered this ship before moving into the 200,000-cubic-foot, dual-controlled ships of the TC class.[33] The Air Service had bad luck with its early TCs. The TC–1 arrived at Scott Field in April 1923 and was wrecked in a storm at Wilbur Wright Field, Ohio, in June. Accidents soon took three more TCs.[34]

In 1925 the school accepted a big, new, semirigid airship—the RS–1. Built by Goodyear, the RS–1 featured a gas capacity of 700,000 cubic feet, a length of 282 feet, and four Liberty engines. The Air Service built its hangar at Scott Field large enough for the rigid ship it wanted but never received. First Lieutenant Orvil A. Anderson tested the RS–1 and found it of no military value. With no money for a new and better ship, the Army dropped its semirigid project.[35]

Technical School

The Mechanics School created at Kelly Field, Texas, in 1917 operated after the war. In the summer of 1920 its staff totaled nearly eighty officers, enlisted men, and civilians; its student body about four hundred. Courses for airplane and engine mechanics enrolled the most students. There were also courses in aircraft armament, auto repair, parachute rigging, and Army paperwork and stenography. Classes for electricians, instrument repairmen, blacksmiths, and welders were being organized. Kelly Field became so crowded that the Mechanics School moved to Chanute Field, Illinois, early in 1921.[36]

The Air Service wanted to send all new recruits directly to the Mechanics School to be tested and trained for the work they seemed best fitted for. However, there was not enough money for transportation to carry out the plan. Most men enlisted by the various units stayed with them. Many other recruits went straight to units for duty and did not get to the Mechanics School until much later—if ever.[37]

The Air Service saw the resumption of recruiting following the congressional ban on enlistments in 1921 as a unique opportunity to train recruits before assigning them for duty. The Mechanics School received a quota of 976 recruits and instructions to go out and get them. Still, the order was not accompanied by money for a recruiting campaign. Parties consisting of an officer, two enlisted men, and a truck driver from Chanute Field opened offices at Danville, Kankakee, Champaign, Decatur, Springfield, and Peoria, Illinois; and Bloomington, Indiana. They put up advertising boards and solicited the help of newspapers, chambers of commerce, the American Legion, postmasters, and others. Planes from Chanute Field flew over in formation, dropped circulars printed by "goodhearted citizens," and gave demonstrations of stunt flying and parachute dropping. In one month, recruiting parties interviewed more than 3,000 men, disqualified many for physical or educational reasons, and accepted 700. Although new classes started every few days, several weeks passed before all of the recruits were in training.[38]

Upon graduation of the large group recruited from Chanute Field late in 1921, the Air Service again faced the old problem of getting men to and from school. Funds still being in short supply, the Air Service turned to its own peculiar resources for transportation. It assigned a Martin bomber to the school as a transport. Other stations with Martin bombers (which could haul 4 or 5 passengers) sometimes flew their men to Chanute for training or came for them after graduation. On occasion they used DHs, though each could take but 1 passenger. A counting in November 1924 showed that during the year 165 men had come to school or returned to duty by airplane. This amounted to only 14 percent of the number arriving and departing. Yet, these were men who could not have attended if required to travel by train. Moreover, the movement of students to and from Chanute Field helped develop air transportation.[39]

Two other schools, the Communications School at Post Field and the Photographic School at Langley Field moved to Chanute Field in 1922. There they merged with the Mechanics School to form a new Technical School under Maj. Frederick L. Martin's command.[40] From time to time the school added a new course, organized or dropped an old one. One notable addition during 1922–26 was a course for crewchiefs, which graduated more students than any other in Fiscal Year 1926. The Technical School, like the flying schools, trained students from the Officers' Reserve Corps, National Guard, Marine Corps, and foreign services. The school's chief mission, however, was the training of enlisted mechanics for the Air Service.[41]

Tactical School

During the war the training of most officers for staff or command positions in aviation was through experience. After the war the Air Service still depended heavily upon experience to prepare officers to command squadrons and fill higher staff and command jobs. But it afforded further training in a Field Officers School that opened at Langley Field in November 1920 with Maj. Thomas DeW. Milling in charge. The bulk of the course dealt with tactical employment of aviation. The seven officers in the first class also studied navigation, meteorology, communications, photography, armament, history of the Air Service, Army regulations, field service regulations, military law, and hygiene and sanitation. The course, scheduled for nine months, was cut short in the spring of 1921, when faculty and students became part of the 1st Provisional Air Brigade for bombing experiments against ex-German war vessels.[42]

The number of graduates from the school (renamed Tactical School in 1922) varied between twelve and seventeen in the next five classes. Flying was added in 1923 and it grew more and more important as the school developed. Elimination of technical subjects in 1925 freed time for courses in aviation tactics and techniques. Students generally spent mornings in lectures and conferences and in working on illustrative problems. They used textbooks, manuals, and other materials prepared by the faculty. Maj. William C. Sherman wrote the first major text, on air tactics, in 1921.[43]

Students usually devoted afternoons to flying and to the practical application of classroom theories and techniques. They also took part in exercises and went on inspection trips. They attended maneuvers, and beginning in 1923 each class visited McCook Field to learn firsthand of new developments in aeronautical equipment.[44]

Engineering School

In 1919 an Air School of Application was set up at McCook Field, Dayton, Ohio, to give Air Service officers an understanding of technical matters relating to airplanes and motors and their maintenance. The objective of the school, proposed by Col. Thurman H. Bane (commander of the experimental station at Dayton), was to improve operations of Air Service flying stations. The first class of six officers entered on November 10, 1919. In 1920 the name changed to Engineering School.[45]

Students selected for the school were rated pilots and graduates of the U.S. Military or Naval Academy or recognized technical colleges. High school graduates could be admitted if they were well versed in fundamental sciences and familiar with calculus, chemistry, physics, and theoretical mechanics. The one-year course covered mechanics, business administration, armament, materials, electricity, powerplants, and theoretical aeronautics. As a rule the classes had no more than a dozen students. Most graduates returned to general duty, but now and then one stayed on a while to work with the Engineering Division. A few went to the Massachusetts Institute of Technology (MIT) to take aeronautical engineering. The outstanding student of the period was 1st Lt. James H. Doolittle. He entered MIT in October

1923, got a master's degree in June 1924, and became a Doctor of Science in June 1925.[46]

In addition the Air Service let a few officers attend other civilian colleges and universities: Yale for radio communications, Harvard for business administration, and Columbia for contract law. It also sent a few to the Command and Staff School, the Army War College, the Army Industrial College, the Chemical Warfare School, and the Signal Corps Radio School.[47]

School of Aviation Medicine

In the spring of 1919 the Air Service began training doctors to replace wartime flight surgeons returning to civilian practice. Doctors selected from officers of the Army's Medical Corps went to the Air Service's Medical Research Laboratory at Hazelhurst Field, Long Island, New York. Established in January 1918, the laboratory's mission was to study the medical aspects of aviation, then shifted from research to training. (*Map 2*)

Three classes graduated before the Air Service abandoned Hazelhurst Field and moved the school to Mitchel Field in November 1919. A fire on March 19, 1921, destroyed teaching materials, equipment, and administrative records. Even so, the Commandant, Maj. Louis H. Bauer, managed to keep the school going until new facilities were ready. The course given to MDs in the grade of major, captain, or lieutenant included practical work in New York City hospitals as well as lectures and demonstrations at the school. Besides, the doctors were taught how to perform the "609" examination required for flying.[48]

By June 30, 1921, forty-six flight surgeons had been graduated and assigned to stations throughout the United States. Their principal job, as stated first by General Menoher and then by General Patrick, was "the prevention of loss of life and property through accidents attributable directly or indirectly to the physical condition of pilots." This they did by physical examination and close personal observation of the flyers, and by investigation of airplane accidents from the medical point of view.[49]

In 1922 the Medical Research Laboratory and its School for Flight Surgeons became the School of Aviation Medicine. As attested by the new name, its main job was educational but it retained a research mission. The school's investigations encompassed the effects of cold on respiration, circulation, and body temperature; the amount of oxygen needed for altitude flights; and lenses for goggles. One study probed the temporary deafness of flyers from aircraft engine noise. Many flyers wore earplugs, but plugs of hard rubber, paraffin, or wax were unsanitary and hard to keep clean. Cotton left fibers in the ear canal, causing irritation. The school recommended that a

MAP 2

lady's powder puff (wool, or wool and cotton) be sewn inside each of the ear flaps on the helmet.[50]

At first the course for flight surgeons lasted for six weeks, but later extended to three months. Medical officers of the Organized Reserve and National Guard could take a basic course of six weeks and qualify as physical examiners. By pursuing an advanced one of six weeks, they could become flight surgeons. Or, after the basic study, they could achieve the same end by completing a correspondence course.[51]

Flight surgeons during the war held flying status and earned flight pay. They continued to draw the extra money until mid-1920, when cut off by Army regulations. General Menoher's recommendation that all medical officers serving in the Air Service be designated for flying duty was twice rejected. Doctors still could fly but not be ordered to do so. General Patrick was able to change this policy in 1922.[52]

Most Air Service doctors flew as passengers but some became pilots. Out of forty-eight flight surgeons on duty in the summer of 1926, seven were rated (four as pilots and three as junior pilots).[53]

At the time of the Mitchel Field fire in 1921, Major Bauer rejected the suggestion that the school move. He thought it should be near the medical facilities and libraries of New York City. The suggestion came up again in the mid-1920s as General Patrick and others grew concerned over the many students washing out of flying training. Perhaps the medical school could help correct the matter if its work was more practical and less theoretical. It needed to be located where pilots trained. So the School of Aviation Medicine moved to Brooks Field in June 1926.[54]

Each of these schools required specific authorization from the War Department. Once a school had been established, responsibility for running it rested with the Chief of Air Service. He had to operate within the administrative framework laid down in Army regulations, and under the monetary and personnel limits imposed. But in technical training matters in the Air Service schools, he was in control. The Air Service Chief determined the kind and extent of the training to be given and set standards for successful completion. This was true even when Air Service policies, such as those on qualifications for flying training, affected the combat capabilities of the Army's air arm. Unlike individual training, unit training for tactical operations did not come under his direction. There he could only advise and recommend, as will be seen in the next chapter.

Chapter V

Tactical Units

The National Defense Act of 1920 did not end the controversy over aeronautical organization. Arguments continued for removing the Air Service from the Army and creating either an aeronautical department comprising all federal aviation, or a defense department combining the air services of the Army and the Navy. Within the Army itself, there was general agreement on the division of aviation into "air service" and "air force." The former consisted of units assigned as integral parts of divisions, corps, and armies to work with ground units. The latter embraced aviation assigned to general headquarters and used by the commander in chief in the field to further his strategical and tactical objectives. Depending upon the situation in the theater of war at any particular time, he might use his air force to support his ground forces, or send it on operations not directly related to the battle on the ground.

In the Army, disagreement arose over what types of aviation should be assigned to the air service, and over the apportionment of available resources between air service and air force. But this was a matter pertaining more to mobilization planning than to the peacetime Army. In the 1920s, corps, armies, general headquarters, and general headquarters reserve (including the air force) existed only on paper as units to come into being in an emergency. For peacetime organization and training, field forces (including both air service and air force units) came under the control of Corps area commanders. This chapter looks at the organization, training, and equipment of the Army's air arm.

Organization

The interim organization the War Department authorized for the Army's air arm after the Armistice was 2 wings, 7 groups, and 27 squadrons of airplanes. All had been formed by the autumn of 1919.[1] In the lighter-than-air branch, the War Department authorized 42 companies (including school detachments), but a shortage of people soon shaved that number to 32. Two airship and 10 balloon companies, plus several school units, had been formed by mid-1920.[2]

A General Reorganization Board that prepared plans for carrying out the National Defense Act of 1920 based its aviation recommendations on this principle: "All aviation in an Army should be employed for participation in the battle, and all strategic bombardment and reconnaissance should be done by aviation in G.H.Q. Reserve."[3] Approving the board's report, the War Department adopted a plan giving divisions, corps, and armies their own observation aviation, assigning attack and pursuit units to armies, and setting up a General Headquarters Reserve composed of all bombardment units and airships as well as some observation units for strategic reconnaissance.

The War Department injected the same organization principle into war planning. An initial mobilization would be based on 2 million men with 6 field armies, 18 corps, and 54 divisions. The Regular Army was to furnish 9 divisions; the National Guard, 18; and the Organized Reserve, 27. To carry out this scheme, the General Staff replaced 6 geographical departments of the United States with 9 corps areas.[4]

The Air Service part of the mobilization plan entailed about 100,000 officers and men and 4,000 airplanes. Aircraft allocations determined tactical organization. The General Staff set as first priority the planes required for observation aviation for divisions, corps, and armies; then planes for 1 attack wing (1 attack and 2 pursuit groups) for each of the 6 field armies and for GHQ Reserve. This toted up to 118 observation squadrons, 26 balloon groups (104 companies), 14 pursuit groups (56 squadrons), 7 attack groups (28 squadrons), 24 airship companies, and 1 bombardment group (4 squadrons).[5]

The theory behind Army organization and mobilization planning was that the peacetime structure set by the 1920 act would be the foundation for mobilization. But as we have seen, Congress commenced slashing the size of the Army early in 1921, eroding the mobilization base. War Department policy was to keep the overseas garrisons as close to authorized strength as possible in peacetime, while taking the cuts at home. Consequently, the Army in 1921 inactivated aviation units in the United States but formed new air units for overseas service.[6]

Further reductions in 1922 brought inactivation of one observation squadron and all balloon companies, and slimmer manning for most observation squadrons in the United States.[7] At the same time, tactical units in the United States relocated.

In the summer of 1919, Kelly Field had become the home of the Air Service's pursuit, bombardment, and surveillance groups. Soon afterwards, however, the surveillance group deployed along the Mexican border. When the surveillance group headquarters returned to Kelly and the group became an attack unit in 1921, the 1st Pursuit Group moved to Ellington Field near Houston. With the creation of the Advanced Flying School at Kelly in 1922, the 2d Bombardment Group departed for Langley Field, leaving the 3d Attack Group the one tactical unit at Kelly. The 1st Pursuit Group relocated at the same time from Ellington to Selfridge Field, Michigan. The 2d Bombardment Group left its planes at Kelly when it moved, but the 1st Pursuit Group flew fourteen SPADs, two SE–5As, and five DH–4Bs to its new post.[8]

After these and further adjustments, Army aviation in the United States in the spring of 1923 consisted of 1 wing, 3 combat groups (pursuit, bombardment, and attack) of 4 squadrons each, and 1 group headquarters and 11 observation squadrons. Each overseas department had 1 composite group of 3 squadrons (observation, pursuit, and bombardment). An exception was Hawaii, with an additional group headquarters and 2 extra squadrons (pursuit and bombardment). The only major changes over the next 3 years inactivated 1 group headquarters in Hawaii and 2 attack squadrons in the United States.[9]

Asking for more people in 1921, General Menoher called Secretary of War John W. Weeks' attention to "two distinct classifications of military air power—'air service' and 'air force.' " The General Reorganization Board recognized two such forms of aviation in 1920, but the Chief of Air Service did not agree with the way the board apportioned aviation between the two.

General Menoher explained that the air service consisted of observation units which carried on visual and photographic reconnaissance, located targets, adjusted artillery fire, and performed contact patrols and other liaison with the infantry. This aviation did not act offensively but engaged in combat only when attacked by hostile aircraft.

On the other hand, air force was purely offensive and comprised pursuit, bombardment, and attack aviation. Pursuit destroyed the enemy's aircraft, protected friendly ones, and harassed ground troops and enemy activities in the theater of operations. Bombardment destroyed military objectives both in the theater of operations and in the enemy's zone of interior. Attack aviation used heavily armored planes to harass enemy ground troops from low altitude with machinegun fire and fragmentation bombs.

For a proper balance, the Chief of Air Service said, 80 percent of the strength should be in air force or combat aviation and the remaining 20 percent in observation or air service. As things stood, air service took 40 percent of the total. General Menoher wanted to form 2 more pursuit groups, 1 more bombardment group, and the necessary auxiliary units.[10]

The War Department accepted and embodied in Army regulations the general concept of aviation being divided between air service and air force.[11] But the department's acceptance of the general proposition did not bring reapportionment of available strength between air force and air service at the level desired by General Menoher and the Air Service.

General Patrick used the war plan to attack the imbalance between air service and air force. He wanted more men and planes for mobilization, less observation, and more pursuit and bombardment. He would take observation from the divisions and concentrate in the corps. He would give everything else—pursuit, bombardment, attack—to Army GHQ and place it under an air force commander for use as the military situation required.[12]

On March 17, 1923, Secretary of War Weeks appointed a board, headed by Maj. Gen. William Lassiter, Assistant Chief of Staff (ACS), G–3 (Operations and Training), to study General Patrick's plan.[13] Appearing before the board, Patrick presented detailed schedules for expanding the peacetime force to 4,000 officers, 2,500 flying cadets, 25,000 enlisted men, 1,680 airplanes, 11 airships, and 24 balloons over a 10-year period. He also produced a plan for mobilizing 12,880 officers, 109,712 enlisted men, 5,194 airplanes, 24 airships, and 104 balloons in an emergency. In this and other meetings, the board gave him full opportunity to explain his plans. The discussion disclosed disagreement on many important points, but the committee easily concluded that something needed to be done to strengthen Army aviation.

The Lassiter Board found the Air Service "in a very unfortunate and critical situation." As it then existed, the service bore "no relationship to the war requirements." If the country was confronted with an emergency, "the Air Service would not be able to play its part in meeting it." The board recommended legislation to strengthen the peacetime Air Service along the lines proposed by General Patrick, and adoption of his plan for mobilization.[14]

General Patrick's proposals on organization did not fare so well. The majority of the Lassiter Board thought aviation should be organized on these principles:

> (a) Observation Air Service should be an integral part of divisions, corps and armies with a reserve under General Headquarters.
>
> (b) An Air Force of attack and pursuit aviation should be an integral part of each Field Army, with a reserve under General Headquarters.
>
> (c) An air force of bombardment and pursuit aviation and airships should be directly under General Headquarters for assignment to special and strategical missions, the accomplishment of which may be either in connection with the operation of ground troops or entirely independent of them. This force should be organized into large units, insuring great mobility and independence of action.[15]

The War Department put these principles in Army regulations.[16] It also approved the Lassiter Board's suggestions on mobilization strength for war plans. However, the board's recommendation that Congress be asked to approve and fund a program for expanding the peacetime Air Service got lost in the Joint Board, due to a disagreement between the Secretaries of War and Navy on appropriations for the two departments.[17]

A committee of the House of Representatives headed by Florian Lampert of Wisconsin took up the Lassiter Board's report while investigating military aviation during 1924–25. Committee members could not understand why the legislation recommended by the Lassiter Board had not yet been sent to Congress. Believing the nation could "have no adequate national defense without an adequate air force," the Lampert Committee urged a five-year aviation program with the War and Navy Departments each getting at least $10 million a year to spend for new flying equipment.[18]

The Lampert Committee had not yet issued its report when President Coolidge, on September 12, 1925, appointed a board headed by Dwight W. Morrow, a New York banker, to study the use of aircraft in national defense. Holding hearings and reviewing testimony taken by various congressional groups, the Morrow Board, like the Lampert Committee, found great conflict in matters of fact as well as of opinion. Witnesses did not agree, for example, on how many usable planes the Army Air Service owned, or whether the United States had more or fewer than Japan.

The Morrow Board rejected the airmen's repeated assertion that air power would be decisive in an armed contest. Its more conservative view was that "the next war may well start in the air but in all probability will wind up, as the last one did, in the mud." The board believed the nation's policy on military aviation should be based on the general military policy of the United States and on the air strength of foreign nations that could menace U.S. security. National policy called for air power as a defensive agent. Protected by broad oceans, the nation need not fear direct invasion from overseas by way of the air. Nor could an enemy move an air force into position to strike the United States so long as the U.S. fleet remained undefeated.

Having considered the personnel situation in the Air Service, the Morrow Board's report of November 30, 1925, recommended the authorization be increased by two brigadier generals—one to head procurement and the other to run the flying schools at San Antonio. It further recommended appropriations to train more cadets and Reserve officers. The board saw the Air Service in transition with new planes ready for production to replace older ones. While special appropriations for fresh aircraft were worthy of consideration, it would be unwise to plan many years ahead, as the Lassiter Board suggested. Such a plan, the Morrow Board held, should not exceed five years. The board also deemed the designation "Air Service" confusing because it seemed to conflict with some of the duties. "The distinction between service rendered by air troops in their auxiliary role and that of an air force acting alone on a separate mission is important," the board said. It suggested the name be changed to "Air Corps."[19]

A number of aviation and Air Service bills appeared in the 69th Congress which began in December 1925. Each dealt in its own way with the strength and tactical structure of Army aviation. Moreover, most treated organization and control of military aviation at the departmental or national level, the proposals being as a rule along two different lines. One plan, advanced by General Patrick, transformed the Air Service into an Air Corps whose relationship to the Army would be similar to that of the Marine Corps to the Navy. Another, championed by Mitchell and suggested by the Lampert Committee, would create a separate Air Force on a level with the Army and Navy in a Department of Defense.[20]

A bill drafted by the War Department contained recommendations of the Morrow Board, and other bills rested on the Lassiter and Lampert reports. But the House Committee on Military Affairs voted them down and came up with its own Air Corps bill. The Senate made so many changes that the bill went to conference. Differences resolved, the amended Air Corps bill became law on July 2, 1926.[21]

The new law changed the name but not the status of Army aviation. More important, it authorized a program for expanding and strengthening the Air Corps over five years.[22] Yet doubts clouded General Patrick's hopes for the future. First, the act did not furnish the air force he thought the nation needed. Second, it failed to guarantee the Air Corps would expand as scheduled. Congress had not let the Air Service attain the strength authorized by law in 1920, and the same thing could happen again. If carried through, however, the five-year program would produce "an effective well equipped Air Force" which, Patrick said, would "greatly strengthen the air defense of the country."[23]

Tactical Training

Units of the 1st Wing formed at Kelly Field, Texas, in the summer of 1919 commenced training as soon as the disruptive effects of demobilization and border duties permitted. The schedule of the 1st Pursuit Group at Kelly, and afterwards at Ellington Field, Texas, offered formation and cross-country flying, pursuit tactics, acrobatics, gunnery, low-level bombing, and reconnaissance. Pilots attended lectures on navigation, mapreading, topography, meteorology, rigging, flying rules, tactical organization, and other subjects. They spent many hours experimenting with tactics and maneuvers. And they improvised constantly to make best use of what they had. To rebuild bomb racks to fit bombs or to devise a sight to use in bombing practice was just part of a normal day's work.[24] Other units had similar programs fitted to their peculiar needs and circumstances.

At first the various groups and separate squadrons enjoyed considerable latitude in formulating training programs. They received guidance and information from manuals, pamphlets, and other documents produced by the Training Division, Training and Operations Group, and Office of the Chief of Air Service.[25] In June 1920 the division began publishing circulars detailing experiences gained during the war. Lt. Col. Harold E. Hartney, who had commanded the 1st Pursuit Group in combat, wrote one on air tactics. Maj. William C. Sherman provided "Notes on Recent Operations," which he had written in France as a "Tentative Manual for the Employment of Air Service."[26]

Guidance from the Office of the Chief of Air Service grew as time went on. General Patrick believed this necessary to achieve the best possible results. By 1923 his staff drew up a training program for all aviation, but because of his limited authority he could not require all units to use it. He therefore arranged for The Adjutant General, War Department, to distribute it as "a directive for commanding officers of exempted stations under the control of the Chief of Air Service," and "as a guide to assist corps area and department commanders in formulating their training programs."[27]

The program for 1925 covered observation, bombardment, pursuit, and attack during four training periods. Running for about three months, the first period gave officers ground instruction an hour a day, five days a week. During this time the officers studied the theory and practice of bombing, gunnery, navigation, night flying, photography, use of parachutes, supply and maintenance, and methods of operations and organization. They heard lectures by medical officers and others on flying at high altitude and on the use of oxygen. Students usually spent their mornings flying, using camera obscura and dummy bombs to learn bombing, and camera guns to practice aerial gunnery. Flying instruction encompassed navigation, cross-country flying, night flying, and aerial photography. Enlisted men took courses in airplane mechanics, armament, communications, and administration.

The second period of about four months was devoted to unit training. All types of units participated in formation and cross-country flying. In aerial gunnery they had to husband their ammunition because the allowance was small—only five thousand rounds to each pilot and observer for the year. Bombardment units practiced from various altitudes and with different kinds of bombs and fuzes, and made raids on simulated targets at a distance from the home airdrome. Attack and pursuit units trained for low-level bombing and machinegun attacks on ground targets. Pursuit also trained for aerial combat, patrol work, and protection of bomber and observation aircraft. Observation units prepared for work with infantry, cavalry, and artillery.

The training for the third period lasted around two months. Regular units of the Air Service ran summer camps for the Organized Reserve, the Reserve Officers' Training Corps, and the National Guard. They spent the fourth period of three months in field training, work with other branches of the Army, and Air Service maneuvers.[28]

Observation units spent much time in exercises with ground forces. The 12th Observation Squadron, for instance, was attached to and worked regularly with a cavalry division in Texas. The 88th at Wilbur Wright Field, Ohio, went to Fort Knox, Kentucky, from time to time to operate with field artillery. Other units participated with ground troops at Fort Monroe, Virginia; Camp Bragg, North Carolina; Camp Benning, Georgia; Fort Sill, Oklahoma; and elsewhere.[29]

Units in Hawaii, the Philippines, and the Canal Zone directed tactical training mainly toward coastal defense. Exercises and maneuvers with ground and sea forces were a regular and important part of their mission.[30]

Balloon companies both at home and in overseas departments used captive balloons to train with field or coast artillery for adjustment of fire. The 19th Airship Company at Langley Field, Virginia, not only worked with coast artillery but experimented with bombing and trained for visual and photographic reconnaissance. Both balloon observers and airship pilots spent time sailing spherical balloons. Demanding a firm grasp of aerostatics and meteorology as well as great skill, free ballooning made a man a better observer or pilot. Despite the hazards, it was great sport. Every unit hoped to produce the winner of a national or international balloon race.[31]

Cross-country flights of entire groups were part of tactical training for pursuit and bombardment. The 2d Bombardment Group at Langley Field made one such flight in October 1924. A year earlier the group had flown to Bangor, Maine, as a test of mobility.[32] This time it would fly to Mitchel Field, New York, at night. The objective, General Patrick said, "was to demonstrate the practicability of changing the base of a bombardment group overnight and the feasibility of operating the group the following morning."[33]

Nine NBS–1s took off from Langley in three formations just before sunset on October 10, 1924. Each plane carried a pilot, navigator, mechanic, and radio operator. Motor trouble over Chesapeake Bay forced Capt. Early E. W. Duncan to return to Langley and change planes. On the second attempt, motor trouble put him down at Lakehurst, New Jersey, where he stayed overnight. Another plane, piloted by Maj. John H. Pirie, the flight commander, came down at Cape May, New Jersey, for motor repairs. Before landing, he told his radio operator, Capt. Harold M. McClelland, to notify the other planes and instruct Capt. Willis H. Hale to assume command. The trouble corrected, Pirie continued to Mitchel Field, arriving about an hour after Hale and the others. At Mitchel the bombers were serviced to be ready to fly the following morning.[34]

One of the flights by the 1st Pursuit Group was from Selfridge Field to

Miami, the aim being to cover the entire distance (about thirteen hundred miles) in a single day. Maj. Thomas G. Lanphier, the group commander, led the flight of twelve PW–8s. He planned to leave before dawn, stop at Fairfield, Ohio, and Macon, Georgia, arriving at Miami around 1800.

Landing at Wilbur Wright Field at Fairfield early Saturday morning, February 28, 1925, 1st Lt. Ennis C. Whitehead broke his landing gear. Major Lanphier decided to hold the flight until a replacement plane came. Two other planes had engine trouble, so the flight did not reach Macon until late afternoon. Because they could not get to Miami before dark and the weather was bad, Lanphier decided to do the safe thing and stay at Macon. Sunday being devoted to tuning the planes, the flight did not touch down at Miami until Monday.

The flight failed, General Patrick said, owing to "the reluctance of the flight commander to eliminate planes which developed trouble and to proceed with the remainder." Nevertheless, the Chief of Air Service believed the flight "demonstrated the practicability of transferring the pursuit component of our Air Force from the cold weather of the extreme northern part of the country to the warmer southern climate in a single day, if properly managed."[35]

Beginning in 1924, the Air Service held bombing and gunnery matches annually at Langley Field. Tactical units in the United States and the Panama Canal Zone participated. Competition was particularly keen in 1926 when the National Guard and Marine Corps together with Air Service squadrons in

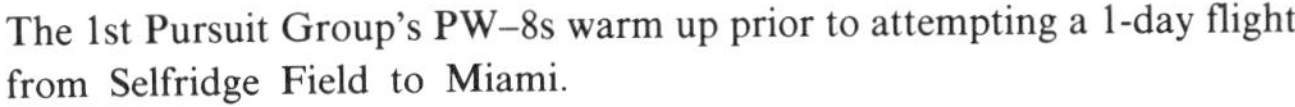

The 1st Pursuit Group's PW–8s warm up prior to attempting a 1-day flight from Selfridge Field to Miami.

the United States took part. Eight events appeared on the program that year—2 for fixed machineguns, firing at ground and towed targets; the same for flexible guns; 3 for bombing from airplanes at low, intermediate, and high altitudes (minimums of 300, 5,000, and 8,000 feet); and 1 for bombing from airships (minimum altitude, 3,000 feet).

It appeared for a time that 1st Lt. Lawson H. M. Sanderson of the Marine Corps would place first in the competition among pursuit pilots. He scored 706 in gunnery against ground and towed targets and in low-altitude bombing. But the Air Service's "ace in the hole," 2d Lt. Louis M. Merrick, came through with 730. The competition among pilots of biplace aircraft became a contest between two members of the 3d Attack Group, with 2d Lt. Earle E. Partridge scoring 646 points to 2d Lt. Hoyt S. Vandenberg's 638. A team made up of 1st Lts. Harold L. George and Ernest E. Harmon from the Office of the Chief of Air Service won the heavier-than-air bombing match. In the lighter-than-air bombing competition, 1st Lt. Alfred I. Puryear and Capt. William E. Kepner came in first and second.[36]

General Patrick considered tactical training incomplete unless air force units trained together from time to time as an air force. He accordingly obtained authority from the War Department to concentrate all available pilots and planes of the 1st Pursuit, 2d Bombardment, and 3d Attack Groups at Mitchel and Langley Fields for maneuvers in October 1925. The air force, assembled on Long Island while the air races were going on at Mitchel Field, consisted of forty-five planes—all the up-to-date fighting ships the Air Service could muster. Under the command of Brig. Gen. James E. Fechet, who had succeeded Brig. Gen. William Mitchell as Assistant Chief of Air Service, the

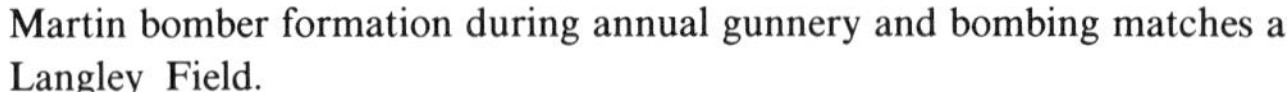
Martin bomber formation during annual gunnery and bombing matches at Langley Field.

NBS–1 during Air Force maneuvers at Fairfield, Ohio, in April 1926.

air force defended the coast against a theoretical attack by a hostile fleet of four airplane carriers and four hundred planes. After the races, when the enemy established a base in Chesapeake Bay, the air force flew off to Langley Field to do battle. Finding these maneuvers valuable for training, General Patrick recommended that similar events be held annually.[37]

The next air force maneuvers took place at Fairfield, Ohio, in April 1926. The Ohio River became the boundary between two nations—Blue to the north and Red to the south. Seeking to annex Ohio, the Red nation mobilized secretly before declaring war on April 17. Blue forces (including simulated infantry, cavalry, and artillery) completed mobilization at Fairfield at 1200 on April 19. General Fechet commanded the Blue air force containing 45 officers, 67 enlisted men, and 44 planes. Included were an observation group (represented by 2 officers and 1 plane) and the 1st Pursuit, 2d Bombardment, and 3d Attack Groups.

Theoretical problems (issuing orders, preparing intelligence summaries, making reports, and compiling war diaries, journals, and other records) took three days, after which the flying began. On April 22 the 2d Group's NBS–1s, supported by the 1st Group's P–1s and the 3d Group's O–2s and DH–4Bs, attacked bridges over the Ohio River at Cincinnati. To avoid the dangers of flying over this thickly populated area, the raid actually was on the town of Circleville, Ohio, about the same distance from Dayton as Cincinnati. The bombers went after bridges again the next day while the attack group struck a railroad ten miles away and pursuit aircraft patrolled the area between. On Saturday, April 24, the target was rail yards at Kenton, Ohio, about sixty miles north of Fairfield. In one tactical problem the following week, pursuit planes detected an attack formation snaking in and out among the trees. They pounced on it in time to prevent serious consequences to ground forces. General Patrick was well pleased with these maneuvers.[38]

Periodic tactical inspections by corps area and department commanders were prescribed by Army regulations and by the training program of the Office of the Chief of Air Service. Maj. Gen. Edward M. Lewis, Commanding General, Eighth Corps Area, and his staff inspected the 3d Attack Group in April 1924, putting it through a number of exercises at Kelly Field. In one the group made diving attacks with machineguns and 25-pound practice bombs on targets on the airdrome. Another involved various combat formations over the field. The group was most pleased with its horizontal bombing. All of the 50-pound bombs struck within fifty yards of the target, demolishing it. General Lewis commended pilots and observers on their excellent showing.[39]

Equipment

The 3d Attack Group had begun training in 1921 with DH–4Bs while awaiting the GA–1.[40] The latter was an armored, ground-attack plane which the Air Service Engineering Division at McCook Field, Ohio, developed for low-level bombing and strafing. A triplane, the GA–1 used two Liberty engines working as pushers, carried a crew of three, mounted eight machineguns, and could carry a 37-mm cannon as well as bombs. Armor protected engines and other vital parts. When the new planes started to arrive early in 1923, the 3d Group quickly discovered they performed poorly. The plane was too heavy. It was slow, required a lot of space to take off and land, had a short range, climbed slowly, and was not very maneuverable. The Air Service reduced its contract with Boeing from twenty planes to ten.[41]

With no new ground-attack plane coming along to take the place of the unsatisfactory GA–1, the 3d Attack Group made attack planes out of DH–4Bs. The group centered on boosting the plane's firepower. After much experimentation, it armed DHs with six fixed machineguns controlled by the pilot and two flexible guns operated by the observer, and with racks for fragmentation, incendiary, demolition, and chemical bombs weighing up to a hundred pounds.[42]

Experience with the GA–1 raised a question whether attack planes should be armored and, if so, how much armor they should carry. The question, not an easy one to answer, would remain open a long time. By 1926, however, students in the Advanced Flying School at Kelly Field were being taught that the proper plane for ground attack was a fast, maneuverable, two-seater. Light armorplate was secondary to offensive power.[43] At that time the 3d Group was replacing its DHs with O–2s adapted to low-level bombing and strafing.

The Air Service first acquired O–2s through a competition in 1924 for a new observation plane with a Liberty engine. With eleven planes entered, Douglas won with the XO–2 and received an initial order for forty-six planes to begin replacing DHs for observation.[44]

De Havilland aircraft had been used by the Air Service's bombardment group for border patrol during 1919–20 and for bombardment training at Kelly Field. At Kelly the group owned large bombers, Capronis and Handley Page O–400s, earmarked by the Air Service for a strategic bombing program that did not get under way before the Armistice. In 1921 the Martin MB–2 (redesignated NBS–1) became the principal bombing plane of the Air Service. The NBS–1 had an empty weight of 7,200 pounds and a gross weight of 12,000, carried a crew of four, and mounted five machineguns. Powered by two Liberty engines, it possessed a top speed of 99 miles per hour, a cruising speed of 91; a range of about 550 miles; a service ceiling of 8,500 feet, and a maximum ceiling of 10,000. Although the Engineering Division tested other planes of the same class, it found none superior to the NBS–1.

The Engineering Division sought a bomber that could carry a heavier load over a greater distance. General Mitchell had high hopes for a giant plane, the XNBL–1, designed by Walter H. Barling, a civilian employee of the Engineering Division. A triplane, the Barling bomber was driven by six Liberty engines and had an empty weight of about 27,000 pounds and a gross weight of more than 42,000. A factory in New Jersey built one in sections and shipped them to Fairfield for assembly. First Lieutenants Harold R. Harris and Muir S. Fairchild, pilots, and Douglas Culver, engineer, took the plane up for its first flight on August 22, 1923. The big bomber attracted a lot of attention at the National Air Races in 1923 and 1924. It set duration and altitude records for lifting useful loads of around 4,400 and 6,600 pounds. Still it was too slow, with a top speed of 95.5 miles per hour. A failure, the Barling bomber spent most of its life parked at Fairfield, Ohio, before being dismantled and burned.[45]

The Air Service made better progress with pursuit aircraft. The 1st Pursuit Group in 1920 envisioned the pursuit ship of the future as an all-metal monoplane with a high-powered, radial motor, carrying two .30-caliber and two .50-caliber machineguns, climbing to 20,000 feet in 10 minutes, flying well over 200 miles per hour, and cruising for 7 hours. The group thought such a plane "within the scope of present day possibilities."[46]

Above: Martin MB–2 (redesignated NBS–1), one of fifty built by Curtiss as part of Army policy of dividing orders among hard-pressed manufacturers during postwar period; below: "Barling Bomber," or NBL–1, designed by Walter Barling of the Engineering Division.

At that time, the 1st Pursuit Group's primary aircraft was an open-cockpit biplane made of wood, wire, and fabric. It used a 180-horsepower Vee-8 engine, climbed to 6,500 feet in 8 minutes, had a top speed of 121 or 122 miles per hour, a range of about 280 miles, and could mount two .30-caliber machineguns. This was the British-designed SE–5A, of which the U.S. Air Service obtained 57. From among new planes being developed by the Engineering Division and American airplane builders, the 1st Pursuit Group received some Orenco Ds and Thomas-Morse MB–3s in 1921. In addition the Air Service withdrew some French-built SPAD XIIIs from storage as the group used up the SE–5As.[47]

After moving to Selfridge Field in 1922, the 1st Pursuit Group flew SPADs for a while before receiving new Boeing MB–3As. Because of the vibration of the plane's powerful (300-horsepower) Wright engine, maintenance was a problem during the two years the MB–3As were standard equipment. A new pursuit ship, the PW–8, developed from the Curtiss racer that won the Pulitzer Race in 1922, became available in 1924. A modification of the PW–8 was the prototype for the P–1s, the first of the Curtiss Hawks, accepted by the 1st Pursuit Group in 1925.

The P–1 was an open-cockpit fabric-covered biplane, the fuselage framed with metal tubing, the wings with wood. It carried one .30-caliber and one .50-caliber or two .30-caliber machineguns. It used a Vee engine (a Curtiss D–12 of 435 horsepower) with duralumin propeller. The P–1 climbed to 5,000 feet in 3.1 minutes and to 20,000 feet in 24, and had a service ceiling of

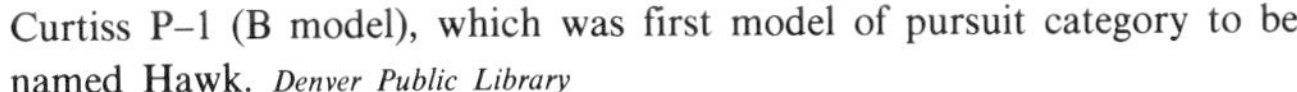

Curtiss P–1 (B model), which was first model of pursuit category to be named Hawk. *Denver Public Library*

22,500 feet, a top speed of 163 miles per hour, and a range of about 400 miles. While not the plane the group dreamed about, its performance mirrored the many advancements in pursuit aircraft development.[48]

During the 1920s the War Department classified aviation as "air service" and "air force." The former, auxiliary to ground forces, comprised observation units assigned to divisions and corps, and observation, pursuit, and attack units assigned to field armies. The latter consisted of bombardment, pursuit, and observation organized in large mobile units and assigned to a general headquarters for employment on strategic or special missions, either independently or in cooperation with ground forces. This organizational scheme served as the basis for mobilization and war planning. The War Department designated active observation units of the Regular Army for peacetime service with Regular divisions. Corps, armies, and a general headquarters were not organized for peacetime service. So other observation units, as well as bombardment, pursuit, and attack, came under the control of corps area and department commanders. From time to time, however, the War Department placed units under the Chief of Air Service for training exercises and maneuvers.

It became obvious early in the 1920s that the active Air Service was too small to be effective in an emergency. The Chief of Air Service proposed, and the War Department approved, a ten-year program for strengthening the air arm. This program fell before Army-Navy rivalry over aviation appropriations.

Unable to enlarge the air arm, the Air Service bolstered the effectiveness of field forces. It developed better unit training programs, held gunnery and bombing competitions, conducted maneuvers, and secured better equipment for tactical operations. Observation units received new planes to replace DH–4s. Pursuit squadrons got ships of greater speed, faster climb, higher ceiling, and longer range. The Air Service did not fare as well with bombardment and attack aircraft. Though dissatisfied with the speed, ceiling, range, and bombload of the Army's bombers, it could find nothing better. Furthermore, it failed to develop a satisfactory plane for strafing and low-level bombing.

Chapter VI

Reserves

In a national emergency requiring a greater force than the Regular Army, the War Department planned to order National Guard units of the various states into federal service. Planning under the National Defense Act of 1920 presumed a Guard manned, organized, equipped, trained, and available for field operations. If the situation called for a still larger force, the War Department would mobilize units of the Organized Reserve composed of members of the Officers' and Enlisted Reserve Corps. Partially manned and equipped in peacetime, Reserve units would not be ready for combat for some time after being mobilized—not until they had been filled by enlistment or conscription, equipped, and trained.

Lack of money prevented the War Department from keeping the Regular Army at the strength envisioned by the 1920 Defense Act. A similar state arose in the National Guard. From 450,000 men, the number on which the act rested, the War Department lowered its sights to 250,000. Actually, the Guard grew to about 180,000 in the 1920s and then leveled off. The Organized Reserve suffered even more, as will be seen in this chapter which deals with the Air Reserve, the Reserve Officers' Training Corps, and aviation units of the National Guard.

Air Reserve

Although the Air Service hoped to retain a powerful peacetime Regular force, it also wanted a large reserve of trained flying officers instantly on hand in an emergency. While it expected to have a reserve of enlisted men, it also counted on drawing men from civil and commercial aeronautics and from airplane plants.[1]

In 1919 the Air Service commenced creating a reserve from trained men leaving the service. The program included both officers and enlisted men, but centered on flyers, chiefly pilots. To induce the discharged pilot to join the Reserve, the Air Service let him continue flying at government expense. He would have to go up with an instructor and prove he could handle a plane. After that he could fly by himself. In an emergency, he would enter the service as an officer without further training.[2]

To attract pilots to the Reserve and enable them to keep in practice, the Air Service permitted them to fly its planes when not on active duty. Such flying depended upon availability of equipment. Otherwise the Air Corps in the beginning imposed just two restrictions. It prohibited cross-country flights and insisted that Reserve flying not interfere with the regular work of the field. Later, it directed commanders to require that each person applying to take up a government plane provide proper identification, be passed by a flight surgeon, and be checked out by a qualified instructor. This would prevent the needless destruction of government property, and see that over-enthusiastic pilots in poor physical condition or out of practice did not unnecessarily endanger their lives.[3]

Of 8,415 flying officers discharged up to May 10, 1919, only 14 percent accepted Reserve commissions. Of 5,429 nonflying officers discharged, 18 percent entered the Reserve. By June 30, 1919, there were 2,434 officers in the Reserve, 1,301 rated and 1,133 nonrated. A year later the number rose to 7,339, of whom 5,046 were flyers.[4]

The Air Service laid plans for Reserve training centers and Reserve units while awaiting congressional action on postwar military policy. Meantime, some Reservists took advantage of the Air Service's offer to let them fly at government fields. Reservists in a flying club at Detroit wanted to fly at nearby Selfridge until they could get an airstrip of their own. The commander at Selfridge, Maj. Norman J. Boots, said that for these men "the lure of the air is great." But he was limited in what he could do. Selfridge, a temporary storage depot, had no flight surgeon to examine Reservists, and only two pilots, Major Boots and 2d Lt. Jerome B. Machle, to accompany them on flights. During one week in the spring of 1920, eleven Reservists made practice flights at Selfridge. Others could not go up since neither Boots nor Machle could spare the time from regular duties.[5]

Other bases were better suited to accommodate Reservists. In May 1920 the Air Service designated Bolling, Carlstrom, Kelly, Langley, March, Mitchel, and Post Fields for Reserve flying. Over one week in July, eleven Reservists flew at Bolling. About the same time, Mitchel Field reported that thirty-four Reservists had passed physical examinations, and hardly a sunny day passed without several Reservists seeking flights.[6]

The ordinary ex-service pilot felt sure he could step into a plane and fly off alone. As a rule he got along fine in the air, but often had trouble landing. After being away from flying for a while, he was "extremely hazy" in judging distance and closeness to the ground. It was clear Reservists needed sufficient flying to keep in condition and be ready to serve in an emergency.[7] The program for voluntary flying, characterized by General Menoher as a "more or less negative arrangement," was not enough.[8]

The Army's Reserve program, which had been based on prewar legislation,[9] was revamped during the general reorganization under the 1920 Defense Act. The new law provided for an Officers' Reserve Corps with an Air Service section (later known as the Air Reserve). Eligible were Army officers, graduates of the Reserve Officers' Training Corps (ROTC), as well as warrant officers and enlisted men with wartime Army service. Appointments, made by the President, were for five years. Those in force at the beginning or during the war would extend to six months after the war ended. If Congress furnished the money, the President could order Reservists to active duty at any time for any period. Save in a national emergency, however, no Reservist could be kept on active duty longer than fifteen days without his consent. The government paid only for active duty, a Reservist receiving the same pay and allowances as a Regular Army officer with the same grade and length of service.[10]

From 7,339 officers (5,046 rated) in 1920, the Air Reserve grew to 7,641 in 1921 and 7,995 in 1922. With 8,249 (5,640 rated) in mid-1923, the Air Service accounted for more than 10 percent of the Army total. Then postwar commissions began to expire, and some officers refused reappointment. The number in the Air Reserve sank to 6,709 in 1924, but rose slowly to 6,985 by mid-1926, when the Air Service had about 6.7 percent of the Army's Reserve officers.[11]

The 1920 act also authorized an Enlisted Reserve Corps with a normal enlistment of 3 years. However, persons with war service could be enlisted for 1 year and were entitled to discharge within 90 days after making application. Enlistments in force at the outset of or during a war would remain in force until 6 months after the war ended. Rules on active duty and pay were similar to those for officers.[12] The Air Service had no trouble commissioning large numbers of Reserve officers but found it very hard to recruit and keep men in its Enlisted Reserve Corps. At the end of Fiscal Year 1922 it counted just 39 (about 8 percent of the Army total). The number climbed to 123 in 1923, 536 in 1924, 821 in 1925, and 1,029 (nearly 18 percent of the Army total) in 1926.[13]

The Act of 1920 enabled the President to form Reservists into tactical units like those of the Regular Army. In the spring of 1921, the War Department allotted aviation units to corps areas for the air service of the divisions, corps, and armies of the Organized Reserve.[14] Included were observation, pursuit, attack, and bombardment squadrons; balloon and airship companies; photographic sections; intelligence offices; repair and service units; and headquarters for groups and wings and for corps and army air services. The Air Service prepared plans and made recommendations for organizing units. Control of units, however, rested with corps area commanders until 1925, when Air Officers of the corps areas were given control of the Air Reserve, except division aviation.[15]

To launch the program, the Air Service assigned one or two Regular officers to the corps areas to stimulate interest among Reserve officers and to assist in creating units.[16] By late 1921, units were formed at New York City, Boston, Philadelphia, St. Louis, Omaha, Tulsa, San Francisco, and elsewhere. In many cases, formal organization stemmed from informal meetings of interested Reservists. The 316th Reserve Squadron (Observation), for example, had its beginning when Reservists from around San Francisco met at nearby Crissy Field in the fall of 1921. The Commander of Crissy Field, Maj. George H. Brett, and his staff developed training schedules for four squadrons. First Lieutenant Robert E. Selff became Officer in Charge, Instruction of Reserve Squadrons. A new hangar having been built at Crissy for the airmail service, the old one was turned over to the 316th Reserve Squadron in January 1922. The squadron then took enlistments. As soon as it accepted its first planes (five Jennies), the pilots (the squadron now had twenty-three officers) showed up for flying. By February 1922 the 316th Reserve Squadron was a going organization.

During the next few weeks, the Ninth Corps Area detailed a staff sergeant and four enlisted specialists of the Regular Air Service to temporary duty with the 316th Squadron. Reservists met Monday nights for instruction and on alternate weekends for classes and flying. By May 1922 the unit had thirty-three of the thirty-five officers authorized, and some pilots had already soloed. Twenty-four enlisted men were in training, and other enlistments were being sought.

The 316th Photo Section, also at Crissy Field, had its officers and was about ready to begin enlisting. Both the 316th Squadron and the photo section were attached to the 91st Division, Organized Reserve, whose Air Officer was Capt. Armin F. Herold from the Regular Air Service. Meantime, Lieutenant Selff had been organizing a pursuit unit, the 440th Reserve Squadron, at San Jose, California. In 1923 another pursuit squadron, the 447th, was formed at Crissy Field.[17]

General Patrick reported 539 units of all types in the Air Service Organized Reserve on June 30, 1923. Of these, 213 had their full quota of officers; assignment of officers to the other 326 was about 50 percent complete.[18] Neither officers nor enlisted men were paid for inactive duty training at night and on weekends. Nor were they entitled to retirement pay and many other benefits enjoyed by Regular Army members. Almost the only incentive for a Reserve officer to join a Reserve unit was the flying, and that was not enough to keep all units filled with eager flyers. Even the 316th Squadron, with its auspicious start at Crissy Field in late 1921, reported a few months later that just 19 of its 33 officers were actively flying.[19]

The situation in the Enlisted Reserve Corps was worse. Reserve units needed enlisted specialists, chiefly mechanics to take care of the planes. But about the only thing the Air Service could offer was a summer "vacation" in camp with pay—if the money was available.[20] Of necessity the Air Service assumed most of the maintenance needed to keep its Reserve units in business.

The Air Service adopted the general policy of using a single system of facilities for Regular Army, Organized Reserve, and National Guard units. Some Reserve units utilized existing Air Service facilities; others required special arrangements. Municipal governments and civic organizations generally cooperated. At Cleveland, for instance, the Chamber of Commerce set up a committee headed by Glenn L. Martin to find a site for a municipal airport that could be shared with the 414th Pursuit Squadron, Organized Reserve. Another pursuit squadron, the 462d, used a field operated by the Chamber of Commerce of Kokomo, Indiana. The Army created airdomes at Fort Benjamin Harrison, Indiana; Fort Douglas, Utah; Vancouver Barracks, Washington; and other Army installations.[21]

In addition the Air Service built facilities at Boston; Pittsburgh; Columbus; Cincinnati; Louisville; Kansas City, Missouri; Santa Monica, California; and Seattle. In each case, it leased the land for one dollar a year, furnished steel hangars from its surplus, and contracted through the Quartermaster Corps for erecting hangars and installing gas and oil facilities. The Air Service supplied a few Jennies, some equipment and tools, a few mechanics (mostly Regular enlisted men but occasionally a civilian or two), and a Regular Army officer as commander.[22]

The Air Service attempted to place these fields on airways that would become a basic part of national defense. It let other federal departments share them, and as a stimulus to U.S. aviation opened them to civilian planes. Thus, early in 1925 the Boston airport accommodated commercial aircraft, the U.S. Navy, the Air Service Organized Reserve (two DHs and five JNs), and the National Guard (four JNs). The commander of the field at that time was 1st Lt. Robert J. Brown, Jr., Air Service; 1st Lt. Aaron E. Jones, Air Service, was on duty with the National Guard. Other Regular Army members stationed there included a flight surgeon and his noncommissioned officer assistant, nine enlisted men, and two civilians in motor repair.[23]

Summer training camps for the Organized Reserve began in 1922. Corps area commanders put about 350 officers on active duty for 2 weeks that year at Mather Field, California; Langley Field, Virginia; Maxwell Field, Alabama; and other fields. The program was supposed to arouse interest in new technical and tactical developments, and to provide as much flying as possible in service planes. Pilots first underwent instruction from Regular Air Service pilots in dual trainers. Work at the Fairfield depot in Ohio that spring concentrated on readying JN–4Hs for shipment to Mitchel Field, New York, before the opening of camp on July 18. At Chanute Field, Illinois, most of the 60 or so Reservists in camp were checked out in the first 3 days; afterwards they soloed in SE–5s, DHs, SPADs, and Fokkers. Nearly all of the 26 Reserve officers in training at Brooks Field, Texas, soloed and flew DHs, SEs, and SPADs before camp ended.[24]

About 500 Reservists attended camp in the summer of 1923, and around 1,000 in each of the next 2 years.[25] Camps were held in 17 places in 1924, with several Air Service stations running more than 1 session. Reservists attended camp in their own corps area. Four pursuit squadrons and a pursuit group headquarters of Fifth Corps Area trained at Selfridge Field, Michigan, in 1924; other pursuit squadrons trained at Mitchel Field; Langley Field; Kelly Field, Texas; Rockwell Field, California; and Fort Bliss, Texas. The attack wing and group headquarters and two attack squadrons trained at Chanute Field. Three other attack squadrons trained at Mitchel Field; Post Field, Oklahoma; and Kelly Field. Two bombardment squadrons used Rockwell Field. Ross Field, California, had a balloon group headquarters and two balloon companies. Scott Field, Illinois, had two balloon companies. Observation aviation was scattered across the country. In certain cases, unit representation was small. While some pursuit squadrons had 20 officers in camp, the number in others ranged from 10 to 3.[26]

Problems hampered the development of all Reserve units, not merely those of the Air Service. There was the unwillingness of certain Reservists to attend summer training, a lack of equipment, and a shortage of Regular Army instructors. But the biggest problem was insufficient money for the program. Aviation unit designations like "pursuit" and "attack" were misnomers. The so-called tactical squadrons consisted almost entirely of officers, most of whom were pilots or would-be pilots. Many units were short of their officer quotas; none had airplanes of its own. A great deal of the flying was in old training planes. Active duty training in summer camps helped keep pilots interested, and gave them a chance to retain a degree of proficiency in service planes. But there was no learning how to function as a unit, the organization existing largely for administrative purposes. General Patrick observed that the Air Service Officers' Reserve Corps was "a 'force' in name only and is in reality nothing more than a pool from which officer material suitable for training may be drawn."[27]

Reservists did the better part of their flying on their own time. In Fiscal Year 1926, for example, they flew 27,800 hours—12,000 hours on inactive status, 10,400 during 15-day training periods, and 5,400 on extended active duty. Of the total hours, around 70 percent was in training planes. Only Reservists on extended active duty logged more hours in service planes than in trainers. In fact, during 15-day tours they did just 30 percent of their flying in service planes.[28]

A Reservist could become a junior airplane pilot by completing either the primary flying course or a special course for Reservists and Guardsmen at Brooks Field. When he went on extended active duty, however, he could not be assigned to a tactical unit until he qualified as an airplane pilot.[29]

General Menoher proposed in 1921 to give Reservists a major role in the Regular Air Service. He thought this would solve a serious problem he expected to arise because of the shortness of the flying life of military pilots. The Army's personnel system, Menoher explained, assumed that an officer's efficiency increased from the time the man received his first commission until he retired. This might be valid for other branches but not for the Air Service. A flyer attained his peak efficiency at an early age and kept it for but a few years, before losing his usefulness as a wartime flyer. It was "recognized and accepted," the general said, "that the maximum period during which a flying officer retains his efficiency as a wartime pilot is about eight years." The flying period in peacetime ran a little longer, probably 10 or 12 years, the maximum age for flying duty being about 35. This meant that 75 percent of all Air Service officers should be lieutenants and captains between ages 21 and 35.

What was to be done with these officers when no longer fit for flying duty?[30] They would still be captains and lieutenants at age thirty-five. Low on the promotion list, they would have little chance of advancement into the few field grade positions available. Some could fill administrative slots but the majority would be without jobs.

General Menoher looked to the Reserve as a source for flyers who could be retained during their peak usefulness and then disposed of by discharge or transfer. He suggested keeping five hundred Reserve officers on active duty at all times. After five years of active duty, they would return to the Reserve to be available for active service in an emergency. Menoher realized the Air Service could not expect ambitious young men to devote five years to the Air Service at Regular Army pay. He recommended legislation that would either give the Reservist much higher pay while on active duty or a generous bonus at the end of such duty.[31]

General Patrick did not take up his predecessor's plan. Perhaps he thought it impracticable, with little likelihood of gaining congressional

students through the complete 4-year course. When his recommendation was not approved, he lowered the goal to 12 units and 500 graduates a year and made plans to train more cadets.[43]

In June 1926 the Air Service had the same 6 units it started with in 1920-21. Enrollment had risen to 1,005, with 463 in the first year's course, 291 in the second, 143 in the third, and 108 in the fourth. As these figures suggest, the dropout rate had improved a bit but was still high, the largest loss being between the first and second years.[44]

Summer camps continued to be held at Air Service stations, with 125 students attending in 1924 and 110 in 1925. Pilot training also proceeded on a small scale. Of a class of 96 in 1924, 24 graduates entered school; 17 completed the course and became junior airplane pilots. In 1925 the pilot course for ROTC graduates expanded from 4 to 6 months; 39 to 88 graduates applied; funds existed for only 16; 13 completed the course. By June 1926 the Air Service ROTC program, in which close to 5,000 students had enrolled over 6 years, had produced some 400 Reserve officers, of whom 44 had been trained as pilots. Forty-four others, from the class of 1926, wanted pilot training, but the Air Service had just enough money to send 35 to flying school in September.[45]

National Guard

The War Department decided against organizing special aero squadrons and other service units in the postwar National Guard. So, when the Militia Bureau asked in September 1919 for information and ideas for forming aviation units, General Menoher said the Air Service had no policies. He saw "formidable obstacles" to these units. Maintenance of flying equipment would be a great expense. Further, the units would have no chance to train with ground forces. They would therefore be unable to work effectively with other branches of the militia.[46]

So much interest developed in aviation for the Guard that the War Department soon reversed its position. Each Guard division would have an air service consisting of an aero squadron, a balloon company, and a photo section. By the spring of 1920, the Air Service and the Militia Bureau had agreed on a plan. The federal government would furnish equipment. The Militia Bureau would arrange with the states for facilities, but units near Regular Air Service stations would be housed and trained there. Guard units would be organized the same way as Regular units and be inspected by officers of the Regular Air Service before being granted federal recognition.[47]

The first aviation unit of the National Guard to pass federal inspection was the 104th Observation Squadron of Maryland, which won federal recognition on June 29, 1921. Four more observation squadrons received recognition before the end of the year, 2 in 1922, 4 in 1923, 3 in 1924, 1 in 1925, and another the following May to make a total of 16 by mid-1926. By that time, photo sections and medical detachments had been organized to complete the air services of 11 divisions. Aviation elements of the National Guard then had less than half of the officers and enlisted men needed at war strength. Among the 271 officers (other than doctors) in the Guard, 184 were rated pilots and 33 rated observers. Of the remaining 54, all but one had passed physical examinations for flying and had been assigned as observers. At war strength an observation squadron required 13 airplanes. Guard units then averaged 5, and these were old trainers rather than service planes. Plans for balloon companies had been abandoned.[48] The addition of units in two states in 1927 and in a third during 1930 completed the organization. This provided an observation squadron for each of the 18 divisions of the National Guard and an extra one designated corps aviation.

Some squadrons found their origins in local organizations of former Army pilots, Reserve officers, and other aviation enthusiasts. The 104th Squadron, for instance, grew out of a flying club active at Baltimore during 1919 and 1920. In Alabama it was the "Birmingham Escadrille," organized by James A. Meissner, an ace. With the help of prominent citizens and the Adjutant General of Alabama, Meissner obtained a National Guard squadron for Birmingham. A local industry made sixty acres of land available for a flying field. The Air Service gave old hangars to Guard units, but the escadrille raised money to prepare the land and erect the hangars. This took most of 1921. By late November the people of Birmingham donated $5,000. Clearing, grading, and building commenced in earnest, with members of the escadrille doing much of the manual labor. So the Birmingham Escadrille transformed itself into the 135th Squadron, Alabama National Guard. When federal recognition came on January 21, 1922, Major Meissner was commanding 25 officers and 120 enlisted men. All of the officers had flown with the Army, Navy, or Marines. Most of the enlisted men were mechanics with experience either at aviation repair depots or at the Air Service Mechanics School.[49]

A few National Guard officers went to Brooks Field in January 1923 for pilot instruction. Flyers of the recent war underwent refresher training while others took the regular course. Eight of the ten officers entering graduated to become junior airplane pilots. The Air Service suggested, and the Militia Bureau adopted, a policy of giving men flying training before commissioning them in the Guard. A few Guard members attended other service schools.[50]

The Air Service detailed an officer to the Air Service of the National Guard of each state as an instructor. Regulations required members of National Guard units to assemble for drill at least forty-eight times a year. The federal government paid officers and enlisted men for up to eight armory drills a month and sixty a year, if each drill lasted at least an hour and a half. The federal government also paid Guardsmen for field training in summer.[51]

Members of five squadrons (83 officers and 375 enlisted men) assembled at the first summer camps for Guardsmen in 1922. Reporting on its activities, the 104th Squadron said it did not take its Jennies to Langley Field since that would be "like taking a ham sandwich to a banquet." At Langley the pilots flew real planes, service planes. Within a week every pilot soloed in a DH. Before the end of the 2 weeks, more than half were flying SE–5s. The next summer, 99 officers and 607 enlisted men from 7 states went to camp. Attendance was high. In 1924, for example, 11 squadrons with a total enrollment of about 200 commissioned and 1,100 enlisted personnel sent 100 officers and 918 men to camp.[52]

Soon after the National Guard's aviation program got under way, General Patrick harbored doubts about pursuing it. He did not think units could handle the day-to-day maintenance and repairs needed to keep the planes in the air. The Militia Bureau helped dispel his doubts by supplying the money to keep a few enlisted mechanics on continuous duty with each squadron.

In mid-1923, General Patrick described the National Guard's Air Service as "an effective and efficient force available for practically immediate field service in an emergency." A year later he reported that the officers were "of a very high type" and that esprit de corps was excellent. "Some of the older units," he said, "could take to the field for active duty after a very short period of intensive training." In 1926 he commented, "With a little additional training and with proper equipment, several of these squadrons could take the field with their divisions and serve them effectively."[53]

When General Patrick spoke of "proper equipment" for the National Guard, he was thinking about the old Jennies the Guardsmen flew. However, newer and better planes were on the way. In mid-1926 the National Guard's aircraft comprised 112 JNs and 7 recently acquired TW–3s. The depot at San Antonio was reconditioning 11 more TW–3s for the Guard, and the Militia Bureau had already transferred funds to the Air Service to purchase 14 Douglas O–2s.[54]

Like the Regular Army, the Reserve forces in the 1920s could not carry out the duties envisioned in the National Defense Act. This was true in aviation as in other branches of the Army of the United States. So far as the Enlisted Reserve was concerned, there seemed slight incentive for men to join; Congress provided little money; the War Department showed scant interest. In an emergency, mechanics and other technicians and specialists would have to be recruited or conscripted for aviation duty.

The Air Reserve was better off with respect to officers. Many men commissioned during the war later joined the Officers' Reserve Corps. But as time went on and they became engrossed in civil pursuits, many lost interest in the military—at least in active participation. The Air Service counted on the Reserve Officers' Training Corps to replace officers lost from the Reserve. Unfortunately, tight congressional purse strings prevented the program from meeting this objective. Further, due to the money shortage, the Air Service could provide pilot training to only a few of those who entered the Reserve by way of ROTC. For the same reason, the Air Service secured but a fraction of the Reserve pilots it wanted from the flying cadet program. Few Reservists flew enough to stay proficient. Organized into units for administration, the Air Reserve was basically a relatively modest pool from which the Air Service could draw individual pilots for further training in an emergency.

The National Guard came closer to achieving its goals but fell well short of being the mobilization force intended by the National Defense Act of 1920. Better pay and other incentives made the Guard generally more attractive than the Organized Reserve. Unlike the latter, Guard units trained as units for wartime service. Having begun with old training planes, the Guard was beginning in 1926 to receive service planes for observation in support of National Guard ground forces.

The test of the Guard and Organized Reserve would await mobilization of the Army of the United States. Yet, some elements would be used from time to time in situations short of a national emergency. While still in the throes of demobilization at the end of World War I, the Air Service, Regular Army, took to the field for defensive operations on the Mexican border.

Chapter VII

Defense

The Army of the United States had as its chief function defense of the nation against attack from without. It protected the nation's land frontier, overseas possessions, and shared with the U.S. Navy responsibility for the seacoast. The war's end in 1918 found the Army's Air Service busy with plans and preparations for defense against attack at sea. During demobilization, the Regular Army and its air arm answered a call to defend the southern border against raids from Mexico, and to halt smuggling of aliens and dope into the United States and arms into Mexico.

Border Patrol

Revolution and disorder in Mexico and trouble along the U.S.-Mexican border in March 1913 brought on the hurried organization of the 1st Aero Squadron, the U.S. Army's first tactical unit equipped with airplanes. In 1916 the squadron took part in General Pershing's punitive expedition into Mexico in pursuit of Mexican revolutionist Pancho Villa. Difficulties along the border continued while the United States was at war in Europe. Mexican bandits often raided American ranches to secure supplies, cattle, and horses, and in doing so sometimes killed the ranchers. U.S. troops stationed along the border shot raiders as they pursued them into Mexico. The biggest clash

Officers of the 1st Aero Squadron, with Capt. Benjamin D. Foulois standing the fourth from left.

came in August 1918, when more than 800 American troops fought some 600 Mexicans near Nogales, Arizona.

Border patrol was one of the many activities being considered for the postwar Air Service. However, no aviation units had been assigned to duty on the Mexican border, when a large force of Villistas moved northward in June 1919 toward Ciudad Juarez, Chihuahua, Mexico (opposite El Paso, Texas), garrisoned by Mexican government forces. Maj. Gen. DeRosey C. Cabell, Commanding General of the Southern Department, received orders to seal off the border if Villa took Juarez. If the Villistas fired across the border, Cabell was to cross into Mexico, disperse Villa's troops, and withdraw as soon as the safety of El Paso was assured. The general ordered Air Service men and planes from Kelly and Ellington Fields, Texas, to Fort Bliss, near El Paso, for border patrol. (*Map 3*)

American troops under Brig. Gen. James B. Erwin, Commander of the El Paso District of the Southern Department, were on alert when about 1,600 of Villa's men attacked Juarez during the night of June 14/15, 1919. Stray fire from across the river killed an American soldier and a civilian, and wounded two other soldiers and four civilians. Around 3,600 U.S. troops crossed into Mexico, quickly dispersed the Villistas, and returned to the American side.[1]

MAP 3

Air Service personnel with DH–4 aircraft began arriving at Fort Bliss on June 15. Maj. Edgar G. Tobin, an ace who had flown with the 103d Aero Squadron in France, inaugurated an aerial patrol on the border on the 19th. By mid-September the force grew to 104 officers, 491 enlisted men, and 67 planes from the 8th, 9th, 11th, 90th, and 96th squadrons.

In the summer of 1919, the Army planned to build at least nine aero squadrons and one airship company for surveillance of the entire border from the Gulf of Mexico to the Pacific Ocean. The plan called for two observation squadrons (the 9th and 91st) of the Western Department to patrol eastward from Rockwell Field, California, to the California-Arizona line. Three surveillance squadrons (the 8th, 90th, and 104th) and four bombardment squadrons (the 11th, 20th, 96th, and 166th) of the Southern Department were to be distributed along the border from Arizona to the Gulf of Mexico.

On July 1, 1919, the three surveillance squadrons organized into the Army Surveillance Group headquartered at Kelly Field. (This group became the 1st Surveillance Group in August 1919.) In September the four bombardment squadrons formed the 1st Day Bombardment Group, also with headquarters at Kelly. In addition the 1st Pursuit Group and its squadrons (27th, 94th, 95th, and 147th) moved from Selfridge Field, Michigan, to Kelly at the end of August to be available if needed. The three groups (surveillance, day bombardment, and pursuit) comprised the 1st Wing at Kelly. Commanded by Lt. Col. Henry B. Clagett, the wing became responsible for aerial patrol of the border in the Southern Department. Also in August, work started on a large steel hangar for an airship station at Camp Owen Bierne, Fort Bliss.[2]

The Army soon scaled down the plan for border patrol. Although minor incidents continued to occur, Pancho Villa never succeeded in rebuilding his force. The major threat had been dispelled by the time aerial patrol began. From January 1920 on, the patrol in the Southern Department was handled by the 1st Surveillance Group which had moved its headquarters to Fort Bliss and gained an extra squadron, the 12th. The group's squadrons operated in two flights, each patrolling a sector on either side of its operating base. From the Gulf of Mexico westward, the deployment was as follows: McAllen and Laredo, Texas, 8th Squadron; Eagle Pass and Sanderson, Texas, 90th Squadron; Marfa and El Paso, Texas, 104th Squadron; Douglas and Nogales, Arizona, 12th Squadron. Most of the time only one squadron, first the 9th and later the 91st, patrolled in the Western Department.[3]

The 8th Balloon Company moved from Brooks Field to Camp Owen Bierne in December 1919 to set up the airship station. Parts for a twin-engine airship (the C–1) commenced to arrive in May 1920, and on September 28 1st Lt. John W. Shoptaw and 1st Lt. Don L. Hutchins took the ship on its first flight. The C–1 made many flights around El Paso, but never played a key part in border patrol.[4]

The patrol bases were hurriedly created. One of the young lieutenants who flew from Marfa in the summer of 1919 remembered the flying field as a pasture at the eastern edge of town. Its five hangars were canvas. A double row of ten or twelve tents served as officer and enlisted quarters and sheltered flight headquarters and supply. The lieutenant, Stacy C. Hinkle, recalled his tour of duty on the border as "a life of hardship, possible death, starvation pay, and a lonely life without social contacts, in hot, barren desert wastes, tortured by sun, wind, and sand." The boredom was as bad as the physical hardship and discomfort, the sole recreation being drinking and gambling. Even so, Hinkle thought the airmen better off than the poor fellows at cavalry outposts up and down the border.[5]

The patrol started with DH–4s and Jennies, both eventually replaced with DH–4Bs. Most of the first planes were not properly equipped for field service. Not knowing what turn events on the border might take, the Army wanted the planes ready for any eventuality. Col. James E. Fechet, Air Service Officer at the Southern Department, found it no easy task to obtain bomb racks, machinegun mounts, cameras, and other equipment. There was a delay, for example, in installing synchronized Martin guns because parts supplied with the guns did not fit the planes on the border. The radios on some planes could send only in code and could not do that very well. Compasses were unreliable, maps sketchy and of little use. The country over which the men had to fly was wild and rough and sparsely populated, with few places for safe emergency landings.[6]

The aerial patrol searched along the border for bands of men and reported to the nearest cavalry post how many men they were, where they were, which way they were heading, what they were doing, and how many horses and cattle they had. The timing of the patrols varied so raiders would not know when the next plane would appear.[7]

The men generally seem to have done all that might reasonably have been expected of them, and sometimes more. Take, for example, the flyer (regrettably unnamed in the story released by the Air Service) sent one morning to find some horses and mules spirited across the border the previous night. Seeing a group of Mexicans and horses in a corral, the flyer sought to notify the cavalry patrol in pursuit of the bandits. Unfamiliar with the use of airplanes with cavalry, the troopers could not comprehend the signals. Flying back across the river, the airman landed at an American picket station, borrowed a horse, swam the Rio Grande River, and chased the cavalry to tell them what he had seen. He then rode back, jumped into his plane, and flew to the corral. The Mexicans apparently had seen the plane the first time and had turned the horses loose and driven them away. When the plane reappeared, the Mexicans scattered, with the plane pursuing three of them who were on horseback. The Mexicans ran under a cottonwood tree and kept it between them and the plane while the flyer circled for a better look. If he found they were bandits (how he expected to do this is not clear), he intended to shoot them. He soon ran low on gasoline, however, and turned back to El Paso. This incident, the Air Service said, illustrated not only "the great service the airplane can render cavalry troops in pursuit of bandits," but also "the necessity of having better liaison between the Air Service and the cavalry."[8]

As time went on, air border units spent less time on patrol and more in training with the infantry, artillery, and cavalry. Air Service personnel further practiced aerial gunnery and formation flying, experimented with radio and other signaling systems, located and marked emergency landing fields, and worked to upgrade facilities and equipment.[9]

Pilots flying along or near the border were under orders not to cross. But they often got lost and strayed into Mexico. At times they went over deliberately, apparently on the spur of the moment. Occasionally, they crossed to carry out a special assignment.

Addressing the National Congress of Mexico on September 1, 1919, President Venustiano Carranza said U.S. military planes had crossed the frontier several times. While his government had protested, the incursions had been repeated.[10] The Mexican president was probably not aware that one of the flights violating Mexico's sovereignty had been made by the ranking pilot of the U.S. Air Service. Inspecting the border patrol in July 1919, General Mitchell had taken Col. Selah R. H. (Tommy) Tompkins, 7th Cavalry Commander, for a reconnaissance.[11]

The day President Carranza addressed the Mexican Congress, Ygnacio Bonillas, Mexican Ambassador to the United States, protested the flight of two planes over Chihuahua City, during the afternoon of August 28. James B. Stewart, American Consul in Chihuahua, had already reported the incident. Soon Stewart was back with another dispatch and Bonillas was protesting again—more American planes had flown over Chihuahua on September 2. Two more planes showed up in the morning of the 5th. When Stewart said these incidents embarrassed members of the American colony, Acting Secretary of State William Phillips replied: "War Department promises to issue strict orders against repetitions."[12]

Not long afterward, Ambassador Bonillas complained that the crew of a U.S. Army airplane had fired a machinegun several times while flying over Nogales, Arizona. Some of the shots hit a dwelling across the border in Nogales, Sonora, luckily without injuring anyone. The Mexican government wanted the guilty persons found and punished. Several weeks later the State Department responded that an Air Service lieutenant was being tried by general court-martial for the shooting.[13]

Another incident protested by the Mexican government began with two Americans getting lost while on a routine flight in the Big Bend area of Texas on Sunday morning, August 10, 1919. A flyer might easily get lost on patrol. Lts. Harold G. Peterson, pilot, and Paul H. Davis, observer-gunner from

Marfa, Texas, found it could happen while following a river on a clear day. Their mission was to patrol along the Rio Grande from Lajitas to Bosque Bonito and then land at Fort Bliss. Coming to the mouth of the Rio Conchos at Ojinaga, Chihuahua (opposite Presidio, Texas), they mistook the Conchos for the Rio Grande and followed it many miles into Mexico before being forced down by engine trouble. Thinking they were still on the Rio Grande, the airmen picked a spot on the "American" side of the river to land. The terrain was rough and the plane was wrecked. Having buried the machine-guns and ammunition to keep them out of the hands of bandits, Peterson and Davis started walking down the river, thinking they would come to the U.S. Cavalry outpost at Candelaria, Texas.

When Peterson and Davis did not arrive at Fort Bliss on Sunday afternoon, the men there assumed they had either returned to Marfa or made a forced landing. When they were unaccounted for on Monday, a search was begun. Flying over the patrol route, 1st Lts. Frank Estell and Russell H. Cooper surmised that Peterson and Davis might have mistakenly followed the Conchos into Mexico. The region along the Conchos almost as far as Chihuahua City was added to the area covered by search planes. Tuesday afternoon Peterson and Davis saw a plane flying up the Conchos, but they were in thick brush and could not attract the crew's attention. The search continued until Sunday, August 17, 1919. Then Capt. Leonard F. ('Two-Gun') Matlack, commanding Troop K, 8th Cavalry, at Candelaria, received word Peterson and Davis were being held for ransom.

The flyers had been taken prisoner on Wednesday, August 13, by a Villista desperado named Jesus Renteria. The bandit sent the ransom note to a rancher at Candelaria, along with telegrams which he forced the airmen to write to their fathers and the Secretary of War, the Commanding General of the Southern Department, and the commanding officer of U.S. forces in the Big Bend District. Renteria demanded $15,000 not later than Monday, August 18, or the two Americans would be killed.

The War Department authorized payment of the ransom, but there remained the matter of getting $15,000 in cash for delivery before the deadline. Ranchers in the area quickly subscribed the full amount, which came from the Marfa National Bank. Negotiation through intermediaries resulted in a plan for Captain Matlack to cross the border Monday night with half of the ransom money for the release of one of the Americans. The meeting took place on schedule, and within forty-five minutes Matlack came back with Lieutenant Peterson. Matlack then took the remaining $7,500 to get Lieutenant Davis. On the way to the rendezvous he overheard two of Renteria's men talking about killing him and Davis as soon as the rest of the ransom money was paid. At the rendezvous, Matlack pulled a gun, told the Mexicans to tell Renteria to "go to hell," and rode off with Davis and the money. Avoiding the ambush, Matlack and Davis safely crossed into the

United States. Questioned by Col. George T. Langhorne, Army Commander in the Big Bend District, Peterson and Davis maintained they had been captured on the American side of the border and had not crossed into Mexico.

At daybreak on Tuesday, August 19, 1919, Captain Matlack once again crossed the border, this time leading Troops C and K, 8th Cavalry, in pursuit of Renteria and his gang. Air Service planes scouted ahead of the cavalry seeking to spot the bandits. They also gathered information on the condition of the trails and the location of waterholes, and conveyed it to the troops by dropping messages.

While flying some twelve or fifteen miles west of Candelaria late Tuesday afternoon, Lieutenants Estell and Cooper saw three horsemen in a canyon and went lower for a better look. When the men on the ground fired on the DH–4, Estell made another pass with his machineguns blazing. Then Cooper opened up with his Lewis guns and killed one of the men, reportedly Renteria.

The search for members of Renteria's gang continued until August 23. With the Mexican government protesting the invasion of its territory, American forces returned to the United States.[14]

A few months later another plane landed in Mexico after its crew followed the wrong railroad tracks. Patrolling on Monday, February 2, 1920, 1st Lts. Leroy M. Wolfe and George L. Usher intended to pick up the El Paso and Southwestern Railroad west of Douglas, Arizona, and follow it to Nogales. Visibility was poor and the compass did not work properly. Sighting a railroad, Wolfe and Usher followed it for some time until it ended. Lost and having engine trouble, they landed and were taken into custody by Mexican officials. The tracks they had steered by ran due south instead of west, and had led them to Nacozari, Sonora, seventy-five or eighty miles below the border. Though treated well, Wolfe and Usher were not set free until February 24. They waited three more days for release of their airplane, shipping it to Douglas by train.[15]

About the same time, a plane on patrol of the lower Rio Grande ran out of gas over Mexico and landed some twenty miles west of Guerrero, Nuevo Leon (opposite Zapata, Texas). The airmen, Lts. E. E. Davis and Gerald E. Grimes of the 8th Surveillance Squadron, were quickly released. With permission from Mexican officials, 1st Lt. Rex K. Stoner took gas and oil into Mexico and flew the plane back to the 8th Squadron's post at McAllen, Texas.[16]

Earlier, in October 1919, two planes from Rockwell Field, California, ended up in Lower California when the flight leader miscalculated in trying to navigate by the sun. In that case, the four men got out safely.[17] Two other American airmen who came down in Baja California were not so fortunate.

On border patrol with the 9th Corps Observation Squadron, Lts.

Frederick Waterhouse and Cecil H. Connolly disappeared after taking off from Calexico, California, bound for Rockwell Field, on August 20, 1919. A search begun the next morning gradually extended farther and farther south in Baja California. When three weeks passed with no trace of the missing men, the search ended. A month later it was learned their bodies had been found near Bahia de Los Angeles on the coast of the Gulf of California, 225 miles south of Calexico.

From the evidence that could be gathered, it appeared Waterhouse and Connolly became lost in a rainstorm and hugged the coast of Baja California southward, thinking they were headed north along the Pacific Coast. They landed safely on the beach about twenty miles north of Bahia de Los Angeles. Their sole chance for survival seemed to be staying with the plane until found. Tortured by heat, thirst, and hunger, they waited seventeen days, but the search never reached that far south. Finally two fishermen came along and took them in a canoe to Bahia de Los Angeles. There the Americans were murdered, apparently for the little money they had. Their bodies, buried in the sand, were discovered within a day or two by an American geological survey party and rediscovered a week later by an American mining engineer. The news, however, did not reach Rockwell Field until October 13. Three days later, a Navy ship, USS *Aaron Ward*, sailed from San Diego with a group of Army officers to recover the bodies.[18]

One of the largest manhunts conducted by the Air Service in the 1920s was organized by Maj. Henry H. Arnold, commanding officer of Rockwell Field, when one of his planes disappeared on a flight late in 1922. First Lieutenant Charles L. Webber had left in a DH on Thursday morning, December 7, to fly Col. Francis C. Marshall on an inspection trip of cavalry posts and camps. Thick fog appeared to be breaking up when Webber and Marshall took off from San Diego at 0905, their destination Fort Huachuca, Arizona, with Nogales an alternate.

Receiving word the men had not reached their destination, Major Arnold sent out every available plane on Friday to search. He also sent messages along the route to secure information. Reports received during the day showed the plane had flown more than one hundred miles into Arizona. Otherwise, no word came concerning the missing plane and men.

Having but a few pilots and planes available at Rockwell, Arnold sought help. Three planes on a cross-country flight from Brooks Field, Texas, had landed at Rockwell on Thursday. These men, including Maj. Ralph Royce, the Brooks Field Commander, joined the search. Crissy Field, California, contributed thirty-two men and sixteen planes. Arnold further secured the help of U.S. Navy flyers from the naval air station at San Diego. Maj. Leo G. Heffernan, Commander of the Air Service, 1st Cavalry Division, Fort Bliss, Texas, brought five planes and pilots to help. By Saturday afternoon, December 9, the Air Service was well organized for an aerial search along the

Maj. Henry H. Arnold, commanding officer of Rockwell Field, organizes extensive manhunt after disappearance of one of his planes.

entire route between San Diego and Fort Huachuca. The addition of planes and pilots from Kelly Field eventually brought the number of planes engaged in the hunt to forty-two. Infantry and cavalry likewise participated.

After ten days the men from Fort Bliss and from Brooks, Kelly, and Crissy Fields gave up and went home. First Lieutenant John P. Richter, Webber's roommate and close friend, had been working from Nogales under Major Arnold's orders. He stayed on for several days to follow any clue.

Arnold was not satisfied that Webber and Marshall had reached eastern Arizona. The evidence did not appear conclusive. No one in the Imperial Valley saw the plane. It seemed to him doubtful whether the plane had gotten through the clouds over the mountains east of San Diego. Dividing this mountain area into sections, Arnold sent fifteen planes to search each section in detail. The whereabouts of the missing plane and men remained a mystery.

Arnold followed every lead. One rumor led to a man in Los Angeles with a reputation for seeing things, past and future. Arnold sent 1st Lt. Frank W. Seifert to talk to him. Afterwards, the lieutenant reported by telegram that the seer was "crazier than 7,000 jackrabbits."

Unwilling to give up, Major Arnold obtained permission to send a party by automobile to try to trace Lieutenant Webber's course by talking to and checking the statements of the various witnesses. Twenty-nine persons from places scattered from San Diego to fifty miles east of Nogales, and from one hundred miles north of Yuma to far south in Mexico, claimed to have seen a DH on December 7. The Air Reserve helped. Maj. Theodore Macauley, who knew the area from his transcontinental flights, went on active duty to head an expedition consisting of another Reserve, Capt. H. A. Erickson, and four Regular Army men from Rockwell Field, Lieutenant Richter, 1st Lt. Virgil Hine, and two privates. The group's equipment and supplies included a Dodge touring car, a light delivery truck, camping gear, and rations for six weeks. Major Macauley and his men left San Diego on January 15, 1923, and did not return until February 23. From their investigation it appeared that Webber's plane crossed the mountains east of San Diego, flew over the Imperial Valley of California, and in Arizona passed south of Yuma and Wellton. But Macauley and his group could track it no farther. A memorial service for Colonel Marshall and Lieutenant Webber was held in Washington on the afternoon of February 28. Major Arnold ordered work suspended at Rockwell Field for two minutes that day in tribute to Lieutenant Webber.

It was not until May 12, 1923, that the plane was found. A man hunting stray cattle discovered it in the mountains, just a few miles east of San Diego. Colonel Marshall and Lieutenant Webber apparently hit Cuyamaca Peak in the fog within thirty minutes after taking off from Rockwell Field five months earlier.[19]

Regular patrol of the border ended some time before. At first, units tried to cover their sectors every day. Later, the number and seriousness of border violations by Mexicans decreased, and the patrols tapered off. In the autumn of 1920, the schedule for the 1st Surveillance Group called for flights twice a week. When exercises with ground forces or other activities interfered, patrols might be canceled for days or even weeks at a time. Brig. Gen. William Mitchell's need for men and planes from the border for bombing tests against naval vessels off the Virginia Capes in June 1921 brought border patrol to an end.[20]

Coastal Defense

In coastal defense, the Army long held responsibility for helping to prevent invasion, a job it shared with the Navy, and for defeating in ground combat any force which an enemy succeeded in putting ashore. To discharge the first of these tasks, the Army's area of operations reached seaward the range of artillery. Aircraft added a new dimension.

Before the end of World War I, the Air Service viewed its mission as embracing patrolling the coast, helping to defend harbors and shores against enemy attack, and assisting coastal batteries by finding targets, determining range, and observing fire. In fact, when the war ended, the Air Service was establishing coastal defense stations on the east and west coasts and in the three overseas departments. The service estimated 15 airplane squadrons and

10 balloon companies would be needed for 10 stations at home, and 15 squadrons and 9 companies overseas.[21]

Col. William Lay Patterson and Lt. Col. Leslie MacDill handled the project for the Air Service, coordinating with the Coast Artillery. By the end of April 1919, they had received Coast Artillery approval for a station on Staten Island, and the Air Service was preparing for coastal defense operations at Langley Field.[22] In June 1919, Maj. Gen. Frank W. Coe, Chief of Coast Artillery, agreed to eight more stations, the precise locations to be determined later. By the end of the year, sites had been chosen on the east coast at Portland, Boston, Narragansett Bay, and eastern Long Island, and in the west at San Francisco and Puget Sound. General Coe, however, opposed permanent construction "until such time as the service of coast batteries is more fully developed."[23]

The Air Service moved the 14th and 24th Balloon Companies from Fort Omaha, Nebraska, to San Francisco in April 1920 to work with coastal batteries in formulating operational procedures. The chief problem in adjusting fire lay in tracking moving vessels. The 14th and 24th tried triangulation, with two balloons about 13,000 yards apart, the observer in each balloon measuring by sextant the angle between the opposite balloon and the target at sea. This was not very accurate. So from an old azimuth instrument and a small telescope the men created a device for an observer to measure the angle. Employing procedures and instruments stemming from experimentation, and communicating by telephone, observers in the balloons could track a moving target at sea, spot the splash of the shells, and report the deviation to the battery. This system was first successful on November 24, 1920, when long-range guns guarding the Golden Gate used only data from balloon observation to fire on a pyramid target towed by a tug 14,000 yards at sea.[24]

A detachment of the 91st Corps Observation Squadron at Crissy Field also worked with coastal batteries at San Francisco. Commanded by 1st Lt. Lowell Smith, the detachment equipped each of its DH–4Bs with two radio transmitters and two sending keys. Either the pilot or observer could send, and if one radio failed a second was on hand. Artillery batteries displayed panels to communicate with the aircraft. When the detachment received a request for a plane, the pilot flew to the battery, where the observer asked by radio for a panel. When "ready to fire" appeared, the pilot proceeded to the target, some 17,000 yards at sea. From 3,000 feet the airmen could see the splash of the shells. Comparing the distance of the splash from the target with the length of the towline, the observer figured the deviation and radioed it to the battery. The pilot then returned to the battery to await another panel.[25]

A combination of planes and balloons produced the best results. They were so good, according to reports from the detachment at Crissy Field, that coast defense officials had "declared artillery obsolete without the aid of balloons for tracking and planes for observing." The detachment therefore began experimenting with flares dropped from airplanes to illuminate targets for tracking from balloons at night.[26]

Nevertheless, the coastal defense project was doomed. Air Service enthusiasm may have cooled somewhat when General Coe tried to place coastal defense aviation under the administration and tactical control of the coast defense commander.[27] Coordination with the Coast Artillery, Coe's insistence on having a system of operations before going far with construction, and other delays prevented much being accomplished before the project fell victim to government economy. With fewer people and less money, both the Air Service and the Coast Artillery had to curtail their programs.[28] Still, coast defense continued to be the principal mission of Air Service units in overseas departments.

While the Air Service was seeking to construct coastal defense stations in mid-1919, the Aeronautical Board was defining aviation functions of the Army and Navy.[29] The board, composed of equal numbers of officers from the Army and Navy (Menoher being the senior member at this time), recommended on August 23, 1919, that Army aircraft conduct offensive and defensive operations with the various arms of the Army, and furnish fire control for coastal defense. Navy aircraft from coastal stations should be employed for convoy, reconnaissance, and patrol. Those operating from ships and bases should carry out reconnaissance and spotting as well as offensive operations against enemy vessels and naval bases.

Upon receipt of the board's statement for publication, the General Staff referred the matter to the Joint Army and Navy Board, whose senior members were the Army Chief of Staff, Gen. Peyton C. March, and the Chief of Naval Operations, Adm. Robert E. Coontz. The Joint Board found the Aeronautical Board's policy too restrictive of the operation of Army aircraft and virtually prohibitive of joint Army-Navy operations. The board recommended on December 18, 1919, that Army aircraft operate as an arm of the mobile army, against enemy aircraft in defense of shore installations, and alone or with other arms of the Army or Navy against vessels attacking the coast. It further proposed that Navy aircraft be employed not only as an arm of the fleet for overseas scouting and against enemy shore establishments, but to protect coastal shipping and against enemy vessels attacking the coast.[30]

Pointing to duplication of functions between Army and Navy aviation in defense against enemy ships, the Joint Board laid out a plan for cooperation and coordination based upon "paramount interest." If an enemy force approaching the coast could be engaged by a U.S. Navy force of approximately the same strength, the U.S. Navy assumed paramount interest and coordinated operations of Army forces with its own. On the other hand, if the enemy force was vastly superior to U.S. naval forces available to use against it, the Army held paramount interest and coordinated operations of the U.S. Navy with those of the U.S. Army. War Secretary Newton D. Baker and Navy Secretary Josephus Daniels approved this scheme, thus giving it effect.[31]

That, however, did not settle the matter. Over the objections of both secretaries, Congress in June 1920 divided aviation differently. It gave the Army control of aerial operations from land bases, and the Navy control of aerial operations of the fleet, and at naval stations when the operations were for instruction, experimentation, or training.[32]

Tests conducted against warships in 1921 and technological advances caused General Patrick in mid-1923 to suggest changes in the aviation functions of the Army and Navy. The use of naval aircraft from coastal stations for overseas scouting and protection of coastal shipping was uneconomical, and failed to secure effective protection for the nation's coasts. This work, he said, should be the sole responsibility of the Army.[33]

Patrick renewed the proposal in testifying before the Lampert Committee in 1925. Both Army and Navy planes might be scouting in the same area at the same time, the Navy to protect coastal lines, the Army in working with coastal defenses. Such duplication should be eliminated. The Army should undertake the air defense of the nation's coasts. How far to sea should the Army's responsibility extend? General Patrick thought two hundred miles under the existing state of aircraft development. Finding that the Army and Navy had "never agreed on a definite air policy," the Lampert Committee suggested Congress "settle by legislation the respective fields of operation of the Army and Navy."[34] But Congress did not act.

The Air Service responded promptly when ordered to the Mexican frontier in June 1919 to help stop raiding and smuggling. Its chief job was air surveillance, to keep the cavalry informed of conditions and activities along the border. It welcomed the assignment because it would afford men valuable training through useful work. However the main threat, Pancho Villa's army, had been eliminated by the time aviation units began operations. Further, the mere presence of planes on the border, and the threat of surveillance, tended to discourage raiders and smugglers. The need for aerial operations diminished. Patrols became fewer and less frequent, and after two years ceased. Service on the border gave the Air Service experience in operating under field conditions, in aerial observation, and in cross-country flying over difficult terrain. It revealed the need for better equipment and training, especially for navigation and communications, and for closer liaison between air and ground forces.

On the sea frontier, the Air Service found its high hopes and elaborate plans frustrated by government economy, coordination problems with the Coast Artillery, and want of a clear, and favorable delineation of the functions of Army and Navy aviation.

Chapter VIII

Planes Versus Ships

Both the Army and the Navy planned to employ aircraft against enemy shipping. Still, neither service at the end of the First World War possessed much information as to the effects of aerial attack on war vessels. Both wanted to ascertain how surface ships would stand up under assault from the air. The quest engendered a controversy more heated and longer enduring than that on the creation of an independent air force. In the fracas, Brig. Gen. William Mitchell quickly captured and long held the spotlight.

Project B

Soon after joining General Menoher's staff in the Office of the Director of Air Service in March 1919, General Mitchell proposed a test of planes versus ships. A year and a half later none had yet been made. During that time Mitchell became more and more vocal on the subject of aviation. His claims as to the importance and effectiveness of air power grew progressively stronger. He plied every means—conversation with fellow officers, testimony before congressional committees, public speeches, magazine articles, interviews with reporters—to publicize his ideas. He held aviation to be at least as vital to the military establishment as ground and naval forces. The nation needed a separate Department of Aeronautics or a Department of Defense with an Air Force coequal to the Army and Navy. Many people agreed, many did not. Those opposed included Army generals and Navy admirals who saw aviation in a supporting role to ground forces or the fleet.[1]

Mitchell was sure aviation rendered battleships obsolete, but he had not enough examples of aerial attacks on warships to prove it. So he examined "the characteristics of the two—relative speeds of airplanes and battleships, the utter vulnerability of the battleship to attacks from above, the futility of attempting to conceal seacraft"—to see what he could learn. He concluded "that seacraft in war lay utterly at the mercy of airplanes."[2]

While urging that a test be made, Mitchell worked on tactics for employment of aircraft against ships. In July 1920, for instance, he went to Langley Field to direct maneuvers which assumed a hostile fleet attacking the Atlantic coast. Airship *Zodiac–1* began patrolling the entrance to Chesapeake Bay at 0200. Airplane pilots and observers, organized into flights represent-

Brig. Gen. William Mitchell

ing pursuit, attack, and bombardment, reported to the field at 0300 to await information on the enemy. At 0400, *Zodiac–1* reported the enemy fleet approaching. Pursuit ships took off at once to seize control of the air. Next came attack aircraft to strafe and bomb from low altitude, followed by bombardment planes with heavy bombs.[3]

In October 1920, Mitchell and three other Army officers accepted an invitation to witness Navy tests on an old battleship, the USS *Indiana*, in Chesapeake Bay. The Navy placed bombs on and near the ship and exploded them. The only bombs dropped on the *Indiana* were dummies filled with sand. The *Air Service Newsletter* reported that planes from Langley Field observed the bombing, but the event did not become a major news item until a London paper published a picture of the twisted wreckage of the *Indiana* in December. American papers picked up the story, and a great public controversy ensued. Some argued that airplanes made battleships a thing of the past. Others contended that battleships could not be destroyed or knocked out completely by aircraft. Skillfully playing this situation, Mitchell succeeded in getting a test of aircraft against warships.[4]

When Secretary of War Baker asked the Navy for ships for tactical experiments, Secretary of Navy Daniels invited the Army to join the Navy in tests against ex-German warships. Baker accepted the invitation to participate in tests under Navy control. While these exchanges took place in February 1921, Mitchell started preparing for the tests, which became Project B.[5]

The next few weeks were busy ones as men were trained and assembled. The 258th Heavy Bombardment Squadron at Aberdeen Proving Ground tested bombs, fuzes, releases, flares, machineguns, and other equipment. Officers from Mitchel Field went to Aberdeen to learn bombing. The depot at Fairfield, Ohio, gave top priority to work on SE–5s, and on Martin, Handley Page, and Caproni bombers. Kelly Field sent 22 SE–5s, 1 Handley Page, and 2 Capronis to Langley. The 1st Pursuit Group transferred 20 officers and 11 cadets to the 2d Bombardment Group for training. Flyers at Fort Bliss practiced bombing a battleship outlined on the ground.

Early in May 1921, Maj. Thomas DeW. Milling, Commandant of the Air Tactical School at Langley Field, commenced forming officers and men into the 1st Provisional Air Brigade for the naval ordnance tests. Other officers and men soon began arriving: 20 cadets and 1 flight surgeon from Carlstrom Field, Florida; 1 lieutenant from Douglas, Arizona; 2 lieutenants from Marfa, Texas; several from McAllen, Texas; 6 officers and a large detachment of men from Mitchel Field; and 65 officers, 43 cadets, and 290 enlisted men of the 49th and 96th Bombardment Squadrons from Kelly Field.[6]

General Mitchell assumed command of the brigade at Langley Field on Friday, May 27.[7] After inspection on Saturday, he left to return to Bolling

Field in his SE–5A. Six other planes departed for Bolling about the same time. All were caught in a violent storm, the worst Mitchell had ever been in. If he tried to land, he probably would crash; if he turned around, he might miss Langley and be blown out to sea. Deciding to try to go around or between storms, he arrived safely at Bolling. For 1st Lt. Stanley M. Ames, pilot of a Curtiss Eagle, and the six men with him the story was different. All seven were killed when Ames tried to land.[8]

The accident received a lot of publicity, with demands for an inquiry. The War Department investigating board found the accident caused by the severity of the storm. Lieutenant Ames took the proper action in the interest of his passengers; he tried to land with the least possible delay.[9] General Mitchell, however, blamed the accident on the lack of regular routes, landing facilities, radio service, and weather bulletins for aviation, all of which could be provided by centralized control over aviation.[10]

This was more than General Menoher, Chief of Air Service, could bear. Mitchell talked too much. For months he had been airing his opinions on air power and aviation organization, opinions usually not shared by General Menoher, the General Staff, or the Secretary of War. He had antagonized the Navy and had quarreled with Rear Adm. William A. Moffett, Chief of the Navy's Bureau of Aeronautics. In securing publicity for himself and his ideas, he had brought unfavorable publicity on his superiors. His remarks after the crash of the Eagle led General Menoher to ask for his removal. Secretary of War John W. Weeks was not prepared to oust the popular Assistant Chief of Air Service. Thus, tests in June and July 1921.[11]

Mitchell tried to give his men as much training as possible. First came instruction on bombsights, bomb racks, radios, bombs, armament, and the theory of bombing. Next came practice with the camera obscura, then bombing with dummy and live bombs of various weights. The men practiced on still targets on Mulberry Island (in James River)[12] and on moving targets towed by a subchaser in Chesapeake Bay. As training progressed, the men operated in small formations and as a brigade. For final training in bombing, the brigade used a Navy target, the *San Marcos*, in Chesapeake Bay.[13]

Vice Adm. Hilary P. Jones was officer-in-charge of the bombing tests scheduled for June 1921. Capt. Alfred W. Johnson, USN, commanded air forces from the USS *Shawmut*. The Navy's plan, designed for scientific tests of aerial bombs and naval gun fire against warships, scheduled bombing to permit inspection by a Board of Observers at intervals during the tests.[14] General Mitchell was of a different mind, desiring to test tactics as well as ordnance. Above all, he wanted to sink ships as quickly and spectacularly as possible. The Navy refused to alter its plan.[15]

The first test came on June 21 on an ex-German submarine, U–117, anchored about fifty miles off Cape Charles Lightship. Eight divisions of naval aircraft (twenty-six airplanes) were to attack at twenty-minute intervals. Then it would be the Army's turn. The first division sank the sub.[16]

The next test, set for June 29, was a search for and attack on the ex-U.S. battleship *Iowa*. Controlled by radio from the battleship USS *Ohio*, the *Iowa* was to be from fifty to a hundred miles off shore at zero hour, seeking to reach the coast before being seen and assaulted by aircraft. So as not to sink the *Iowa*, the Navy planned to drop dummy bombs. During the planning conference in May, Mitchell refused to commit his bombers to this kind of an operation. When he changed his mind, the Navy would not let him take part.[17]

Mitchell's turn came on July 13, the target being the destroyer G–102 anchored fifty miles off Cape Charles Lightship. As for other ex-German ships, the operation was to continue until the ship went down. Naval air forces were in reserve in case the Army's bombs did not sink the destroyer. If Navy planes failed, U.S. destroyers would take over. Then battleships. Finally, if the G–102 still floated, a wrecking party from the USS *North Dakota* would sink her with depth charges.

Mitchell directed operations from a DH–4B. Airships and planes from Langley Field photographed the bombing. First, SE–5s attacked with machineguns and bombs from about 200 feet, the strafing simulated but the bombing real. Each plane carried four 25-pound personnel bombs. In four passes the SE–5s registered twenty-five hits, four of them duds. First

Direct hit on G–102 by SE–5 aircraft.

Lieutenant Thomas K. Matthews put a good bomb down the forward stack. The 1st Brigade reported it "an assured fact that no human being on the unprotected decks of the G–102 could have lived through the hail of iron and lead delivered in this attack."

That was merely the beginning. Next came 16 Martin bombers led by Capt. Walter R. Lawson.[18] As the planes passed over the target in succession at 1,500 feet, each dropped two of its six 300-pound demolition bombs. Lawson's first 2 bombs straddled the ship, each missing by 75 feet. The other planes followed at 45-second intervals. One scored 2 hits. The G–102 began to sink by the stern. Two minutes later a plane got 2 hits. When the formation returned to the attack, the destroyer was settling fast. One plane scored 2 more hits, and another struck just as the bow raised for the final plunge. The *Shawmut* signaled: "Return to base." No naval planes, destroyers, battleships, or wrecking parties were needed.[19]

On July 18 the Navy and Army took turns bombing an ex-German cruiser, *Frankfurt*, anchored in the test area off Cape Charles Lightship. During the morning the Navy dropped 250-pound bombs, the Army 300-pounders. In the afternoon, after the Board of Observers inspected the ship, the Navy dispensed more 250-pounders and some 550-pounders. A second inspection disclosed that the only damage thus far had been to the superstructure. The stability of the cruiser was unaffected; she was not taking water.

Captain Lawson and a flight of Martins with 600-pound bombs arrived

Bomb hit on the ex–German battleship *Ostfriesland.*

while the board was inspecting the ship. They had left Langley Field without awaiting Captain Johnson's signal to take off. When Lawson reported he was running short of gas, the board cut the inspection short. At 1611 Johnson signaled for Lawson to attack. Three and a half minutes later, Martin 3 (Lawson's plane) dropped the first bomb 200 feet off the starboard side. As the attack continued, Lawson came around again and put two bombs close enough to throw tons of water over the ship. Immediately afterwards, Martin 4 bracketed the funnels with two hits amidship. Another bomb, from Martin 2, exploded alongside, "fairly lifting *Frankfurt* out of the water." The ship started to sink by the bow. Two more bombs fell before the attack ended at 1625. At 1645 the *Frankfurt* took a pronounced list to port and at 1650 she disappeared beneath the water.[20]

The tests climaxed on July 20–21, 1921, with the bombing of the ex-German battleship *Ostfriesland*. Secretary of War Weeks, Secretary of Navy Edwin Denby, Generals Pershing and Menoher, senators and representatives, foreign observers, and many other distinguished guests as well as reporters went aboard the transport USS *Henderson* to witness the operation. The Atlantic Fleet and Admiral Jones' flagship, USS *Pennsylvania*, lay off the target to observe the action.

Because of heavy seas, the Navy postponed the operation until "further orders." General Mitchell thought this some kind of Navy trick. He requested but did not receive permission to attack. When the bombing commenced, the Navy went first. Mitchell impatiently dispatched his bombers without waiting for orders from the *Shawmut*, so the flight had to wait until the observers completed inspection of the ship. Afterwards, both the Army and Navy planes attacked, but the operation ceased due to an approaching storm. The Board of Observers reported little damage on the top side but considerable underwater injury. The battleship was listing to port and had settled by the stern.

The following morning, Army and Navy planes were on the scene when Captain Johnson ordered the attack. The rules called for individual attacks. Each plane commander was to make certain that the "All Clear" was out on the forecastle of the *Shawmut* before attacking. Further, bombing was to be discontinued for inspection after each hit.

The Army led at 0832 with 1,100-pound bombs. First Lieutenant Clayton L. Bissell scored a hit on the forecastle with the first bomb. The *Shawmut* hauled in the "All Clear" and repeated the "cease bombing" order by radio. The 1st Brigade reported that "due to the speed with which the attack was pressed four more bombs were dropped before the signal had been removed." Two of the four were hits. The Board of Observers found no vital damage to the ship or main battery, but her fighting efficiency had been affected by a large hole in the starboard side which was taking in water. By noon the ship was down about four feet by the stern and one foot by the bow.

Then came the test with 2,000-pound bombs. The Army was allowed a maximum of three bombs and two direct hits, with a pause after the first hit for inspection of the ship. General Mitchell dispatched Captain Lawson with eight Martin bombers and three Handley Pages. With the force on its way, Mitchell notified the *Shawmut* that the attack would go on until the Army got its two hits. But Mitchell was not after hits. He ordered his men to try for near-misses, believing them more damaging. The second bomb hit the side armor, glanced off, and exploded close, twenty to twenty-five feet from the port side. The stern settled fast. In a few minutes the *Ostfriesland* turned over and disappeared.[21]

General Mitchell accomplished what he had determined to do—sink a battleship. He threw a big party that night at the Langley Officers' Club.[22] Captain Johnson prepared a report on the "Disregard of Orders by Army Air Units in the Bombing Experiments on the *Ostfriesland*."[23] Years later, however, this naval officer who had commanded the air forces for the tests would write: "I don't see what else Mitchell could have done except keep on dropping bombs until the ship sank. If the ship had not sunk soon he would have been the object of ridicule because of his pre-bombing pronouncements. The operation would make or break him." As Admiral Johnson said: "It made him. In the public eye he became the infallible prophet on aviation."[24] Thus Mitchell emerged a national hero—with an obsession that Navy officials had conspired to try to prevent him from sinking the *Ostfriesland*.[25]

What did the tests prove? As ordnance tests, Mitchell said, they showed "that seacraft of all kinds, up to and including the most modern of battleships, can be destroyed easily by bombs dropped from aircraft, and further, that by far the most effective means of destruction are bombs." They "demonstrated beyond a doubt that, given sufficient bombing planes—in short an adequate air force—aircraft constitute a positive defense of our country against hostile invasion."[26]

The tests convinced the Navy's Board of Observers "that the airplane is a powerful weapon of offense." The board noted that the ships sunk by the bombing had been at anchor, carried no crews to rectify damage and keep them afloat, and had no antiaircraft fire or planes of their own to protect them. Further, a row of destroyers guided the bombers to their targets, the weather was good, except when operations were postponed because of fog, wind, or sea, and the bombers struck from very low altitude. But the fact remained, the board said, that "the ships so attacked, whether submarine, destroyer, light cruiser, or battleship, were eventually sunk, and by airplanes with bombs alone."[27]

The Joint Army and Navy Board found the battleship "still the backbone of the fleet and bulwark of the Nation's sea defense." Aircraft added to the dangers battleships faced but did not make them obsolete. It was "imperative as a matter of national defense to provide for the maximum possible development of aviation in both the Army and Navy." The test also proved the need for aircraft carriers and effective antiaircraft armament.[28]

Following the tests, Brig. Gen. William Mitchell led a mock raid against New York City on July 29, 1921. Seventeen Martin bombers, a Handley Page, and a Caproni flew northward in a great "V." Changing to battleline, they passed the Battery at 8,000 feet, bombed lower Manhattan, flew up Broadway to Central Park, turned and attacked again, and landed on Long Island. Hit by twenty-one tons of demolition, gas, and incendiary bombs, the city lay in ruins. The people were dead or in flight—or so the reporters learned when they talked with the airmen at Mitchel Field that afternoon. Mitchell put his planes on public exhibition and gave his men leave to visit the city they had theoretically destroyed. Returning to Langley Field on Monday, he bombed Philadelphia, Wilmington, Baltimore, and the Naval Academy at Annapolis.

The stated purpose of the maneuver had been to ascertain the radius of action of the current equipment, instruct personnel in the theory and practice of bombing, and familiarize them with the Atlantic seaboard (nothing being said about possible effects on public opinion). Mitchell said the operation "proved conclusively that even with present day equipment, unless adequate measures are taken, enemy planes could be landed on our shores and practically demolish the Atlantic Seaboard cities in a period of a few days."[29]

On August 29 General Mitchell sent his report on the naval tests and the bombing of the cities to his chief, General Menoher. Mitchell said, "National defense should be revised at once." The Army should defend the land. The Navy should operate on the high seas. An Air Force should be furnished for frontier and coastal defense against airplanes and ships, with the defense zone extending two hundred miles to sea. "Aviation," he said, "can only be developed to its fullest extent under its own direction and control. An efficient solution of our defensive needs will not exist until a Department of National Defense is organized."[30]

Noting his disapproval, General Menoher forwarded the report to Secretary of War Weeks. That was the end of that—until the report leaked to the press on September 13, 1921. The *New York Times* called it a "sensational chapter" in the "aircraft versus capital ships" story.[31] General Menoher decided either he or Mitchell had to go. Secretary Weeks decided to keep Mitchell, at least until after the *Alabama* tests, scheduled to start in a few days.

Many people expected the position of Chief of Air Service to go to Mitchell. Instead it went to Maj. Gen. Mason M. Patrick, who made it clear he would be chief and Mitchell would have to operate within set limits. Mitchell threatened to resign but quickly backed down when Patrick did not object.[32]

USS *Alabama*

Having agreed to give the Air Service an old battleship for bombing tests, the Navy in early September 1921 towed the USS *Alabama* into position near the USS *Indiana*, about seven miles southwest of Tangier Island in Chesapeake Bay. The Air Service had asked that the ship be in operating condition so it could discover what type of attack would put her out of action. However, neither the Navy nor the Army could afford the $450,000 needed to place the *Alabama* in commission.

The 1st Provisional Air Brigade meantime collected bombs and other equipment, practiced on targets on Mulberry Island, trained for night operations, created emergency landing fields and radio stations between Langley Field and the testing grounds, and completed other preparations. The Navy made a subchaser and a launch available to carry members of the brigade to and from the target and to observe the tests. Maj. Joseph T. McNarney went to Remo Field, about fifty miles north of Langley, to take charge of operations at the target.

The tests got under way on Friday morning, September 23, 1921. At 0923 two Martin bombers began laying a smokescreen windward of the *Alabama*. Then came two DHs, each with four 25-pound phosphorous bombs, and afterwards four DHs with tear gas bombs. Shortly after 1300, two Martin bombers dropped eight 100-pound phosphorous bombs, scoring six hits. Phosphorous flames and fumes covered the entire ship. General Mitchell then departed from the plan to direct an additional attack by two Martin bombers each carrying fifteen 25-pound phosphorous bombs. The Martins got six hits; the other twenty-four bombs "were well-placed about the battleship." The next scheduled attack was by two Martins with tear gas shortly after 1600. Four of the sixteen 50-pound bombs hit; the others landed alongside.

The weather being favorable, Mitchell ordered the night attack that had been planned for the tests. About 2240 a DH dropped flares to illuminate the target. Three DHs followed, each with one flare and 100-pound demolition bombs. Next three Martin bombers, each with two flares, attacked with 300-pound bombs. The brilliant light of the flares impaired bombing accuracy. The Martins got two hits, the DHs none.

After DHs put down a smokescreen the next morning, seven SE–5s attacked with machineguns and small personnel bombs, while four Martins dropped 300-pound demolition bombs. Four other Martins took off with 1,100-pound armor-piercing bombs, but only three attacked. Motor trouble forced the fourth down in the York River, the crew being rescued by a motorboat. The day's work ended with Martins launching a torpedo attack, simulated because the brigade lacked torpedoes and torpedo racks. The

Martin bomber flies away after dropping a 25-lb. phosphorous bomb on the USS *Alabama.*

Alabama had been damaged and her fighting efficiency impaired. Yet she floated which was according to plan.

The 1st Brigade got down to the business of destroying the *Alabama* when operations resumed on Monday, September 26, 1921. All of the seven Martin bombers dispatched carried a 100-pound bomb for a sighting shot. Four also carried 2,000-pound bombs, the other three two 1,100-pound bombs apiece. The first bomber dropped its sighting shot at 1159 and returned at once to release its 2,000-pounder. "This bomb," the brigade reported, "shook the warship from stem to stern, [and] threw smoke, mud, water and flames several hundred feet in the air."[33] The other bombs followed one after the other. Several drove holes in the ship, destroyed the funnels, cut off a fire-control mast, or did other damage. The onslaught sent the *Alabama* to the bottom of the bay.

The tests yielded useful information and interesting conclusions concerning aerial operations against seacraft. Airplanes could employ a smokescreen as cover for day bombing. Flares might be useful at night to alert friendly aircraft to a target but not for illuminating it for attack. The best conditions for night assault were when the moon was out behind the target. Air-delivered phosphorous and gas bombs could put seacraft out of commission by destroying the crew. Light bombs could destroy the superstructure and upper decks. No battleship in use or contemplated for the near future could withstand heavy bombs. And finally, the argument that the Air Service was limited in its operations by any condition or possibility had been proven a fallacy—or so Mitchell maintained.[34]

The tests over, the 1st Provisional Air Brigade disbanded.[35] General Mitchell resumed his duties as Assistant Chief of Air Service under the new Chief of Air Service, General Patrick.

USS *New Jersey* and USS *Virginia*

Patrick gave the Assistant Chief many special assignments, several of them keeping him away from Washington for long periods. In November and December 1921, Mitchell served on the staff of the American delegation to the Washington Conference for the Limitation of Armament and as a member of the subcommittee on aircraft. That duty was cut short, however, by an assignment to Europe to see what countries there were doing in aviation.

Later, in the summer of 1923, Patrick put Mitchell in charge of more operations against warships. A number of people saw this as reopening the controversy over the effectiveness of aerial bombs against seacraft. The Air Service disavowed any such intention: "The proposed exercises are simply in the nature of training to increase the efficiency of the bombardment personnel of the Army Air Service."[36]

Ships were being scrapped under the terms of the Five-Power Naval Treaty signed at the Washington Conference. Consequently, Congress specified that not less than $50,000 of the amount appropriated for the Air Service for Fiscal Year 1924 be used for bombing tests against obsolete naval craft.[37] The Air Service asked for two battleships. Planning to simulate battle conditions insofar as possible, Mitchell wanted ships equipped for radio control and turned over to him with steam up and the magazines full. He suggested the tests be held near the Diamond Shoals Lighthouse off Cape Hatteras. By setting up an airdrome on Cape Hatteras, he would give Air Service personnel experience in operating from an advance base.

Mitchell aimed to sink one ship in the shortest possible time. He would begin with an aerial torpedo launched against one of the radio-directed ships. He then would attack with a single plane and a two-ton bomb. This would give him a chance to test the Owl, a twin-fuselage, trimotor plane built by L. W. F. (Lowe, Willard, and Fowler Engineering Co., Inc.).[38] The torpedo and bomb should sink the ship, but other planes would be on alert. If the ship

Above: USS *New Jersey* listing after first attack; below: a closeup of the USS *Virginia* wreckage resulting from the bombing tests of 1923.

stayed afloat, the attack would go on with 1,100- and 2,000-pound bombs. Mitchell planned to subject the second ship to a succession of tests to try out smokebombs, lay a smokescreen over a maneuvering ship and, among other things, determine the effects of gas on animals aboard the ship.

The tests, conducted on Wednesday, September 5, 1923, did not turn out quite the way Mitchell wished. The two battleships, USS *New Jersey* and USS *Virginia,* lay in the vicinity of the Diamond Shoals Lightship, but at anchor, and without animals to be gassed. Moreover, the War Department on August 31 had specified that the tests commence with a bombing attack from 10,000 feet.

The directive posed a serious problem for Mitchell and the 2d Bombardment Group, which provided most of the people and planes. The standard NBS–1 could not bomb from 10,000 feet. The group had assembled six new NBS–1s with superchargers at Langley Field in mid-July for testing but found them unsatisfactory. The motor ran hot, and with motors throttled to try to keep them cool, the bomber climbed slowly. A supercharged NBS–1 required on the average 127.25 minutes to reach 10,000 feet with twelve 100-pound bombs, a full load of gas, and a crew of three.

First Lieutenant Carl A. Cover moved the radiators of one supercharged NBS–1 from beside to beneath the motors. With twelve 100-pound bombs, the plane could go to 10,000 feet in 54 minutes. With two 1,100-pound bombs, it took 85 minutes. This was the 2d Group's one plane capable of bombing from 10,000 feet when the War Department order arrived.

The men at Langley Field worked night and day, including Sunday and Labor Day, to modify other bombers. They completed four by Tuesday, September 4, 1923, and secured two more from Aberdeen Proving Ground. The group, save the supercharged planes under the command of 1st Lt. Charles B. Austin, moved to Cape Hatteras. That evening, Mitchell talked briefly to the men on the importance of the tests and what he expected. The attack plan for the following morning assumed friendly pursuit had gained temporary superiority in the air, and attack aviation had successfully completed low-altitude attacks against personnel on the battleships. Lieutenant Austin's flight from Langley Field would go first.

Fog at Langley Field on Wednesday lifted a bit about 0600, letting Austin take off. Each of his flight's six planes carried four 600-pound demolition bombs. One plane having been delayed, five attacked the *New Jersey* about 0840 and landed at Cape Hatteras. A flight of seven bombers, commanded by Capt. Lloyd L. Harvey, mounted the second assault from Cape Hatteras. All of these planes packed one 2,000-pound demolition bomb. After this run from 6,000 feet, Mitchell flew close to the *New Jersey,* saw she had settled considerably, and concluded she would eventually go down. The 2,000-pound bombs should have sunk her but, as Mitchell explained, the bombs had stuck in their racks, and either the fuzes had deteriorated or the bombs had not exploded at the proper depth.

Mitchell diverted the next attack from the *New Jersey* to the *Virginia*. Fitted with 1,100-pound bombs, all seven aircraft, under the command of 1st Lt. Harrison G. Crocker, struck from 6,000 feet. Mitchell said the bombs "literally tore the ship to pieces." Within thirty minutes the *Virginia* was under water.

Captain Harvey was standing by with a flight armed with 2,000-pound bombs. Mitchell sent him to bomb the *New Jersey* from 3,000 feet. The bombs stuck in the racks and overshot the target. Only one dropped close enough to do damage.

The gasoline supply ran low when a barge failed to arrive on schedule. Gas drained from airplanes was enough for two planes loaded with 1,100-pound bombs. The first plane put a bomb about ninety feet from the *New Jersey*; its next one fell alongside, a dud. The second plane dropped a bomb squarely on the ship. Before the pilot could come back to drop the other, the *New Jersey* turned over.[39]

Mitchell proclaimed that, as a result of the tests, "the problem of the destruction of seacraft by Air Forces has been solved and is finished."[40] But the public had begun to lose interest, and Mitchell was unable to exploit the tests to the fullest. For one thing, the Army Chief of Staff, General of the Armies John J. Pershing, curtailed publicity so as not to antagonize the Navy. For another, Mitchell got married in October 1923 and departed on an extended honeymoon and inspection tour of the Pacific and Far East.[41]

Court-Martial

When General Mitchell returned home in July 1924, he commenced a vigorous campaign to win acceptance for his ideas on air power. His chief, General Patrick, had adopted many of them and was working through normal military channels to secure acceptance for others. But this did not satisfy Mitchell, who wanted his program approved completely and immediately. Unable to make headway in the usual military way, he presented his case to the people to force the politicians to act. He gained a large following but made many enemies. His speeches, writings, and testimony before Congress antagonized President Coolidge, Secretary of War Weeks, the General Staff, and top Navy officers. General Patrick had supported and apologized for Mitchell as best he could, but that was becoming harder and harder to do. In March 1925, Secretary Weeks informed the President he could not recommend reappointment when Mitchell's four-year tour as Assistant Chief of Air Service expired on April 26, 1925. The job, and rank of brigadier general, went to James E. Fechet. Mitchell reverted to his permanent rank and became Air Officer, Eighth Corps Area, at Fort Sam Houston, Texas.

Colonel Mitchell continued his campaign and even stepped it up after the disappearance of a Navy plane, the PN–9, on a flight from San Francisco to Hawaii on August 31, 1925, and the crash of the Navy dirigible *Shenandoah* in Ohio on September 3. On September 5 he charged that these accidents were "the direct result of the incompetency, criminal negligence and almost treasonable administration of our national defense by the Navy and War Departments." This statement spurred the President to appoint the

"Billy" Mitchell court-martial. *USAF Art Collection (Robert Schaar, artist)*

board headed by Dwight W. Morrow to investigate aviation. The President also decided to give Mitchell the court-martial he seemed to be seeking.

The trial began in Washington on October 28, 1925. The charges, preferred by the President, were that Mitchell had made statements which were insubordinate, contemptuous, disrespectful, and prejudicial to good order and military discipline. Pleading not guilty, he succeeded in getting the court to go into the whole business of air power and the management of national defense, prolonging the trial to seven weeks. Found guilty on all counts, Mitchell was sentenced to suspension and forfeiture of pay and allowances for five years. The President approved the sentence on January 26, 1926, but granted Colonel Mitchell full subsistence and half pay. When he quickly offered his resignation to be effective February 1, 1926, the War Department instantly accepted.[42]

In the tests at sea in 1921 and 1923, General Mitchell and the U.S. Army Air Service proved conclusively that airplanes could sink unmanned, unarmed, undefended warships attacked while sitting still in the water in clear weather. The tests did not demonstrate what airplanes might be capable of doing under different circumstances. They did not forthwith render battleships obsolete, as Mitchell would have it, but rather spurred development and growth of naval aviation and of antiaircraft defense of the fleet, and further strained relations between the Army and Navy. Not only did the tests fail to garner substantial support for Mitchell's campaign for a powerful and separate air force, they put him on a course that eventually led to court-martial and resignation.

Chapter IX

Civil Affairs

Besides its role of defending the United States against attack and invasion, U.S. Army aviation performed a variety of civil functions. It patrolled national forests to spot and report fires, took aerial photographs for the U.S. Geological Survey and various other government agencies, assisted in several scientific experiments and, among other things, went to the aid of persons in distress. On one occasion, it furnished a squadron as part of Regular Army forces called out amid labor disturbances in the West Virginia coal fields.

Forest Fire Patrol

The use of airplanes in detecting and reporting forest fires had been advocated by American foresters as early as 1909 and tried in the state forests of Wisconsin in 1915. After a disastrous fire swept over ten thousand acres in northern Minnesota during October 1918, Chief Forester Henry S. Graves of the U.S. Forest Service, Department of Agriculture, asked the Air Service for help. Approving the request in March 1919, Secretary of War Baker authorized General Menoher to furnish Air Service personnel, equipment, and facilities for experimental patrols. The Air Service and Forest Service agreed California was the place to begin. Menoher therefore advised Graves to work out the details with the commanders of the Air Service stations in the area.

The Air Service began operations on June 1, 1919, to discover, locate, and report fires to the Forest Service. JN–4Hs at March Field, California, flew two patrols daily, starting at 1000 and 1300. One route was south of Riverside, over the northern part of Cleveland National Forest. The other was northward over Angeles National Forest. JN–6Hs from Rockwell Field, California, covered the southern part of Cleveland National Forest. JN–4Ds from Mather Field near Sacramento patrolled Tahoe, El Dorado, and Stanislaus forests. The Balloon School at Ross Field, Arcadia, California, kept a balloon flying at three thousand feet from 0700 to 1430 each day, providing lookout service for Angeles Forest from La Canada to San Dimas Canyon. A detachment of balloon students studying observation at Mount Wilson served as lookouts. In addition, the school at Ross Field made available ten enlisted men, a 1½-ton truck with a 50-gallon tank, shovels, axes, and fire extinguishers, to help fight fires at the call of the Forest Service.

To report the location of fires, men on patrol carried maps marked with a grid. Some wireless equipment was available for telegraphic messages from patrol planes to ground stations, but radio communication proved so unsatisfactory and unreliable that other means were frequently used. The aerial observer discovering a fire often dropped a message at the nearest town, asking the finder to inform the Forest Service. On occasion a pilot, if he could find a suitable place, landed so the observer could report by telephone. Flyers at times carried pigeons to release with a message. Repeatedly, however, the report waited until the plane returned to base.

The patrols soon grew so valuable and promising that forestry officials in California wanted to extend the area. Governor Ben W. Olcott and forestry officials of Oregon asked Col. Henry H. Arnold, Air Service Officer, Western Department, for help. Early in August after General Menoher's approval, Arnold established patrols with JN–4D and JN–4H aircraft from bases at Salem and Roseburg, Oregon. Late the same month, Arnold took charge of all forest patrol in the Western Department.

One of Arnold's first acts in his new role consolidated the Rockwell and March patrols at March Field. Now, one patrol went to Rockwell in the morning and back in the afternoon, while another covered a large area north and northwest of March Field. Arnold also merged operations in Oregon at Eugene, and changed all patrols to DH–4s for better range.

Arnold used the JN–4Ds in Oregon to create a new patrol from Redding, California, on August 31, 1919, but soon replaced them with DH–4s. After inspecting Redding, he shifted the patrol to Red Bluff, California, to gain better facilities. He further formed a patrol at Fresno, California, at the end of August, using JN–4Ds until DH–4s became available about three weeks later.

When the forest fire season ended in October 1919, Air Service crews had flown more than 2,800 hours and over 235,000 miles. They reported 570 fires, 27 having first been reported by plane. Eight major accidents caused one fatality, that on the first day of operations. Patrol was not an easy job. As Colonel Arnold pointed out in his review of the season's work, the men flew over dense forests and mountains, emergency landing places were few, and smoke and fog made flying hazardous. Despite the difficulties, the men showed a "willing spirit" and did "excellent work" in checking the number and extent of forest fires. Commanding pilots and mechanics alike, Arnold commented the latter often flew as observers, and all who had done so had been placed on flying status.

The Forest Service supervisor at Los Angeles thought aerial patrol "the most efficient system of fire detection . . . at their disposal." The district forester said results "far exceed the expectations." Forestry officials stressed that the airplane, besides its contribution to fire protection, had a significant educational and psychological effect on the public. A plane on patrol reminded people that the national forests were being watched, and greatly increased public cooperation in fire protection and prevention. The foresters wanted the help of the Air Service in the future.[1]

Before the 1919 season closed, the Air Service and Forestry Service planned for 1920, both with the idea of expanding the aerial patrols. Colonel Arnold wanted to organize 5 patrol squadrons (each with 32 officers, 132 enlisted men, and 19 airplanes) for operations in California, Oregon, Washington, Idaho, Montana, and Wyoming. Supported by western forestry and conservation interests and approved by the War Department, the plan went to Congress. Efforts were made to authorize additional Air Service personnel for forest patrol. Instead, Congress appropriated $50,000 to the Forest Service for aerial patrols. Under the circumstances, the two services agreed patrols should be confined to California, the area of greatest fire hazard.

Colonel Arnold's plans for 1920 also envisioned a training conference. This fitted in with the approved plans of both services to use foresters as aerial observers, and to station a forester at each of the bases and sub-bases utilized by patrols. The conference was held at March Field for a month beginning February 19, 1920. The Air Service provided four instructors, headed by Lt. Col. Barton K. Yount, the March Field Commander, to give courses in flying, meteorology, mapreading, first aid, and radio. Six Forest Service instructors taught economics of forestry, organization and work of the Forest Service, fire protection and suppression, and safeguarding of timber areas outside national forests. There were likewise meetings to discuss patrol routes, landing fields, ground markings, panels, radio code, pigeons, aerial photography, law enforcement, and other matters.

The sole Air Service squadron that could be spared in the west for forest patrol in the spring of 1920 was the 9th Corps Observation Squadron at Rockwell Field. Relieved from border patrol, the unit moved to Mather Field in April to prepare for the new patrol season in California. Patrol bases were founded at Red Bluff, Fresno, and Mather Field. In addition, landing fields (each with gasoline, oil, spare parts, and an enlisted mechanic) were started at Covelo, Alturas, Cooperstown, and Bakersfield. To communicate with DH–4s on patrol, the Signal Corps set up and operated radio stations at Alturas, Red Bluff, Mather, Sonora, Fresno, and Hot Springs. (*Map 4*)

The 9th Squadron commenced operations in mid-May 1920 with seven patrols daily—two from Mather Field, two from Fresno, and three from Red Bluff. Later it canceled various patrols owing to a gasoline shortage. The building of automobiles in the United States was burgeoning, and though more gasoline was being refined than ever before, the demand was great. Shortages occurred due to distribution problems, and a number of communities instituted gasless Sundays. Production was pushed and more oil imported. In the face of predictions that the nation's oil supplies would run out by 1940, engineers sought to enhance the efficiency of refineries and of automobile engines.

The shortage of gasoline hampered the Air Service more and more during the summer and fall of 1920. The 90th Surveillance Squadron at Sanderson, Texas, reported it was carrying out instructions to conserve gasoline. The 1st Pursuit Group at Kelly Field had to abandon several cross-country flights. At France Field in the Panama Canal Zone only short flights were authorized, to let officers and enlisted men on flying status get in the required number of flights before the gas supply ran out. At Mather Field the 9th Corps Observation Squadron could get regular but not always high-test gasoline. Since the regular gas was too poor to be used safely in DH–4s, the squadron at times could not fly scheduled patrols.[2]

For the patrols from Mather, Red Bluff, and Fresno, the 9th Squadron supplied the planes, pilots, and mechanics, and stood the cost of operating the planes. The Forest Service provided the observers and paid their expenses along with the cost of telephones, telegraph, and transportation.

After Congress refused to authorize more people for forest patrol, the Air Service decided to use cadets recently graduated from the Primary Flying School at March Field. Beginning May 20, 1920, cadets flew two patrols daily from March, one to Santa Barbara, the other to San Diego and return.

About the time operations got under way in California, the fire hazard in Oregon grew serious. The governor, the state forester, and the U.S. district forester asked for Air Service assistance. Arnold—who on July 1, 1920, dropped to his permanent rank of major but continued as Air Service Officer for the Western Department—established patrols from Eugene and Medford, Oregon. The men and planes came from the 91st Corps Observation Squadron at Rockwell Field.

The detachment at Eugene reported 169 new fires during the last half of August 1920 and a record 36 on September 2. In one week in August, the 9th Squadron's flight at Red Bluff detected 100 fires. On one patrol from March Field, observers found 5 fires within 30 minutes. Reported by wireless from the patrol planes, all 5 were extinguished without great damage, saving thousands of dollars worth of valuable timber. On another occasion, the 9th Squadron flew firefighters into the Lassen National Forest. Train and trail would have required 48 hours, the planes needed just 3. That made a big difference in suppressing the fire and brought predictions of a time to come

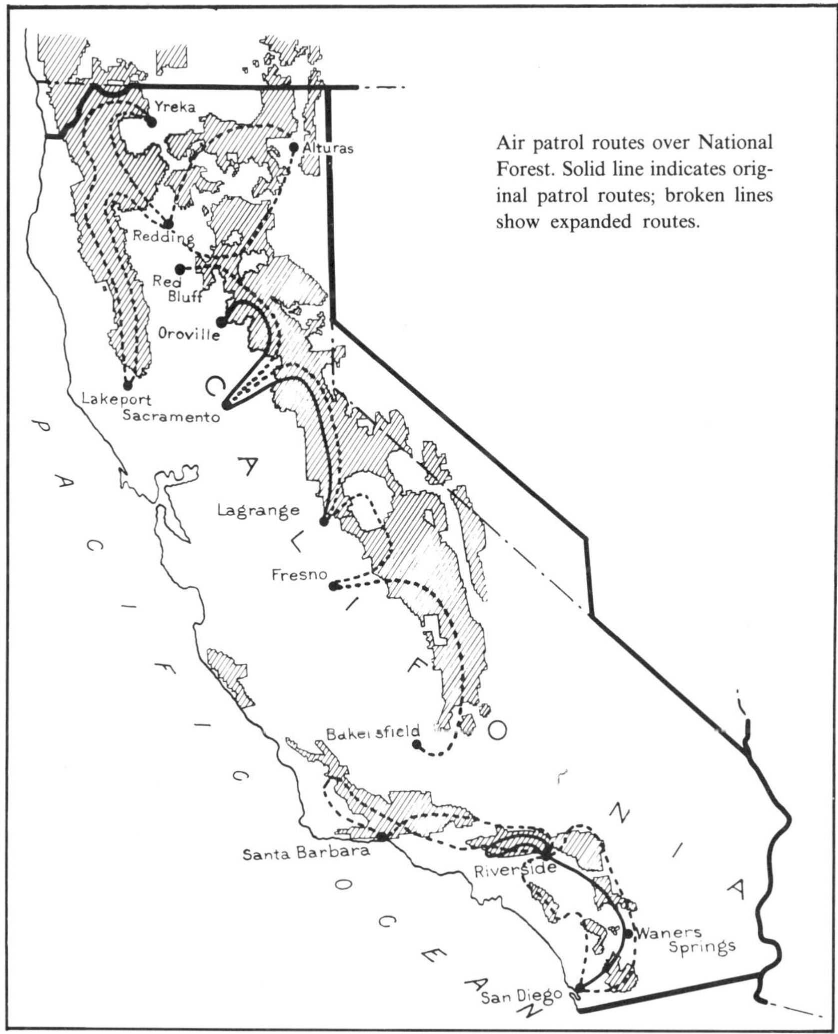

Air patrol routes over National Forest. Solid line indicates original patrol routes; broken lines show expanded routes.

MAP 4

when men and equipment would be flown to the scene of a fire and dropped by parachutes while airships rained down fire-extinguishing chemicals.

During the 1920 season, the Air Service had 37 planes, 35 officers, and 149 enlisted men on forest patrol. The pilots flew nearly 4,000 hours and over 476,000 miles in patrolling more than 16 million square miles of forest in California and Oregon. On these patrols, 1,632 fires were discovered, 818 being reported first by the aerial patrol.

Committing three times as many planes, the Air Service completed triple the flights and flew twice as many miles on forest patrol in 1920 as in 1919. Moreover, vastly improved radio communications rendered the aerial patrols more effective than in the previous year.[3] Even so, the role of the Air Service in protecting the nation's forests declined after 1920.

Aerial forest patrol had been undertaken by the Air Service in 1919 as an experiment. Although requested by the Forest Service, the work in 1919 was directed by Air Service officers and carried out with Air Service personnel, equipment, and funds. In 1920 the Forest Service assumed control, the role of the Air Service being that of lending pilots and planes to the Forest Service. For 1921, Major Arnold again recommended five squadrons. This time, however, he proposed that the patrols be flown by cadets and Reserve officers as pilots and observers, with Regular officers as squadron commanders and flight leaders. Governor Olcott urged Secretary Baker to extend the patrol over the entire Pacific Northwest. Forestry associations and private timber owners also wanted the patrol expanded to take in more territory. Nevertheless, the only aid forthcoming from Congress was another $50,000 for the Forest Service. In a period of general retrenchment throughout the federal government, the Air Service, short both personnel and funds, could provide little help.

For the most part, the plan for 1921 resembled that for 1920 with the addition of a patrol over the Olympic Peninsula in Washington. The Air Service contributed pilots, mechanics, and planes and assumed the normal expenses of aircraft operation. The Forest Service furnished observers and paid the expenses of the patrol bases and the cost of telephone and telegraph. The Air Service supplied radio equipment and recruited licensed amateur operators paid by the Forest Service. They operated stations at headquarters in the national forests to receive messages from patrol planes and link together the various bases. The Forest Service also paid for much of the gasoline and oil used in the patrol planes, when a cut in the Quartermaster's appropriation severely reduced the quantity available to the Air Service. In fact, Oregon and Washington bought gas and oil for patrols in June, before Forest Service money for Fiscal Year 1922 became available on July 1.

Cadets flew patrols in central, northern, and southern California. Detachments from the 91st Corps Observation Squadron at Rockwell Field handled patrols in Washington and Oregon. The gas shortages, which delayed the beginning of patrols in California until July, continued to hamper operations through the summer of 1921. The Air Service also was handicapped by personnel shortages stemming from the general reduction of the Air Service and Army.

Of necessity the activity in 1921 was considerably smaller, the area covered being reduced by one-fourth, patrols by half, and flying hours by a third. The results were generally good, and the airplane had achieved a significant place in fighting fires. Nonetheless, by the end of the 1921 season, the Air Service and Forest Service saw that aerial patrol could supplement but not replace other surveillance. Aerial patrols appeared to be of most value at the time of the greatest danger of fire, such as immediately after electrical storms. Aerial observers could also perform valuable services during a forest fire, by keeping foresters posted on its course and the progress of teams in bringing it under control.

Major Arnold placed high value on the training and experience patrol flights afforded Air Service personnel. He therefore recommended patrols be continued in 1922. Paul G. Redington, Chief Forester in District 5 (California), wanted the work to go on and even be increased in especially hazardous areas. Yet, he thought daily patrols involved too much unnecessary flying. He suggested planes, pilots, and observers be furnished for special reconnaissance as needed. For the fiscal year commencing July 1, 1922, Congress failed to grant a special Forest Service appropriation for aerial patrols. Due to reduced Army appropriations, the War Department insisted expenditures be limited to the most important military activities. Consequently, the department ruled Army planes would not fly forest patrol in 1922.

Persistent requests persuaded Secretary of War Weeks in mid-July to permit the Air Service to fly patrols in Oregon. Major Arnold at once sent eight DH–4Bs and pilots to Eugene, where they remained at the call of the Forest Service until mid-September. At the end of July, he set aside two more planes for similar flights over the Olympic Peninsula. And he assured California of Air Service aid in an emergency.

Lt. Col. William E. Gillmore, who succeeded Major Arnold as Air Service Officer of Ninth Corps Area (the former Western Department) in the fall of 1922, wanted a definite policy covering the Air Service's role in the campaign against forest fires. Favoring continued cooperation with the Forest Service, Gillmore disliked handling patrols on a yearly basis, and sought approval of a permanent program. General Patrick and the Forest Service supported him. But the War Department turned down the plan Patrick submitted, taking the view that assistance to the Forest Service would eventually have to be discontinued.

In 1923 the Air Service once more reserved planes and pilots for the Forest Service. Forest conditions that season, however, were such that the fire hazard was very low. The Air Service made just seventy-five flights, total

Plane on forest patrol.

flying time being slightly over two hundred hours. With the danger of fire far greater in 1924, the Air Service used but two planes, for patrols in Oregon.

After three years without a special appropriation for aerial patrols, the Forest Service received $50,000 for the fiscal year starting July 1, 1925. Secretary Weeks refused the Forest Service's request for assistance. Air Service would have to bear the cost of pilots' time and of wear and tear on planes, and continued participation of the Air Service in forest patrol would interfere with training. The Army undertook the patrols originally as an experiment to determine their value. The time had come, Weeks said, to turn over the work to commercial aviation. Supporting the Secretary's position, General Patrick suggested the Forest Service either hire commercial aviation or take on the work itself. The Air Service could spare some planes if the Forest Service wanted to set up its own patrol.

Having received ten DH–4Bs from the Air Service, and having hired Reserve officers as pilots and ex-enlisted men with Air Service training as mechanics, the Forest Service flew its own patrols in 1925. After two more years of operations, however, the Forest Service gave the work to civilian contractors.[4]

Mapping

Soon after the Armistice the U.S. Coast and Geodetic Survey sought the help of the Army and Navy with experiments in applying aerial photography to mapping. Cooperating with the survey, the Air Service in June and July 1919 took pictures of the coast in the vicinity of Atlantic City. It tried three cameras. The first was an L-type plate camera adapted from a British model and manufactured by Eastman Kodak Company. The second, an Eastman K–l automatic camera, used a 75-foot roll of film to produce ninety 7- by 9-inch exposures. The third, designed by Maj. James W. Bagley of the Corps of Engineers, had three lenses with one directed downward for vertical photographs and the other two inclined for oblique pictures. The test showed the value of cameras in mapping level country, and aerial photography could probably be developed for employment in mountainous terrain.[5]

Later in 1919, Capt. Albert W. Stevens and members of Army Air Service Photo Section 7 photographed Camp Gordon, Georgia, for the General Staff. During the first half of 1920, various units mapped the peninsula above Langley Field, Virginia; areas around Yuma, Arizona, Camp Bragg, North Carolina, and Fort Leavenworth, Kansas; and the coast of Mariveles Bay in the Philippines.[6]

On the basis of these and other projects, the Assistant Chief of Air Service, General Mitchell, announced in the fall of 1920, "With proper organization we could make [an] accurate survey of the entire country from the air in three years." Such mapping could be done at one-tenth the cost and in one-hundredth the time required by any other method. The Air Service had the camera (he thought the K–1 "second to none") and knew how to do the job. "The need now," Mitchell said, "is to coordinate all these requirements and get a single organization which can carry the necessary measures into effect."[7]

The Air Service did photographic work for a number of government agencies during the first half of the 1920s. It made an aerial survey of the Grand Canyon for the Department of Interior, photographed an area in Mississippi for the U.S. Public Health Service, mapped five thousand square miles of Tennessee for the U.S. Geological Survey, and helped the Geological Survey update several quadrangles in the state of New York. Among other projects were the photographing of two thousand square miles along the U.S. Canadian border near Duluth, Minnesota, and the making of mosaic maps of areas proposed for the Shenandoah and Great Smoky Mountains National Parks. The Air Service extended photographic support to a University of Rochester expedition to southern Panama, to determine the suitability of the region for rubber plantings and to hunt for a tribe of blond Indians. It furnished thirty-five planes at Mitchel Field, New York, to take pictures and assist Professor David Todd of Amherst College and other scientists with observation of a solar eclipse on January 24, 1925. Military projects included photographing water routes and docking facilities for the Board of Rivers and Harbors, and making mosaics of Army stations, reservations, and training areas. Finally, the Air Service worked with the Field Artillery in making fire-control maps for aerial photographs.[8]

The U.S. Geological Survey tried without success for several years to gain congressional approval for a general topographical survey of the United States. In February 1925, Congress finally authorized a 20-year project and supplied funds for work through Fiscal Year 1926. The law permitted use of existing services and government facilities. The Director of the Geological Survey indicated less than half of the 3 million square miles of the United States had been mapped and most of what had been done needed resurveying. That agency's chief topographical engineer estimated the job would take 100 years using ground survey methods. He and the director wanted the Army to photograph 45,000 miles in Fiscal Year 1926 then over 120,000 square miles a year. But the Air Service lacked resources for such a large undertaking. Projects completed for the Geological Survey in 1926 totaled about 22,500 square miles. The following year the War Department imposed a fee of $51.75 per flying hour (later raised) for photographic work outside the department.[9]

Officers having key roles in Air Service photography included 1st Lt. George W. Goddard, Captain Stevens, and Major Bagley. The leading Air Service civilian engaged in photography was Dr. Sam M. Burka of McCook Field, Ohio. Stevens spent most of his time in the field doing photographic work. In fact, there were complaints he was too long away from his duties as head of the Photographic Laboratory at McCook Field. Though some of his projects were self conceived, they had approval of General Patrick or his staff. One entailed a two-month expedition in 1923, with 1st Lt. John A. Macready as his pilot, to photograph the cities and natural wonders of the West. For about nine months during 1924 and 1925 Stevens was on leave to accompany Dr. Alexander Hamilton Rice on an expedition to the upper Amazon River. Afterwards, he and Macready made another extended tour through the West.[10]

Captain Stevens, Lieutenant Goddard, and members of the Engineering Division at McCook Field were in the forefront in developing and testing new equipment, materials, and techniques. They spent a lot of time with the Eastman Kodak Company and the Fairchild Aerial Camera Corporation. The K–2 camera featured a device for spacing exposures evenly, the K–3 a faster shutter and other refinements. Major Bagley added transparent spirit levels to his tri-lens (T–1) camera so the position of the bubble registered on the film to help evaluate and interpret mapping photographs. To create the T–2, Bagley inserted a fourth lens trained backward to show the relative position of the other exposures. The T–3, with one vertical and four oblique lenses, could cover in one exposure eight times the terrain of the T–1. Other experiments at McCook Field dealt with photography at night and at high altitudes. The depot at Fairfield, Ohio, did its part by reworking DH aircraft

Dr. Sam M. Burka (left), physicist at McCook Field, and Lt. George W. Goddard prepare for photographing of solar eclipse, January 24, 1925; below: Capt. Albert W. Stevens with camera suspension on Salmson plane.

to produce DH–4BP–1s, which became the standard photographic planes of the Air Service.[11]

Helping Farmers

When C. N. Neillie, entomologist of the city of Cleveland, had difficulty in applying insecticides to the leaves of closely planted trees in 1921, he conceived the idea of dusting the trees by airplane. Working through ex-Secretary of War Baker, he secured the assistance of the Air Service's Engineering Division at McCook Field, Dayton, Ohio. The Ohio Experimental Station at Wooster, Ohio, assigned J. S. Houser to the project. Houser and E. Dormoy of McCook Field designed a system for dispensing insecticide from a hopper attached to the side of an airplane. Houser tested the device in August 1921 on the farm of Harry A. Carver near Troy, Ohio. Lieutenant Macready flew the plane, a JN–6, about 20 or 25 feet above a grove of catalpa trees ravaged by caterpillars. Houser, in the rear cockpit, turned a crank to scatter arsenate of lead. Within three days ninety-nine percent of the caterpillars were dead. H. A. Gossard, the state entomologist who witnessed the test, reported the Air Service "seems not only willing but anxious to cooperate with anybody wishing to test out the value of the airplane for this use."[12]

The following summer the Air Service sent two airplanes, two pilots, and three enlisted men to Tallulah, Louisiana, to join with the U.S. Department of Agriculture on experiments in dusting cotton. The work went on until December 1924 under varying conditions of weather, altitude, and time of day, and with different kinds of insecticide. These experiments led to the establishment of a crop-dusting industry. The Air Service temporarily released 1st Lt. Harold R. Harris to assist Huff-Daland Dusters, Incorporated, get started in the business. The company dusted some fifty thousand acres in 1925. By 1927, commercial dusters covered ten times as many acres.[13]

The Air Service further aided the Department of Agriculture on experiments to exterminate mosquitoes in Louisiana and gypsy moths in New England. At another time it spared a plane to the Department of Agriculture for a campaign against the alfalfa weevil. After 1st Lt. Oakley G. Kelly left the Regular service and became executive officer of the 321st Observation Squadron of the Organized Reserve, he lent a hand to pomologists in Oregon dusting apple orchards to eliminate scab and coddling moths.[14]

Among similar projects undertaken by the Air Service was one in the Philippine Islands. The Mindoro Sugar Company wanted help in killing locusts that were wasting its cane fields. The Philippine Department sent 1st Lt. Harry Weddington with a detachment from Camp Nichols at Manila to

San Jose on the island of Mindoro. Weddington created six landing fields in a semicircle around San Jose, airplanes or carabao being the sole means of transporting supplies and insecticide to them. The crews flew a DH–4B to observe and photograph the cane fields and a JN–4D to dust with calcium arsenate.

Weddington's experimentation was characteristic of many such projects. Seeking the best dispenser for dusting, he first tried a "Troy" hopper built to specifications supplied by the Engineering Division at McCook Field. It proved unsatisfactory. The Engineering Department at Camp Nichols came up with a new cone-shaped hopper which was installed in the front cockpit of a Jenny and cranked by the pilot. Though this improved hopper was better than the Troy, Weddington found cranking hard while piloting the plane a few feet above the ground. The sugar company's engineer made a dispenser operated by a venturi, but it was discarded after testing. Good features of the various systems were then incorporated in a new one driven by an electric motor. But the motor, generator, and battery weighed too much. The hopper was then redesigned to be powered by a wind-driven fan, coupled by worm gear to a paddle wheel inside the hopper. Other experiments focused on finding the best time and the most effective way of dusting.

From operations at San Jose in the fall of 1923, Weddington concluded the airplane could kill locusts and do it more economically than any other means. In 1924 the Philippine Bureau of Agriculture purchased a JN–4D for dusting, and hired a pilot tutored at Camp Nichols by 1st Lt. Milo McCune of the U.S. Army Air Service.[15]

For several summers, beginning in 1921, the Air Service assisted the Department of Agriculture with studies of wheat rust. Flying with Air Service pilots, Department of Agriculture observers collected rust spores at various altitudes and in different parts of the country to trace the spread of the disease.[16] Air Service pilots and planes helped government officials survey storm damage in a national forest in Oregon,[17] count elk in California,[18] reforest areas of Hawaii by scattering seeds from the air,[19] and, among other things, report crops being grown in sundry places.[20]

Relief Missions

The Air Service always appeared ready to reach out to people in distress. Planes from Kelly Field, Texas, dropped food and supplies to those stranded by a flood along the Mexican border in September 1919. When the Rio Grande flooded again in 1922, planes from Kelly flew patrols to report on flood conditions and locate stranded people and cattle. Air Service planes helped during floods at Dayton, Ohio; Pueblo, Colorado; and elsewhere.

Planes from Aberdeen Proving Ground, Maryland, bombed an ice jam on the Delaware River in March 1923 and another on the Susquehanna in February 1926. NBS–1s from Chanute Field, Illinois, and DH–4Bs from Fort Riley, Kansas, saved Union Pacific bridges and rail lines by bombing an ice jam on the Platte River in March 1924.[21]

When people in a motorboat were reported missing after a storm on Lake St. Clair, First Lieutenant Jerome B. Machle flew out from Selfridge Field, Michigan, and found them on an uninhabited island. A speedboat from Selfridge took them off. First Lieutenants Eugene H. Barksdale and Newton Longfellow of Mitchel Field, New York, had no such good luck in their search for a yacht blown out to sea in a storm. First Lieutenant Arthur L. Foster and his observer, Maj. Walter W. Vautsmeier, of March Field, California, could not find a two-year-old child who had wandered away from home. Nor were planes from the 8th Attack Squadron able to locate two

Above: JN–6 used in spraying grove of catalpa trees, near Troy, Ohio; right: spraying dust to kill locusts in the Philippine Islands, 1924.

youths supposed to have drowned in Corpus Christi Bay, Texas. On that search, 1st Lts. Peter E. Skanse and Edwin R. McReynolds carried inflated innertubes in case they had motor trouble over the bay.[22]

Several organizations were caught up in rescue operations during April 1923 after reports came in of people marooned and starving on South Fox Island in Lake Michigan. The only plane available at the moment at Selfridge was a Jenny. First Lieutenant Ennis C. Whitehead flew it to Northport, Michigan, on April 19 with Harold V. Willcox of the *Detroit News* as passenger. When they landed on ice on Grand Traverse Bay, the plane broke through. Whitehead and Willcox were rescued, but the plane went down. Whitehead found a landing field and asked for a plane. First Lieutenant Russell L. Meredith and a reporter from the *Detroit Times* arrived in a DH the next day. In landing the plane was damaged beyond field repair. Reporting the accident, Whitehead asked Selfridge not to send another plane until he obtained information from Fox Island.

A plane from Chicago, carrying representatives of the *Herald and Examiner*, landed at Northport earlier on April 20 and took off for South Fox Island. The same day, 1st Lt. Henry E. Wooldridge, who had flown from Chanute Field to Gaylord, dropped several hundred pounds of food on the island. A mail plane, arriving at Charlevoix the evening of the 20th, took two hundred more pounds of food to the island the next day. Then word came that the people on South Fox Island were not starving, just short of some items like butter, sugar, and coffee.

While Lieutenant Whitehead was salvaging the Jenny from the bay and preparing to ship it and the DH to Selfridge by train, the people of Northport asked him for aid in establishing a landing field. On the way back to Selfridge, he stopped at Traverse City to help the Chamber of Commerce find a field and mark it.[23]

In 1920 DH–4Bs from Post Field, Oklahoma, and Fort Leavenworth, Kansas, assisted rescue work in a Colorado mine disaster. In 1922 planes from Crissy Field, California, flew rescue workers to a mine where forty-eight men were trapped five thousand feet down. In April 1926, when Mauna Loa on the island of Hawaii erupted, three planes commanded by 1st Lt. Harold R. Rivers responded. They photographed the volcano and lava flow for Dr. Thomas A. Jaggar, volcanologist at Kilauea Observatory.[24]

First Lieutenant John D. Corkille of Brooks Field, Texas, saw a house on fire while out flying one day. He circled down for a closer look. Not seeing anyone around the place, he landed, ran to the door, knocked, and said (politely), "Mister, your house is on fire."[25] Similarly, 1st Lt. Harold A. Bartron of Wilbur Wright Field, Ohio, saw a fire, landed, and helped the woman remove furniture and valuables from her house.[26]

First Lieutenant James E. Duke, Jr., of the Air Intermediate Depot at San Antonio flew to Winchester, Texas, to drop smallpox serum. When a

Panamanian army officer at David, in western Panama, needed serum for tetanus, 1st Lt. Homer B. Chandler of France Field reacted. He flew across the isthmus to Fort Clayton, picked up the serum, carried it back to France Field, and gave it to another pilot, 1st Lt. Charles B. Austin, to deliver. On another occasion, France Field received a call to assist Mrs. F. N. Gage, the wife of a University of Michigan professor collecting snakes and insects in Panama. She had developed tetanus after being accidentally shot in the hand. Two DHs and two Martin bombers went to David. After a plane dropped serum to the party, Mrs. Gage was taken eighteen miles by horseback to the field, where she was placed in a litter aboard a bomber. Professor Gage and Capt. Andrew W. Smith, a flight surgeon who had studied under the professor, accompanied her on the flight to a hospital at Balboa.[27]

General Mitchell was inspecting Selfridge Field when word came that a boy on Beaver Island in Lake Michigan had been seriously injured and required medical attention. Ice floes prevented help being sent by boat. The boy's father had called General Patrick, who told him to get in touch with the Commander at Selfridge. Mitchell let Lieutenant Meredith take his DH, the *Osprey*, for the mission. Landing on ice in the harbor at Charlevoix, Michigan, Meredith picked up Dr. R. M. Armstrong and flew through a snowstorm to the island, where the doctor operated on the boy and said he would recover.[28]

Assisting Civil Authority

Between the bombing of the German warships in June 1921 and the tests on the *Alabama* the following September, the 1st Provisional Air Brigade assisted civil authorities in West Virginia. John L. Lewis and the United Mine Workers of America were trying to unionize coal miners in Mingo County. Miners had clashed several times with mine guards and police, and in late August they were gathering in the mountains of Logan County to march into Mingo and force recognition of the union. Lacking National Guard units (the state having not yet organized militia for the postwar period), Governor Ephraim F. Morgan asked President Harding for federal troops to maintain order.[29]

Brig. Gen. Harry H. Bandholtz, Commander of the District of Washington, went to West Virginia to investigate. So did General Mitchell. Questioned by newsmen shortly after landing at Kanawha City on Friday, August 26, 1921, Mitchell told a reporter how air power might be used in West Virginia:

> "All this could be left on the air service," he said. "If I get orders I can move in the necessary forces in three hours."
>
> "How would you handle the masses of men under cover of gullies?"

> "Gas," said the general. "Gas. You understand we wouldn't try to kill these people at first. We'd drop tear gas all over the place. If they refuse to disperse then we'd open up with artillery preparation and everything."[30]

On Saturday, General Bandholtz talked with a group of miners and threatened martial law if they did not disband and go home. Satisfied with the reaction, he returned to Washington. Mitchell flew back to Bolling Field.

Five miners died and three deputy sheriffs were captured in a pitched battle on Sunday. With reports the miners were reassembling to march on Mingo, President Harding granted the governor's request for troops. On Thursday, September 1, 1921, a squadron of DH–4Bs from the 1st Provisional Air Brigade was ordered to Charleston, under the command of Maj. Davenport Johnson. The planes mounted machineguns and carried ammunition. Two Martin bombers transported extra ammunition, medical supplies, and a flight surgeon. Additional personnel and equipment, as well as bombs, went by rail.

Shortly after departing Langley Field, Major Johnson had trouble with his DH and returned for another plane. The squadron landed on a muddy field at Roanoke to refuel and spend the night. On Friday morning, planes had been jacked up to get them out of the mud. In taking off, 1st Lt. Valentine S. Miner hit a corn shock and wrecked his plane. His passenger, Cadet Virgil D. Lovell, was cut on the left arm. Capt. John J. Devery, Jr., broke an axle and blew a tire when engine trouble forced him down in a rough field near Beckley, West Virginia. Two planes, piloted by 1st Lts. Donald R. Goodrich and Edgar A. Liebhauser, became lost in the fog and spent Friday night at Mooresburg, Tennessee. Heading for Charleston on Saturday, both planes wrecked when forced down by a storm. A Martin bomber piloted by 1st Lt. Leslie P. Arnold was blown off course on Friday and crashed while en route from Aberdeen Proving Ground to Charleston. Another Martin bomber crashed while returning from Charleston to Langley Field on Saturday, killing four men and severely injuring another.

From Charleston, General Bandholtz sent the squadron to find the miners' camp and report on activities and conditions in the disturbed area. The fighting grew more intense but ended swiftly when Bandholtz dispatched columns of troops along mountain roads with full field equipment, including rifles, machineguns, artillery, ammunition carts, baggage wagons, and rolling kitchens. The Army did not need its guns or tear gas. By Monday, September 5, 1921, everything was quiet. Thursday the squadron returned to Langley Field with only one minor accident.

Afterwards, General Mitchell pointed out that the movement of a squadron to Charleston in an "incredibly short space of time" had been achieved by "the same organization and same planes which had sunk the battleships far out at sea; now they had crossed the Alleghanies and landed in the midst of the mountains." The Mingo War, Mitchell said, afforded "an excellent example of the potentialities of air power, that can go wherever there is air, no matter whether they may be over the water or over the land."[31]

Believing military aviation would benefit from the development and growth of civil aviation, the U.S. Army Air Service eagerly undertook numerous civil projects that advanced aviation in general. In addition, it flew mercy missions and performed other civic functions which brought it much goodwill. The Air Service detected, reported, and fought forest fires, and it applied aerial photography to civil and military mapmaking. Added to these were the transporting of medicines in an emergency, doing relief work during floods and other natural disasters, and moving the sick and injured. Equally significant were the scientific studies and experiments and the eradication of mosquitoes, locusts, and other pests. The Air Service would willingly have done much more but it had not the means. In forest patrol and crop-dusting, for example, it could only help to start the business then withdraw, permitting others to carry on.

Chapter X

Aviation Facilities and Equipment

The Air Service at an early date saw the need for a nationwide network of airways and landing fields to permit rapid movement of aviation units across country for military purposes. It also foresaw the necessity for such facilities for civil aviation. Air Service flights like those of Maj. Albert D. Smith and Maj. Theodore Macauley across the continent, and Lt. Col. Rutherford S. Hartz around the rim of the United States, helped to open cross-country operations. The first transcontinental reliability and endurance test likewise yielded much valuable information. Development of airways and landing fields became a cooperative venture involving several federal agencies as well as state and municipal governments, business and industry, and civic groups. The Air Service made a major contribution in establishing and operating a model airway. As part of that project, or in work closely related to it, the Air Service pursued development and improvement of navigational instruments, night-flying equipment, air-to-air and air-ground communications, weather service, and safety devices.

Landing Fields and Airways

Airfields were scarce and widely scattered in America at the end of the First World War. A cow pasture might serve as a landing place for a gypsy flyer or for an Air Service pilot on a recruiting trip. If his motor quit in the air, a flyer picked the best place he could see within gliding distance. Clearly a nationwide network of airways, airports, and emergency fields seemed essential to the advancement of military aeronautics and commercial aviation. Capitalizing on public interest in flying, the Air Service encouraged cities and towns across the nation to build landing fields for use by all flyers.

Many towns already showed interest. Thomasville, Georgia, was one. Mr. Edward R. Jerger, editor of the *Times-Enterprise*, and three other men motored in April 1919 to Souther Field in Americus, Georgia, for help. The Commander, Maj. Earl S. Schofield, sent Lt. John McRae back with them to advise on the selection of a field. The policy of the Air Service was to grant such requests and lend expert flyers to help locate proper landing facilities.[1]

General Menoher used a Southeastern Aeronautical Congress at Macon, Georgia, in May 1919 to unveil a plan for municipal airports. He told delegates the Air Service and Post Office Department were working with cities and towns to create flying fields. For the time being, the program was limited to thirty-two communities where stations were needed for mail service or for cross-country flying by the Air Service. The cities would pay for establishing and maintaining the fields and erecting the steel hangars furnished by the government. General Menoher urged other cities and towns to construct fields to government specifications, so they could be fitted into the network in the future.[2]

The Air Service devoted considerable effort to the program during the next year and a half. Representatives of Pennsylvania cities heard Lt. J. H. Sullivan talk on the need for and value of municipal fields. Afterwards, the convention adopted a plan for each city to form an aviation committee, select a field, devise means for its upkeep, and arouse public interest. At Selfridge Field, Maj. Norman J. Boots received a request from Charlotte, Michigan, for help in laying out a field. In the west, Col. Henry H. Arnold designated 2d Lts. Ralph M. Kelly and Leland W. Miller to advise the city of Seattle on building an airfield. The same kind of thing went on elsewhere throughout the country.[3]

Airfields alone would not enable commercial aviation to operate safely or allow the Air Service to deploy aircraft quickly from one part of the country to another to meet national defense demands. Routes needed to be laid out and marked, and emergency landing places located. Weather services, communications, maintenance and repairs, lighting—all had to be at hand if aviation was to progress to the point where cross-country flying could be done safely, day or night, good weather or bad.[4]

To launch an airway program in the United States, the Air Service established a model airway between Washington and Dayton, Ohio. At the opening exercise at Bolling Field, on February 12, 1921, Brig. Gen. William Mitchell explained how the airway would serve aviation as roadways served the automobile. The Air Service leased land near Moundsville, West Virginia, and opened a landing field stocked with gasoline and supplies to afford an intermediate stop between Washington and Dayton. It invited towns along the route to provide emergency landing fields and mark them. Although the airway was used, a shortage of money and people prevented the Air Service from making any real progress in its development during 1921. In fact the situation became so bad the Airways Section, which had been created in September 1920 in the Office of the Chief of Air Service, was discontinued.[5]

President Harding supported the development of transcontinental airways. He wanted the Air Service to work with other government agencies to set up routes, and with the states to start airdromes. But that did not furnish the wherewithal for the task. After General Patrick became Chief of Air Service in October 1921, he diverted funds from other activities to airways. On December 1, 1921, he reinstated the Airways Section with Capt. Burdette S. Wright as chief.[6]

One of the section's chief tasks was collecting and disseminating information about airdromes. The Air Service had already assembled in Washington many facts gleaned from reports of cross-country flights and other sources, but they needed to be checked, updated, and expanded. Capt. St. Clair Streett of the Airways Section made a five-week, four thousand-mile survey flight in the spring of 1920 to gather data in Indiana, Kentucky, Illinois, Missouri, Kansas, Iowa, Wisconsin, Ohio, New York, Massachusetts, and Connecticut.[7]

General Patrick instructed several Air Service stations to investigate sections of a nationwide system of airways.[8] Capt. Lowell H. Smith and Sgt. William B. Whitefield surveyed Idaho, Wyoming, Nevada, Utah, Oregon, and Washington. First Lieutenant Delmar H. Dunton investigated the route from Kelly Field, Texas, to Scott Field, Illinois. Pilots of the 22d Observation Squadron's detachment at Pope Field, North Carolina, mapped routes to Savannah and Macon, Georgia, and Louisville, Kentucky. So it went in order that all parts of the country might be covered. By mid-1925 the Air Service possessed information on nearly thirty-five hundred landing places, including more than twenty-eight hundred emergency landing areas, in the United States.[9]

On March 1, 1923, the Air Service started publishing a pamphlet, *Airways and Landing Facilities,* describing all known landing fields. Later that year, the Air Service began issuing information on fields in looseleaf form for easy revision. Using Air Service data, the Corps of Engineers prepared and printed a map of the United States showing fields and proposed airways.[10]

The Air Service sent flights from fifteen stations in 1923 to scan segments of airways and depict routes in both directions. It published this information in bulletins, the first issued on September 15, 1923, for the route from San Francisco to Sacramento. Further, the Air Service asked the

Geological Survey and Corps of Engineers to prepare strip maps, each covering a section about 200 or 220 miles long and 80 miles wide on a scale of 1:500,000. Copies of these publications went to the Navy and Post Office Department, and to the U.S. Superintendent of Documents for sale to the public.[11]

The Air Service meanwhile proceeded to build part of the proposed nationwide airways as a model for the whole system. It extended the original airway by adding in 1922 routes north and south from Washington, D.C. to Mitchel Field and Langley Field; from Moundsville, West Virginia, to Selfridge Field; and from Dayton to Scott Field. And in 1923 it added a southern division from Scott to Kelly Field by way of Kansas City, Missouri; Muskogee, Oklahoma; and Dallas.[12] (*Map 5*)

The Air Service wanted towns and landing fields along the airways uniformly marked. The service expected the various communities to do the work, but it supplied information and commenced marking the model airway to test the system and furnish an example. Painting crews marked 106 towns between Dayton and Washington in the summer of 1922, then ran out of money to go on. Reserve and National Guard units as well as the Regular Air Service had scant success in getting chambers of commerce, civic clubs, Boy Scouts, and other organizations to take up the job.[13] Renewing the marking effort in 1925, the Air Service secured the support of the Standard Oil Company of Indiana. On December 17, 1925, the twenty-second anniversary of the Wright Brothers' flight at Kitty Hawk, the company started a program of marking airways in its ten-state area of operation. Other oil companies soon joined in. Standard Oil of California, for example, marked 110 of its storage stations between San Diego and Seattle by early 1926, and planned to mark all 650 stations.[14]

The Air Service had begun scheduled flights over the model airway in 1922 to haul passengers and packages. At first it used DH–4Bs, but later the depot at Fairfield, Ohio, tailored DHs to this work. Designated DH–4B4,the airways plane had larger gas and oil tanks as well as the latest devices for the aid and comfort of the pilot. The rear cockpit became a cargo compartment with a streamlined cover, but with a collapsible seat so a passenger could be carried. The Air Service reported that in 4 years it flew over 1.2 million miles on 671 scheduled flights transporting more than 1,200 passengers and 62,000 pounds of freight.[15] The control station for the model airway was at Bolling Field until 1925, when it moved to a more central location at McCook Field, Ohio. Soon afterwards it shifted to Wilbur Wright Field, Ohio. Control officers at the stations made reservations for passengers and kept records of planes in transit. Airborne radio was not regularly used in airways operations during this period, but radio linked stations on the model airway.[16]

The operations office at Mitchel Field became a model others might copy. The building afforded quarters for the officer and five enlisted men who

MAP 5

ran the office as well as space for the flight surgeon, lounging facilities, reading material, lockers, toilets and showers, a bulletin board with weather and flight schedules, and a map of the eastern division of the model airway. A map room contained a complete file of maps, and photographs of landing fields lined the wall. At each end of the field stood a battery of six floodlights controlled from the operations office. The wind vane on the field was lighted. From an observation platform atop the operations building, one could see the entire flight line and much of the surrounding country. On top of the building a beacon light flashed "M–I" in Morse code.[17]

To make a long flight over the model airway or elsewhere, a pilot requested permission from the Office of the Chief of Air Service. Approval usually carried restrictions. For example, the pilot might be told not to stop any place for more than two days unless delayed by mechanical trouble or bad weather, or he should be back within a certain time. At times he was given use of an airplane but not granted per diem. He was always required to send a full report of the trip to Washington for review by the Airways Section for information concerning landing fields and routes.[18]

Flight and Navigational Instruments

A pilot flying over familiar territory with known landmarks in sight needed just a few instruments—a tachometer and gauges for gas, oil pressure, and water temperature—to inform him of the engine's performance. If he did not know the country or could not see the ground because of fog, clouds, or darkness, he had to have instruments to tell him the plane's attitude, heading, and altitude. Magnetic compasses and aneroid barometers had been adapted for aircraft, and simple bank and incline indicators were available at the end of the war. Pilots, however, put little faith in such instruments. The Engineering Division of the Air Service always had projects for driftmeters, compasses, airspeed indicators, altimeters, flight indicators, sextants, and other instruments to provide better aids to navigation. The man in charge of this area in the early 1920s was 1st Lt. Albert F. Hegenberger, "one of the foremost exponents of the use of navigational instruments in flying."[19] Bradley Jones, America's leading authority on aerial navigation, worked with him at McCook Field on an earth-inductor compass.[20]

Magnetic compasses were influenced by the aircraft's engine, wiring, and other metal parts. They required adjustment to the plane. Even if properly set, a magnetic compass could not be depended upon, too often starting to spin just when needed most. Interested in the new earth-inductor compasses which the Bureau of Standards and the Pioneer Instrument Company were developing in 1923, the Air Service bought some for testing.[21]

Sensor for Earth Inductor Compass mounted in a DH–4 aircraft.

Lt. Albert F. Hegenberger, an advocate for the use of navigational instruments in flying.

Hegenberger and Jones made many flights to try out the earth-inductor compass and other instruments. They went to Boston, for instance, on September 6, 1923, in a DH with large gas and oil tanks. The rear cockpit was fitted for Jones to make sextant observations. It had a driftmeter in the floor, a magnetic compass, and the dial of the inductor compass which Jones turned to set the course. Designed by Hegenberger, the front cockpit's instrument board featured vertical scales for easy reading. The instruments included a flight indicator, an airspeed meter, and a dial pointing to zero so long as Hegenberger stayed on the course set by Jones on the indicator compass, but moving one way or another if the plane swerved to left or right.[22]

When Hegenberger and Jones departed McCook Field for Boston, the large patches of clouds grew as the plane bore eastward. Using the driftmeter to measure drift to the north from a crosswind, Jones rectified the compass setting. About 1040 they saw the Ohio State University stadium at Columbus. The clouds now quite dense, the men caught only a glimpse of the Ohio River. The cloud layer reached from 300 to 7,000 feet with more clouds above 9,000 feet. Hegenberger climbed through to 10,000 feet and crossed Pennsylvania without seeing the ground. Thinking he must be close to the Hudson River, he descended. At Hartford, Connecticut, the men got the first check of the course in 400 miles. Boston lay just ahead.[23]

Hegenberger soon afterwards transferred to Hawaii. In January 1924, Jones went to Langley Field to instruct men in navigation who would soon begin an around-the-world flight in planes carrying both magnetic and earth-inductor compasses. Work on instruments progressed at McCook Field. On one trip to New York, 1st Lt. Eugene H. Barksdale and Bradley Jones flew above clouds and on instruments most of the way. With a 50-mile tailwind they covered the 575 miles in a record 3 hours and 45 minutes.[24] Another notable flight went from McCook Field to Albany, New York. Neither the pilot, 1st Lt. Hugh C. Downey, nor Bradley Jones had been over the route before. Their maps, taken along for an emergency, were sealed. They made most of the trip above clouds, at 10,000 feet. Despite a strong crosswind they came down within 5 miles of their destination.[25]

Interested in radio signals as a navigation aid, the Engineering Division commenced testing an interlock system early in 1924.[26] Tuned in on a radio beacon, a pilot heard a dot-dash signal (a Morse code "A") if he was to the right of the course, a dash-dot ("N") if to the left, and a continuous sound if on course. Tests showed this entailed considerable concentration and possible error in interpreting the signal. The engineers ran the signals to lights on the instrument board—white for on course, green to the right, and red to the left. Pilots employing signals from Wilbur Wright Field came home on the beacon from as far away as two hundred miles. The tests extended into 1926, when the Air Service was helping the Air Mail Service establish and test a radio beacon at Monmouth, Illinois.[27]

At midpoint in the 1920s, Air Service pilots still did nearly all their flying by the landmark method. A magnetic compass, an altimeter, and a flight indicator were the instruments most commonly used for navigation on cross-country or night flying.[28]

Night Flying

At the end of the First World War, few Air Service stations had searchlights or other electrical equipment to light landing areas. Where available, the lighting was applied chiefly to aid pilots suddenly overtaken by darkness. At other stations, bonfires, flares, gasoline poured on the ground and ignited, burned cans of gasoline-soaked waste, or a combination of these helped a pilot find the field to land. If a pilot flew at night (which was seldom), he liked to pick a moonlit one. He might take along parachute flares and put wingtip flares on the plane. Though especially useful for an emergency landing, flares were often faulty. Wingtip flares were likewise hazardous, sometimes igniting grass, weeds, or brush on the field or setting the plane on fire if it crashed.[29]

The Air Service during the early 1920s made good progress in night-flying equipment. First Lieutenant Donald L. Bruner and his men at McCook Field accounted for numerous improvements. Powerful landing lights were developed in streamlined cases for installation on airplane wings. A maroon, nonglare finish cut reflection of light from a propeller. The glare caused by the flames from Liberty engines was reduced by extending the exhaust pipe past the rear cockpit. Streamlined housings for running lights were devised. Instrument dials were painted with luminous paint, and instrument panels were lighted. With the Ordnance Department, more powerful and more reliable parachute flares were developed. Moreover, airfield lighting was improved with electrical boundary lights, obstacle lights, beacons, and searchlights.[30]

The first night flight over the model airway took place without the benefit of the new equipment. First Lieutenant Clayton L. Bissell flew from Washington to Dayton on the night of August 5/6, 1922, and returned the next night. He thereupon questioned the need for putting a lot of expensive equipment on planes just for night work. By 1923, however, the Air Service was installing the new electrical devices on its aircraft and landing fields.[31]

Lieutenant Bruner formulated an airways plan giving emergency fields and terminal fields boundary and obstacle lights. Each field would also have an illuminated wind indicator and a horizontal, rotating, flashing light. He placed the beacons so a pilot flying the airway could always see at least one, and most of the time two. The Air Service let him install equipment and test his plan in 1923 on the section of the model airway between Dayton and Columbus, Ohio.[32]

Earlier that year the Air Service granted night flying a regular place in training and operations. It launched a project to put night equipment on eighty DH–4Bs, sixty-five NBS–1s, and twelve MB–3As. The depot at Fairfield, Ohio, shipped the gear to the fields with installation instructions. By October 1924 the 2d Bombardment Group at Langley Field had equipped its planes and gained enough experience for a night flight to New York. An Air Service press release on the flight called attention to the two electric landing lights on the wings, and to the running lights (red and green on wingtips and white on the tail). The story noted each bomber carried four parachute flares and four wingtip flares. At Aberdeen Proving Ground the Air Service experimented in night bombing. The Advanced Flying School made night flying part of its course. The Hawaiian Department added night missions to annual maneuvers. Thus the Air Service injected a new dimension into its operations.[33]

Communications

When the Air Service founded the border patrol in 1919, General Menoher informed the Chief Signal Officer the Air Service would require two-way, air-to-ground radiotelephony. It would in addition need air-to-ground radiotelegraphy for planes operating beyond the range of radiotelephone communication, and in certain instances want radiotelephony for plane-to-plane communication. The aviation depot at San Antonio would furnish some equipment, but General Menoher asked the Signal Corps to supply other items, handle the installation, and operate ground stations as called for by directives governing radio work in the Army.[34]

Equipment existed for the types of communication mentioned by General Menoher. But most of it was not up to par, and Air Service personnel were without training and experience in its use. Then, too, the Signal Corps had few men it could detail to install and maintain radios on airplanes or operate ground stations. Radio therefore saw little use in border patrol.[35]

A handwritten note dropped in a tube or pouch attached to a parachute, or with just a streamer to attract attention, had proven during the war to be more reliable than radio for sending a message from an airplane to the ground. Airmen working with ground forces gave prearranged signals with flares or aerial maneuvers. Ground troops displayed panels in sundry patterns or resorted to flares, rockets, smoke, or lights to convey prearranged messages to airplanes. All of these methods persisted through the first half of the 1920s.[36]

Seeking better communications, the 12th Squadron at Nogales, Arizona, experimented with wigwag from ground to air. Aerial observers in Hawaii tried wigwag to communicate between planes on interisland hops. Troops in the Panama Canal Zone used shutter panels. Several units experimented with broadcasting messages from planes by Magnavox Telemegaphone. Pilots on cross-country flights on the Mexican border or in the Canal Zone often carried pigeons to release if the plane went down in uninhabited country.[37]

As soon as it became apparent the Signal Corps would offer little help with radios for border or forest patrol, General Menoher sought control over all aviation radio activities. Early in 1920 the War Department shifted to the Air Service responsibility for installing, maintaining, and operating radio apparatus of its units and stations. It did not allow the Air Service to take over radio equipment development, but the Signal Corps set up an Aircraft Radio Laboratory at McCook Field for closer coordination between the two services.[38]

With a lift from the Signal Corps while it trained its own men, the Air Service improved radio communication for forest patrol in 1920 and 1921.[39] The bombing tests of 1921 applied radio to control aircraft within formations, direct operations from the flagship USS *Shawmut*, and report forced landings. Bombing planes of the 1st Provisional Air Brigade carried voice sets

Radio SCR–134 installed in an XO–25A.

to talk between planes in formation. Each formation's control plane also had a spark set of longer range to communicate with stations at Langley Field, the naval base at Norfolk, the *Shawmut*, and destroyers along the line of flight to and from the target.[40] Radio came to be standard equipment on NBS–1s. Gradually observation units adopted two-way radio for adjusting artillery fire. Experience with radios, together with modifications of apparatus and procedures, steadily enhanced communications.[41]

The Signal Corps worked on new radios with superheterodyne receivers for pursuit, observation, and bombardment. The pursuit set (SCR–133) assured voice communication between planes to a distance of five miles. The observation and bombardment sets (SCR–134 and –135) enabled voice communication, the 134 reaching out thirty miles and the 135 somewhat farther. These two sets additionally offered both tone and continuous wave (CW) telegraph for greater distances. CW, having the longer range, could transmit only. The radioman operating the bombardment set kept in touch with the aircraft commander by interphone (SCR–155). The commander was usually on interphone talking with other members of the crew, but he could operate the radio himself if he wanted to. An interphone system (SCR–160) tied in with the observation set (SCR–134).[42]

Many problems remained with the 130-series of radios. Electrical interference caused reception trouble. Ignition systems needed shielding, amounting in some cases to complete rewiring of aircraft. Wooden planes had to be metallized by adding wire and metal strips to wings and fuselage for adequate grounding. All metal required bonding to prevent absorption of radiated energy, eliminate danger of sparks between metal parts, and reduce receiver noise. These steps demanded vast experimentation. Field tests starting in the spring of 1925 revealed other problems. In the end the results proved disappointing.[43]

Weather Service

The swift growth of cross-country flying after the war engendered fresh requirements for weather information. The Signal Corps, responsible for Army meteorological work, could not organize stations fast enough or come up with the kind of service to satisfy the Air Service. Enlisted observers at flying fields gathered information useful for local flying. The information also went to Washington for use by the U.S. Weather Bureau in preparing forecasts for the Army and Navy. These forecasts, telegraphed to flying fields, aided in planning cross-country flights. So did the special aviation forecasts regularly written by the bureau for broadcast by the Navy's powerful radio station at Arlington, Virginia.[44]

While the support of the Signal Corps and Weather Bureau was helpful, it did not meet Air Service requirements. The service, General Menoher said, had to have a weather unit of its own, one operated by people familiar with flying needs. Still, neither General Menoher nor General Patrick succeeded in securing War Department approval for a meteorological section in the Air Service.[45]

Unable to get much help from the Signal Corps in a period when funds were short for nearly all activities, Patrick in 1922 obtained agreement with the Weather Bureau for cooperation. Air Service pilots on training flights visited Weather Bureau stations within three hundred miles of their home fields, became acquainted with the meteorologists, learned of prevailing weather conditions, and discussed problems of the two services. Weather Bureau personnel accepted invitations to lecture at Air Service stations on the bureau's work and area weather. Air Service stations could receive forecasts from the bureau by paying for a telephone call or telegram. If a pilot landing away from his home station requested a forecast by telegram, the bureau paid for the reply.[46]

By 1925, the Signal Corps had created new weather stations and every Air Service flying field had a Signal Corps weather detachment. Under the supervision of an Air Service officer the detachment made observations, prepared and transmitted reports, received reports from the Weather Bureau and other weather stations, posted weather data for flyers, and furnished special reports for cross-country flights.[47]

At the beginning of 1924, the Engineering Division at McCook Field, Ohio, instituted a Meteorological Branch headed by a Signal Corps officer, Maj. William R. Blair, formerly chief meteorologist for the American Expeditionary Force. The branch concerned itself with better ways of collecting data on upper air currents, finding means for determining the height and depth of clouds and, among other things, ascertaining the chances of an airplane or balloon being struck by lightning.[48]

The Air Service also assisted the U.S. Weather Bureau in research. The Balloon and Airship School scheduled fifteen balloon flights from Scott Field, Illinois, for Dr. C. LeRoy Meisinger, who had gained experience with balloons and meteorology as an Air Service officer during the war. Working for the Weather Bureau in 1924, he studied the path of air currents, the amount of dust in the air, sky brightness, and the size of cloud droplets. The project ended with the tenth flight, on June 2, 1924, when lightning struck the balloon, killing both Dr. Meisinger and his pilot, 1st Lt. James M. G. T. Neely.[49]

One of the projects the Air Service took an interest in was for dispersing fog. General Patrick furnished Air Service planes and pilots to work with Dr. Francis Warren of Harvard University, who wished to experiment with dropping electrically charged sand on clouds. Tests at Phillips Field,

Maryland, and later at Bolling Field produced what Dr. Warren called "uncanny manifestations" and convinced him his theories were correct. He could do better with a plane that could carry more sand than the DH he was using. Two large planes, he believed, could disperse fog covering an area the size of the city and harbor of New York. So experiments proceeded on a project destined to be around in one form or another for a long time to come.[50]

Parachutes

Members of the U.S. Air Service serving as aerial observers during World War I had parachutes to escape from captive balloons when the enemy attacked. Pilots and observers in airplanes carried none. After the Armistice, the Equipment Section of the Engineering Division at McCook Field pursued work begun during the war on parachutes for use with airplanes. Maj. Edward L. Hoffman headed the section; James Floyd Smith was his assistant for parachute development.[51]

For balloons the parachute attached to the balloon basket and pulled out by a static line when the person jumped. Attempts to adapt this type of chute for airplanes were not very successful. A number of people, including Smith and another civilian at McCook Field, Leslie L. Irvin,[52] were fashioning chutes to be carried and operated by the jumper. On April 28, 1919, Irvin tested one developed by Smith. Leaving the plane at 1,500 feet, Irvin descended safely but broke an ankle on landing. After further tests and modifications of Smith's chute, the Air Service ordered four hundred on June 25, 1919, from a company Irvin had formed while recovering.[53]

When the new parachutes (U.S. Airplane Type A) were distributed in 1920, the Air Service added a parachute course at the Mechanics School. It next issued instructions requiring chutes to be tested and used under the supervision of course graduates. Persons who wanted to jump should be given the opportunity, but jumping was to be entirely voluntary. Two chutes, at least one being of the free-fall, manually operated type, were specified for all except emergency jumps. Moreover, none but emergency jumps were to be made from lower than 1,500 feet.[54]

Kelly Field, Texas, recorded forty jumps in the two months after the course started. Besides a backpack, each jumper carried a reserve chest chute.[55] Among officers completing parachute training was lst Lt. Harrison G. Crocker. Upon returning to his squadron at Laredo, he found his companions eager to learn to jump. About the same time, 1st Lt. Cyrus Bettis demonstrated the use of the parachute at El Paso. At Carlstrom Field, Florida, parachute "hopping" appealed to "hardy souls" who had become so "blasé with flying" they were looking for new thrills.[56] At Post Field, Oklahoma, Sgt. Encil Chambers jumped from 5,000 feet, opened his chute at 4,000, cut loose at 3,000, dropped another 500 feet, opened his second chute, and executed a "neat" landing in the middle of the field. SSgt. Gilbert A. Shoemaker, chief instructor in the parachute course, stood on the upper wing of a DH–4B, pulled the ripcord, and was pulled off by the open chute as a Pathe News cameraman took pictures from another plane.[57]

Competition to see who would jump from the greatest height quickly pushed the record up to the point where the men needed planes with oxygen equipment and superchargers. Capt. Albert W. Stevens carried an oxygen flask when he jumped from 24,000 feet after conducting photographic experiments at Dayton, Ohio, on June 12, 1922. "The parachute," he said, "was simply another way of getting down after the real work was done." The chute opened with a jerk. The 11-pound oxygen bottle strapped to Stevens' leg broke loose. Stevens grabbed it, tucked it under a shoulder strap, and took a few puffs from the tube. It soon slipped out and disappeared, but he descended so rapidly there was no further need for it. In a strong wind the

Parachute lift-off from Martin bomber at McCook Field, Dayton, Ohio.

chute rocked and tossed so much Stevens became "miserably seasick." In landing he broke bones in his foot.[58]

The safety record for parachute jumping was good, and the Air Service wanted to keep it that way. After Pvt. Earl W. Moon drowned when he landed in Chesapeake Bay during a demonstration at Aberdeen, Maryland, in October 1920, the Air Service changed regulations to require a jumper to wear a life preserver if in danger of landing in water. At the same time, it tightened directives to make sure jumpers had proper equipment and instruction.[59] Later, General Patrick required approval of his office for training, test, or exhibition jumps entailing unusual risks. Taking note soon after of jumps with delayed releases to test personnel and equipment, Patrick ordered that approval be secured from him before making any jumps not in the normal course of training.[60]

Parachutes purchased for training and demonstration were unsuited for general use. By early 1920, test pilots of McCook Field adopted seatpacks. Lap packs were developed for gunners and photographers. Some time passed, however, before parachutes were regularly used on Air Service flights.

On April 10, 1921, the *Chicago Tribune* stated flyers generally did not wear parachutes. Thirty percent of the aviators killed in crashes could have been saved, the *Tribune* said, if they had worn them. Maj. Follett Bradley took issue. A pilot should try to save his plane. If he did not, he was "guilty of misconduct." To require a pilot to wear a parachute and encourage him to employ it would foster "faintheartedness" and lead to many crashes that could have been safe landings. Bradley thought there were just three conditions—fire, collision, and collapse of the plane or loss of an essential part like a wing—under which a pilot would be justified in taking to his chute. Major Hoffman thought there were other occasions, such as engine failure at night over rough country in a plane without landing lights. He said pilots at McCook Field considered parachutes indispensable. They did not suffer from the faintheartedness predicted by Bradley.[61]

First Lieutenant Harold R. Harris became the first member of the Air Service to save his life by parachute. A terrible vibration arose in the Loening PW–2A he was testing at McCook Field on October 20, 1922. Seeing part of the left wing or aileron come off, Harris opened his belt, left the cockpit, and descended safely. Talking of Harris' jump, two Dayton newsmen, Morris Hutton and Verne Timmerman, and an employee of the Engineering Division, M. H. St. Clair, came up with an idea for a club composed of people saved by parachute when forced to leave aircraft in flight. The Caterpillar Club thus came into being as a "membership" list compiled by the Parachute Unit at McCook Field.[62] Another Air Service pilot, 1st Lt. Frank B. Tyndall, jumped when the wings came off the MB–3A he was testing at Seattle on November 11, 1922. Two pilots having been saved within a period of three weeks, the Air Service issued orders requiring pilots and passengers in Army aircraft to have parachutes for all flights.[63] It would be some time, however, before enough parachutes would be available to carry out the regulation at all stations.[64]

No Air Service man joined the Caterpillar Club in 1923. Eighteen qualified between April 1924 and the end of June 1926. Included were Horace M. Hickam and Harold Geiger (whose planes collided), John A. Macready, Frank O'D. (Monk) Hunter, and Eugene H. Barksdale. The first person to be saved by parachuting twice from an airplane was Charles A. Lindbergh, once as a cadet and again as a Reserve officer. By mid-1926 two other Air Service men, Hunter and Barksdale, had also jumped twice.[65]

The requirement for carrying parachutes, improvements in airplane design and construction, better equipment and facilities, and stress on flying safety reduced the fatalities from airplane crashes. Fifty-one men died in 1920, 57 in 1923, 34 in 1924, and 128 in 1925. Flying hours per fatality rose annually, from 1,018 hours in 1921 to 4,063 in 1925. Miles flown per fatality climbed from 73,631 to 297,375 during the same period. The Air Service investigated each accident to find the cause. It attributed about half of the fatal accidents to errors of judgment or mistakes of pilots, with engine failures the next most frequent cause. Only a few accidents were due to structural failure or weather conditions.[66]

The U.S. Army Air Service contributed much to flying safety by pioneering the development and use of free-fall parachutes. Its work with radio for air-to-air and air-ground communication moved slowly, that on instruments and aerial navigation somewhat faster. A significant project was the building and operation of an airway as a model for a nationwide system. The Air Service drew plans and specifications for landing fields, devised a standard scheme for marking fields and routes, furnished information and advice, and encouraged towns and cities to create and operate airports. The service worked with the Post Office Department's Air Mail Service and with individuals, organizations, and state and local governments. It meshed its efforts with the Army's Signal Corps on radios, the Signal Corps and U.S. Weather Bureau on meteorological services, the Bureau of Standards on instruments, and with industry on equipment of all kinds. Besides, help for the airways project came from farmers, businessmen, Boy Scouts, Reservists, and others. Through this cooperation, military and civil aviation advanced side by side during the 1920s.

Chapter XI

Higher, Faster, Farther, and Longer

Whatever the performance of an airplane, it seldom satisfied the airmen. They wanted to fly higher, faster, farther, and longer, this being as true of U.S. Army flyers as of others. Those in the Army needed altitude to reduce the chance of detection and to lessen vulnerability to ground fire, speed to engage and defeat an enemy in aerial combat, distance and duration to spy on the enemy from the air and to attack him far within his own territory. Moreover, the Air Service required these things to permit rapid military moves of aviation units. So it devoted much effort to achieve more altitude, speed, range, and endurance. Encouraging the competitive spirit of its members, the Air Service welcomed competition between Army and Navy flyers, military and civilians, Americans and foreigners. For competition stimulated interest in aviation, aided the development and improvement of aircraft, and afforded opportunities to test and evaluate equipment. Besides, the Air Service liked to see Army names on the official list of aviation records certified by the Federation Aeronautique Internationale (FAI).

Altitude Records

Maj. Rudolph W. Schroeder was one Air Service officer who got his name on the FAI's roster of record holders. He was chief test pilot of the Air Service's Engineering Division at McCook Field, which in 1919 was trying to solve problems of high-altitude flight. Oxygen systems used during the war to assist breathing in the thinner air at higher levels needed further development and improvement. In addition, the lighter air at higher altitudes cut the efficiency of an aircraft's internal combustion engine. So some means had to be found to supply sufficient air for mixing with the gasoline in the carburetor. Attacking the latter problem, Dr. Sanford A. Moss of the General Electric Company had devised a supercharger. Exhaust gases drove a turbine that compressed air for the carburetor. On September 6, 1919, Major Schroeder tried out the Moss turbo-supercharger on a Le Pere-Liberty 400 at McCook Field. With Lt. George A. Elfrey as passenger, he took the Le Pere

Left: Maj. Rudolph W. Schroeder; below: Le Pere plane used to set world altitude record on Feb. 27, 1920, with Schroeder at far right.

to 28,500 feet, setting a two-man altitude record. Continuing the test, Schroeder and Elfrey went to 30,900 feet on September 24 and to 31,821 feet on October 4.[1]

On February 27, 1920, Major Schroeder flew the Le Pere alone to see how high he could go and to explore the trade winds (later called the jet stream) known to exist at upper altitudes. He had the warmest flying clothing obtainable, an automatic oxygen system good for three hours, a reserve tank of oxygen with manual control, and special gasoline developed by Thomas Midgley, Jr., who was working on antiknock fuel for the Dayton Wright Airplane Company.

At 18,000 feet the automatic oxygen system failed. Schroeder turned on the reserve and noted the temperature was 67 degrees below zero, Fahrenheit. An inch-thick coat of ice (condensation from the exhaust) covered the plane's center section. The exhaust poured carbon monoxide over his head. At 33,000 feet he wondered what the ceiling would be. The Liberty engine was working fine with the help of the supercharger and there was still enough gas for an hour and a half. He pushed on, recording the temperature every 50 feet and paying particular attention to the wind, so strong it pushed the plane backwards. Suddenly the oxygen ran out. The reserve was gone and the automatic system still did not work. Gasping for breath, Schroeder tore off his mask and goggles. The carbon monoxide hit him. He managed to put the machine into a dive and switch off the motor before he passed out. The plane fell 5 miles with Schroeder unconscious. Reviving just in time, he straightened out over Dayton and glided to a landing. The men at McCook Field found Schroeder slouched in the cockpit, covered with ice, limp and helpless, his head drooped to one side, his eyes frozen wide open.

Schroeder had carried two barographs to record altitude. When his instruments were calibrated and calculations made, he received credit for 33,143 feet, more than enough to capture the world record. Even so, he spent time in the post hospital recovering from the effects of lack of oxygen, carbon monoxide poisoning, rapid descent, and frozen eyelids.[2]

Why did Schroeder do it? What did he hope to accomplish? Many people were asking questions. Newspapers commented on the uselessness of such a venture. One editor called it a "suicidal altitude flight." The Air Service tried to explain. In the future, antiaircraft fire would become more and more deadly, forcing military aircraft higher and higher. Whatever the Air Service could learn about altitude flying would benefit commercial as well as military aviation.[3]

First Lieutenant John A. Macready assumed the altitude task after Major Schroeder left the service in late 1920. Testing superchargers, propellers, and other equipment, he sought several times to shatter Schroeder's record. He finally succeeded on September 28, 1921, with an altitude of

First Lieutenant John A. Macready stands in front of plane after beating Schroeder's high-altitude record on September 28, 1921.

34,508 feet. However, a Frenchman, Sadi-Lecointe, reached 36,555 feet in 1923. Though Macready could not beat that, he set a new American record of 35,900 feet on January 29, 1926, six months before he resigned from the service.[4]

Races

Before Major Schroeder left the Army to join Aviation Engineer Underwriters Laboratories, he represented the U.S. Air Service in the James Gordon Bennett Airplane Race in France in September 1920. He flew a racer, the VCP–R, designed by Alfred V. Verville of the Air Service Engineering Division, and built at McCook Field. In the race the 12-cylinder, 600-horsepower Packard engine, the most powerful yet made in America, overheated and forced Schroeder to drop out. Sadi-Lecointe won with an average speed of 168.26 miles per hour for three laps over a 50-kilometer, straight-line course.[5]

On Thanksgiving Day 1920, 1st Lt. Corliss C. Moseley flew the VCP–R (with a larger radiator) in the first Pulitzer Race, held at Mitchel Field. He won but was disappointed because he averaged just 156.54 miles per hour. After several trials with the VCP–R to break the world speed record of 192 miles per hour, he could do no better than 186.[6]

The Harding administration's campaign for economy prevented the Army and Navy from entering the second Pulitzer Race at Omaha in 1921. Nevertheless, General Patrick gave Lieutenant Macready leave to fly a racer (MB–6) built (but not yet delivered) by the Thomas-Morse Aircraft Corporation for the Air Service. Macready came in third, the winner being Bertram B. Acosta, a well-known Curtiss test pilot. Acosta had previously taught a number of men how to fly including George C. Kenney, prior to his joining the Air Service.

Though there was no letup in the drive for economy, both the Army and the Navy found money for the National Air Races in October 1922. The meet that year was held at Selfridge Field under the auspices of the Detroit Aviation Society. A change in Air Service policy permitted Army flyers to accept cash prizes.[7] This meet, like others in the 1920s, offered a variety of events to attract participants from all branches of aviation, and draw a large crowd for the show. Seven planes took part in a race, "On to Detroit," calculated to stimulate civilian aviation. Flying to Selfridge from several parts of the country during the week of October 8, contestants received points based on the horsepower of each plane, elapsed time, miles flown, and average speed. Walter H. Beech of the Laird Airplane Company, Wichita, Kansas, won with a Laird Swallow.

Navy and Marine Corps pilots competed for the Curtiss Marine Flying Trophy on Sunday, October 8. The next race, held the following Tuesday, was for the *Detroit News* Aerial Mail Trophy, created by William E. Scripps to inspire performance in the postal service. The rules confined the contest to multimotored planes carrying payloads of eight hundred pounds or more. None of the Post Office Department's planes at Detroit could qualify. The Air Service entered five Martins—four bombers and a transport. First Lieutenant Erik H. Nelson won the first prize of $1,200 with the transport.

On Friday the Aviation Country Club of Detroit tendered a trophy and $2,000 in prizes for a speed contest open to airplanes seating three or more passengers, including a pilot. Four planes entered and three completed the ten laps around the twenty-five-mile course. First Lieutenant Harold R. Harris of the Air Service won in a redesigned DH–4–400 (Honeymoon Express). Another Air Service pilot, 1st Lt. Robert S. Worthington, in a Fokker T–2, tied for second place with Charles S. Jones of the Curtiss Aeroplane and Motor Company flying a Curtiss Oriole. The same day, 1st Lt.

Theodore J. Koenig won the Liberty Engine Builders' Trophy presented by the Packard Motor Car Company in a race for military personnel in biplace observation airplanes. The entries, all from the Air Service, were six DH–4Bs, two XB–1As, and Koenig's Le Pere.

On Saturday, six pilots of the 1st Pursuit Group competed for the John L. Mitchell Trophy, proffered by Brig. Gen. William Mitchell in honor of his brother who was killed in France. The race went four laps over a triangular, fifty-kilometer course based on Selfridge, Gaukler Point, and the USS *Dubuque* anchored in Lake St. Clair. First Lieutenant Donald F. Stace won with an average speed of 147.8 miles per hour.

Then came the main event, the Pulitzer Race. Months earlier the Air Service asked various airplane manufacturers to design new pursuit planes. The sole restrictions were that the planes be for military work and fly more than 190 miles per hour, some 30 miles per hour faster than any service plane at the time. The Navy asked several builders to develop fast maneuverable planes for duty on ship or shore. The Navy put 4 planes in the Pulitzer Race. The Army Air Service entered 10: 3 Verville-Sperry R–3s, 2 Loening R–4s, 2 Thomas-Morse R–5s, 2 Curtiss R–6s, and an old R–1, stream-lined, refined, and redesignated VCP–1.

The Air Service's R–6s came in first and second, 1st Lt. Russell L. Maughan with an average speed of 205.8 miles per hour, and 1st Lieutenant Lester J. Maitland with 198.8. The Navy's year-old Curtiss racers took third and fourth places. First Lieutenant Eugene H. Barksdale brought his R–3 in fifth, Lieutenant Moseley came sixth in the VCP–1.[8] The tremendous speeds and the daring way Maughan and Maitland turned the pylons made it a thrilling race. Maughan told reporters he "was stunned more or less at the fifteen turns." Once turning Gaukler Point he became unconscious for three or four seconds. Maitland reported similar sensations.[9]

Evaluation of the new planes went on at Selfridge Field the ensuing week. After the Curtiss R–6's showing in the Pulitzer Race there seemed slight doubt the Air Service would adopt the plane as its standard pursuit ship. What would the Curtiss racer do in straight flight? Sadi-Lecointe recently flew 212.035 miles per hour on a 1-kilometer course. Lieutenant Maughan broke the record with his racer on a 1-kilometer course at Selfridge Field on Monday, October 16, when he averaged 229 miles per hour for 8 laps, 232.22 for 4, and 248.5 on 1. The record he set did not become official, however, because no representative of the FAI observed the flight and reported the results.[10] General Mitchell arranged for an official observer when he tested the two R–2s 2 days later on the 1-kilometer straightway at Selfridge. The official speed would be the average of the times for 4 passes, 2 each way over the course. Using Maitland's plane, Mitchell averaged 211.34. Shifting to Maughan's, he averaged 219.78 on the first try and 222.969 on the second.[11]

First Lieutenant Russell L. Maughan (left), winner of the Pulitzer Trophy at the 1922 Detroit Aviation Meet; and 1st Lt. Lester J. Maitland, second prize winner at the Pulitzer Race.

Lecointe broke Mitchell's record with a speed of 233.013 miles per hour on February 15, 1923. The Air Service made more speed trials with the R–6 over a one-kilometer course at Fairfield with officials present to certify the results. On March 26, 1923, Maughan averaged 236.587 miles per hour. On the 29th, Maitland averaged 239.95, but his record was disallowed because he did not maintain level fight over the entire course.[12]

For the Pulitzer Race at St. Louis in 1923, the Army used old racing planes (R–3s and –6s) with different pilots (General Patrick wanted to give more men a chance).[13] Two Navy flyers, Lts. Alford (Al) J. Williams and Harold J. Brow, took first and second places.[14] Afterwards at Mitchel Field, the Navy's pilots attacked Lieutenant Maughan's record. Starting with a long dive to gain speed, Lieutenant Brow took the record on November 2, 1923, with a speed of 259.115 miles per hour. He lost the record two days later to Lieutenant Williams, who flew 266.583 miles per hour.[15]

The Navy, busy preparing for the Schneider hydroplane race,[16] did not participate in the Pulitzer contest during the International Air Races at Dayton, Ohio, in 1924. That left only the Army in the running. General Patrick designated Capt. Burt E. Skeel and 1st Lts. Alexander Pearson, Jr., and Wendell H. Brookley for the Pulitzer, with 1st Lt. Harry H. Mills as alternate. Later, 1st Lt. Rex K. Stoner was added. The Army acquired Lieutenant Brow's plane from the Navy for Pearson to fly. The wings collapsed and Pearson died while practicing a diving start a few days before the contest. On the day of the race, Captain Skeel crashed in view of the stands when the wings came off his R–6 during a steep dive to gain speed for the start. Brookley, in another R–6 close behind, dodged flying debris, circled for another start, and flew the designated four laps at an average speed of 214.41 miles per hour. Mills, unaware of what happened, came in first in an R–3, with a speed of 216.55 miles per hour. Stoner finished third in a PW–8A.[17]

In March 1925, a committee appointed by General Patrick to survey the Engineering Division recommended the Air Service give up racing. The committee, headed by Dexter S. Kimball, dean of the College of Engineering at Cornell University, thought racing had been a good way to attract attention and interest to aeronautics. Now, however, the Air Service was well enough established "to stand on its own real worth and dignity." The continued expenditure of large amounts of money for racing could no longer be justified, not to mention the outlay of "nervous energy" in preparations. The annual races, the committee thought, "may very well upset the whole year's work."[18] By the time Kimball submitted the committee's report, the Air Service and Navy already had plans for the Pulitzer and Schneider races in 1925. The services wanted to recapture for America the maximum speed record that a French officer, Bonnett, took from Al Williams on December 11, 1924, with a speed of 278.48 miles per hour. Cooperating, the Army and

Navy bought three R3C–1 racing planes from Curtiss.[19] Each service entered one in the sixth (and last) Pulitzer Race at Mitchel Field on October 12, 1925. The Army's pilot, 1st Lt. Cyrus Bettis, won by averaging 248.975 miles per hour.[20] Two weeks later, flying Bettis' plane fitted with pontoons, 1st Lt. James H. Doolittle won the Schneider race and set a seaplane record of 245.713 miles per hour.[21]

Each year the U.S. Air Service took part in the National Balloon Race, the winners becoming eligible for international competition in the annual James Gordon Bennett Balloon Race. Air Service balloonists won the national contest twice, in 1922 and in 1923. Maj. Oscar Westover (pilot) and 1st Lt. Carlton F. Bond (aide) won at Milwaukee in 1922 but had bad luck in the international race from Geneva, Switzerland, the following August. With the balloon over Hungary, sailing close to the ground in a favorable wind, peasants caught the dragrope, pulled the basket down, and turned the Americans over to the police. After offering a satisfactory explanation, Westover and Bond gathered up their equipment and returned to Geneva by

First Lieutenant James H. Doolittle, (left) winner of Schneider Cup Race, with 1st Lt. Cyrus Bettis, winner of Pulitzer Trophy Race, both events in October 1925.

train.[22] In 1923, 1st Lts. Robert S. Olmsted and John W. Shoptaw flew Army balloon S–6 to victory in the national contest held at Indianapolis. The weather being bad for the race from Brussels on September 23, 1923, some contestants dropped out, but Olmsted and Shoptaw were among those who started. Lightning hit the S–6 over Nistelrode, Holland. Olmsted was killed outright, and Shoptaw when the balloon fell.[23]

Distance and Duration

One of the historic flights of the early 1920s took Air Service planes from New York City to Nome, Alaska, and back. Brig. Gen. William Mitchell, Chief, Training and Operations Group, Office of the Director of Air Service, wanted to lay out an air route so if conditions dictated, air units could be moved to Asia by direct flight. In addition, he wanted to collect data for improvement of the DH–4B and to photograph unmapped areas of Alaska. Mitchell and his group set about planning late in 1919, and by spring 1920 had selected a route. General Menoher and Secretary Baker approved the flight the Air Service billed as "The Year's Greatest Aerial Event."

The expedition consisted of four planes and crews: Capt. St. Clair Streett, flight commander, and Sgt. Edmund Henriques, mechanic, in Plane No. 1; 1st Lt. Clifford C. Nutt, pilot and second in command, and 1st Lt. Erik H. Nelson, navigating and engineering officer, No. 2; 2d Lt. Clarence E. Crumrine, pilot and photographic officer, and Sgt. James D. Long, mechanic, No. 3; 2d Lt. Ross C. Kirkpatrick, pilot and information officer, and MSgt. Joseph E. English, mechanic, No. 4. Taking off from Mitchel Field at noon on Thursday, July 25, 1920, they set a course for Erie, Pennsylvania. (*Map 6*)

Running into fog and rain, the planes spread out to avoid collision. Captain Streett climbed to get above the clouds; the others pressed steadily ahead. At 9,000 feet Streett encountered hail. To save his propeller, he cut the motor and glided down. Coming out under the clouds and recognizing the country, he landed at Elmhurst near Scranton, Pennsylvania. Benny Troop's hayfield was not as smooth as it looked from the air. The result was one broken axle and the loss of a day in receiving a replacement from Mitchel Field. Then the truck bringing gas and oil stuck in the mud. So after laying over a second night, Streett caught up with the others at Erie on Saturday. Delayed several days by rain and mud, the flight finally got under way once again—only 4,000 more miles to Nome! The route led over Lake Erie, across Michigan, Wisconsin, Minnesota, North Dakota, Saskatchewan, Alberta, and British Columbia to Wrangell, Alaska, then northward across the Yukon to Fairbanks and Nome. The men flew through sunshine, fog, and clouds; over open plains, wooded hills, and glaciers; through mountain gorges and around

1920 Flig

1934 Flig

MAP 6

Brig. Gen. William Mitchell welcoming Capt. St. Clair Streett and his crew on return to Bolling Field from Nome, Alaska.

mountain peaks. Sergeant Henriques took the controls while Captain Streett went out on a wing to put out a fire when oil overflowed on a hot exhaust pipe. Henriques flew the plane again while Streett worked a handpump mile after mile to maintain a flow of gas when dirt got in the gas tank's pressure relief valve. Sergeant Long rode the tail of No. 3 to hold it down for a landing; when a tire blew, he was thrown off headfirst but was not seriously injured. Lacking a spare, Lieutenant Crumrine wrapped the rim with rope and tied on the tire casing. At one time or another, the men repaired broken wing skids and bent ailerons, fixed a tire cut by glass in landing on a field that had once been the town dump, persuaded a farmer to put a runway through his oat field, and recruited a cabinetmaker and a tailor to repair a broken wing. The crews shook hands with the many people who came out to greet them or see them off, lunched with Board of Trade members, dined with the City Club, and ate sandwiches and drank coffee brought to them as they prepared the planes for the next hop. So it went until 40 days, 4,502 miles, and 50 flying hours after leaving Mitchel Field they landed on an old parade ground at Fort Davis on the Nome River, the mission accomplished.

Leaving Nome on the last day of August, the men arrived at Mitchel Field on October 20. The next morning they flew to Washington to accept the congratulations of Generals Menoher, Mitchell, Pershing, and others. The Air Service's public relations staff compared this flying exhibition with the Navy's NC–4 hop in May 1919 over the Atlantic and found it ranked high. Captain Streett and his men had pulled off a feat surpassing even Capt. John Alcock's and Lt. Arthur W. Brown's nonstop Atlantic flight. The trip to Nome and back was "one of the most hazardous and stupendous aerial events yet attempted in any country."[24]

One of the great goals of the Air Service was to reduce the time for

Above: 1st Lt. Alexander Pearson, Jr. in front of his DH–4B which he pilots on Atlantic to Pacific flight; and 1st Lt. William D. Coney poses before his flight from the West to the East.

deploying Army aircraft from one part of the country to another. Accordingly, General Menoher announced in January 1921 that on Washington's Birthday the Air Service would seek to fly from Jacksonville, Florida, to San Diego (2,079 miles) in less than twenty-four hours. Actually, two men would try simultaneously: 1st Lt. Alexander Pearson, Jr., flying from east to west in three hops, stopping at Houston and El Paso; 1st Lt. William D. Coney going in the opposite direction, making just one stop, at Love Field near Dallas. The Chief of Air Service expected these flights to produce performance records that would advance both military and commercial aviation.[25]

Lieutenant Pearson, stationed at Douglas, Arizona, had engine trouble while on the way to Jacksonville in his DH–4. Forced down in the Big Bend area of the Rio Grande, he was missing nearly a week while making his way to the patrol post at Sanderson, Texas. So ended his half of the project.[26] Lieutenant Coney got away from Rockwell Field on schedule but motor trouble brought him down in Texas. He therefore took 2 days, 9 hours, and 24 minutes to cross the continent, but his flying time was a record 22 hours and 27 minutes. Believing he could fly coast to coast within 24 hours, he tried again, leaving Jacksonville on March 25, 1921. Lost in fog and having motor trouble, he hit a tree while landing. Taken to a hospital at Natchez, Mississippi, he died there 5 days later.[27]

The next challenger for the coast-to-coast record was 1st Lt. James H. Doolittle. Before he started, he flew the entire route keeping a detailed diary of the performance of his DH–4B and its Liberty engine. He then went to Dayton to consult engineers at McCook Field, and completely rebuilt his plane to make it more durable and extend its range. He planned a single stop at Kelly Field for service. After several months work, he was finally ready to go on August 6, 1922, but broke a propeller and wing taking off from Jacksonville. His second try came on September 2, 1922. He flew coast to coast in 22 hours and 35 minutes, which included an 85-minute stop at Kelly Field.[28]

Meanwhile in late 1921, 1st Lts. Oakley G. Kelly and Muir S. Fairchild became targets of good-natured ridicule at McCook Field, Ohio, when they talked of a nonstop coast-to-coast flight. The idea was preposterous. No plane could carry enough fuel. No man could stand up under the strain. Kelly and Fairchild wanted to try if they could find a plane. After studying the cruising speed, load-carrying capacity, fuel consumption, reliability, and other characteristics of several aircraft, they decided on an F–IV transport built in the Netherlands by Anthony H. G. Fokker under contract with the U.S. Air Service. General Patrick approved the flight during a visit to McCook Field on August 10, 1922.

The Fokker transport (designated T–2 by the Air Service) was a monoplane built to take a Liberty 12 engine and to accommodate eight passengers in a cabin behind the pilot's open cockpit. It carried 130 gallons of fuel, sufficient for about six hours. Additional gasoline tanks would have to be installed and other modifications made before the T–2 would be ready for the transcontinental hop. First Lieutenant Ernest W. Dichman took charge of the engineering work at McCook Field. He enlarged the gasoline storage to 725 gallons by adding tanks in the wing and cabin. Among other changes were more capacity for oil and water, an auxiliary radiator, an oil radiator, larger and stronger wheels, a door between cockpit and cabin, and a set of controls in the cabin.

Weighted with gas for the nonstop transcontinental flight, the plane could cross the Alleghenies. And with departure from Mitchel Field, the plane would consume enough gas by the time it reached the Rockies to permit it to cross without difficulty. The Weather Bureau advised, however, that a west wind of some twenty miles per hour could be expected at 5,000 feet during September and October, when the flight was being planned. Desiring the advantage of a good tailwind, Lieutenant Kelly set out from McCook in a DH–4 in mid-August to find a low-altitude route through the mountains in the west. Since Fairchild had not fully recovered from injuries sustained in an airplane crash, Dichman went along on the survey.

The route laid out by Kelly and Dichman was a tortuous one of 1,000 miles through valleys and canyons and around mountains. Beginning in California at Rockwell Field, the route ran northeast through Temecula Pass to Banning and south to the Salton Sea, southeast to the Southern Pacific Railroad, east along the rail line to Tucson, Arizona, and in New Mexico to Deming, northeast to Rincon, north to Carthage and Estancia, east to Santa Rosa and Tucumcari. Through the western range the route was under 3,000 feet. An altitude of 4,000 feet would be needed at Tucson and 6,500 at Carthage, but fuel consumption should lighten the plane sufficiently so it could reach those heights. After Tucumcari, altitude would not be a problem through Wichita, St. Louis, Indianapolis, Dayton, Pittsburgh, to New York City.

Back at McCook Field by the end of August, Kelly and Dichman pushed preparations for the flight. Because such a venture demanded someone more experienced than Dichman, 1st Lt. John A. Macready took his place. Leaving McCook on September 19, Kelly and Macready arrived in the T–2 at Rockwell on the 24th. Two expert mechanics, Charles Dworack and Clyde Reitz, arrived from McCook the next day to help ready the plane. An overhauled Liberty motor was installed and several minor changes were completed in the T–2. The back of the pilot's seat was made detachable to allow more room for movement between cockpit and cabin. A continuous-cord message conveyer replaced the unsatisfactory speaking tube for communication between crewmembers. Weeds and sand were cleared from a long runway that had not been used for some time. Having been tested, the T–2 was serviced on October 4 and placed at the end of the runway poised for takeoff. At 2030 a message from the Washington Weather Bureau forecast almost ideal conditions for the 5th.

Kelly and Macready were at the plane by 0515 on Thursday, October 5, 1922, to take care of last-minute checks and other details. A flip of a coin gave Kelly the cockpit for takeoff. After the engine warmed up, Kelly signaled for pulling the blocks and opened the throttle. The plane, with a gross weight of 10,695 pounds, hesitated, then moved forward very slowly. Gradually gathering speed, the plane lifted off after a run of about six-tenths of a mile. The time was 0553. Airborne, the T–2 gained so little altitude Kelly had to turn left over the Pacific to avoid Point Loma. Now flying downwind, the heavily laden plane commenced to settle until it came dangerously close to the water. Kelly circled the field twice before he reached an altitude of 200 feet and headed northeast toward Temecula Pass.

Twelve miles out, the plane was up to 500 feet; at 30 miles, 1,200 feet; at 50 miles (after 35 minutes flying), 1,700 feet. It would have to rise another 1,000 feet in the next 35 miles to clear the hills at Banning. But Kelly and Macready did not get that far. At San Jacinto they flew around for an hour waiting for fog to break or dissipate. With this delay, they would not be out of the mountain passes of New Mexico before dark. The extra consumption of gas impaired their chance of reaching New York. Kelly turned the plane back to Rockwell Field. Hoping to salvage something from the flight, Kelly and Macready decided to go for an endurance record. This would also afford valuable data on gas, oil, and water consumption for use on a subsequent transcontinental hop. Over Rockwell, Macready dropped a message to Capt. Robert G. Ervin, the commanding officer, telling him of their plans and asking him to notify the Aero Club of America and take steps necessary to authenticate a world's endurance record. The men stationed at Rockwell and people in San Diego watched as Kelly flew the T–2 around and around within sight of the field.

Kelly's turn as pilot ended at noon. Just before that Macready ate sandwiches, drank hot beef broth from a thermos, and swallowed strong, hot coffee to prime himself for six hours in the cockpit. When the time came, Macready took the controls in the cabin. Kelly opened the little door to the cabin, removed the back of the pilot's seat and dropped it and the parachute cushion down the hole. After lifting one side of the hinged seat, he crawled through the hole into the cabin. By speaking loudly, he could converse with Macready. The two men changed places and Macready crawled through the door into the cockpit and took control. Kelly put the parachute and the back of the seat into position in the cockpit and closed the door. Five times the men changed places before the flight ended. Though there was a bench in the cabin, the off-duty pilot slept hardly at all. There were so many things to do, like relieving the pilot briefly, flying the plane while the pilot made a minor repair or adjustment to the engine (which was on the pilot's right and

First Lieutenant Ernest W. Dichman (right), engineer of attempted non-stop transcontinental flight; center: 1st Lt. John A. Macready (left) and 1st Lt. Oakley G. Kelly, pilots in the transcontinental flight; bottom: Fokker transport T–2 used in the flight.

accessible to him), checking gas, oil, and water consumption, or dropping messages.

With the fuel supply dwindling, Macready landed the T–2 Friday evening. Having stayed aloft for 35 hours, 18 minutes, and 30 seconds, Kelly and Macready beat the world's record by almost 9 hours. Their record was not officially recognized, however, because the flight had not been observed and timed by the rules of the Aero Club of America and the Federation Aeronautique Internationale.[29]

Kelly and Macready hoped for a transcontinental hop in a few days, but maintenance troubles and bad weather intervened. Conditions did not become favorable until November 3, 1922. Crossing the mountains safely and surviving terrible storms over Oklahoma and Kansas during the night, the T–2 was forced down at Indianapolis the following morning by a cracked cylinder jacket.[30] Given a new engine, the T–2 flew to McCook Field where Kelly and Macready prepared for another assault on the world's endurance record. They closed in the pilot's cockpit, piped heat from the engine into the cabin, and among other things installed a high-compression Liberty engine. Since Wilbur Wright Field was far larger than McCook, they planned to use it to take off the heavily loaded plane. The Corps of Engineers laid out a fifty-kilometer, triangular course around the water towers at Wilbur Wright and McCook Fields and a pylon placed near the town of New Carlisle. Orville Wright agreed to serve as official observer for the Aero Club of America.[31]

A thousand people gathered to see Kelly and Macready begin their flight Friday morning, March 2, 1923, but the T–2 bogged down in the soft field. Another flight on the 30th was not much more successful, for engine trouble ended it after 7 hours and 55 minutes. The next try, on April 16 with a regular Liberty 400 engine, succeeded. Besides setting a new endurance record of 36 hours, 4 minutes, and 34 seconds, Kelly and Macready posted world records for distance (2,516.55 miles) and for speeds over distances of 1,500, 2,000, 2,500, 3,500, and 4,000 kilometers.[32]

Some of their records did not last long. In fact they lost two speed records before they finished their endurance flight. First Lieutenant Harold R. Harris, Kelly's boss at McCook Field, beat them in a DH–4L with an extra gas tank in the rear cockpit. He started his flight on April 17, when Kelly and Macready had been up almost 24 hours and had already taken the 1,500- and 2,000-kilometer records. Flying the same course as Kelly and Macready and using the same timers and observers, Harris captured the record for 1,500 kilometers with a speed of 114.35 miles per hour, and for 2,000 kilometers with a speed of 114.22. Seeing Harris that night after the T–2 landed, Kelly roared: "You're a h—l of a chief. Here I work hard to get a record and you take it away from me before I get out of the air."[33]

On Wednesday, May 2, 1923, Kelly and Macready embarked on a nonstop transcontinental flight once more. This time they flew in the opposite direction, starting from Roosevelt Field on Long Island. On takeoff the plane was still rolling when it reached the edge of Roosevelt Field. There the ground dropped off about twenty feet to the adjacent Hazelhurst Field. When the plane went over the edge, it settled but never quite touched the ground. Kelly fought to clear trees, poles, and wires. As Macready told it: "We scraped along the housetops and hillsides with our Liberty motor running absolutely full power, and for hours we felt as though we could stick out hands and grab a handful of daisies off the fields." About thirty minutes after takeoff the voltage regulator registered discharge from the batteries. With Macready flying the plane from the cabin, Kelly took off the regulator and adjusted the points. That evening Macready contended with clouds and rain. Later, with Kelly again at the controls, the aircraft came out of the overcast into the moonlight. They flew by dead reckoning to Spearman, Texas, where a positive check showed them on course and on time. The takeoff had been arranged so dawn would come before the plane entered the most treacherous part of the route at Tucumcari, New Mexico. With a great deal of the gas gone, the T–2 had no trouble crossing the mountains.

The people of San Diego—Macready's hometown—awaited the T–2. When the plane could be seen a few minutes after noon, sirens sounded, women waved handkerchiefs, and men tossed hats into the air and shouted. Macready flew on, over the city and across the bay. A roar went up from North Island when the transcontinental express touched down. Kelly and Macready had accomplished the impossible—a nonstop flight across the American continent. Their flying time was 26 hours, 50 minutes, 38.4 seconds. The flyers, begrimed and spattered with oil, and looking drawn and tired, were at once surrounded by the jubilant crowd. Pretty young girls put flowers in their arms. Newspaper reporters and cameramen struggled to get close enough to record the great moment. Soon the flyers would be showered with telegrams of congratulations from General Patrick, General Pershing, and well-wishers across the country.

Explaining to the American people the significance of the nonstop flight, the Air Service said the distance traveled by Kelly and Macready might just as well have taken them across the Atlantic from Halifax to Liverpool, or across the Pacific from San Francisco to Honolulu. For a businessman, it meant that in the time it would take him to travel from New York to Chicago by train he could go to the Pacific coast by plane. From the military standpoint, it meant that in an emergency men, ammunition, and supplies could be moved from one coast to the other in one day.[34]

The Air Service's statement was typical of a time when enthusiastic airmen let fantasies of the future obscure realities of the present. It is not necessary to say Kelly and Macready's flight did not demonstrate any of the things the Air Service claimed. It merely showed that two crack pilots after long and elaborate preparations, and with a plane especially fitted for the job,

First Lieutenant Frank Seifert (holding hose) and 1st Lt. Virgil Hine in De Havilland plane that was used during refueling flight of June 27, 1923.

could do what they did—on the third try. Saying that in no way detracts from the achievement, for this first nonstop hop from coast to coast on May 2–3, 1923, marked a milestone in American aviation history. Moreover, the event spurred airmen to lend substance to their dreams.

On June 27, 1923, 1st Lts. Lowell H. Smith and John P. Richter of Rockwell Field, California, attacked the endurance record recently achieved by Kelly and Macready. They used a DH–4B with a gas capacity of less than one third of the recordbreaking T–2's, but the DH was fitted for inflight refueling. A 50-gallon tank with a wide opening, installed back of the rear cockpit, gave the plane a fuel capacity of 200 gallons. Another DH, flown by 1st Lts. Virgil Hine and Frank W. Seifert, was equipped as a servicing plane. Its 40-foot hose connected to the bottom of the main fuel tank, with special fitting for transferring gas to the rear end of the endurance plane. As Hine flew over the endurance plane, Seifert let the hose down through a hole in the floor. Richter grasped the two handles on the hose, put the end in the 50-gallon tank behind his cockpit, and opened the valve. When the 50-gallon tank filled up, contact was broken briefly to allow Richter to pump into the main tank, then contact was reestablished. During the transfer the servicing plane operated on its emergency tank. A second hose was furnished for oil, and a rope used to lower food or messages.

Two refueling contacts were completed on June 27 before a burnt-out generator forced the endurance plane down after six hours and thirty-eight minutes. Trying again the next day, Smith and Richter flew around a triangular, fifty-kilometer course all day and all night. Fog, which prevented refueling, forced them down next morning. Although the flight set no record, Maj. Henry H. Arnold, the Rockwell Field Commander, thought it contributed greatly to aviation development. It showed that midair refueling could keep a plane up "until such time as either the plane wears out mechanically or the pilot and observer are subjected to such strain that they can no longer maneuver it."

The *Air Service Newsletter* said Army flyers were "firm believers in the maxim—'If at first you don't succeed, try, try, again.' " Smith and Richter demonstrated that belief at 0504, Monday, August 27, 1923. This time they used two refueling planes alternately, one manned by Hine and Seifert, the other by Capt. Robert G. Ervin and Lt. Oliver R. McNeel, ORC. The only problem Smith and Richter had was with their Liberty motor. Several times on Monday it cut out completely. Smith thought the trouble lay in a fuel fitting (the Lunkenheimer valve) in the cockpit. Dirt in the valve's strainer apparently blocked the gas flow. By hitting the fitting with a wrench he could clear the blockage for a time. So the endurance flyers went around and around the pylons all day Monday, Monday night, and Tuesday with Smith banging on the Lunkenheimer whenever the motor started to miss. At the end of 106 laps, Smith headed for Rockwell Field and landed at 1819. In addition to a new endurance record of 37 hours and 15 minutes, Smith and Richter set a new distance record (5,300 kilometers) and six speed records over distances from 2,500 to 5,000 kilometers.[35]

The speed attained by the Curtiss racer in the National Air Races in 1922 and during tests at Dayton gave 1st Lt. Russell L. Maughan the idea of

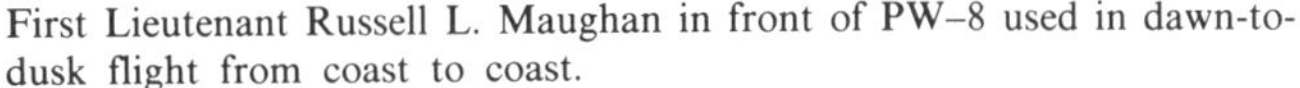
First Lieutenant Russell L. Maughan in front of PW–8 used in dawn-to-dusk flight from coast to coast.

trying to fly from coast to coast between dawn and dusk. He figured that by going westward with the sun in midsummer, he would have 20 hours for the flight.[36] By averaging 160 miles an hour he could cover the 2,670 miles from New York City to San Francisco in 16.7 hours. If he removed the armament and some other equipment from a Curtiss PW–8 (which was patterned after the R–6), he could add gas and oil tanks to extend the range so he would need just four stops of 30 minutes each, at Dayton; St. Joseph, Missouri; Cheyenne; and Salduro, Utah.

The project approved, the route surveyed, the plane prepared, and spare parts sent to the stopping places, Maughan took off from Mitchel Field at dawn on July 9, 1923. On the way to Dayton, he wandered off course in fog and fell two hours behind schedule. Ten miles beyond his second stop, at St. Joseph, a clogged gasline forced him down in a pasture. Turning sharply to avoid hitting a cow, he damaged the landing gear. So ended that flight. Trying again on July 19, Maughan got as far as Rock Springs, Wyoming, where an oil leak ended the flight. "Each failure," the *Air Service Newsletter* said, "augurs for better success on the next attempt, for it is only by a thorough test of any equipment that defects are found and constructors are enabled to take proper steps to correct them." As the days became shorter, however, the odds against a successful dawn-to-dusk flight increased. General Patrick vetoed an attempt in 1923.[37]

When Lieutenant Maughan tried once more on June 23, 1924, he used a new PW–8 modified to extend its range. The weather forecast for Sunday night, June 22, being favorable, Maughan opted to go the next morning. Taking off at 0258,[38] he had rain and fog between Pittsburgh and Zanesville, but arrived safely at McCook Field at 0710 Eastern Standard Time. Mechanics gassed and oiled the plane in 20 minutes. Eager to send Maughan on his way in record time, one of the men put an 18-inch monkey wrench on the gas valve to tighten it. He broke it off, costing Maughan an hour while the valve was removed, soldered, and replaced. Between Dayton and St. Joseph, Maughan had to fly around three local thunderstorms, but otherwise his only trouble was trying to stay awake as the plane droned along mile after mile on a long hot day. At St. Joseph a soft field prevented taking on a full load of fuel, causing an extra 20-minute stop at North Platte, Nebraska.

West of North Platte, Maughan lost time due to strong headwinds. Leaving the refueling stop at Salduro at 1751, he knew he would be hard pressed to reach San Francisco before dusk. Flying grew more and more difficult. Clouds cut off light from above, fog shut it off from below. Fortunately he knew the country from Reno to San Francisco. He recognized Sacramento from the Travelers Hotel and identified Mare Island (in San Pablo Bay) from ships in the bay. The lights of San Francisco appeared as a dull glow barely visible through the fog. Then he spotted the revolving light on the federal penitentiary on Alcatraz Island in San Francisco Bay.

Relieved, he hoped "never again to be so glad to see prison lights." Getting below the fog, Maughan passed over Crissy Field at 2140—one minute before the official time of dusk. Lights on the field revealed a large crowd awaiting him. A hundred people saw Maughan off that morning; fifty thousand came to see him land. He circled several times until sure no one was in the landing area. When the plane set down, the crowd broke loose and rushed headlong onto the field to greet the "brave and daring airman."[39]

A race to see who would be the first to fly around the world was developing among aviators of several nations when General Patrick made plans to send U.S. Army planes around from east to west in 1924. Donald W. Douglas built the planes, Douglas World Cruisers (DWCs), which were open-cockpit, two-seater biplanes with dual controls. Powered by a Liberty motor, a DWC cruised at eighty miles per hour and had a range of about 830 miles. It could be fitted with either wheels or pontoons for operation from land or water. The Air Service Engineering Division at McCook Field, Ohio, selected motors from government stock and sent them to the Douglas plant at Santa Monica, California, for installation. The division also made the propellers: oak for use with pontoons, walnut with wheeled landing gear. The Air Service depot at Fairfield, Ohio, packed extra engines, propellers, wheels, pontoons, spare parts, tools, and other supplies and shipped them to points along the route. The U.S. State Department obtained clearances from foreign governments. The U.S. Navy, the Coast Guard, and the Bureau of Fisheries moved supplies and stood by to assist airmen in trouble at sea. Air Service officers went ahead over parts of the route to collect information, select landing places, and arrange for supplies and services. The Signal Corps furnished a meteorologist to help the flyers across the North Pacific. With the assistance of these and others, the U.S. Army Air Service World Flight became a great cooperative venture.

General Patrick picked Maj. Frederick L. Martin, Chanute Field Commander, to head the World Flight. Martin and his mechanic, SSgt. Alva L. Harvey, had their DWC christened the *Seattle*; 1st Lts. Lowell H. Smith and Leslie P. Arnold, the *Chicago*; 1st Lt. Leigh Wade and SSgt. Henry H. Ogden, the *Boston*; and 1st Lts. Erik H. Nelson and John (Jack) Harding, the *New Orleans*. They started their flight at Seattle on April 6, 1924.

The *Seattle* crashed on a mountain on the Alaska Peninsula, but Major Martin and Sergeant Harvey survived and returned home. Command fell to Lieutenant Smith. Delayed by weather in the Aleutians and North Pacific, the flight did not reach Japan until May 23. After overhauling the planes with new engines and pontoons, the men continued along the coast of Asia by way of Shanghai, Hongkong, Saigon, Bangkok, and Rangoon, to Calcutta, arriving there on June 26. There the men overhauled the planes again, this time with new wings and engines, and with wheels in lieu of pontoons.

Their route took them overland via Karachi, Constantinople, and Paris,

Sailors refueling Douglas World Cruiser at Houton Bay, Scotland, and pilots for Round-the-World Flight (l. to r.): 1st Lt. John Harding, 1st Lt. Erik Nelson, 1st Lt. Leigh Wade, Maj. Frederick Martin, 1st Lt. Leslie Arnold, 1st Lt. Lowell Smith, and Lt. L.D. Schulze. (Schulze and Arnold are alternates)

to London which they reached on July 16. After another complete overhaul and a change back to pontoons, the men were ready for the North Atlantic crossing via Iceland and Greenland to Newfoundland. The *Boston* went down at sea en route to Iceland but a Navy cruiser, the USS *Richmond*, rescued Wade and Ogden. They rejoined the flight at Nova Scotia, where General Patrick had another World Cruiser waiting for them. Patrick himself flew to Maine to meet the *Chicago*, *New Orleans*, and *Boston II* and escort them to Boston. After a leisurely flight across the United States, the World Flyers returned to Seattle on September 28, 1924, after being gone 175 days.[40] (*Map 7*)

Gifts and honors were showered upon the flyers,[41] the significance of their achievement being highlighted by the failures of others that same year. A British pilot wrecked his plane in the northern Pacific, two Frenchmen crashed near Shanghai, two Portuguese flyers got as far as Macao, two Argentineans crashed at Hanoi, and an Italian plane went down in the northern Atlantic. But none of these made the supply arrangements and organized the support the U.S. Air Service did for its World Flight.[42]

Members of the U.S. Army Air Service set many flying records in the 1920s. In fact at one time they held over two-thirds of the world records for land-based airplanes, certified by the Federation Aeronautique Internationale. The Air Service found competition advanced aviation through improvement of equipment and techniques. Seeing competitive events stimulating interest in aviation, it sought to use them for mustering public opinion and support. The Curtiss Hawks—those speedy little ships Army pursuit pilots flew in the twenties and thirties—stemmed from the racers Lieutenant Maughan flew to win the Pulitzer prize in 1922 and set a world speed record in 1923. Major Schroeder and Lieutenant Macready's altitude work had a direct bearing on air power for it led to superchargers, oxygen systems, and other equipment that let Army flyers carry out bombardment and pursuit operations at higher altitude.

These advances and improvements became subjects for refinement and perfection, a process that proceeded unceasingly. A particular flight's contribution was sometimes not very direct and immediate, Smith and Richter's 1923 endurance flight being a case in point. Many more flights would be flown and many years would pass before aerial refueling became a practicable and routine means for extending a plane's time aloft—something that never would have come about without a Smith and Richter along the way. In some Air Service projects technological advancement played but a minor role; honor and glory, and a good press, counted more. A prime example was the first round-the-world flight by the Air Service in 1924.

MAP 7

PTEMBER 28, 1924

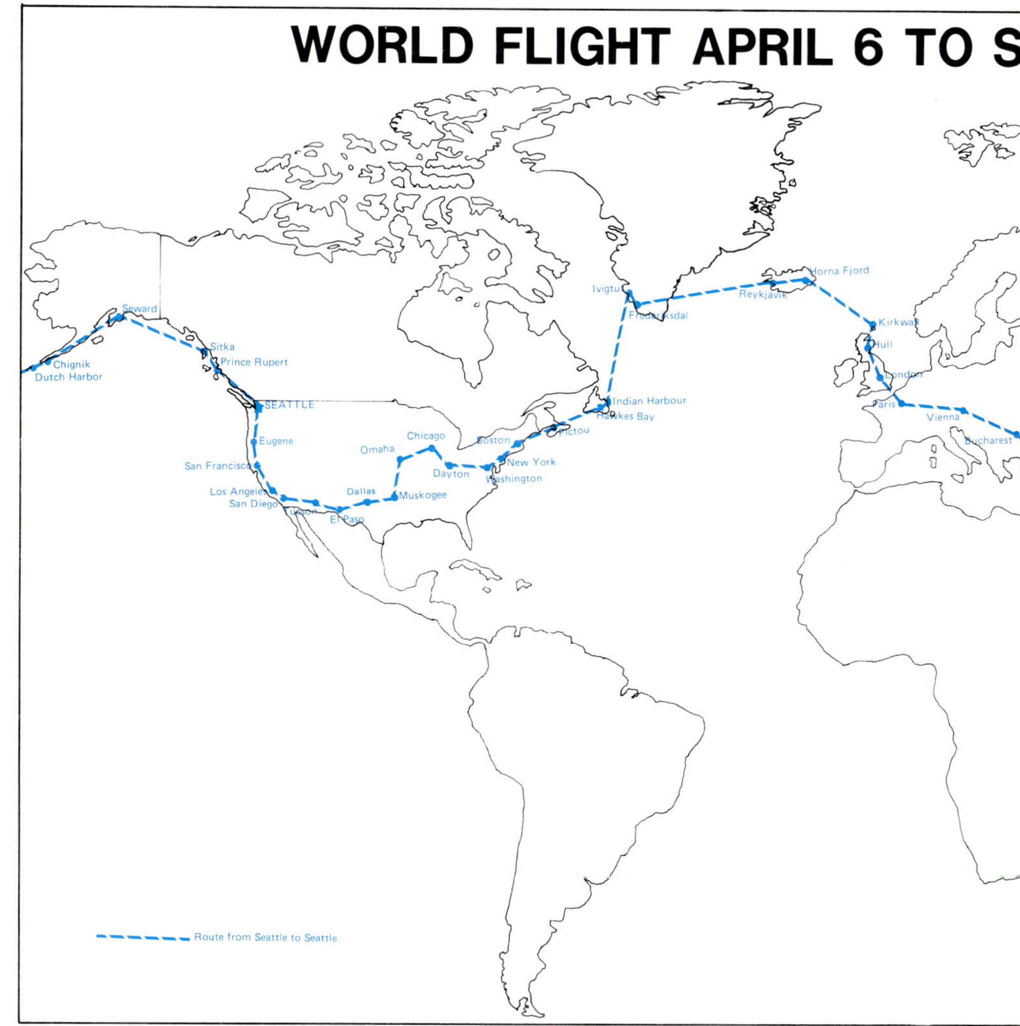

WORLD FLIGHT APRIL 6 TO S
Seward
Chignik
Dutch Harbor
Sitka
Prince Rupert
SEATTLE
Eugene
San Francisco
Los Angeles
San Diego
El Paso
Dallas
Muskogee
Omaha
Chicago
Dayton
Boston
New York
Washington
Pictou
Indian Harbour
Hawkes Bay
Ivigtut
Reykjavik
Horna Fjord
Kirkwall
Hull
London
Paris
Vienna
Bucharest
Route from Seattle to Seattle

Part Two

Air Corps 1926–1933

Chapter XII

The New Air Corps and the Five-Year Program

The Air Corps Act of 1926, which grew out of Air Service and War Department General Staff efforts to strengthen Army aviation, authorized expansion of the Army's air arm over a five-year period. The 1926 act changed the name of the air arm from Air Service to Air Corps, created a position for an Assistant Secretary of War to deal with aviation matters, directed establishment of Air Sections in the General Staff, and provided two more general officers to serve as Assistant Chiefs of Air Corps. The act also authorized temporary promotions for Air Corps officers, permitted the use of Reserve officers on extended active duty to supplement the Regular Army officer corps, and gave extra pay to enlisted men qualified as air mechanics. Administrative changes occasioned by this act came easily, but the five-year program proved difficult.

The New Air Corps

The Air Corps Act took effect when signed by President Coolidge on July 2, 1926. Maj. Gen. Mason M. Patrick, Chief of Air Service, then became Chief of Air Corps, and the President at once nominated F. (Frederick) Trubee Davison to fill the new position of second Assistant Secretary of War

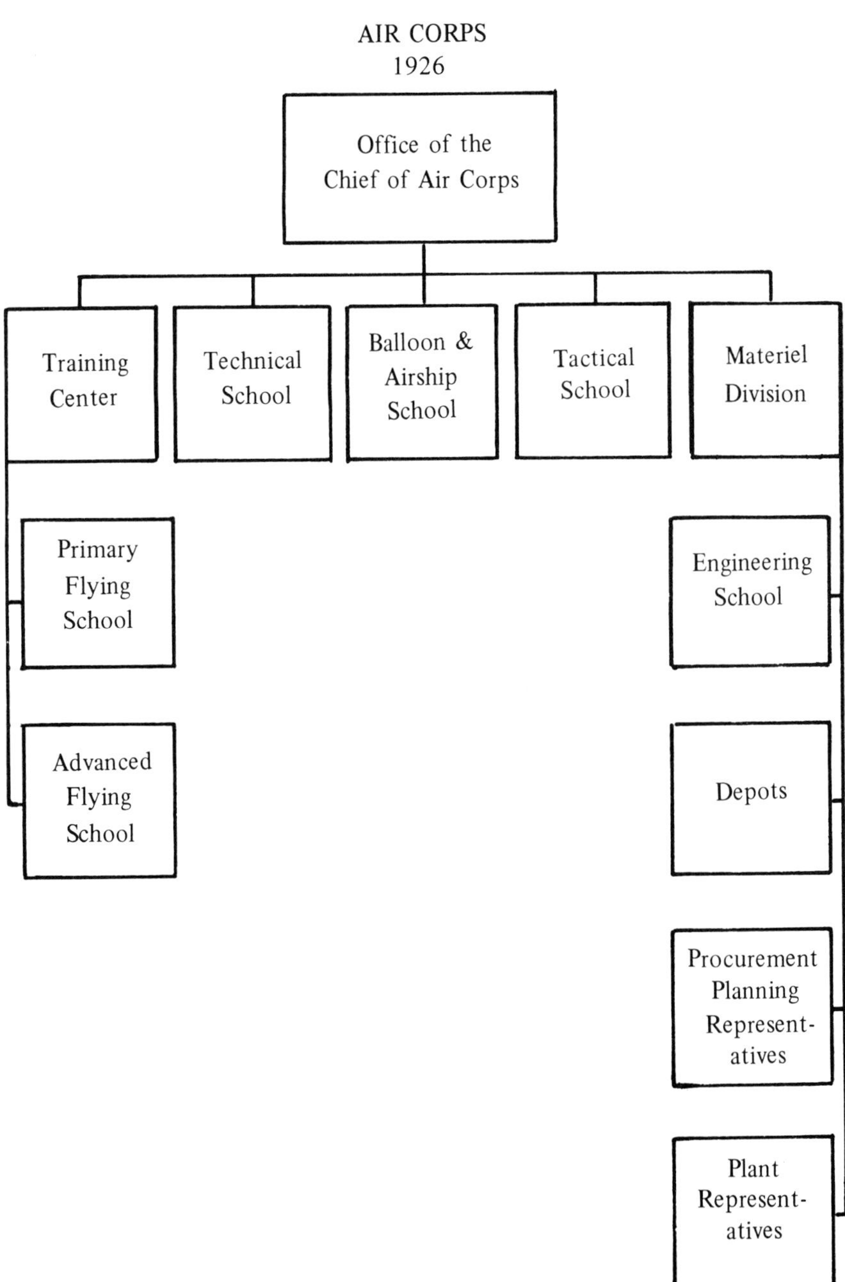
AIR CORPS
1926
Office of the Chief of Air Corps
Training Center
Technical School
Balloon & Airship School
Tactical School
Materiel Division
Primary Flying School
Advanced Flying School
Engineering School
Depots
Procurement Planning Represent-atives
Plant Represent-atives

F. Trubee Davison,
Assistant Secretary of War (for Air)

(for Air) established by the Air Corps Act. The son of a partner of J. P. Morgan and Company, Trubee Davison had been educated at Groton and Yale. He drove an ambulance in France before the United States entered the war. Later he joined the U.S. Naval Air Service and was lamed for life in a seaplane crash. An endowment of $6 million left by his father enabled him to make a career in politics. He took a law degree at Columbia University and in 1922 became a member of the New York Assembly. Retaining an interest in aviation, he served as a director of the Daniel Guggenheim Fund for the Promotion of Aeronautics. Now he would assist Secretary of War Dwight F. Davis "in fostering military aeronautics."[1]

Secretary Davison's chief duty became supervision of the Air Corps five-year program. This often took him into the area of operations, where he worked through the General Staff. Given responsibility for Air Corps procurement, he also worked with the first Assistant Secretary of War, who was in charge of all procurement for the Army. However, he thought his most important function to be one he took upon himself. He tried to heal wounds left by the Mitchell affair, to help bring the airmen and ground Army together.[2]

On July 17, 1926, the day after Davison was installed in his new office, the Air Corps got two new brigadier generals. The Air Service previously had just one, to serve as Assistant Chief, a post held first by William Mitchell and then by James E. Fechet. The Air Corps Act having furnished two additional assistants,[3] two lieutenant colonels received promotions—Frank P. Lahm to be Commanding General of the Air Corps Training Center at San Antonio,

Brig. Gen. Frank P. Lahm

Brig. Gen. William E. Gillmore

Texas, and William E. Gillmore to be Chief of the Materiel Division to be created at Dayton, Ohio.

Another innovation of the Air Corps Act was the creation of Air Sections in the various divisions of the War Department General Staff for three years beginning July 1, 1926. Air Corps officers had been detailed from time to time for duty with the staff. Two majors, Millard F. Harmon, Jr., and Rush B. Lincoln, were already with G–3 (Operations and Training), having been sent after their graduation from the Army War College in 1925. Nevertheless, the Air Corps view had not always been represented. Now the law required each division to include an Air Section, headed by an Air Corps officer, "to consider and recommend proper action on such matters as may be referred to such division."[4] Major Harmon stayed with G–3 to head its Air Section; Major Lincoln went to G–4 (Supply). Lt. Col. Roy C. Kirtland, recently graduated from the Army War College, went to G–1 (Personnel); Maj. John D. Reardan, of the same class, to the War Plans Division; Maj. Joseph T. McNarney, who just finished Command and Staff College, to G–2 (Intelligence).

The Air Corps Act contained a provision assuring that flyers would hold most of the key positions in the Army's air arm. It required the Chief of Air Corps, at least two of the Corps' three brigadier generals, and at least ninety percent of its officers below the rank of brigadier general to be "flying officers." By statutory definition this meant in peacetime pilots qualified to fly service aircraft, and in wartime pilots or rated observers. Like the 1920 act, the Air Corps Act required assignment of flying officers to command flying units. Other provisions of the law, including those relating to mechanics ratings, temporary promotion, and extended active duty for Reservists in peacetime, will be treated later in this chapter.

Planning for Expansion

The five-year program authorized in the Air Corps Act of 1926 held some hope of real progress in Army aviation. General Patrick had labored long for a large increase in the size and strength of the Air Service. As a result, the War Department board headed by General Lassiter in 1923 suggested a program, approved by Secretary Weeks, to enlarge the peacetime Air Service over ten years, but that program did not begin because the Secretaries of War and Navy could not agree on the division of money for aviation. When the Morrow Board proposed a smaller, five-year program in late 1925, The Adjutant General asked General Patrick for suggestions to carry out the recommendation.[5]

Patrick disliked the Morrow Board's proposal and said so. Since 1923,

War Department policy for expanding the Air Service followed the Lassiter report.[6] He deemed it a mistake to depart from that policy. He could not agree to "any less ideal solution." Because he had to offer a plan, he suggested a 5-year one based on the 10-year plan; afterwards a second 5-year one could complete the job. Patrick wanted to compress to 5 years the entire 10-year schedule for developing heavier-than-air tactical units. The Air Service then had 32 tactical squadrons, many undermanned and only partially equipped. The Lassiter plan called for 62, Patrick proposed 63. To secure that many in 5 years, he would curtail growth in other areas, especially in school squadrons and lighter-than-air companies.[7]

Evaluation of Patrick's suggestion fell to G–3, headed by Brig. Gen. Hugh A. Drum, a member of the Lassiter Board. In G–3 the head of the Air Section, Major Millard F. Harmon, Jr., got the job. The report ultimately approved by Drum regarded Patrick's plan as too ambitious for accomplishment. Drum proposed a smaller program of fifty-two tactical squadrons. Other divisions of the General Staff agreed. Secretary Davis approved Drum's report "as a study" and asked that the details be worked out as quickly as possible. That was in early March 1926.[8] Congress already had several bills concerning aviation and the Air Service. One supported by the War Department would expand the Air Service as suggested by the Morrow Board.[9]

The War Department study went back to the Air Service for complete information and schedules covering organizations, personnel, aircraft, facilities, and costs.[10] Maj. Herbert A. Dargue, in charge of the War Plans Section of the Air Service, gave Major Harmon detailed plans for personnel and airplanes before the end of March.[11] By mid-April all the information was in G–3 for review and coordination in the General Staff. On May 21, 1926, Maj. Gen. Malin Craig, successor to Drum in G–3, asked the Chief of Staff, Maj. Gen. John L. Hines, to approve the project with the understanding that the increase in the Air Service would not be at the expense of any other branch of the Army.[12] General Hines did not act until after Congress passed the Air Corps Act, approving expansion along the general lines of the War Department's five-year plan. He then noted that the yearly program for carrying out War Department policy as embodied in the 1926 act depended upon the budget and would be considered before and during work on budget estimates each year.[13]

The Air Corps came into being with 919 officers, 8,725 enlisted men, and 1,254 airplanes.[14] The law authorized a buildup to 1,650 officers, 15,000 enlisted men, and 1,800 airplanes over 5 years. The number of tactical squadrons would grow from 32 to 52, pursuit from 8 to 21, bombardment from 8 to 12, and attack from 2 to 4, with observation squadrons staying at 14. Ten of the new squadrons would become part of the combat forces in the United States. Nine would strengthen the garrisons in Hawaii and the Canal

Zone. The other new squadron (a composite unit with bombardment, pursuit, observation, and transport planes) would be stationed in Alaska.[15] Completion of the five-year program would give the Army in the United States an air force consisting of two bombardment wings and an extra pursuit group for employment by General Headquarters in an emergency. The wings, one for the east coast and the other for the west, would each have a bombardment and a pursuit group of three squadrons. The extra pursuit group of three squadrons would be situated in the north-central part of the country, close to the transcontinental airway, for rapid deployment in an emergency. An attack wing comprising an attack and a pursuit group, each with three squadrons, was designated army aviation. This wing was to be stationed so as to be ready for defense of the southern frontier. Nine observation squadrons would afford corps and division air services.[16] For administrative, logistical, and strategical reasons it would be better to group these units in some central place in the United States, but political and economic reasons made this impossible. General Patrick said the distribution adopted placed near each border the type of aviation best for an emergency.[17]

The War Department estimated the Air Corps needed direct appropriations of about $31 million a year during the 5 years of expansion and $26 million a year thereafter. The biggest item in the budget would be procurement of new airplanes, ranging from $16.5 million the first year down to $13.5 million the fifth year, then $11 million a year. With indirect appropriations for pay, construction, ordnance supplies, and other things added, the total each year during the expansion ranged between $68 million and $76 million.[18]

Delay

The Air Corps expected to start expanding at once, with the five-year program running from July 2, 1926 to June 30, 1931 (Fiscal Years 1927 through 1931). But money for expansion was not instantly available. The Air Corps Act itself carried no money. Funds would therefore have to come from annual or supplemental appropriations. The Air Service appropriation of $15.3 million for Fiscal Year 1927 had been made in April 1926 before Congress enacted the Air Corps bill.[19] In the Air Corps Act, however, Congress authorized submission of a request for a supplemental appropriation for Fiscal Year 1927 for the program's first increment. Always seeking ways to economize, President Coolidge directed the program be deferred until Fiscal Year 1928. Still, he authorized the War Department to request a supplemental appropriation for Fiscal Year 1927 to let the Air Corps prepare for the first increment of the five-year program. The Air Corps asked for $8.6

million, which Secretary Davis approved on August 18. The Air Corps also revised estimates previously prepared for Fiscal Year 1928 to provide for the first year's expansion.[20] When President Coolidge inquired about the program on October 4, 1926, Secretary Davis told him supplemental estimates were pending before the Bureau of the Budget, but nothing could be done until money became available. Davis added that the Air Corps Act specified the Air Corps buildup not reduce other branches of the Army. Though no such requirement was spelled out in the law, Congress evidently intended that other branches would not have to give up officers and enlisted men to expand the Air Corps. And that of course was the way Davis, the General Staff, and the other arms and services wanted it.[21]

Assuming the Bureau of the Budget would approve the supplemental estimate for 1927 and Congress would appropriate the money, General Patrick asked permission on October 28 to do a number of things to get the five-year program off the ground. Davis granted permission the same day.[22] But Patrick soon found his assumption wrong—the Bureau of the Budget denied more money for 1927. The Deputy Chief of Staff, Maj. Gen. Dennis E. Nolan, told General Staff members on November 4, 1926, that if the program was to be started that year it would have to be with money taken from other Army activities. This naturally elicited protests from those who stood to lose. Patrick assured Nolan he had "never advocated the increase of the Air Corps at the expense of other branches of the Army." It would "be greatly regretted," he said, "if no other method of bringing about this increase can be found." Referring to his previous request to start the program, Patrick said he would regard Secretary Davis' approval as not having been given.[23]

Reexamining the Air Corps Act, General Patrick found a way to complete the five-year program on time even though it commenced a year late. The section of the law spreading the expansion in personnel and equipment over five years said growth the first year would not exceed one-fifth of the total, and the remainder would be in four approximately equal increments. Thus, after taking into account whatever might be accomplished during Fiscal Year 1927 (no matter how little), the remainder could legally be divided into four parts to complete the program on June 30, 1931, as planned. If General Patrick seriously thought this solution, which he proposed to Secretary Davison, would be approved and carried out, he was due more disappointment. The five-year program would run five years, from July 1, 1927, to June 30, 1932.[24]

Money Problems

Money shortages persisted during the entire five-year program. The introduction of aviation had added much to the cost of the Army, but other new or improved weapons and equipment (including tanks, armored cars, motor transport, antiaircraft artillery, gas and gas masks, smoke projectors, radios, and sound-ranging equipment) helped push military costs above prewar levels. Military expenditures rose in the last half of the 1920s as a result of further technical developments. Expansion of the Organized Reserve, Reserve Officers' Training Corps, National Guard, Chemical Warfare Service, Air Corps, and other activities boosted costs still more.

For several years the Army lived on materiel left over from the war, but by the mid-1920s the old equipment was wearing out and stocks were being used up. Of all the branches of the Army, the Air Corps had the biggest program for modernization and expansion.[25] The Air Corps tried hard to get money for the five-year program. But an economy-minded government saw no urgent need for military preparedness in a day when isolationism and pacifism ran strong among the American people and no foreign power posed an immediate threat to the nation's security. President Coolidge, trying to keep government costs down, did not oppose the Air Corps expansion so long as it did not cost a lot more.

And when President Hoover entered the White House in 1929, he sought ways to shave expenses. He stepped up his efforts when the administration faced deficits stemming from the Great Depression following the stock market crash in the fall of 1929. With the nation's economy in shambles and federal revenues declining, it was harder and harder to get money to keep the Army going, much less pay for expanding the Air Corps.

The War Department's instructions for the annual budget called for progressive accomplishment of the five-year program. The budgeting process leading to the annual appropriation of monies for the Army began many months before the opening of the fiscal year when the funds were to be available. The War Department first asked each branch for preliminary estimates of the amount needed. The War Department's Budget Advisory Committee studied the estimates in light of military policy and with the idea of preserving balance among the various arms and between personnel and materiel. The estimates and the committee's recommendations next went to the Chief of Staff, who issued guidelines for final estimates. These, after review by the advisory committee and approval by the Chief of Staff and Secretary of War, became the department's estimates, which went to the Bureau of the Budget in the September preceding the beginning of the fiscal year the following July.

The Bureau of the Budget, created in the Treasury Department in 1921, coordinated executive department estimates, weighing them individually and

President Herbert Hoover seeks spending cuts in the new Air Corps budget.

collectively in relation to the administration's financial policies. The bureau usually set a dollar ceiling for each department, and at times revised or deleted items. After the departments completed necessary changes, the bureau prepared a budget for presidential approval and submission to Congress. Legislative hearings and debates often produced further amendments before appropriations passed.[26]

For Fiscal Year 1928 the Air Corps put in for $28.6 million to support the present establishment for a year and to pay for the first expansion increment. Congress appropriated $20.6 million, a third more than appropriations in recent years. At the end of Fiscal Year 1928, General Fechet, successor to General Patrick in December 1927, reported the first expansion increment completed.[27] Secretary Davison, who regarded "fulfillment of the 5-year program . . . the first objective of the Air Corps," stressed that sufficient funds were a must if the full program was to be achieved "smoothly and efficiently."[28] For Fiscal Year 1929 the Air Corps asked for $36.5 million and received $24.6 million. While it made considerable progress, it did not meet all objectives of the second increment. General Fechet reported shortages of housing, equipment, and senior officers.[29]

For Fiscal Year 1930 the Air Corps requested over twice the amount than for 1929. It needed to take care of deficits in earlier appropriations and deficiencies discovered in the program. It wanted $77.5 million and received $34.7 million.[30] At the end of the year, General Fechet once more reported failure to attain five-year program objectives. Secretary Davison said, "In the broad sense, progress had been gratifying and results satisfactory," but he found "stumbling blocks" removable solely by adequate appropriations.

Patrick J. Hurley, now Secretary of War, told President Hoover progress had been "normal" and the program had been budgeted to completion in Fiscal Year 1932.[31]

In the spring of 1929 the Air Corps sought $54 million for Fiscal Year 1931. Desiring a minimum of military preparedness, President Hoover ordered a survey of all military activities to find ways to trim expenditures. Gen. Charles P. Summerall, War Department Chief of Staff, asked the chiefs of the arms and services for recommendations. The chiefs generally defended their own programs and activities. Fechet wanted more money for the Air Corps, and the General Staff urged completion of the five-year plan.[32] The Bureau of the Budget limited the Air Corps to $35 million which, General Fechet said, left the five-year program in "a haze of uncertainty." He recommended abandoning it for "a new balanced program which can be achieved within the limited funds available."[33]

A week earlier the great bull market collapsed on Wall Street, sending stock prices plunging. No one then foresaw the extent of the disaster. But the Great Depression was just beginning to be felt when Congress appropriated money for Fiscal Year 1931. The Air Corps received $35.8 million.[34] The new fiscal year had scarcely begun before President Hoover called for spending cuts. Because of the depression, he needed to shave expenses to prevent a deficit. He wanted to know what the War Department could defer. The department figured it could withhold $20 million out of the $409 million appropriated. The President wanted to delete $65 million. Army officials objected but, expecting the nation's economy to shortly take a turn for the better, were inclined to cooperate with the President. The Air Corps resisted. It could not carry out its mission; the aviation industry would collapse. In the end the Air Corps was permitted to spend about $36 million, $2 million diverted from other arms and services.[35]

For the fifth year of expansion, the Air Corps put in for $64 million and got $31.5 million. Before the fiscal year began, however, it was clear the Treasury would run deeply into the red. Rigid economy became the rule. Secretary Davison aimed to economize "without seriously impairing the basic structure of the Air Corps."[36] The Air Corps sharply curtailed flying. Purchasing officers shopped for better prices. These and other savings enabled the Air Corps to send more than $2 million back to the Treasury.[37]

The Air Corps did not complete the expansion program in 5 years. Short of officers, enlisted men, airplanes, and tactical units, it blamed these deficiencies on the War Department and the Bureau of the Budget, which pared more than 45 percent from the amount the Air Corps asked for.[38] The Air Corps' calculation of the size of the cut rested on the amounts it requested from year to year while the program was under way, the total being $260.6 million. The sum appropriated over the 5-year period ($147.2 million) actually was but 5.8 percent short of the amount the Air Corps originally told

Congress the program would cost ($156.2 million). Davison referred to the higher, recomputed estimates when he observed that "the cut taken [annually] by the Corps usually greatly exceeds a proportion based on the amount of the Air Corps estimates as compared with the estimate for the entire Army."[39]

The depression of course was one of the stumbling blocks to successful completion of the five-year program on schedule.[40] Still, the two appropriations made after the stock market crash were far higher than those of the program's first two years. Shortages of money in indirect appropriations, including pay of officers and enlisted men, also curtailed expansion, particularly in the fifth year of the program. The Air Corps had not yet felt the full force of the depression. Worse times loomed ahead. All the Air Corps could do was hope, as Davison said, that "when the necessity for such rigid economy no longer exists, first priority will be given to the completion of the 5-year program in all its phases."[41]

Sixteen Hundred and Fifty Officers

When the 5-year program began on July 1, 1927, 919 men held commissions in the Air Corps, Regular Army. The Air Corps needed to add 146 officers a year to reach the goal of 1,650 by June 30, 1932. It could secure them in two ways: by transfers of officers from other branches of the Army, and by commissioning Reserve officers, enlisted men, and civilians in the Air Corps, Regular Army. Both methods were adopted. The Air Corps Act rested on a War Department allotment to the Air Service of 1,247 officers from the 12,000 Congress authorized for the Army in 1922.[42] To close the gap between allotment and goal, Congress authorized 403 additional officers for the Army.[43] But the 403 vacancies did not become available in the beginning.[44] President Coolidge ordered Air Corps expansion to be kept within the limitation of the 12,000 officers Congress authorized for the Army in 1922. Colonel Kirtland of G–1's Air Section perceived that if the President's directive remained in force, the expansion would "be indefinitely extended over a long period of years."[45]

In the second year the Air Corps received 200 of the 403 vacancies authorized. Having filled about half by the end of the fiscal year, it counted on filling the rest with cadets graduating from the Advanced Flying School in October 1928. Calling attention to the improvement in officer manning, General Fechet deemed future prospects bright.[46] But the Air Corps did not receive the remaining 203 vacancies. Consequently, it could not accept graduates of pilot training who desired commissions in the Regular Army. During the last 2 years of the program all new officers came by transfer.[47]

Besides authorizing more Regular officers, the 1926 act permitted the Air Corps 550 Reserve officers on extended active duty.[48] The allotment was 110 for the first year of the expansion. Of those ordered to duty, 2 were killed, 1 relieved due to physical disability, 1 discharged, 15 released to take civilian flying jobs, and 39 appointed in the Regular Army.[49] The program specified 220 Reservists on active duty the second year. The Air Corps was funded for just 110 but called 192, the difference being taken care of by losses during the year. Shortages of equipment and money precluded the Air Corps from filling its quota of Reservists over the remaining years. The number on duty nevertheless rose, the average being 241 during 1931 and 271 in 1932. The upturn stemmed in part from the dearth of Regular Army vacancies, which kept the Air Corps from commissioning Reservists in the Regular Army.[50]

The Air Corps acquired fewer officers by transfer from other branches than it hoped for. Those wanting to change had to apply, the number being smaller than anticipated. Several, like Col. Albert E. Waldron, a 54-year-old graduate of the Military Academy with more than 28 years of service when he applied for transfer from the Corps of Engineers, failed the physical. Some were unacceptable owing to unfavorable records or sundry reasons. Second Lieutenant Glenn O. Barcus and others the Air Corps accepted were detailed to flying school. Many washed out. In one group of 29 entering primary training, 27 failed; one lasted only 12 days. Those, like Maj. William O. Ryan, who finished advanced training were transferred to and commissioned in the Air Corps. During the 5-year period the War Department detailed 592 officers and transferred 220.[51] Many officers wanting flying training were second lieutenants fresh out of West Point. Since the Army did not commission Military Academy graduates directly in the Air Corps, these officers had to apply. The Air Corps gained 18 by detail from the Class of 1926, 30 in 1927, 70 in 1928, 110 in 1929, and 86 in 1930. Thus the Military Academy was a major source of new Air Corps officers. When 86 from the class of 1931 underwent flying training, the Air Corps figured not more than 50 would finish. The number earning their wings came closer to 40.[52]

The washout rate went higher with older officers detailed for flying training. General Fechet, who viewed flying as "a young man's game" (he began pilot training at age 40), invited Secretary Davison's attention to the remarks of a lieutenant colonel recently dropped from primary training at Brooks Field. Like Fechet, the colonel had enlisted in 1898 and had served in Cuba and the Philippines. He won an Infantry commission in 1902, was awarded the Distinguished Service Medal for achievement in World War I, and graduated from the Army and the Navy War Colleges before entering flying training in March 1927. After he washed out in October, someone asked him what the age limit should be at the school. He replied: "You ought not to have anybody come into the Air Corps unless he is a youngster. That is to say, a man certainly under thirty." The 45-year-old lieutenant colonel thought it "a waste of time to fool with anybody else." Older men could competently fill staff and technical positions despite being only "fair" pilots: "But this theory of sending a man down there far over forty to become a crack pilot is the most ridiculous thing in the world."[53] An officer did not suffer disgrace from washing out of flying school. His detail to the Air Corps quickly ended in another assignment, and eventually he might even become a four-star general, as Lt. Col. Walter Krueger did.

Having gained 521 officers and lost 186, the Air Corps closed the 5-year period 396 short of the goal of 1,650. About a third of the losses were from aircraft accidents, and another third from resignations of officers who thought the future looked brighter on the outside. Resignations were felt Air Force-wide, in the training organization and in the tactical force. The Engineering School at Wright Field shut down for the school year 1927–28, and the Technical School at Chanute Field canceled several courses. However, the urgent need for engineers and technicians impelled the school at Wright to reopen and the one at Chanute to resume full operation the following year. A different situation existed at Scott Field. Interest in lighter-than-air aviation lagged. Needing money and people elsewhere, the Air Corps curtailed lighter-than-air operations and in 1928 closed the school at Scott indefinitely.[54]

Pilot Shortage

With a goal of 1,650 officers, the Air Corps needed at least 1,485 commissioned airplane pilots at the end of the fifth year of expansion.[55] This meant an increase of 759 over the number on July 2, 1926. Considering the percentage of students who would not complete training, and the number of pilots who would be lost by resignation, fatal accidents, and other causes, the Air Corps figured on 500 cadets entering training each year.[56] The 5-year program also envisioned more enlisted pilots. The Air Corps still wanted all pilots commissioned, but Congress insisted on saving money by using pilots who drew less pay. The 1926 act required that after July 1, 1929, not less than 20 percent of the pilots in tactical units be enlisted men unless the Secretary of War determined it impracticable to secure that number. In July 1926 the Air Corps counted 50 enlisted airplane pilots. To meet requirements of law, it planned for 225.[57]

The Primary Flying School at Brooks Field, Texas, expanded but still could not accommodate all primary students. Needing another school, the Air Corps reopened March Field, California. Ultimately March would house the new bombardment wing to be stationed on the west coast, but meantime it served as a primary flying school.[58] To free officers and men to run the enlarged training program, the Air Corps temporarily inactivated certain tactical squadrons, kept the fewest pilots in other tactical units, and did not replace some officers returning from foreign service.[59]

Reorganizing pilot training, the Air Corps created a Training Center at San Antonio with Brig. Gen. Frank P. Lahm in charge. He opened headquarters at Duncan Field[60] on September 1, 1926. As an Assistant Chief of Air Corps, he commanded the Primary Flying School and the School of Aviation Medicine at Brooks Field, and the Advanced Flying School at Kelly.[61] He wanted closer cooperation between the two flying schools to ease students' transition from one school to the other. He likewise desired unsatisfactory students eliminated earlier, with more weeded out during primary training and fewer at the advanced school. His objective was more Kelly Field graduates without lowering standards.[62]

Patrick dispatched Fechet to San Antonio to discuss changes with Lahm. Afterwards, Fechet recommended the school at March Field be placed under Lahm's command, and transformation training (wherein the students changed from primary trainers to service aircraft) be moved from the advanced to the primary course.[63] Fechet's plan provided two "basic" flying schools, at Brooks and March Fields, and a "specialized" flying school at Kelly Field. It called for a 12-month course, 8 in basic and 4 in specialized training, with classes entering July 1, November 1, and March 1. It cut academic instruction to the bone and divided instruction at the basic schools between primary flying in primary training planes and basic flying in observation planes. The plan required all students in the specialized school to be trained as observation pilots and observers and in one other branch of aviation—attack, bombardment, or pursuit. Fechet proposed 100 students for each class entering basic training, with the new program to begin on July 1, 1927. Estimating that 30 percent would graduate to the specialized school and elimination would then be negligible, Fechet forecast the graduation of 180 airplane pilots a year.[64] Patrick and the General Staff approved.[65] Brooks and March were still named Primary Flying Schools, but with their course divided into a primary and basic phase. Graduates went to the Advanced Flying School at Kelly Field for specialized training.[66]

During Fiscal Year 1928, 592 students (500 cadets, 91 officers of the Regular Army, and 1 noncommissioned officer) entered the primary flying schools; 149 finished basic training; 128 completed advanced training and received the airplane pilot rating.[67] Colonel Kirtland in G–1 said the Air Corps was short 239 pilots (114 Regular Army officers, 85 Reserve officers on extended active duty, and 40 enlisted men) from the number planned for the end of the first increment of expansion. Believing the Air Corps could not meet its 5-year goal, he suggested revision of the program "along lines which are clearly susceptible of execution."[68] General Fechet did not agree. He thought the percentage of cadets graduating could be increased by raising qualifications for pilot training, stimulating the morale of students and instructors, and improving teaching methods. It would help if adequate equipment could be supplied the schools, not an easy task in view of the general shortages throughout the Air Corps.[69]

Fechet further proposed to have more active duty pilots by requiring flying cadets to serve 3 years—1 as a cadet and 2 more as a Reserve officer unless sooner commissioned in the Regular Army. The cadet winning his wings entered the Reserve with a valuable tuition-free education, the government getting nothing in return in peacetime. U.S. laws compelled cadets of the U.S. Military Academy to sign up for 8 years. The solution: Change the Air Corps Act to make flying cadets serve at least 3 years.[70]

Maj. Robert E. M. Goolrick, commanding 2d Cavalry Division Aviation at Fort Riley, Kansas, had a similar idea in April 1929. He was losing too many pilots. Three second lieutenants resigned the previous year, 2 had applications in the War Department, and 1 planned to submit his in a day or two. Others were thinking about resigning in the next few months. All were graduates of the Advanced Flying School, had held Reserve commissions, had applied for and received Regular commissions, and had been Regular Army second lieutenants from 3 to 12 months. It cost the government a lot of money to send these men through flying school. And to become useful officers they needed a year or 18 months more training in flying and military duties with a tactical unit. Goolrick suggested second lieutenants appointed to the Regular Army from cadet classes be made to serve at least 2 more years.[71]

Fechet received Goolrick's recommendation while heading a War Department board to consider changes in the five-year program. The board concluded a change in law was unnecessary, for the War Department already possessed authority to prevent second lieutenants of the Air Corps from resigning.[72] The War Department acted quickly. On August 14, 1929, it altered the articles candidates signed for appointment as aviation cadets. Now applicants needed to serve 3 years, 1 in flying school as a cadet and 2 either as a Reserve second lieutenant on active duty or as a Regular Army officer. The first class affected entered in November 1929. Some applicants withdrew when the articles changed but most signed for the extended period.[73]

Another major change was the relocation and consolidation of primary pilot training in 1931. Soon after taking command of the Training Center, General Lahm acted to bring together all flying training at a new field. Brooks and Kelly Field lay too close to San Antonio which was spreading out. The government held land in Florida and California it might use, but Lahm and many other Air Corps officers preferred San Antonio. The Chamber of Commerce did not want to lose such a large government operation. A number of cities offered land, but the Air Corps eventually accepted twenty-three hundred acres donated by the city of San Antonio. The

War Department named the new field, situated fifteen miles northeast of the city, in honor of Capt. William M. Randolph, a native Texan and Adjutant of the Advanced Flying School, who died in the crash of an AT–4 on February 6, 1928.

Construction commenced in the summer of 1929. However, only a small part of the new station was completed before its dedication on June 20, 1930. This marked the end of General Lahm's tour of duty at San Antonio. The "Father of Randolph Field" again became a lieutenant colonel with assignment at the Presidio of San Francisco. Brig. Gen. Charles H. Danforth commanded when training got under way at Randolph Field. The headquarters of the Training Center and the primary schools at Brooks and March Fields moved in September 1931 to the new station, which soon became famous as the "West Point of the Air." The advanced school stayed at Kelly Field. The 12th Observation Group, previously at Fort Sam Houston, took over Brooks Field. March Field became the home of the 1st Bombardment Wing.[74]

Few enlisted men became students at the Training Center. It cost the government at least $25,000 to teach a man to fly. Assuming the enlisted graduate elected to stay in the Air Corps, he had but a few years (perhaps just a few months) before he could no longer pass the physical examination for flying. An officer at this point could still serve in administrative and staff positions. But an enlisted pilot who could no longer fly was of no more use

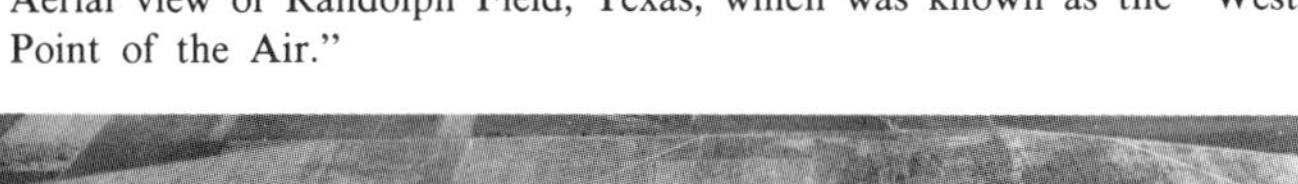
Aerial view of Randolph Field, Texas, which was known as the "West Point of the Air."

than an ordinary enlisted man. Enlisted pilots often sought more lucrative jobs available in civilian life. Their Reserve commissions created a problem. While some Reserve officers served as enlisted men, others served as officers.[75] A board of officers reviewing the five-year program suggested enlisted pilots be used chiefly as transport and ferrying pilots rather than in tactical units as required by the Air Corps Act.[76]

Though the law remained unchanged, the number of enlisted pilots went down instead of up. Few enlisted men met pilot training requirements. Those with wings frequently left for jobs with other government agencies or in civil aviation, but such jobs became harder to find during the depression. The Air Corps itself arranged for enlisted pilots to go into other positions in the service, calling some to extended active duty as Reserve officers and commissioning others in the Air Corps, Regular Army. Therefore, enlisted men on duty as airplane pilots dropped from 50 at the end of Fiscal Year 1926 to 38 in Fiscal Year 1930.[77]

During the 5-year period of expansion, 3,486 flying cadets, Regular Army officers, and enlisted men entered primary flying schools; 1,350 completed basic training, and 1,286 finished advanced. At the end of the fifth year the Air Corps counted 1,158 Regular Army officers and 33 enlisted men with the airplane pilot rating. Thus the Air Corps fell some 300 commissioned and 200 enlisted pilots short of the 5-year goal.[78]

Fifteen Thousand Enlisted Men

The 5-year plan scheduled expansion of the enlisted force at the rate of 1,248 a year, with the allotment climbing from 8,760 to 15,000. Congress added 6,240 men to the Army's authorization of 125,000.[79] After the Army appropriation for Fiscal Year 1928 provided for only 118,750 enlisted men, President Coolidge ordered the War Department to take 1,248 men from other branches for the first increment of Air Corps expansion. Six weeks into the fiscal year, the Air Corps received vacancies for 115 sergeants, 118 corporals, and 1,015 privates.[80] The allotment of additional men being delayed the second year, Capt. Charles B. B. Bubb of the Air Corps War Plans Section tried to show that by law the War Department had to furnish the men. Construing that the Air Corps Act gave "authorize" a "mandatory" meaning, Bubb asserted the act "demanded" the air arm expand to 15,000 men. The Air Corps could not increase or decrease the number—nor could the War Department. That figure had to be met in 5 years regardless of whether the Army grew, shrunk, or stayed the same. Examining the history of the act, Bubb found that Congress did not intend Air Corps expansion to be at the expense of the rest of the Army. But since Congress did not supply sufficient money for expansion, "the War Department has no alternative," Bubb declared, "but to continue to increase the Air Corps at the expense of other branches."[81]

Bubb's colleagues could easily embrace his interpretation, but the Air Corps could not throw it at the Chief of Staff or Secretary of War with a demand for men. The War Department had already arrived at a contrary view. Moreover, the President did not want the Air Corps buildup to add to the cost of government. Consequently, the 536 enlisted men allotted to the Air Corps for the second year of expansion also came from elsewhere in the Army.[82] The policy of taking from other branches to bolster the Air Corps abided after Hoover assumed the presidency. Under James W. Good (Secretary of War from March 6, 1929, until his death on November 18, 1929), and under Patrick J. Hurley, his successor, the Army went on giving up enlisted men to provide the Air Corps the men still due from the second year and to supply those for the third and fourth increments of expansion.

Over four years, other branches parted with 4,992 men to the Air Corps. To obtain 1,248 men for the third year, for example, the War Department inactivated five infantry battalions and a regiment (less one battalion) of field artillery, and made numerous small adjustments in the Cavalry, Corps of Engineers, Chemical Warfare Service, and Ordnance Department. Secretary Hurley warned: "If the policy of building up the enlisted strength of the Air Corps by transfers from the other arms is continued, the Army will soon be unable to properly perform its many missions."[83]

The transfers ended in 1931. Because of the demand for stricter economy during the depression, the Air Corps allotment did not grow for the fifth year of expansion. A small gain in the number of men on duty the last year brought the enlisted strength to 13,060 on June 30, 1932. The Air Corps thus fell more than 1,900 short of the 5-year objective.[84]

Air Corps allotments of enlisted men were by grades reflecting seven levels of military rank from private to master sergeant, and by ratings for privates and privates first class as specialists of six classes.[85] Specialists received extra pay in recognition of their skills and as an inducement to reenlist. A private, for instance, drew $3 besides his base pay of $21 a month if rated specialist sixth class. His pay went up to $51 when he advanced to specialist first class.[86] As incentives for mechanics to stay in, the Air Corps Act authorized air mechanic ratings for enlisted men in grades of sergeant and below. First class mechanics received the pay of technical sergeants ($84 a month), second class the pay of staff sergeants ($72).[87]

The Air Corps came into existence with 2,440 specialists out of a total allotment of 8,342 enlisted men. It was allotted 1,258 air mechanics for Fiscal Year 1927.[88] In September 1926, however, General Patrick learned the number of specialists and mechanics might be trimmed due to budget cuts.[89] He insisted the Air Corps had to have specialists and mechanics, and could not achieve the results Congress expected from the five-year program if constantly handicapped by personnel reductions and other obstacles. He could not concur in "curtailing the activities and usefulness of the Air Corps in order that other branches of the service may be spared embarrassment." In seeking a solution for such problems, he said, both the spirit and the letter of law should be considered.[90] The Air Corps succeeded in retaining and filling its allotment of air mechanics. In fact the number allotted advanced to 1,517 in 1929, to 1,692 in 1930, and to 1,866 in 1931. In 1932 the number on the rolls averaged 1,816.[91]

The Air Corps did not fare so well in securing other specialists or noncommissioned officers in the upper grades. The enlisted men allotted for the first increment consisted mostly of privates and included no specialists or grades above sergeant.[92] General Fechet complained that only a few eligible candidates could be promoted to technical or master sergeant.[93] Not enough specialist vacancies were to be had. Tracing the source of the trouble, Secretary Davison discovered the sole grades and ratings available for Air Corps expansion came from disbanded units of the Infantry, Cavalry, and Field Artillery. Those branches could not furnish the grades and ratings needed by the Air Corps.[94]

The Air Corps was given 118 more specialists the second year and 308 the third. These, along with an upturn in the number of noncommissioned officers, made possible the promotion of a number of worthy men.[95] Noting considerable improvement the fourth year, Secretary Davison said it was "unfortunately at the expense of the rest of the Army." Owing to the shortage of money, allotments of specialists and noncommissioned officers did not increase the fifth year.[96]

Eighteen Hundred Airplanes

The Air Corps Act authorized a buildup to 1,800 "serviceable" airplanes by the end of the 5-year program. General Patrick had used that figure in January 1926 in plans for expansion. That number encompassed just operational planes, not war reserves, operational replacements for planes out of commission for repair or overhaul, or planes for the National Guard and Organized Reserve.[97] General Drum, War Department G–3, recommended 2,500 aircraft, embracing those on order as well as war reserves and planes for the Guard, and Reserve.[98] As G–3 and the Air Corps refined the plans, the number became 2,200 not counting aircraft on order. Secretary Weeks gave the Chairman of the House Committee on Military Affairs that figure but said it took in craft on order.[99] Using 2,200 a little later, General Patrick divided it into 1,672 active, 110 in reserve, and 418 on order.[100] In authorizing 1,800 serviceable planes, Congress included equipment for the National Guard and Organized Reserve but not obsolete planes, those awaiting salvage, or those undergoing experiment or service tests. To maintain 1,800 serviceable aircraft, Congress permitted procurement of 400 a year to replace obsolete and unserviceable items.[101]

Seeing it was impossible to secure the force he thought necessary, General Patrick worked with the General Staff to get what he could into the 5-year program. He pointed out the inadequacies to the Assistant Secretary of War 2 weeks after Davison took office. The Air Corps required 1,388 airplanes at war strength. In addition, it needed 695 more as replacements for those undergoing overhaul, and 907 approved by the War Department as war reserves for Panama and Hawaii. That brought the total to 2,990. General Patrick called the 110-plane reserve "totally inadequate."[102] Sixty aircraft in operation in the United States required 40 in reserve, 15 at the various stations, and the other 25 at depots for distribution. Each overseas garrison needed a reserve of sixty, forty overseas and twenty in United States depots, to keep forty in operation.[103]

The Air Corps studied several ways to solve the problem, one being to earmark a certain proportion (say 25 percent) of the 1,800 airplanes as replacements. Patrick's staff prepared a plan for reallocating planes to supply a station reserve of 106 and a depot reserve of 344, leaving 1,350 operating aircraft. This plan gave pursuit and attack squadrons 10 in lieu of 16 operational ships, bombardment squadrons 5 instead of 9, and observation squadrons 8 rather than 13.[104] Patrick's staff also devised plans based on 2,200 and 2,393 airplanes. The latter allowed 1,738 in operation, 140 in station reserves, and 515 in depot reserves.[105]

On November 30, 1927, two weeks before he retired, General Patrick recommended changing the authorization from 1,800 planes to 2,400 and doing away with the limitation on procurement of replacements.[106] Davison judged the proposal sound but desired not to change the law so soon. General Fechet inherited the problem when he moved up to Chief of Air Corps on December 14, 1927.[107] He waited until November 1928 before requesting changes in the law: Keep the figure 1,800 but describe it as being "service" (i.e, tactical) rather than "serviceable" airplanes. Besides the 1,800, authorize as many more aircraft as the Secretary of War thought necessary for the National Guard, Organized Reserve, and schools. Remove the restriction on the number of replacements.[108]

In February 1929, Davison named General Fechet head of a board to revise the 5-year program and draft legislation to carry it into effect. The board found that an authorization of 1,800 planes left these shortages: 330 (tactical units), 374 (schools), 103 (National Guard), 79 (Organized Reserve), and 8 (miscellaneous activities). Add 387 for depot maintenance and the aircraft needed became 3,081, divided as follows: 840 pursuit, 229 bombardment, 178 attack, 802 observation, 103 cargo, 546 primary training, and 383 advanced training. The board suggested changes in the Air Corps Act similar to those Fechet proposed the previous year. Davison urged the War Department to rush study of the report so legislation could be secured in the next session of Congress, starting in December 1929.[109]

While trying to change the law, the Air Corps attacked the problem from other angles. Using a direct approach, it asked for money to buy airplanes to replace those out of operation for overhaul. The request being rejected, Davison warned that unless aircraft undergoing overhaul were excluded from the 1,800 authorized by law, tactical units could not be brought up to the strength specified by War Department Tables of Organization.[110]

During the following year the number of planes wanting overhaul averaged 12.5 percent. Davison noted this meant the Air Corps had to have 2,058 airplanes to keep 1,800 in operation. He said the Air Corps required 515 replacements annually (25 percent of 2,058), but this might change with better materials and improved maintenance.[111]

The Air Corps sought to capitalize on the wording of the 1926 act, which distinguished between "serviceable" and "obsolete" aircraft. The Air Corps classified several of its craft, including DH–4s and NBS–1s, as "limited standard," meaning they served some purposes but were unfit for military operations. The Air Corps wanted to keep and fly these planes while classifying them "obsolete" so they would not count against the 1,800 authorized by law.[112]

Secretary Davison, who thought this worth investigating, discovered the Army's definition of "obsolete" prevented the Air Corps from flying "limited standard" airplanes after reclassification. Davison's executive officer, Maj. Delos C. Emmons, and the Air Corps staff produced a definition to fit the situation, but the General Staff insisted on applying the present one.[113] The Air Corps then set out to see what it could do with "serviceable," a term not in common use in the Air Service before the 1926 act. What did "serviceable" mean? Did it cover aircraft being overhauled, for instance, or those on order? The Air Corps asked the General Staff for a definition. Davison added a question: Did the term as used in the law include training planes? The Judge Advocate General gave an opinion: The Air Corps' training planes, as well as those of the National Guard and Organized Reserve, counted as part of the 1,800, but planes undergoing overhaul did not. Secretary of War Hurley authorized the Air Corps to exclude equipment out of commission for depot maintenance in preparing estimates for 1,800 serviceable airplanes.[114]

The Air Corps budget estimate for Fiscal Year 1932 included money to purchase 234 aircraft for an operational reserve, which would give the Air Corps a total of 2,034. This made the overall estimate for airplanes more than $32 million for the year. The Bureau of the Budget approved $13.6 million.

Congress appropriated $15.3 million.[115] The Air Corps then returned to basing its estimates on 1,800 planes.[116] Over the 5-year period, Congress appropriated money for more than 2,000 new tactical and training aircraft and related equipment, parts, and accessories. On June 30, 1932, the Air Corps counted 1,814 planes on hand for the Regular Army, National Guard, and Organized Reserve, with 210 (11.6 percent) of these temporarily out of commission for repair or overhaul.[117]

The period of expansion was one of transition in airplane construction. In 1926 the era of stick-and-wire aircraft for military use was drawing to a close. New craft still came covered with fabric, but used metal instead of wood in the fuselage. The Air Service had produced a "modernized" de Havilland (DH–4M) by putting the wings, tail, and engine of the old wood-and-fabric DH–4 on a fuselage of welded steel tubing. The PW–9, standard for pursuit in 1926, used steel tubing in fuselage and empennage. Engineers were making good progress in developing all-metal planes of monocoque or semimonocoque design. The Air Corps bought some all-metal ones in the five-year program. By 1933 the trend in the Air Corps was away from metal and fabric biplanes toward all-metal monoplanes.[118]

For the first increment of expansion the Air Corps bought 16 additional PW–9Ds, and for the second, 33 more P–1Cs. The second year it also ordered 18 P–6s equipped with the 600-horsepower Curtiss Conqueror (V–1750), a new engine cooled with Prestone.[119] The P–6 likewise had oleo hydraulic struts on the landing gear and toe-operated wheel brakes, both adopted by the Air Corps as standard for its aircraft. After experimenting with many changes in the P–6s, the Air Corps in 1932 ordered 46 P–6Es, which came with wheel spats and with tailwheels in lieu of skids.[120] A formation of the fast, little ships dashing across the sky above Selfridge Field, Michigan, made a wondrous sight. The chrome-yellow the Air Corps adopted in 1927 for wings and tail to give greater visibility contrasted sharply with the old olive drab still being used on the fuselage. The national insignia—a red dot, inside a five-pointed white star, in a circumscribed blue circle, adopted in 1921—stood out clearly on the wings. "U.S. ARMY," lettered in black, appeared between the stars on the under surface of the lower wing. The rudder, bearing a blue vertical stripe and 13 red-and-white horizontal stripes, resembled "Old Glory."[121] The fuselage aft of the cockpit bore the insignia of the 17th Pursuit Squadron—a great snow owl swooping down on its prey, the illusion being heightened by talons painted on the spats.[122] The 27th Pursuit Squadron at Selfridge further put on a good show with its P–12s, the Air Corps' best acrobatic plane during this period. The government placed its initial order for 10 with Boeing in November 1928. Capt. Ira C. Eaker got the first plane in February 1929, for a flight from Brownsville, Texas, to France Field, Panama Canal Zone. In June 1929 the Air Corps ordered 90 P–12Bs, the biggest single airplane order the Army had ever placed in peacetime. Orders for later series raised the number of P–12s acquired by the Air Corps to 366, with the last item of this model delivered in May 1932.[123]

The standard pursuit ships of the Air Corps carried one man and two forward-firing machineguns. In the late 1920s, however, the Corps became interested in a two-place pursuit plane carrying a gunner and an additional gun on a flexible mount. Such a ship, the Air Corps thought, might be better than the standard one-place aircraft in protecting bombers against hostile craft. Fitted with bombs, it could also serve as an attack plane. In maneuvers in California in April 1930 the Air Corps used two-place attack aircraft to try out tactics for two-place pursuit protecting bombers.

Six months after the maneuvers, the Air Corps tested an experimental ship, the XP–16, a gull-wing, open-cockpit biplane made of metal tubing covered with fabric. In addition to two forward-firing .50-caliber machineguns, it carried a .30-caliber flexible gun operated by a gunner who rode in the rear cockpit facing backwards. The Air Corps ordered twenty-five from the Berliner-Joyce Aircraft Corporation for the 1st Pursuit Group to test. The 94th Pursuit Squadron began receiving P–16s in mid-1932. The Air Corps bought two copies of a metal, two-place, low-wing monoplane from Consolidated, one to try for pursuit (Y1P–25), the other for attack (XA–11). Both crashed during tests at Wright Field, Ohio, in January 1933. Refined, the Y1P–25 became the P–30 and eventually the P–30A, fifty being ordered by the Air Corps in 1934.[124]

Unable to develop a successful plane specifically for attack, the Air Corps asked Curtiss to revise the O–1B observation plane for attack. It acquired 76 of these planes (A–3s), the first in 1928, followed by 78 of an improved version (A–3B) in 1930. The observation aircraft mounted two machineguns in the nose and two more on a Scarff ring in the rear cockpit. The chief change in conversion to attack added two guns, one in each of the lower wings beyond the propeller arc. Carrying 200 pounds of bombs under the wings, these attack ships cruised about 110 miles per hour and had a range of around 630 miles.[125]

The search for a new bomber to replace the NBS–1 led to the B–2 Condor, which Curtiss developed from the NBS–1 by way of the XNBS–4. The Air Corps ordered 12 in 1928 and 1929. But it bought most of the bombers for the 5-year program from the Keystone Aircraft Corporation, the successor to the Huff-Daland and Company, Inc., producer of the LB–1. The Air Corps ordered nearly 200 planes from Keystone from 1927 to 1932. They came in several models and series, the earlier ones designated light bombers (LB–5, LB–5A, LB–6, LB–7, and LB–10A). The distinction between light and heavy being eliminated in bomber nomenclature, some LB–10As became B–3As; the remainder, with different engines, B–5As. Later, Keystone included B–4As and B–6As. All were slow and of short range. Carrying three .30-caliber guns and 2,500 pounds of bombs, the B–4A went 121 miles per hour at top speed, cruised at 103, operated at altitudes up to 14,000 feet, and could go about 850 miles. The B–6As performed about the same except for less range.

Speed rose steeply in bombers the Air Corps ordered from Douglas (seven Y1B–7s) and Boeing (seven XB–9s and XB–9As) in 1932. The Boeing, the Air Corps' first all-metal, low-winged monoplane, flew 188 miles per hour and cruised at 165. The Douglas, an all-metal, gull-wing monoplane, was just slightly slower. Both models had open cockpits and retractable landing gear, used two engines, mounted two .30-caliber guns, and carried a crew of four (pilot, bomber, and two gunners). The service ceiling of both was about 20,500 feet. The Boeing could fly farther than the Douglas (540 miles against 410) and haul a bigger bombload (2,260 pounds against 1,200).[126]

The Air Corps purchased observation aircraft in many models and series including Curtiss O–1ES, O–1Gs, O–11s, and O–39s; Douglas O–2Hs, O–2Ks, O–25As, O–25Cs, O–32As, O–38s, and O–38Bs; Thomas-Morse O–19Bs and O19–Cs; and a few Loening amphibians, OA–1s and OA–2s. Moreover, it bought Douglas O–2Hs and Consolidated O–17s to replace National Guard Jennies, withdrawn from training in September 1927.[127]

The Air Corps also acquired a variety of ships for hauling passengers and cargo: 9 C–1s, 8-passenger version of the Douglas World Cruiser; 17 C–1As, with room for 9 people; 3 C–2s, and a 3-engine monoplane built by Atlantic; 13 Ford Tri-Motors (C–3 and C–3A, C–4 and C–4A); 11 Sikorsky sesquiplane amphibians (C–6s and C–6As), and 20 single-engine C–14s made by General Aviation. The Douglas planes had an open cockpit for 2 pilots seated side by side ahead of the cabin, the C–14s an open cockpit behind the enclosed cabin. In other models, the crew area was enclosed. The Air Corps obtained 8 7-place cabin monoplanes (YF–1s) from Fairchild to test as photographic planes. It redesignated these and 6 F–1As as C–8 and C–8A cargo planes, but continued to use them chiefly for photography.[128]

The Air Corps proceeded to buy PT–1s in Fiscal Year 1927 for primary training. The next two years it ordered PT–3s and PT–3As, similar to PT–1s but with different engines. Later it secured a few PT–11s and PT–12s, both modifications of the PT–3. The engines of the PT–12s were too powerful for primary training, so the Air Corps designated the planes basic trainers (BT–7s).[129] In the training program the transition from primary trainers to service aircraft took place at the Advanced Flying School until July 1927, when the Air Corps divided flying training into three parts— primary, basic, and advanced. The Training Center used de Havilland aircraft, especially DH–4M–1Ts and DH–4M–2Ts, in transition training until it replaced them with basic trainers commencing in 1930.[130] The first basic trainers were forty Douglas O–2Ks and thirty O–32As converted to BT–1s and BT–2s in 1930. The following year the Air Corps ordered from Douglas 146 BT–2Bs and 20 BT–2Cs, the former with tailwheels in lieu of skids and the latter with a new

headquarters at Fort Crockett on June 15, 1932, but did not get all of its units until April 1, 1933. The move of the 3d Attack Group and wing headquarters to Barksdale Field was delayed until February 1935.[142]

Throughout five years of expansion, tactical units for the most part approached authorized strength in enlisted men but not in commissioned officers. At the period's close, units at March Field had 118 officers (67 Regular and 51 Reserve) of the 218 set forth in tables of organization; Langley Field, 140 of 238; Hawaiian Department, 103 of 249; Fort Crockett, just 52 of 134. Selfridge Field was a notable exception, having one more officer than the 84 authorized.[143] The officer shortage stemmed primarily from the Air Corps' failure to secure its full quota. A contributing factor was the many officers diverted to other duties—running schools; studying in Air Corps, Army, or civilian institutions; managing depots; representing the government at Douglas, Boeing, Curtiss, and other aircraft factories; working with National Guard, Organized Reserve, ROTC units, the General Staff and the Office of the Assistant Secretary of War, at Army posts, and in U.S. embassies abroad. Forty-four served in the Office of the Chief of Air Corps. The Air Corps did not take into account some of these requirements in drawing up the 5-year program.[144] General Patrick had scant success in seeking to eliminate detachments to gain officers for training programs and tactical units.[145] About 50 percent of the Air Corps' officers, and 70 percent of its enlisted men, served in the tactical organization at the end of the fifth year of expansion.[146]

The Air Corps suffered not only from a shortage of officers but from low rank in many important positions. The problem was most acute in tactical squadrons, which were supposed to be headed by majors but often were commanded by junior officers. The root cause was poor promotion opportunities. In mid-1930, for example, 502 second lieutenants, 495 first lieutenants, 161 captains, 93 majors, 15 lieutenant colonels, and 5 colonels held commissions in the Air Corps, Regular Army. About three-fifths of the second lieutenants had less than 2 years of service; more than four-fifths of the first lieutenants had over a dozen years. The Air Corps needed senior officers with specialized training for staff positions and special assignments. Hence lieutenants were put into many positions calling for captains and majors.[147]

The War Department failed to get Congress to solve the promotion problem, plaguing all branches not just the Air Corps.[148] Even so, the Air Corps did not take advantage of a statute enabling it to provide proper rank for officers in many key positions. The Air Corps Act of 1926 authorized temporary promotions of one or two grades for officers holding command or staff positions when no one was available in the authorized grade.[149] General Patrick saw the disadvantages outweighing the advantages. Since this section of the law was permissive, he suggested it not be used. He said the Air Corps assigned officers according to their qualifications and availability. Consequently, it could not treat all men exactly alike. One first lieutenant might command a squadron; another, equally qualified and equally deserving, might teach in a flying school. Under the Air Corps Act the former, gaining an "accidental advantage," advanced to major; the latter stayed a first lieutenant. Just the same, General Patrick, did not want this section repealed; the Air Corps might want to use it later. Gen. Charles P. Summerall, War Department Chief of Staff, and Assistant Secretary of War Dwight F. Davis agreed. And Congress raised no objections.[150]

Junior officers remained in charge of tactical units. Secretary Davison noted in mid-1930 that majors commanded just five of the Air Corps' 53 tactical, school, and service squadrons. The rest of the squadrons, he said, were "headed by captains (not many), first lieutenants (quite a few), and second lieutenants (several)." He discovered a second lieutenant with less than four years of service commanding a squadron composed of 18 other second lieutenants and 129 enlisted men, with government property worth in excess of half a million dollars. Such a "spectacle," Davison said, "should make defense-minded citizens ponder."[151]

The Air Corps Act of 1926, a milestone in the history of the U.S. Army's air arm, worked fine in some respects but not in others. How well F. Trubee Davison performed as Assistant Secretary of War for Air can be seen in his pleasing neither the War Department General Staff nor the Air Corps. The former complained he assumed functions belonging to them. The latter insisted he was not vigorous enough in supporting aviation when dealing with the General Staff and the Secretary of War. Davison remained as Assistant Secretary until Democrats replaced Republicans in Washington in 1933. The new President, Franklin D. Roosevelt, left the post vacant until April 1940 then appointed Robert A. Lovett. The Air Sections of the General Staff functioned until discontinued in 1930. The Adjutant General thereupon announced that at least one Air Officer would normally be on duty with each division.

Although Congress provided for temporary promotions, the Air Corps, foreseeing other problems, did not use the authorization to solve the difficulties arising from the low rank of officers filling many key positions. However, the promotion provision remained on the statute books for future use.

The expansion program contained problems the Air Corps could not or was unwilling to solve. It could do nothing to change the attitude of enlisted men who wanted to fly but not as enlisted pilots. On the other hand, it would not modify restrictive personnel and training policies to alleviate a shortage of flying officers.

The airplane program was particularly nettlesome, due in part to misconception and confusion concerning statistics used in developing and approving the program. Members of the Air Corps, the War Department, and Congress did not always make clear what they included in a particular figure. Attempts at definition were at times misunderstood or, if understood, soon forgotten. The number "1,800" in the Air Corps Act meant different things to different people. As used in General Patrick's plan, it would have given the Army a lot more planes than the same figure used by Secretary of War Weeks. This matter was further muddied by the absence of sound data for calculating requirements, especially those for operational reserves of airplanes, something not very well understood by the airmen, much less by the General Staff and Congress. Hence, the five-year program was attended by the Air Corps' constant struggle to get the airplane authorization raised, in which it failed.

Lack of money stood in the way of a bigger airplane authorization. It resulted in personnel ceilings, prevented full allocation of manpower spaces, and curtailed training and commissioning. Expansion commenced at a time of American business prosperity, but economy of government was a prime objective of the Republican administration, the United States Congress, and the American people in general. Halfway through the five years came the stock market crash, then an economic depression forcing more stringent economies. Even so, congressional appropriations for the five years were only slightly less than the estimates on which Congress approved the program. The total amount available, however, was far below the total of the larger estimates the Air Corps submitted during the course of the program to take care of deficiencies and new requirements, particularly in procurement of airplanes.

During expansion the Air Corps did fairly well with respect to enlisted men, ending about 10 percent below the goal. But it was 22 percent short of Regular officers and 33 percent in rated airplane pilots. One factor in the pilot shortage was inability to reach the target set for Reserve officers on extended active duty (40 percent short), and the complete failure of the enlisted pilot program.

In numbers of aircraft, the Air Corps met the objective set by the Air Corps Act. During the expansion, the Air Corps acquired new and better equipment in larger quantities and much more rapidly than at any time since the war. Reliability rose in all types of planes. In performance, pursuit ships showed good gains in speed, climb, ceiling, and handling; bombardment aircraft in ceiling and range. Furthermore, the program achieved a sharp increase in the number of tactical planes. Nonetheless, at the period's close, many tactical units were far below peacetime strength in planes and pilots. Moreover, formation of tactical units lagged behind schedule. Any assessment of the expansion program ultimately hinged on whether the emphasis was on gains or deficits. The Air Corps emerged at the end of five years much better off than it was at the beginning, though nothing like what the airmen wanted and hoped to attain.

Chapter XIII

Tactical Training

With the unfolding of the five-year program, the tactical capabilities of the Army's air arm suffered from unit inactivations and personnel transfers. As the program progressed, however, tactical units recouped those losses and gained new strength, even though most ended the period short of the goal. Their training consisted in part of ground instruction in bombing, gunnery, navigation, photography, and night flying. Unit training included formation, night, and cross-country flying, navigation, and work with the ground forces. Squadrons spent little time on the range and seldom used live ammunition. To drop live bombs on an abandoned bridge or an old ship, to test mobility in a long cross-country flight under severe operating conditions, to pioneer in high-altitude tactical flying, or to experiment with a new technique—these were significant events in a unit's life.

The training of tactical units in the mid-1920s was hampered by a lack of people, money, equipment, supplies, and facilities. In 1927, General Patrick reported a well-rounded program of unit training impossible due to the personnel pinch. At the same time too few airplanes caused him to recommend, and Secretary Davis to direct, curtailment of cross-country flying. Regulations required each pilot to make one cross-country trip a month. The new directive let a commander authorize flights in a zone of five hundred miles around his station. It permitted longer flights only under exceptional circumstances or to meet a military requirement other than cross-country training. However, pilots acquired valuable cross-country experience ferrying new aircraft from factories to their units.[1]

General Fechet reported in 1929 that insufficient equipment prevented observation units from making satisfactory progress in unit training. The 2d Bombardment Group at Langley Field, Virginia, owned just eight bombing planes at midyear. The 7th Bombardment Group at Rockwell Field, California, needed hangars for its planes. The 3d Attack Group was without adequate housing for its equipment at Fort Crockett, Texas.[2] But conditions in tactical units improved with construction and repair of facilities, procurement of new aircraft, and assignment of more officers and men. In 1931, Fechet termed tactical training "generally satisfactory," though some units, especially observation squadrons, were not up to expected proficiency in aerial gunnery and bombing.[3]

Owing to inactivations of units and transfers of officers from tactical to school units to start the five-year program, fewer men than usual engaged in the annual bombing and gunnery matches held at Langley Field in May 1927. The scores, however, rose in every event.[4] In 1928 the score in the bombardment match went up, but scores dropped in the pilots match (pursuit), observers match, and pilots match (observation and attack). All heavier-than-air organizations in the United States, the Panama Canal Zone, and Hawaii took part. The winners, including those of previous years, received badges recently approved for distinguished aerial gunners and bombers.[5] The War Department authorized all attack, bombardment, observation, and pursuit squadrons in the United States and the Canal Zone to send representatives to the matches in 1929. Six units in the United States did not participate, the 11th Bombardment and the 1st, 12th, 16th, 22d, and 99th Observation Squadrons having no men proficient enough to be eligible.[6] Some observation squadrons found no officers qualified for matches at Langley Field in September 1930. Disappointed by gunnery scores, General Fechet wanted more attention given to aerial gunnery training, chiefly for observers.[7] The number competing climbed in 1931. Pilots scored a little higher, though the shooting was still below par. The observers did worst of all,[8] but improved in 1932, the last year of the matches.[9]

The Air Corps did not set up a special school to teach aerial gunnery and bombing. That work fell to tactical units, which tried to increase proficiency by practicing with camera guns, dummy bombs, and live ammunition. But they received only small allotments of live ammunition and often wanted a nearby range for practice. General Fechet and his successor as Chief of Air Corps, Gen. Benjamin D. Foulois, complained of too few ranges and about training deficiencies.[10] Pilots and observers of Mitchel Field, New York, with no range of their own, went to Aberdeen, Maryland, the nearest place available to them. The 1st Pursuit Group operated a range 130 miles from its station at Selfridge Field, Michigan. The 2d Bombardment Group at Langley Field was better off than most Air Corps units. It continued to use Mulberry Island in the James River, where the 1st Provisional Air Brigade practiced for the tests against the battleships in 1921. The group also developed a high-altitude range on Plum Tree Island across Back River from Langley and in addition used targets in Chesapeake Bay.

Like other groups, the 2d always welcomed an opportunity to attack a "real" target. It thought its bombing of a highway bridge over the Pee Dee River near Albemarle, North Carolina, in December 1927, "second in importance only to the battleship bombing in 1921." The state of North Carolina built the bridge in 1925 of reinforced concrete with piers sunk in bedrock. Afterwards, it gave the North Carolina Power Company a permit to build a dam that would raise the water over the bridge. To get the permit, the power company built a new bridge a short distance upstream. The state then turned over the older bridge to the Army for tests by the Air Corps, Field Artillery, and Corps of Engineers. On Saturday, December 17, 1927, the 2d Bombardment Group sent 28 officers and 60 enlisted men under the command of Capt. Asa N. Duncan to Pope Field for the tests. Operations began early Monday morning and continued 5½ days. The schedule called for 20 missions a day, with two planes dispatched together at 20-minute intervals. Each mission took about 2½ hours. With but 8 pilot-bomber teams available, the men flew at least 5 and sometimes 7 or 8 hours a day. The group's target, the west span of the bridge and its approach, was about 20 feet

View of Pee Dee River Bridge after bombing.

wide and 400 feet long. Orders cautioned against damaging the new bridge and powerlines north of the target.

Starting with bomb cases loaded with sand, the group scored 1 direct hit on Monday from 8,000 feet. Proceeding with sandloaded bombs on Tuesday and Wednesday morning, it got another direct hit from 6,000 feet. The aircraft used 300- and 600-pound demolition bombs Wednesday afternoon and Thursday. Those landing just a few feet from the bridge did no damage. Direct hits with 600-pound bombs merely blew off the railing and dug a 3-foot hole in the floor. The group tried 3-plane flights in vee formation, each plane releasing two 600-pound bombs in salvo on signal from the leader. The results proved poor, the 3-plane formation being too small to offset any error of judgment on the part of the leading bomber. On Friday the 2d Group made 5 direct hits, 3 with 600-pound and 2 with 1,100-pound bombs. The heavier bombs demolished 3 sections of the approach and damaged the floor of the west span. With 7 bombs on Saturday, the group scored 2 hits on the pier at the west end of the bridge, "tearing and twisting the span beyond repair." The work completed, the group flew back to Langley Field that afternoon (Christmas Eve).

These attacks on the Pee Dee River bridge gave the 2d Bombardment Group some idea of how bombs of different weights affected reinforced concrete. The results likewise interested the tacticians and theorists of the Air Corps Tactical School, which shared Langley Field with the 2d Bombardment Group. The group's operations in North Carolina also showed a need for far greater bombing accuracy, meaning a lot more practice. They further confirmed that several planes in formation and dropping on the lead bomber had a much better chance of hitting the target than did a smaller formation. General Fechet was pleased the 2d Group had accomplished such a large operation, involving flying some twenty-five thousand miles and carrying about seventy thousand pounds of bombs, "without injury or casualty and without any motor failure or airplane trouble."[11] Two incidents, however, might have had serious consequences.

Two Reservists, 1st Lts. W. O. Bunge (pilot) and W. K. Andrews (bomber), on active duty with the 2d Group, landed at the end of a mission not knowing their Martin bomber carried an armed 300-pound demolition bomb in the bay. After landing they discovered that when Andrews had salvoed the bombs, one had hung on the rack. The other incident involved 2d Lts. G. F. Stowell, Air Reserve (bomber) and Herbert C. Lichtenberger, Air Corps (pilot). When Stowell released a 100-pound bomb as a sighting shot, it failed to drop off and held up the 600-pound demolition bomb he released next. The wind in the bomb bay whirled the arming vanes off the 600-pound bomb, leaving it ready to explode at the slightest jolt. Stowell crawled over the top of the fuselage from his bombing position in the nose of the Martin bomber to the rear cockpit. Using a forked stick carried for just such an emergency, he tripped the release. Both bombs dropped and exploded in the river about two miles below the bridge.[12]

In 1931 the 2d Bombardment Group acquired another "real" target, a World War cargo vessel, the USS *Mount Shasta*. The group last dropped live ammunition on a ship in the tests against the USS *New Jersey* and USS *Virginia* in 1923. The *Shasta* had spent the last decade tied up in the James River. The U.S. Shipping Board made the old hulk available to the Army for Langley Field bombers to attack and sink. Maj. Herbert A. Dargue, 2d Group Commander, and Lt. Col. Roy C. Kirtland, commanding Langley Field, told reporters the operations would show how long bombers required

Maj. Herbert A. Dargue, (right) 2d Group Commander, conducts bombing operations against USS *Mount Shasta* (below).

to take off and intercept an enemy ship, and would test marksmanship and the effects of bombs. They cautioned, however, against expecting anything sensational because this was bombing practice. The nine bombers carried no heavy ordnance, just 100- and 300-pound high-explosive, and 30-pound phosphorus bombs.

An Army mine planter, the *General Schofield*, towed *Shasta* out to sea Monday evening, August 10, 1931. The Coast Guard cutter *Mascoutin* followed with newsmen and other observers. At daybreak Tuesday, the sea was so rough and the weather so bad at the rendezvous (60 miles off Currituck Light and 110 miles southeast of Langley) that Lt. Col. Ira Longanecker, Air Corps information officer, radioed from the *Mascoutin* telling Dargue to delay an hour and bomb at 1200 instead of 1100, but Dargue did not receive the message. The bombers left Langley at 0930, the time originally scheduled. Six observation planes and a blimp left earlier with observers and photographers.

Because of the weather, Major Dargue decided to go by way of Currituck Light. This was farther than a direct line from Langley Field, but kept the planes over land longer in case of emergency. The observation planes and blimp flew straight to the rendezvous, arriving about 1100. Expecting the bombers soon, the *Schofield* turned the target loose for the attack, but the bombers failed to find the *Shasta*. The Navy did not help matters by offering to lend planes and flyers to sink the ship. The bombers spotted the target the next day but scored just two direct hits, one a dud, the other a 300-pound bomb doing no damage. When the planes left, the *Shasta* still floated. A crew from the *Mascoutin* could not open the sea cocks to sink her. The *Mascoutin* and another cutter, *Carrabassett*, then fired 1-pound shells and machinegun bullets into the hull below the waterline for two hours before the *Shasta* began to go down.

Trying to explain the "bombing flop"[13] to the public, Major Dargue said: "The Squadron actually went through hell in getting to the *Shasta* and the attack was staged under trying conditions." He pointed out that five miles out from Currituck Light the planes ran into a heavy line squall with driving rain, high winds, and weather so thick the planes became separated. (Even so, the weather cleared enough for the planes to reassemble, and the sun was shining when the time came to bomb.) Major Dargue was also quoted as saying: "We scored several direct hits that would have undoubtedly sunk the vessel promptly if the bombs had been of adequate size." He explained: "We wanted to use larger bombs, but were instructed to use the 100- and 300-pound bombs we use in our regular target practice." A correspondent of *The New York Times* reported talk among newsmen about the inaccuracies of the bombing. The consensus, he believed, was the bombs were too light and 1,000-pounders would have sunk the *Shasta*.[14]

A few months later an old vessel, the *Haines*, being towed to sea by a

Coast Guard cutter to be sunk, went down in a fishing channel off Plum Tree Island, where it was a hazard to boats. The 2d Group's 49th Bombardment Squadron was given a chance to destroy it. The flyers could not easily see the *Haines* under water, but a 10-foot-square float anchored over the wreck gave them an aiming point. Six planes from the squadron flew two practice runs with sand-filled bombs. The next time the aircraft carried 100- and 300-pound bombs loaded with TNT. After the attack, an inspection party found only a few splinters, pleasing the 49th Squadron immensely. During Air Corps bombing matches, the distance of the strikes from the center of the target averaged about 200 feet in 1929, 194 in 1930, and 150 in 1931. The radial error in the bombing of the *Haines* averaged 25 feet, an achievement the 49th Squadron thought "worthy of note."[15]

Another squadron, the 90th Attack at Fort Crockett, interrupted training in 1929 to resume border patrol. During insurrection in Mexico, bullets whizzed across the border as government troops and rebels fought at Naco, Sonora. Maj. Gen. William Lassiter, Commanding General, Eighth Corps Area, Fort Sam Houston, Texas, sent Brig. Gen. Frank S. Cocheu to Fort Huachuca, Arizona, to take charge. Cocheu's warning to both government and rebel forces in Mexico went unheeded. After a rebel airplane dropped bombs on the American side of the border at Naco, Arizona, General Lassiter ordered aircraft to the border. Twelve planes of the 90th Squadron led by Capt. Horace N. Heisen left Fort Crockett with full equipment at 1310 on April 6, 1929, stopped at Marfa, Texas, overnight, and reached Fort Huachuca at 1210 the next day. Six planes of the 12th Observation Squadron had already arrived from Fort Sam Houston. Aerial operations came under Maj. Eugene A. Lohman, 2d Division Aviation Commander. Other officers and men, traveling by land, soon were supporting flying operations.

Major Lohman first staged an aerial parade of all eighteen aircraft between Douglas and Nogales, Arizona, to warn the Mexicans and reassure Americans along the border. After that he set up patrols, two planes departing at dawn and two at three in the afternoon, keeping two other planes and pilots on alert. The parade and patrols produced the desired result. At least no border incidents occurred, and both detachments went home early in May.[16]

Units adjusted training and operations to take advantage of changes and improvements in equipment. The arrival of P–12s, for instance, let pursuit units operate at higher altitudes. On December 13, 1929, 2d Lt. Norman H. Ives of the 95th Pursuit Squadron at Rockwell Field took a P–12 with military load to 30,000 feet. He thought this a record for pursuit planes, but 2d Lt. George E. Price of his squadron soon broke it. Price went to 31,200 feet, this being part of the squadron's "intensive training at service and absolute ceiling."[17] Capt. Hugh M. Elmendorf, the squadron commander, led the 95th "to its zenith" in April 1930. His altimeter registered 33,000 feet, but was probably off by several thousand feet at that height. Consequently, the squadron claimed just 30,000. The squadron's operations showed 26,000 feet to be the highest efficient altitude for a formation of P–12Bs. The 95th trained at that altitude but sometimes went to maximum height for tests and experiments.[18]

The tactical application of high-altitude flying in pursuit craft was shown on June 7, 1930, when the 95th Squadron flew over Los Angeles at an altitude "beyond the vision of the human eye." Engaged in mimic warfare, it protected six B–2s of the 11th Bombardment Squadron that struck Los Angeles from 15,000 feet. Just before the bombers attacked, the pursuit planes "dived earthward, belching out volleys of machine gun fire and releasing light bombs on . . . anti-aircraft nests." Then the pursuit ships "zoomed heavenward," set to dive and assist the bombers if needed. People on the ground basked in California sunshine, but the pursuit pilots, five miles up, found the temperature forty degrees below zero. Against this, each pilot wore fifty pounds of clothing. At high altitude, he sucked oxygen from the tank in his cockpit. The 95th claimed being the "first air squadron in history . . . [to deliver] an attack from so great a height," the only one, "which has reached that altitude in a formation flight."[19]

Obtaining P–12s, the 1st Pursuit Group at Selfridge Field also started high-altitude training. After flying to Muskegon, Michigan, to help dedicate a municipal airport in September 1930, the 94th Pursuit Squadron, commanded by 1st Lt. Harry A. Johnson, flew back to Selfridge at 25,000 feet.[20] Johnson, formerly a test pilot at Wright Field, Ohio, supervised service tests by the 94th Squadron on a new system that automatically regulated the flow of liquid oxygen. A face mask connected to the oxygen supply enabled the pilot to breathe through his nose, which was better and more comfortable than taking oxygen from a tube held in the mouth. The squadron's training at that time consisted mainly of tactical maneuvers at altitudes between 27,000 and 28,000 feet.[21] The 94th Pursuit Squadron undertook a series of flights during October 1931 to test the oxygen system, other equipment, and operating procedures. It wanted to learn how long it took to climb to 20,000 feet and fly to an objective 100 miles away, and the amount of gas the aircraft consumed at different speeds at high altitude. The squadron then was prepared for a long cross-country flight.

Taking off for Washington, D.C., the 12 planes led by Lieutenant Johnson spiraled upward through the clouds and set a compass course at 20,000 feet. After an hour a slight break in the clouds allowed Johnson to check the course. A short distance past Pittsburgh the clouds disappeared, making the rest of the flight easy. Not far from Washington, members of the squadron noticed 2d Lt. George A. Hersam, Jr., "cavorting crazily about the sky and then diving for the earth." Hersam had accidentally detached his oxygen tube when he reached for a map. He passed out but recovered at 7,000 feet, righted his plane, and rejoined the squadron just before it touched down at Bolling Field. All planes but one went the entire distance on auxiliary gas tanks, flying at an average airspeed of 200 miles per hour. The flying suits kept the pilots warm enough the first hour, then the men commenced to chill. Most of the pilots had trouble with their goggles fogging; one lost his 15 minutes after taking off. The flight showed that "should Washington be endangered by attack, a squadron could take off from Selfridge Field, arrive at Washington, drop the auxiliary tanks, and have sufficient gasoline in the main tank to go right into combat at ceiling just 2 hours after leaving Selfridge Field."[22]

As the 27th Pursuit Squadron and other units received the proper equipment, they too began training at high altitude. By late 1932, the 19th Pursuit Squadron, for example, owned six planes (P–12Cs and P–12Es) equipped with oxygen for practicing maneuvers at 20,000 to 25,000 feet. Stationed at Wheeler Field in Hawaii, the 19th thought the excessive cold the biggest objection to such work.[23]

Pursuit and bombardment units also practiced operations wherein commanders gave orders by radio. During the last half of the 1920s, the Air Corps put on a number of demonstrations of radio communication between ground stations and airplanes. One took place at a military tournament at Chicago in June 1927. Station WLS broadcast conversations between students of the Air Corps Communications School, flying in O–2Cs, and persons on the ground. People attending the tournament at Soldier Field heard the conversations over a public address system. The flyers promptly acknowledged and executed orders from the ground to ascend or descend, go in one direction or another, or fly in a certain formation.[24]

At that time, however, tactical units had no equipment suitable for "radio control" of operations. The low-frequency sets in the "130" series proved deficient. Instead of the intended SCR–135 set, bombers used the observation set (SCR–134). The pursuit set (SCR–133) was too heavy for the little planes, was hard for a pursuit pilot to operate, and performed poorly. The Air Corps therefore purchased only a few and seldom used them. To communicate his orders to other pilots in his formation, a pursuit squadron or flight commander generally preferred older methods, like rocking the wings of his plane, waving his arms, or making other visual signals.[25]

The Materiel Division worked with the Signal Corps and industry to find better radio equipment for the Air Corps. It sought lighter, more reliable, easier to operate sets of longer range and higher quality communication. It encountered many obstacles, one being noise in voice transmissions. Despite static and interference from the electrical systems of airplanes, flyers usually could read telegraphic signals. On the other hand, microphones picked up so much noise from the engine and from the rush of the wind in an open cockpit that flyers usually could not understand spoken transmissions. Among other technical problems was the search for suitable aircraft antennas. The masts and other fixed antennas sometimes used provided only short-range communication. More often the antenna consisted of a wire from one hundred to two hundred feet long, weighted at the end, and let out from the plane to trail behind. A trailing antenna gave greater range than a mast, but presented a hazard to other planes and prevented flying in close formation. If the pilot maneuvered suddenly, the antenna snapped off. If he flew too close to the ground, the wire caught on a tree or other object and pulled off.[26]

In 1928, a board of Air Corps and Signal Corps officers, headed by Maj. Horace M. Hickam, concluded that two types of radio communication were required: "command" communication within a pursuit, bombardment, or attack unit in the air, or between units in combined operations; and "liaison" between aircraft in the air and Air Corps or other units on the ground. The equipment on hand did not meet requirements. As an interim measure, the board suggested modifying present transmitters and adoption of a new receiver (BC–152), being developed by the Signal Corps. Compact, light, easy to install, and simple to operate, the BC–152 was compatible with any of the three ground sets then standard for aircraft communication, as well as with the Air Corps' interphone. Three plug-in coils for 250–400, 400–850, and 850–1500 kilocycles afforded a broad frequency range. With one coil the receiver weighed only 10 pounds; with all three, 11.5 pounds. Much smaller and thinner (12 x 8 x 2.75 inches) than previous sets, it became known as the "pancake receiver."[27]

The BC–152, linked with the BC–114 transmitter from the SCR–134, enhanced radio communication for bombardment, observation, and transport planes. Tied in with the transmitter (BC–129) from the SCR–133 (adapted for code only), it marked the first real efforts to employ radio in pursuit operations. The 95th Pursuit Squadron, under Captain Elmendorf, pioneered in radio control of pursuit. The squadron exhibited its methods during a flight of eighteen planes from Santa Monica, California, to San Diego in March 1930. Elmendorf's plane carried a radio receiver and transmitter with a trailing antenna; the three flight commanders had receivers with mast antennas. General Fechet, in California on an inspection tour, directed the pursuit ships by voice radio from a transport plane piloted by Capt. Ira C. Eaker. Receiving Fechet's orders, Elmendorf relayed them by radiotelegraphy to flight commanders, who passed them on to the other pilots by arm signals.[28]

The 95th Squadron tried out its radio equipment and procedures during Air Corps maneuvers at Mather Field, California, in March 1930. Receiving orders from a ground station or from the commander of the bomber formation his squadron supported, Captain Elmendorf had no trouble controlling his flights by radio. The maneuvers disclosed, however, that radio slowed the action and tended "to prevent the sought for dash and slam so desirable in Pursuit operations."[29] Maj. Willis H. Hale, chief of the operations section throughout the maneuvers, reported that the "use of radio exceeded anything . . . ever attempted heretofore in the history of the Air Corps." In bombing operations assisted by attack aviation, the bomber commander directed attack planes against unexpected targets. Radio added flexibility to command in the air. It might eliminate the need for predetermined itineraries, rendezvous points, and times of attack. Moreover, it might keep a mission from failing due to unexpected antiaircraft resistance or a storm. Nevertheless, operations were hindered by trailing antennas on attack planes.

Radio served other purposes in the maneuvers. On one occasion the ground station at Mather Field picked up a message from an observation plane reporting the landing of a hostile group at an airdome eighty-five miles away. The station relayed the message by telephone to bombardment and attack groups, both with armed aircraft on alert. Within seven minutes from the time the observer began transmitting, a bombardment and an attack squadron took off to assault the hostile airdrome.[30]

When a new Fokker C–7A arrived at Rockwell Field, Maj. Carl Spatz, 7th Bombardment Group Commander, took out some of the seats and installed two BC–114 transmitters and two BC–152 receivers. Now, without changing frequency, he communicated with pursuit and bombardment aircraft in voice or code. On October 4, 1930, Spatz explained radio control to the public through Fox Movietone News. In maneuvers at Burbank, California, the same month, he parked his command ship at United Airport, fastened the trailing antenna to the flagpole, and by radio directed pursuit planes defending the airport.[31] Early in 1931 the 7th Group secured BC–SA–167 receivers to use with the old transmitters, thus injecting voice radio in nearly every phase of training. For example, it devoted the last two weeks in March to radio-controlled interception. A bomber acting as a target reported regularly by radio to the ground station, giving its location, altitude, and course. From that information, the ground station guided pursuit to the objective.[32]

Communication officers at Rockwell Field sent information to Selfridge Field, where control grew more complex as the 1st Pursuit Group expanded under the five-year program. The 1st Group discovered that a commander could control no more than six planes with visual signals. The group needed a radio to become an effective combat force—a set "as light as possible, using voice as a means of transmission, be 'fool proof' and easy to operate, and able to withstand the violent maneuvering to which these small, speedy airplanes are subjected."[33] Beyond that, pursuit pilots dreamed of a radio "which can be fitted into a watch pocket and will always work."[34] Meantime, the 1st Group made do with the sets on hand, including a few recently acquired

Maj. Carl Spatz directs flight by radio during Air Corps maneuvers at Mather Field, California.

BC–SA–167s. The Group Commander, Maj. Gerald E. Brower, used radio to direct a mock attack on Cleveland, Ohio, on April 7, 1931. Of the fifty-two planes making the flight, just those of group, squadron, and flight commanders had radios. Station WTAM rebroadcast Major Brower's orders, described the attack, and broadcast an address by the Selfridge Commander, Maj. George H. Brett, from an airplane over Cleveland. From time to time the group put on similar shows elsewhere as it practiced operating under radio control.[35]

In 1931 pursuit units assisted the Materiel Division in testing experimental radios for command communication. Of five sets submitted by various manufacturers, one satisfied weight, space, and performance specifications. This radio (the SCR–183) weighed barely 43 pounds, handled both voice and code, and had a fixed antenna. With plug-in coils, the receiving component (BC–192) covered all required Air Corps frequencies. The set's automatic volume control could be tuned and operated by touch and sound without dependence on sight, a big plus when operated by the pilot. Pleased with the performance in 1932 tests, and believing the SCR–183 and BC–192 could serve bombardment and attack as well as pursuit, the Air Corps decided to order as many as funds permitted.[36] These were the sets the Air Corps used in the 1934 airmail operations.

Tactical training of aviation units encompassed cross-country flights to test mobility. Maj. Hugh J. Knerr, 2d Bombardment Group Commander, led 3 flights from Langley Field to the west coast in 3 successive years. In September 1928 he took 9 LB–5As to Los Angeles for the National Air Races. In light of bad weather, sandstorms, high temperatures, and poor servicing facilities, the trip out consumed 6 days. Home at the end of 3 weeks, the group believed the trip has produced "facts concerning the transcontinental movement of a bombardment squadron which are far more valuable than any paper data which might be obtained."[37]

In 1929, the 2d Group dispatched nine LB–7s with sufficient gasoline for eleven or twelve hours. Departing early on August 5, the planes stopped twice for servicing, and reached Rockwell Field the evening of the 6th. Three of the bombers at once flew a mission over the ocean off Point Loma. Major Knerr viewed the flight as proof of the Air Corps' ability to "move with an offensive or defensive Air Force from one coast to another in less than 40 hours flying time and still be able to undertake defense or attack maneuvers upon reaching the objective." The men returned at a more leisurely pace punctuated by several stops, one at Cleveland for the National Air Races.[38]

The most notable event on the 1929 trip ensued soon after the flight left Denver's municipal airport on the way home. As often the case on long flights, the men tuned their radios to a station offering music and entertainment, in this case the General Electric station at Denver. About 125 miles east of the city, they heard the announcer say:

> This is Station KOA, Denver, calling DO–1, commander's ship of the air fleet that left Denver this morning. The plane remaining in Denver because of motor trouble requests that a spare magneto be returned to the field.[39]

Second Lieutenant A. J. Kerwin Malone, commander of the disabled bomber, could not pinpoint the trouble before the other ships took off. Left behind, he soon traced it to a defective magneto. The sole spare being aboard one of the other planes, he telephoned KOA, explained the problem, and asked that a message be broadcast.

Mrs. J. C. Traw of Flagler, Colorado, turned on her radio at 0915, just in time to catch the announcement and the plane's reply—message received, "and thanks very much, KOA." Then she heard KOA repeat the announcement. She was unaware the planes' transmitters could not reach Denver, but she knew KOA had not heard the reply. Listening to the men in the planes talking to each other, she heard one say, "Maybe someone will phone them." Mrs. Traw made the call—a plane was being sent back with the extra magneto. By noon, Malone's plane and the relief ship were set to rejoin the air fleet. In 1930 the 2d Group once more flew to California, this time to join other units in maneuvers at Mather Field.[40]

The 1st Pursuit Group's training also involved cross-country flights to test mobility under low temperatures. In January 1927, Maj. Thomas G. Lanphier led twelve planes from Selfridge Field to Montreal, Canada.[41] Three years later, Maj. Ralph Royce took the group on an arctic patrol to Spokane, Washington, to test equipment and see if pursuit units could move long distances by air during severe winter weather. Royce's expedition comprised eighteen P–1s, three cargo ships, and an observation plane. The transports hauled twenty mechanics and spare parts. One carried radio sending and receiving equipment to communicate with amateur radio operators along the way. First Lieutenant Ennis C. Whitehead piloted the O–2K, his passenger being Hans Christian Adamson, public relations man on Secretary Davison's staff.[42] The route lay along the northern border, by way of Saint Ignace, Michigan; Duluth, Minnesota; Minot, North Dakota; and Great Falls, Montana. The return course ran a little to the south, through Helena, Montana; Minneapolis, Minnesota; Wausau, Wisconsin; and Escanaba, Michigan. The plan would have the patrol leave on January 8, 1930, arrive at Spokane on the 11th, depart on the 13th, and return to Selfridge on the 16th. But it did not work that way. As a reporter at Selfridge Field wrote: "Jack Frost and Old King Boreas were not kindly disposed to the Pursuiters."[43]

Right: Lt. Ennis C. Whitehead (left) and Mr. Hans Christian Adamson during stopover in Wausau, Wisconsin en route to Spokane, Washington; below: Curtiss P–1C pursuit plane equipped with skis for 1st Pursuit Group's Arctic Patrol.

Bad weather delayed departure until the 10th. The flight was marred by subzero temperatures, blizzards, drifts, snow flurries, and fog; frozen hands, feet, and noses; broken starters, oil lines, pistons, axles, and skis. These made the trip miserable, slowed progress, and scattered the group. Thirteen pursuit ships reached Spokane on the 19th, and three more pursuit planes got there the next day. Royce and his flyers started back on the 22d. He assembled all remaining aircraft at Minneapolis on the 27th, those missing being a pursuit ship wrecked in North Dakota and the O–2K. Whitehead and Adamson fell so far behind on the way west that they gave up and went to Dayton. Sixteen pursuit ships and a transport returned to Selfridge on January 29, the other two transports the next day, and the last pursuit plane on February 2, 1930. A Selfridge Field reporter thought that after the men thawed out, they no doubt would "look back on their long jaunt through the frigid Northwest as the experience of a lifetime."[44] Soon Major Royce and the 1st Group headed westward once again, this time for Mather Field for maneuvers.[45]

Besides the time consumed in routine training conducted generally along lines suggested by the Chief of Air Corps, tactical units spent much training time testing and experimenting with new equipment and techniques that might enhance tactical capabilities. The assignment of fresh pursuit ships with a long-sought increase in altitude was not accompanied by detailed instructions as to how the unit was to employ the ships in combat. The unit knew the plane's performance as determined by Air Corps test pilots at Wright Field, Ohio. Even so, its own pilots needed to ascertain for themselves how high they could go and, more important, discover the practicable limit for tactical operations. Moreover, they had to fit tactics to the best use of the additional altitude afforded by the new planes, and then perfect operations to exploit the new advantage to the fullest. That is what Captain Elmendorf and the 95th Pursuit Squadron at Rockwell Field did with the P–12.

One innovation often led to another, and not infrequently depended on still another for its successful application. To give pursuit planes more altitude did not suffice; pilots had to be able to fly and fight in the rarefied atmosphere of the higher altitude. Engineers came up with an automatic oxygen system, and test pilots checked it. Nevertheless, the system still required testing by tactical units, and training for its use in actual operations.

In the early 1930s, tactical training began to deal with the control of aerial operations through radio communication. Development in this area suffered long from lack of satisfactory equipment. The 130-series of radios at first seemed promising but failed to measure up in field tests. When improvements appeared, the Air Corps endured long delays in securing sufficient equipment for operations. But using what they could get, tactical units worked out and practiced techniques that combined radio and visual communication to direct aerial operations.

Special projects undertaken by units from time to time afforded valuable training as well as assessment of the unit's capabilities. Theorists might calculate bomb damage under various conditions, but the one sure way to measure the damage was to drop a particular bomb on a particular target, as the 2d Bombardment Group did when it attacked the Pee Dee bridge. How well the Army's bombardment group might have performed if called upon in 1931 to defend the nation against enemy ships may perhaps be judged by its performance in the matter of the *Shasta*. A demand to defend the Mexican border brought units quickly to the scene in a show of force, apparently ending the affair. Mobility, such as that shown by the 90th Attack Squadron in the border incident, began in the last half of the 1920s to assume a much larger place in tactical training. Mobility allowed assembly of aviation elements from all over the United States to form a wing, division, or air force for exercises or maneuvers.

Chapter XIV

Exercises and Maneuvers

Annual maneuvers, begun by the U.S. Army Air Service in 1925, brought aviation units of the corps areas together for training and operations, to experiment with and to test organization, equipment, tactics, and logistics. Maneuvers gave Air Corps officers command and staff experience and practice in handling large units in the field. A different military problem was set for each occasion. For the 1927 maneuvers, the War Department authorized a one-sided operation in Texas, a field army and its aviation against a simulated enemy. In lieu of maneuvers the following year, the Air Corps dispatched a composite group around the country demonstrating operations at Army service schools. In 1929, it collected units into a provisional wing for operations in California. The next year it gathered all available men and planes into a provisional air division for the two-fold purpose of exercising as large a unit as possible, and of displaying the Air Corps to the American people. Lack of money prevented concentration of aviation units for maneuvers in 1932.

Planning the Air Corps' annual maneuvers for 1927, General Patrick obtained permission for a one-sided operation of a field army against a simulated enemy. It took place in Texas during May 15–19. Maj. Gen. Ernest Hinds, Commanding General, Eighth Corps Area, led the army, General Fechet the army's air service. The 2d Division of Fort Sam Houston, Texas, furnished the ground troops. The Air Corps assembled 127 officers, 296 enlisted men, 108 airplanes, and 1 airship. Over and above the observation, pursuit, and attack aviation that normally made up the air service of an army,

Secretary of War Dwight Davis (left) and Ass't. Secretary of War for Aviation Trubee Davison at the 1927 Texas Air Corps manuevers.

the Air Corps attached a bombardment group for the maneuvers. Fechet's force included 1st Pursuit Group P–1s, 2d Bombardment Group Martin bombers, 3d Attack Group O–2s, and O–7s from the 12th and 16th Observation Squadrons. Transport planes (C–1s and C–2s) helped move men and equipment to and from the maneuver area. Among the observers were Secretary of War Davis, Assistant Secretary Davison, chiefs of the Army's various arms, military attaches from 8 countries, and newsmen.

The principles behind the tactical problem chosen for the maneuvers mirrored the War Department's concept for the employment of army aviation at that time:

(1) Before contact between opposing ground troops, aerial reconnaissance extended the reconnaissance area. Receiving objectives in general terms, the air service commander enjoyed great liberty of action.

(2) As ground forces came into contact, the army commander exercised closer control over aviation. Mission assignments grew more definite, sometimes specific, but the air service commander retained his freedom as to means and methods.

(3) The air force (if not dissipated) constituted a highly mobile and powerful reserve that could be rapidly concentrated at threatened points to hamper and delay the enemy, whether he be aggressive or retiring.

(4) The air force (if not dissipated) was available for special missions against sensitive points in the battle area, the enemy's supply organization and installations, and the rail and road nets. It could also be used to extend artillery action and directly support ground forces.

(5) The rapidity and power of air action made possible in battle many things previously impossible.

(6) An air force had two major limitations: its inability to conquer and hold terrain alone; and its dependency, to a greater degree than other arms, on atmospheric conditions.

Under these principles, pursuit pilots from Selfridge Field patrolled enemy lines, sought out and engaged enemy pursuit, intercepted enemy bombers, and supported bombardment and attack aviation. Bombing crews from Langley Field attacked enemy positions, troops, airdromes, bridges, and lines of communication. Attack planes from Fort Crockett struck enemy columns on the march, a division in an assembly area, and a brigade moving into position for counterattack. Pilots and observers from Fort Riley and Fort Sam Houston used visual and photographic methods to reconnoiter roads, railroads, airdromes, and the battle area day and night. Some night observation missions failed due to low ceiling or mechanical trouble with flares or cameras. Otherwise, General Fechet reported, aerial missions of all kinds usually "were promptly and efficiently executed." Just two planes failed to take off on scheduled missions, two left a mission and went home, and two landed away from home. Of 430 mission scheduled, Fechet's men completed 98.8 percent.

On the way to San Antonio, four aircraft and crews from each of the 1st, 2d, and 3d Groups put on a two-day display at Fort Benning, Georgia. Attack planes placed live ammunition on targets representing field artillery

Officers attending Air Corps demonstrations at Langley Field, Virginia, in May 1928.

and infantry on the march and under cover. Bombers struck an ammunition dump with 100-pound bombs, and pursuit executed aerial maneuvers and assisted attack ships in spreading smokescreens over infantry and artillery on the march. Other ships and crews from the three groups as well as observation craft and crews from Marshall Field at Fort Riley gave demonstrations at Fort Riley and at Fort Sill, Oklahoma, before arriving at San Antonio. After the maneuvers, the Air Corps simulated an air attack on marching troops of the 2d Division at Camp Stanley, northeast of San Antonio, a smokescreen being laid by airplanes to cover friendly troops.[1]

Instead of maneuvers in 1928, the Air Corps formed a group at Langley Field on May 1 to perform aerial operations at Army service schools. The group consisted of a pursuit squadron (25 P–1s) from Selfridge Field, an attack squadron (16 O–2s and 9 A–3s) from Fort Crockett, and a transport squadron (15 cargo planes) drawn from various stations. The 2d Bombardment Group further took part in a show at Langley for the Air Corps Tactical School. Congressmen, other officials, and newsmen flew to Langley for the event, the climax coming when pursuit, attack, and bombardment aircraft hit ground targets with live ammunition and ended by passing in review at 200 feet. Afterwards the group gave exhibitions at Army posts in North Carolina, Georgia, Oklahoma, and Kansas, before disbanding on May 22.[2]

Dissatisfied with the limited training afforded by one-sided fighting like that of 1927, the Air Corps secured permission for two-sided maneuvers with ground forces in Ohio during 1929. A line north and south across the state between Dayton and Columbus was the boundary between two nations. Red (to the east) and Blue (to the west) went to war at 0001, Thursday, May 16. For ground forces the maneuvers took the form of a staff exercise for Regular Army, National Guard, and Reserve officers of Fifth Corps Area, commanded by Maj. Gen. Dennis E. Nolan. Canvas panels on the ground (white for Blue forces, orange for Red) marked infantry and artillery; mounted on trucks they signified troop movements. Troops from Fort Benjamin Harrison, Indiana, and Fort Thomas, Kentucky, placed and moved panels.

The Air Corps contributed about 275 officers, 225 enlisted men, and 200 airplanes under the command of General Foulois. In addition to the 1st Pursuit, 2d Bombardment, and 3d Attack Groups, tactical forces consisted of the 11th Bombardment and 95th Pursuit Squadrons from the new 7th Bombardment Group at Rockwell Field, California. The 9th Observation Group and the 15th and 16th Observation Squadrons supplied observation. Blue air force, commanded by Lt. Col. Henry C. Pratt, had 18 bombardment, 15 pursuit, 15 observation, and 34 attack planes. Red, under Maj. John N. Reynolds, comprised 6 bombardment, 18 observation, and 50 pursuit aircraft (pursuit also being used for attack). Red air force set up headquarters at Norton Field, Columbus; Blue at Fairfield. War having been declared, Colonel Pratt ordered a night attack by 7 bombers on a transportation junction at Newark, east of Columbus. Bad weather precluded flying, but the umpires (faculty members of the Air Corps Tactical School) permitted simulation and ruled the bombers stopped traffic for 12 hours.

The first mission in which aircraft actually took to the air came soon after daylight on May 16, when Colonel Pratt sent 38 attack and pursuit planes against a Red airdrome at Brice, southeast of Columbus. The raid failed because the Reds, fearing such an attack, had moved their airplanes. But Blue found other targets, rail yards at Columbus and a depot east of the city. Encountering three Red observation craft, Blue pursuit shot them down and a little later got three more. Major Reynolds retaliated by strafing a Blue airdrome north of Dayton at Troy, destroying 10 attack planes. The Blues then launched a second raid on Newark, this time putting the junction out of commission for 2 days. Expecting a raid on the airdrome at Fairfield, Colonel Pratt scattered his aircraft about the field to render them less vulnerable. Even so, 5 Red bombers, escorted by 18 pursuit ships, succeeded in demolishing 9 planes on the ground and in escaping without loss despite the withering machinegun fire Blue put up in defense of the field. Aerial operations the first day closed with a night bombing raid on a Red airdrome spotted that afternoon by a Blue observation plane.

All this fighting occurred without firing a shot. Umpires flying with each mission assessed the results by applying rules drawn up for the maneuvers. An observation plane not encountering hostile pursuit secured any information available from the air. If it met 1 hostile pursuit ship, it still accomplished its mission; if it met 2 or more, it failed. When an umpire with a bomber formation saw interception by hostile pursuit with 2 to 1 superiority, he marked the mission unsuccessful, divided the number of bombers by 3 to get the number shot down, and divided that by 2 for pursuit losses. An attack formation minus pursuit support fulfilled its mission even in the face of equal numbers of hostile pursuit. A division in bivouac lost 1 percent of its people and 10 percent of its animals when hit by an attack group in daylight. A pursuit unit engaging an equal number of pursuit aircraft lost 5 percent of its ships in combat over its own territory, 10 percent if within 5 miles of either side of the front, and 15 percent if over hostile territory. The rules covered almost every conceivable condition. They specified, for example, that a bombardment, attack, or pursuit group required 2 hours to refuel and reload for another mission, and that the group could carry out 2 daylight missions at 75-percent group strength each day.[3]

War correspondents reported "violent air fighting" as Blue ground forces drove eastward toward Columbus on Friday, May 17. The first real casualty of the air war came on Saturday after Blue bombers and attack aircraft raided a Red depot. Escort planes of the 95th Pursuit Squadron met Red pursuit over the north side of Columbus. In maneuvering, two planes from the 95th collided, the prop of 2d Lt. Andrew F. Solter's plane cutting the rear of the fuselage of 2d Lt. Edward L. Meadow's P–12. Meadow died but Solter saved himself by parachute. Learning of the "unfortunate affair," General Foulois told newsmen: "It's all in the day's work of the Air Corps. Although an unhappy occurrence, the accident will cause no change in the maneuver plans, which will be carried out as scheduled."[4] Operations Saturday morning wound up the first phase of the maneuvers. That afternoon the opposing air forces joined in a demonstration for the people of Cincinnati. Thousands watched as pursuit ships battled bombers and attack planes bent on destroying the city. Oscar B. Hanson of the National Broadcasting Company described the battle by radio from an Air Corps transport plane. Afterwards, aircraft went into formation and, with General Foulois leading, flew to Norton Field to pass in review before General Nolan, the maneuver director.

The second phase commenced with a bombing raid on New York. Following the endurance flight of the *Question Mark* in January 1929, Secretary Davison wanted to show the public the practical application of aerial refueling to military operations. He directed that during the maneuvers a bombing plane fly from Dayton to New York and back without landing along the way. First Lieutenant Odas Moon piloted a Keystone bomber, with First Lieutenant Eugene L. Eubank, copilot; Mr. Bradley Jones, navigator; 1st Lt. John P. Richter, refueling officer; and 1st Lt. Charles T. Skow, radio operator. Capt. Ross G. Hoyt piloted the refueling plane with Sgts. Robert A. Brewer and Wilbur J. Simmons operating the refueling equipment. First Lieutenant Leroy M. Wolfe flew a transport carrying a National Broadcasting Company crew.

The plan called for Moon to depart Dayton at 1400, Tuesday, May 21, and to refuel over Washington, D.C., at 1815 while Hanson described the operation over a national network. Wolfe would follow Moon to New York so Hanson could broadcast while the bomber circled the city and dropped flashlight bombs over lower Manhattan. Moon would next head for Atlantic City, dispense a parachute flare, continue to Washington for another refueling about 2300, and be back at Dayton before dawn. On the way from Dayton to Washington, weather stopped the refueling and radio planes at Uniontown, Pennsylvania, but the bomber pressed on to New York. The duty officer at Governors Island heard the Keystone's motors at 2125 but could not see the plane. He told reporters that without searchlights, his antiaircraft guns were useless. ("Of course, we would have [lights] if she had been a real enemy plane.") Moon flew up Broadway to Central Park, turned south, and dropped a flare that lighted the bay, ships in the harbor, and Statue of Liberty. After dispensing two more flares, the bomber circled and disappeared. Mitchel Field and Newark Airport turned on lights in case Moon wanted to land. He had enough gas to go to Bolling Field, where he received orders to stay overnight.

The Air Corps quickly improvised a new plan for the following day. All three planes assembled at Bolling Field and flew to New York. Over the city, the bomber took a load of gas from the refueling plane. To give people a better chance to see how it was done, Moon and Hoyt repeated the performance (without transferring gasoline) four times. All the while Hanson broadcast. The demonstration completed, the planes landed at Mitchel Field to remain overnight before returning to Ohio, where war raged once again.

As the Red Army resisted the Blue drive toward Columbus, aircraft of opposing air forces reconnoitered and strafed troop columns and enemy positions. They attacked supply bases, ammunition dumps, rail centers, bridges, dams, and airdromes, and battled each other in the air. The climax came Saturday, May 25, 1929, when the Air Corps loaded planes with real bombs and live ammunition to show the might of aviation against ground forces. A regiment of soldiers stood guard to keep any of the thousands of visitors from entering the target areas at Fairfield. Cardboard figures represented ground troops; canvas portrayed transport columns. A clatter of machinegun fire signaled the start as planes of the 3d Attack Group dove on the targets. Circling, the attackers wheeled back to drop 25-pound bombs. Three other attack ships swooped low and spread a dense smokescreen over the destruction. The 2d Group's bombers triggered 100-pound bombs on a dummy ammunition dump. The crowd saw a bluish-white pattern of bursts just right of the target and an instant later heard the blast. The attack aircraft swept back over the field, and landed, trailed by the bombers. Fifty or more pursuit planes dived from the clouds, roared across the field, zoomed upward, and made a wide circle before landing. An armistice ended the maneuvers at 0400 Sunday.[5]

Brig. Gen. William E. Gillmore directed field exercises at Mather Field in April 1930. The Air Corps publicized movement of the 1st, 2d, and 3d Groups to California as an example of its mobility. Other units taking part included the 7th Bombardment Group from Rockwell Field and the 91st Observation Squadron from Crissy Field. At March Field they formed a provisional wing. Gillmore said the 114 air force planes (62 pursuit, 24 bombers, 28 attack) "represented the maximum in concentration of air force units in the Continental limits of the United States."[6] These exercises highlighted the value of radio in directing and controlling tactical operations. Tests using attack aircraft in the role of pursuit to protect bomber formations led to purchase of biplace pursuit planes for the purpose. Air transportation received special attention. Maj. Henry H. Arnold, the provisional wing's S–4 (Supply), used airplanes to move supplies from Rockwell to Mather (which, being inactive, provided nothing). He employed 3 C–2As and 1 C–1 (all the transport planes in the Ninth Corps Area) and 1 LB–7 on the job. They completed 36 flights from March 3 to April 1 to haul 36,548 pounds of cargo. When the exercise commenced, these planes and the transports arriving with

Brig. Gen. William Gillmore (right) with Mayor of Sacramento, Calif., and the Adjutant General, State of Calif., at the 1930 Mather Field exercises.

the tactical units formed a transportation squadron. They furnished daily passenger service between Mather and Crissy and, when and where needed, made special flights for passengers and supplies.[7]

Secretary Davison directed that the maneuvers of 1931 be in the form of demonstrations to acquaint the American people with the Air Corps and afford them a clearer idea of the Army's air effort. The Air Corps seized this opportunity to test division organization for the first time. War plans envisioned a division of some 2,200 aircraft, 4,000 officers, and 28,000 enlisted men attached to General Headquarters. General Foulois commanded a provisional division of 667 airplanes, 692 officers, 69 flying cadets, and 643 enlisted men. To come up with that many planes and men the Air Corps committed all save one of the squadrons stationed in the United States,[8] the instructors, students, and planes of the Advanced Flying School, flights from 19 National Guard squadrons, and every transport plane it could lay hands on. This furnished Foulois a pursuit wing (commanded by Maj. George H. Brett), a bombardment wing (Maj. John H. Pirie), an attack group (Maj. Davenport Johnson), two observation wings (Majs. Walter R. Weaver and Ralph Royce), and a transport group (Lt. Col. Augustine W. Robins). The planes were 130 P–12s, 17 P–6s, and 58 P–1s for the pursuit wing; 31 B–3As, 9 B–2s, 4 LB–7s, and a group (three squadrons) of O–38s for the bombardment wing; 70 A–3Bs and some A–3s for attack; 48 cargo planes of 9 models; and 9 models of observation planes ranging from the O–1 to the O–38.

The division assembled at Dayton, Ohio, on Saturday, May 16. The schedule allowed some time to practice forming in the air before going forth to display the Air Corps to the public. Bad weather delayed departure from Dayton, throwing the schedule off 1 day. Accordingly, the division put on its show at Chicago on Thursday the 21st, New York on the 23d, Boston on the 25th, Atlantic City on the 27th, Philadelphia on the 29th, and Washington on the 30th. The performance at each place was twofold. First came 29 aircraft from the 11th Bombardment, 13th Attack, and 95th Pursuit Squadrons in a combat demonstration of "thrilling dives, mimic air battle, and dog fighting." Then the rest of the division passed in review. The bombardment wing was first, followed by an observation wing, the attack group, and the other observation wing. Pursuit planes flew above and on both flanks. Bringing the division together in the air proved a difficult problem not entirely solved until the show at Boston. While at Chicago, General Foulois received a message from the Army Chief of Staff, Gen. Douglas MacArthur, ordering the division to New York at once to play at war. With the Atlantic Fleet sunk, the Panama Canal destroyed, the Pacific Fleet unable to come to the defense of the east coast, it was fortunate the Air Corps was already mobilized. Moreover, Maj. Henry H. Arnold, division G–4 (Supply), had stocks at Pittsburgh; Cleveland; Buffalo; Middletown, Pennsylvania; Aberdeen, Maryland; and Bolling Field to service units as they flew eastward.

In New York, General Foulois got word of an enemy fleet headed for New England. Ordering the division to proceed at once, General MacArthur flew to Boston and set up a command post. With the foe 250 miles east of Boston on the morning of May 25, the Chief of Staff directed the division to attack but the fleet faded into a fog bank. Bringing the division back to New York, MacArthur returned to Washington to control operations against an attacker whose movements dovetailed with the schedule of the division's demonstrations. The planes repulsed an assault at Atlantic City on the 27th and gained victory on the 30th. To the delight of the airmen, General MacArthur took to the air (with Capt. Ira C. Eaker at the controls) to lead the division on its triumphant procession over Washington.[9]

Lt. Col. Ira Longanecker, in charge of publicity, believed the demonstrations highly successful. Support from famous aviators helped put them over. James H. Doolittle followed the division as service representative for the Shell Oil Company. After the show at Boston, he flew pictures to New York to make the evening newspapers, affording publicity for both Doolittle and the Air Corps. In the show at New York, Charles A. Lindbergh flew one of the

on the full name in press releases) gave newspapers a picture made when the city was bombed. The caption read: "A flashlight photograph taken by Captain Albert W. Stevens, famous U.S. Army Air Corps photographer, from a plane flying at a height of 1,000 feet, near the middle of the Hudson River, showing Grant's Tomb, the new Riverside Church and an adjacent part of Riverside Drive."[12] This did not satisfy everyone. Mrs. F. Adler, writing to the editor of the *New York Times*, said it was "an outrage that bombs should land near apartment houses, awakening children and throwing invalids into a state of collapse from which they will suffer for a long time." She wanted an "energetic protest" lodged "to prevent such an outrage in the future."[13]

Congratulating the Air Corps on the "remarkable success of the maneuvers," President Hoover told Secretary of War Patrick J. Hurley they "reflected great credit on our military establishment."[14] Assistant Secretary of War for Air Davison thanked the public, the press, and the broadcasting companies for their support. "Their attitude," he said, "has shown an appreciation of the fact that the purpose of these maneuvers was to solve in actual operation difficult problems which can be solved in no other way." These maneuvers of 1931, he added, showed "the United States in first place in so far as the operation of large units is concerned."[15] General Foulois suggested exercises the following year to develop teamwork among units, improve tactics, and test basic principles of organization.[16] Nevertheless, the Air Corps did not get enough money for a major concentration of aviation units in 1932.[17]

Apart from Air Corps maneuvers over these years, Army air units engaged in numerous training exercises with ground or sea forces. Late in 1929, the Secretaries of War and Navy announced 5 exercises to be conducted jointly during the first half of 1930. All involved Air Corps units. Three dealt with defense of Panama, the Philippines, and Hawaii, the other 2 with harbor defense of Long Island Sound and San Francisco. The services carried out such exercises to test defense, improve tactics, and develop communication and teamwork.[18] One on April 15, 1932, had the 1st Bombardment Wing of March Field attacking Fort MacArthur, guarding Los Angeles Harbor. Maj. Carl Spatz commanded 27 aircraft from 3 squadrons of the 7th Bombardment Group and 3 planes from the 17th Pursuit Group. The 63d Coast Artillery (Antiaircraft) employed 3 sound locators and 5 searchlights, but no antiaircraft guns, to defend the fort. The defense attempted to spotlight the planes, the attackers to bomb and get away without being illuminated.

Flying between 6,000 and 10,000 feet, the three squadrons approached from different directions. They switched off navigation lights 3 minutes before the target. The first bomber of each squadron attacked at 2045; the rest followed at 3-minute intervals. As each came over the objective, it released a flare to simulate bombing, turned, dived 1,000 feet, and withdrew with lights on. At 2100 the bombing was half through, and pursuit planes arrived at 4,000 feet with lights on. Their job was to confuse the sound detectors. One bomber turned back because of engine trouble. Twenty were over the target exactly on time; 6 were from 30 seconds to 2 minutes early or late. Three did not drop flares because the release mechanism malfunctioned. Eleven bombers finished the mission undetected. Searchlights picked up 14 on the way out. Only 1st Lt. John M. Davies was spotted before he got to the objective and the Air Corps thought this "probably a matter of chance."[19]

Work with Army and Navy forces continued to occupy a prominent place in the activities of Air Corps units at overseas stations. The 6th Composite Group at France Field, for example, participated in the maneuvers of the Panama Canal Department in January 1932. When an enemy force landed forty miles east of Panama City, Maj. Lewis H. Brereton moved the 6th Group (Panama Canal Department Air Force) across the isthmus to an airdrome at Fort Clayton. With observation, bombardment, and pursuit forces divided about equally, opposing armies battled "fast and furiously" for ten days.[20] Also, Blue and Red forces often clashed in the Philippines.[21] And during Grand Joint Exercise No. 4 in Hawaii in February 1932, a Blue force composed of U.S. Navy ships and Army transports attacked Oahu. The defending Black force included units of the 18th Composite Wing commanded by Lt. Col. Gerald C. Brant. Just one of the umpires came from the Air Corps; the rest, including the chief umpire, Rear Adm. Frank H. Schofield, were Navy. As the exercise progressed, the "impersonal and impartial rulings" handed down by Navy officers impressed Air Corps officers.

The action began when the aircraft carriers USS *Lexington* and USS *Saratoga* launched 172 planes before daylight on February 7 to attack Luke and Wheeler Fields. At Luke they destroyed 15 ships on the ground, but these turned out to be dummies assembled by the Hawaiian Air Depot from scraps of old planes, the real ones having been removed to outlying airdromes. The 18th Wing flew a few reconnaissance missions but otherwise remained in hiding the first day. The Navy returned the next morning to machinegun and bomb the outlying fields but found no targets. Those of the 18th Wing departed earlier and concentrated high over Kahuku Point. They intended to follow Navy planes back to the carriers but rain and low visibility prevented. During the morning a Black observation aircraft reported 31 Blue seaplanes at anchor in Hilo Bay on the island of Hawaii. Bad weather kept commercial planes on the ground, but the 26th Attack Squadron took off at 1315, flew 230 miles to Hilo, attacked the seaplanes and the airport without opposition except for antiaircraft fire from a cruiser about 10,000 yards away, and got back by 1820.

Brant put his wing in the air before dawn on the 9th. At 0650 a patrol plane from Luke Field spotted the Blue carriers 55 miles off Barbers Point. Receiving orders by radio, the 23d and 72d Bombardment Squadrons immediately changed course for the carriers and put the *Saratoga* out of commission. Vice Adm. Henry E. Yarnell and his staff then transferred to the *Lexington,* which the umpires (not wanting to terminate the exercise) declared uninjured by the rain of 2,000-pound bombs. Aircraft of the 26th Attack and the 6th and 19th Pursuit Squadrons meanwhile engaged an equal number of Navy craft in a dogfight off Barbers Point, with each claiming victory.

As the mock war went on, bombardment and attack planes of the 18th Wing struck 3 transport ships off the north shore of Oahu. Capt. Auby C. Strickland led the 6th and 19th Pursuit Squadrons to machinegun and bomb a carrier. When marines landed on the west coast of Oahu early in the morning on February 12, 18th Wing attack planes assaulted the troops, bombers hit the transports, and pursuit ships battled Navy aircraft in the air. At 0930, Brant and Schofield agreed nothing further could be learned by prolonging the exercise. More than 1,000 Army and Navy officers attended the critique. The *Air Corps Newsletter* reported, "One thing was clearly brought out, and that was the fact that Oahu is a hard nut for an enemy to crack."[22]

The aerial demonstrations given by aviation units on their way to and from maneuvers in Texas in 1927, as well as the performances of the demonstration group on tour the following year, showed ground troops how planes could work with and support them. The Air Corps found the maneuvers of 1927 an opportunity to exhibit aerial operations to ground troops, and to train air and ground forces to work together as a field army. General Hinds, the army commander for those maneuvers, took a similar view but placed a bit more emphasis on the other side. The maneuvers depicted ground operations to air forces and air operations to ground forces, fostered teamwork and cooperation between ground and air forces, and showed both the powers and limitations of aviation.

The operations of two opposing air forces in Ohio in 1929 lent weight to ideas being advanced at the Air Corps Tactical School. The maneuvers confirmed, for instance, that bombardment should normally be used against strategic targets, but also might properly be used on frontline targets not destructible by any other available weapon. Pursuit tactics and training changed after being shown by these maneuvers to be faulty. Attack, for which pursuit was being equipped and trained in addition to its own work, could be done more effectively by attack aviation. The 1929 maneuvers also revealed that the Army needed a two-seat or multiseat ship with more firepower than single-seat pursuit to protect bombardment formations. Among other things, the 1929 maneuvers disclosed poor communication—between pilot and observer, air and ground, and air to air. But the maneuvers that year produced nothing on supply or aerial transportation, both needing attention.

The California maneuvers of 1930, where radio figured prominently, showed solid improvement in communication but far more was needed. Experiments that year using two-seat planes to protect bombers led to procurement of biplace pursuit for that work. The maneuvers of 1930 affirmed organization for supply as correct and efficient. Use of airplanes to move supplies and passengers for the maneuvers indicated that, except for gasoline and ammunition, air units could be supplied by air, if sufficient cargo planes were available. These findings on supply and transportation were confirmed in 1931 during the concentration, operation, and dispersal of the provisional air division.

Maneuvers and exercises yielded technical data, proved or disproved theories, uncovered new ideas, afforded training not obtainable in any other way, and otherwise contributed to the progressive advancement of Army aviation. At the same time, they served Air Corps efforts to elicit public support for military aeronautics.

Chapter XV

Flying

The late 1920s were filled with remarkable flights recording a succession of "firsts" and setting new records. The premier event of the period was, of course, Charles A. Lindbergh's solo flight nonstop from New York to Paris in May 1927. The following year, Amelia Earhart, the first woman to make a transatlantic flight, crossed with Wilmer Stultz. Capt. Charles Kingsford-Smith achieved the first transpacific flight by airplane. In 1929, Hugo Eckener circumnavigated the globe in the *Graf Zeppelin* and Navy Comdr. Richard E. Byrd and Bernt Balchen flew over the South Pole. In 1931, Wiley Post and Harold Gatty went around the world in the *Winnie Mae*, and in 1932 Amelia Earhart flew the Atlantic alone. A complete list would contain the names of a number of U.S. Army flyers who won places for themselves and for the Air Corps in the annals of aviation.

One of the major Air Corps projects was a flight of five amphibians (AO–lAs) through Mexico and Central America to the Panama Canal Zone, then around South America and back to the United States by way of the West Indies. This flight was publicized as a goodwill tour to show the possibilities for establishing aerial transportation and communication in Latin America. The hope was it would place the United States in position to assume a leading role in the development of aviation in that part of the world. General Patrick expected a lot of good publicity for the Air Corps. Led by Maj. Herbert A. Dargue, the flight left Kelly Field, Texas, on December 21, 1926. Breakdowns and accidents delayed the flyers all along the route, a midair collision killing two crewmembers at Buenos Aires, Argentina. Stops and side trips to

Maj. Herbert A. Dargue (left) and Grover Loening, designer of amphibian used on Pan American flight.

deliver messages from President Coolidge to heads of Latin American governments took time. The Pan American Flight dragged on so long the American people lost interest. In some countries the flyers received a chilly reception. In a few places they met anti-American demonstrations despite the friendly attitude of government officials. But they were greeted enthusiastically at Bolling Field, when they wound up their tour on May 2, 1927. Their work, however, was soon eclipsed by Charles A. Lindbergh's flight across the Atlantic on May 20–21.[1] (*Map 8*)

California to Hawaii

Four days after Lindbergh landed the *Spirit of St. Louis* at Le Bourget Airport in France, James D. Dole of the Hawaiian Pineapple Company offered $25,000 for the first flight between the United States and Hawaii. He announced the prize on August 15, 1927, to afford Lindbergh time to come home from Europe and get ready. The Lone Eagle was not interested but others were, some more eager to be first than to take Dole's money.[2] Army

MAP 8

flyers had talked for years about flying to Hawaii. First Lieutenant Lester J. Maitland sought permission in 1919 and repeated his request to General Patrick in 1924. First Lieutenant Albert F. Hegenberger had worked on navigation equipment at McCook Field, Ohio, before being transferred to Hawaii in 1923. Now he wanted to fly back to the mainland. General Patrick said Hegenberger "bored me to death" with letters asking to make the trip. The Chief of Air Service turned down these petitions because "the time was not yet ripe."[3]

Lts. Albert Hegenberger and Lester Maitland (l. to r.), pilots for Hawaiian flight; below: *Bird of Paradise* makes the first nonstop flight to Hawaii from California, June 28–29, 1927.

Conditions changed by late 1926. Lieutenant Maitland served as aide to Assistant Secretary of War for Air Davison. Lieutenant Hegenberger returned to McCook Field. A radio beacon to aid navigation was tested successfully at Dayton, Ohio. The Air Corps acquired a new trimotor transport (the C–2) that could conquer the long trip over water. In December 1926, Maitland secured tentative approval for flight pending outcome of the plane's tests. Hegenberger would accompany him as navigator, radio operator, engineer, and relief pilot.

Upon completion of tests on the plane at Rockwell Field, California, on June 24, 1927, Secretary of War Davis authorized the flight. Richard V. (Dick) Grace, a movie stunt pilot (he flew in the movie *Wings*), was in Hawaii for a solo flight to the west coast. Ernest L. Smith and C. H. Carter were at Oakland, California, preparing their aircraft for a hop to the islands. Announcing the Maitland-Hegenberger flight, Secretary Davison called it "strictly an Army project," one unrelated "to any prize or any other offer made by private individuals for a successful flight across the Pacific." The Army had been considering such a flight for years. That it came at a time when prizes were tendered was, Davison said, "purely a coincidence."[4] Forbidden to accept any prize or award for the flight, the Army flyers turned down $10,000 for their story.

The Army field at San Francisco, California, was too small for the C–2 to lift off fully loaded for the Hawaiian flight. Maitland and Hegenberger therefore joined their competitors at Oakland Municipal Airport. Patrick inspected their plane (*Bird of Paradise*) at Oakland on June 27, 1927, and gave the green light for the next morning. Smith and Carter hoped to start ahead of the Army flyers. General Patrick sent Maitland and Hegenberger on their way at 0709: "God bless you my boys, I know you'll make it."[5] Maitland gunned the *Bird of Paradise* down the 7,000-foot runway. After rolling some 4,000 feet he got airborne, climbed to 2,000 feet, passed the Golden Gate, and headed for a tiny speck in the ocean 2,400 miles away. Smith and Carter departed 2 hours later but returned with a broken windshield. Grace was repairing his plane, damaged while trying to take off.

Hegenberger's job was to hold the *Bird of Paradise* on a course to the Hawaiian Islands in the vast Pacific. The margin for error was slim. If he should be off more than 3½ degrees from the great circle course he plotted, the plane would miss the island entirely. But Hegenberger was probably the best navigator the Air Corps had in uniform, and careful preparations had been made. The Signal Corps Aircraft Radio Laboratory at Dayton set up radio beacons at Crissy Field, California, and near Paia, Maui, in the center of the Hawaiian group. The *Bird of Paradise* had radio receivers and transmitters, an earth-inductor compass, four magnetic compasses, a drift sight, smokebombs for measuring drift, and a sextant. The aircraft further carried an inflatable rubber raft, tinned beef and hardtack, and five gallons of drinking water—just in case the flyers missed the island or for some other reason came down on the ocean.

Out of San Francisco, the *Bird of Paradise* cruised at 108 miles an hour, the air clear, the wind light. Maitland felt a bit uneasy at the controls; for the first time since 1922 he flew without a parachute. Why bother with one when it would be more hindrance than help in case of trouble? Hegenberger tried to check drift by sighting on smokebombs over the tail but could not because of the sun's glare. When high winds and rough water later rendered the bombs useless, he received good readings by using the drift sight in the floor to check white caps. The earth-inductor compass failed the first hour, and Hegenberger could not fix it. Soon the signal from Crissy Field stopped. He tried frequently to get it again but without success.

About 0900 Maitland and Hegenberger came to clouds extending from the water up to about 1,500 feet. After flying just above the clouds for about half an hour they saw water through some small holes. The holes were not big enough, however, for Hegenberger to measure drift. Maitland went down to 500 feet to get beneath. The ceiling decreased to 300 feet. They faced rain squalls and strong winds swinging to the northeast. Haze hid the horizon. Hegenberger navigated by ordinary compass, determining drift from the waves, and shooting the sun when it broke through the clouds. Though confident of his calculations, he wanted to check. He changed course to intercept the *Sonoma*. Knowing the steamship's position at noon, he computed the point and time of meeting. He figured the ship to be 25 miles ahead at 1423. They passed 724 miles out at 1445. Hegenberger put the *Bird of Paradise* on a track parallel to the original course. Rain squalls grew more numerous. Visibility decreased. Figuring they were near the *President Cleveland*, Hegenberger established radio contact at 1910. The *Cleveland* reported her position 1,157 miles from San Francisco Lightship; wind northeast at 30 miles per hour. But the plane's signal came in too weak for the ship to obtain a bearing by direction finder. From time to time Hegenberger relieved Maitland at the controls. During the night they flew above the clouds so Hegenberger could shoot the stars. About 2300 he picked up the headphones once again and heard the signal from the radio beacon. The aircraft was south of the course, instead of north, as both he and Maitland had thought. He adjusted the course to the beacon, but the signal stopped abruptly after about 40 minutes and was not heard again.

It was cold above the clouds, at 10,000–11,000 feet. The center engine ran rough and cut out. The *Bird of Paradise* lost altitude. Near 3,000 feet Maitland got the engine running. He then discovered ice had formed over the carburetor intake at high altitude. Not planning on cold weather they had left air-intake heaters off to save power. Maitland climbed back to 7,000 feet, the best he could do under the circumstances. Still, sufficient breaks appeared in the clouds to permit Hegenberger to sight the stars from time to time. At the rate they were flying, aided now by a strong tailwind, they would reach Hawaii before daylight. Not wanting to land in the dark, they cut their speed to 70 miles per hour. Eventually they saw a light they identified as the lighthouse at Kilauea Point, Kauai, the most northerly spot on the northernmost island in the Hawaiian group. They were about a hundred miles from their destination, Wheeler Field, in the middle of Oahu. It was raining. Rather than try to land in the dark on a field situated between two mountain ranges, they decided to circle Kauai until daybreak. When dawn came about an hour and a half later, they crossed Kauai Channel and reached Wheeler at 0629 (Hawaiian time).[6] They were tired and hungry, having flown 25 hours and 50 minutes with nothing to eat. Hegenberger did not find the chicken sandwiches and thermos bottles of soup and coffee put aboard at Oakland. (He did not look under the tarpaulin beneath his plotting board.)

Grace, the stunt man, came to congratulate the Army flyers. Having tried to take off several times, he had given up after wrecking his plane. Ernest L. Smith started from Oakland on July 14 with a new navigator, Emory B. Bronte. The Army turned on the beacons, but Smith and Bronte received the signal only part of the time. Running out of gas just as they reached Molokai, they crashed in a tree. Fourteen planes entered the Dole race on August 19. Eight started, four turned back, two went down at sea, two reached Honolulu. Arthur Goebel and Navy Lt. William V. Davis, Jr., who used the Army's radio beacons until within sight of Maui, won the Dole prize. But the Army flyers had been first. They wanted to fly home in the *Bird of Paradise* but General Patrick said "no." The Pacific flight measured up well when compared, as it inevitably was, with many other newsworthy and historic flights of 1927. Lindbergh called it "the most perfectly organized and carefully planned flight ever attempted."[7] Patrick attributed the success of the venture to "determination, preparation, and navigation."[8]

Question Mark

On New Year's Day 1929 the Air Corps set out to capture the world endurance record with a trimotor transport refueled in the air. Over 5 years had passed since Capt. Lowell H. Smith and 1st Lt. John P. Richter completed their historic flight of more than 37 hours. Since then Americans had paid scant attention to aerial refueling. How long could an Air Corps crew commanded by Maj. Carl Spatz keep the C–2A (28–120) in the air? The question—which gave name to the plane—was answered on January 7: 150 hours, 40 minutes, 14 seconds.

The flight of the *Question Mark* commenced and ended at Los Angeles Metropolitan Airport at Van Nuys, California. Spatz's strategy called for sustaining slow, even speed to prolong engine life while shuttling between Los Angeles and San Diego. Each morning he made pilot and flight officer assignments for the next twenty-four hours. Capt. Ira C. Eaker, chief pilot and second in command, was responsible for the throttle setting. If either of the pilots, 1st Lt. Harry A. Halverson or 2d Lt. Elwood R. Quesada, was at the controls when conditions demanded a change in the setting, Spatz and Eaker were to be notified at once. They were also to be informed in case of bad weather, uncertain visibility, motor trouble, or other emergency. If the plane had to be abandoned, the flight officer would alert all crewmembers and see they were ready to jump before he left the plane. The pilot was to stay at the controls (unless they were useless) until all others left the cabin. The flight officer kept the log, which he dropped to the ground each day. An observer for the National Aeronautical Association witnessed and certified the flight. Eaker's duties included the important one of winding the barograph, the instrument that continuously recorded time and altitude, furnishing documentary evidence of the flight's duration.

Two C–1s served as refueling aircraft. Capt. Ross G. Hoyt, 1st Lt. Auby C. Strickland, and 2d Lt. Irvin A. Woodring comprised the crew of one; 1st Lt. Odas Moon and 2d Lts. Andrew F. Solter and Joseph G. Hopkins the other. Capt. Hugh M. Elmendorf handled ground operations. Due to the weight of the equipment, the *Question Mark* took off on New Year's Day

Question Mark refuels shortly after taking off from Los Angeles, California.

with less than 100 gallons of gasoline. Moon and his crew soon linked up and transferred 100 gallons before Eaker, at the controls during takeoff and refueling, started for San Diego. Hoyt awaited the *Question Mark* at Rockwell Field, California. Installation of two 150-gallon gasoline tanks in the cabin afforded the *Question Mark* a capacity of 580 gallons. The cabin also contained a 40-gallon tank of oil. A trapdoor in the top of the fuselage let the crew receive gas, oil, water, food, supplies, and equipment from the refueling planes. The C–1s had extra gas tanks and a 30-foot hose that dropped through a hole in the floor to transfer gas to the *Question Mark*. The hose lacked a valve at the lower end, the flow being controlled from the refueling plane. A copper wire ran down the hose for grounding to the *Question Mark* during refueling to prevent static electricity from firing the gasoline.

Eaker took the controls during refueling. Halverson assisted him in the cockpit. Quesada helped Spatz with the equipment for receiving gas. SSgt. Roy W. Hooe operated a wobble pump to transfer gas from the cabin tanks to the wings. To refuel, Spatz donned goggles, a rubber raincoat, gloves, and face mask to avoid being burned if gasoline spattered. Climbing up on a platform, he thrust his head and shoulders through the hole in the top. The refueling plane came in behind, and stayed slightly above and a bit ahead of the *Question Mark*. In Refueling Plane No. 1, Hoyt held the C–1 on a straight-and-level course at 80 miles per hour. Strickland and Woodring lowered the hose and jockeyed it into position. With Eaker keeping the *Question Mark* steady, Spatz caught the hose and put the end in a funnel connected to the cabin tanks by aluminum pipe. Strickland opened the valve and the gas rushed down 75 gallons per minute. Refueling went well the first day. After the initial refueling by Moon, Hoyt made three contacts and transferred 600 gallons. When the *Question Mark* received 200 more gallons shortly after midnight, Spatz was showered with gasoline. He noted in the log that it "was distributed over the face and eyes principally. The first aid kit was used and zinc oxide applied. Nothing serious." The fumes that filled the cabin affected Quesada but he soon recovered. Spatz was sprayed two more times before the flight ended but with no real harm.

During the week the *Question Mark* stayed aloft, fog, rough weather, bumpy air, or darkness at times complicated refueling. Some contacts could not be completed on schedule or needed to be cut short. When conditions were too bad for refueling at Van Nuys or San Diego, the planes met elsewhere, once over Oceanside, and 5 times over El Centro in the Imperial Valley. It was not easy to link up for even the few minutes required. The planes tended to separate as the refueling aircraft grew lighter and the endurance aircraft heavier. Now and then the pilot had trouble holding the correct speed. Eaker could see the refueling craft, but the pilot of the C–1 could not see the endurance ship. If Hoyt went a little too fast, his

Crew in *Question Mark* endurance plane, (l. to r.): Maj. Carl Spatz, Capt. Ira Eaker, 1st Lt. Harry Halverson, 2d Lt. Elwood Quesada, and SSgt. Roy W. Hooe.

crewmember Woodring (who could see) tugged on a string fastened to the pilot's arm. If Hoyt had to speed up a little, Woodring pulled twice. Finished refueling, the *Question Mark* glided off to the left, and the refueling aircraft landed to prepare for the next linkup. In 37 contacts, the C–1s poured 5,600 gallons of gasoline and 245 gallons of oil into the *Question Mark*. Moreover, the refueling planes took up batteries and materials for repairs, including a window to replace one blown out of the cabin.

The plans originally called for radiotelephone equipment so the *Question Mark* could communicate with the refueling planes, the Air Corps radio station at Rockwell Field, and one to be set up at Santa Monica. But because of the weight, a radio was not installed in the endurance aircraft. Spatz communicated with the ground by dropping messages (which ground crews sometimes had trouble finding), by firing flares (at times not understood), or by tying notes to the refueling hose or the rope used to transfer supplies. He received messages by notes delivered by the refueling planes, by means of panels or lights on the ground, and by "blackboard planes." The latter were PW–9Ds with sides painted black so messages could be written on them with chalk. First Lieutenant Archie F. Roth and 2d Lts. Homer W. Kiefer, Norman H. Ives, and Roger V. Williams flew these ships. A flying blackboard told Spatz "Water—250 gals gas coming up," or "Message received—ready at midnight" for refueling he requested. The first blackboard message, at Van Nuys at noon on New Year's day, was a reminder ("Don't forget Rose Bowl") to fly over the stadium at Pasadena when the University of California played Georgia Tech.

Since Spatz and his crew expected to stay up a long time, provisions were made for their welfare and comfort. Each man underwent a physical examination before and after the flight. Flight surgeons drew up a special diet. The men on the *Question Mark* were supposed to have an electric stove to heat food, but they omitted it to lighten the equipment load. The refueling planes brought them hot meals and ice cream; women of a church in Van Nuys sent a turkey dinner on New Year's Day. The refueling craft also brought telegrams, mail, water, a collapsible bathtub, clean towels, and woolen underwear. Off-duty men slept on bunks atop the tanks. A novel combination of propellers reduced noise, which could be most annoying during a long flight. The two wing engines had Westinghouse two-blade Micarta propellers, the center engine a Standard three-blade steel prop.

Oil was to be transferred by hose, but in a test at Rockwell on December 29 the oil flowed too slowly even after heating. The depot built slings to lower 5-gallon cans of oil with the ropes that transferred other supplies. Special equipment on the *Question Mark* allowed Sergeant Hooe to change engine oil (Pennzoil triple-extra-heavy) in flight. Copper tubes ran from the oil tank in the cabin to the engines. Rods connected to indicators in the cabin operated valves in the nacelles. Eaker took control of the plane to change oil. Hooe opened the drain valve to get rid of the warm oil (it made a mess on the fuselage and tail). Closing the drain valve, Hooe opened the feed valve and by wobble pump put new oil in the engine. Other copper tubes went from the engine to a board in the cabin, each tube fitted with an Alemite connector. Using an ordinary grease gun, Hooe greased the engine rockers during flight. The log regularly carried the notation, "Rockers greased," but failure of rockers ultimately ended the flight. Hooe likewise could work on the engines in the air. An extra door in each side of the cockpit opened on a walkway to each of the side engines. The walk, of steel tubing, had a handrail along the rear side. A step on the landing gear gave Hooe a place to stand while working on the engine. Another platform below the nose of the plane afforded him access to the center engine, but to reach it he had to climb up through the cockpit and over the nose. For work outside Hooe wore a parachute and a lineman's safety belt that he fastened to the walk or nacelle. He carried a rubber hook to hold the propeller still after the pilot nosed the plane up to slow the prop on the engine requiring repair. One day gas leaked into the fuselage from valves in lines leading to wing tanks. The refueling plane delivered red lead, soap, and shellac for repairs. Then cylinders started missing. It was night. Eaker kept the *Question Mark* within gliding distance of Metropolitan Airport in case landing became necessary. By opening the throttle he finally cleared plugs that had become fouled. On the seventh day the left engine stopped completely. Eaker opened the throttles on the other engines to keep the plane in the air while Hooe went out on the catwalk. The rockers had failed. Hooe began work but the other two engines could not stand the additional strain. So the flight ended.[9]

"The flight of the *Question Mark*," Spatz wrote in his report to General

Fechet, "demonstrates conclusively that one transport plane can safely refuel another transport in the air." He viewed refueling as "a distinct advantage in military operations." Bombers could take off with lighter loads of gas and heavier bombloads. With refueling, a bomber's radius of action "has scarcely any limit at all." Refueling could be extended to pursuit, attack, and observation planes to enlarge their areas of operations and enhance flying safety. Spatz foresaw commercial planes accomplishing "transcontinental and transoceanic flights without landing." Refueling would reduce time on long trips by eliminating fuel stops. Flying would be safer, for with refueling planes located at frequent intervals across the country, transports could circumvent bad weather and use radio to ask for inflight refueling.[10] The *Cincinnati Enquirer* believed the flight of the *Question Mark* "won the interest not only of the man in the street but that of the world of science in an unusual degree." *The New York Post* deemed it "a new chapter in the history of aviation." *The Washington Star* thought it might lead to a nonstop trip around the world. The *Albany Evening News* said that while the flight might not have an immediate benefit, it should further aviation and give more confidence in flying. Editors praised the crew. "A man who has to listen for days to the unbroken roar of airplane motors," the *Richmond News-Leader* asserted, "is entitled to all the distinction he gets for breaking a record."[11]

The distinction of holding the record did not long remain with Major Spatz and his crew. James Kelly and Reginald L. Robbins chalked up more than 172 hours at Fort Worth in May. And that was just the beginning. That year there were nearly forty attempts in America to set refueling endurance records. Nine surpassed the *Question Mark*. The record at year's end was held by Dale Jackson and Forest O'Brine who had flown the *St. Louis Robin* over 420 hours. That record, too, would soon fall. The Air Corps was out of the contest.[12]

The End of Racing

Absence of international competition and failure of the U.S. military services to enter brought the Pulitzer Race to an end in 1925. A race for standard pursuit ships took its place at the National Air Races at Philadelphia in 1926. Navy Lt. George T. Cuddihy won by averaging 180.5 miles per hour. Second Lieutenant Lawrence C. Elliott, Air Corps, flying a P–2, came in second with 178.6 miles per hour—far below 1st Lt. Cyrus Bettis' 249 miles per hour in the Pulitzer Race the previous year.[13] First Lieutenant Eugene C. Batten, Air Corps, won the pursuit race at Spokane, Washington, in 1927 by averaging 201.2 miles per hour. He flew the XP–6A, built by the Materiel Division for the race from the fuselage of a P–1A, the wings of the XPW–8A with old PW–8 surface radiators, and a 700-horsepower Conqueror engine.[14] The Air Corps did not enter the pursuit event at Los Angeles in 1928 but competed with Navy and civilian flyers in the main event at Cleveland in 1929. At Cleveland a civilian, Douglas Davis, flying Travel Air Model R, won (194.9 miles per hour); 1st Lt. Robert G. Breene, Air Corps, placed second (186.8) in the XP–3A; another civilian, Roscoe Turner, third in the Lockheed Vega (163.8).[15] The Air Corps did not enter the national races again. Thus it did not compete for the Thompson Trophy, offered first in 1930, or the Bendix, established in 1931. It continued, however, to send men and planes to the National Air Races to put on aerial exhibitions, but the race committee sometimes paid the Air Corps' expenses. The Air Corps' exhibition at Los Angeles in 1928 featured an acrobatic team, the "Three Musketeers," composed of 2d Lts. William L. Cornelius, John J. Williams, and Irvin A. Woodring. These officers served together at Selfridge Field, Michigan, before being reassigned in June 1928 to the 95th Pursuit Squadron at Rockwell Field, California. For the air races at Mines Field at Los Angeles, they flew PW–9Ds to demonstrate pursuit maneuvers and perform acrobatics.

Ford Tri–motor plane rounds pylon at the 1926 National Air Races at Philadelphia.

On the afternoon of the third day of the races, the Musketeers did a series of stunts in formation. Then Lieutenants Cornelius and Woodring climbed to 10,000 feet and executed a three-quarters outside loop while Lieutenant Williams entertained with solo acrobatics. Just as Cornelius and Woodring finished their maneuver, Williams roared across the field in front of the grandstand. Flattening out at about 200 feet, he turned the plane on its back. The engine sputtered. He finished the roll. The motor continued to miss. He lost altitude swiftly. The plane pancaked, hit on both wheels, bounced, landed on one wing, bounced again, came down on the other wing and collapsed, pinning Williams in the wreckage. Cornelius and Woodring saw the accident but went on with their act. A Reserve officer among the spectators volunteered to fly for the fatally injured Musketeer. The next day the crowd thrilled to Charles A. Lindbergh's performance.[16]

The Air Corps clung to balloon racing a little longer than to airplane racing, despite the general decline of its lighter-than-air branch and scarcity of money. In 1928, Capt. William E. Kepner and 2d Lt. William O. Eareckson won both the national and international races.[17] Second Lieutenant Wilfred J. Paul and MSgt. (2d Lt., Air Reserve) Joseph H. Bishop won the national race in 1932, but the Air Corps could not find money to send them to Switzerland for the international contest.[18] In the national contest at Denver on July 3, 1936, the Army balloon, with Capt. Haynie McCormick, pilot, and Capt. John A. Tarro, aide, got caught in a down current, crashed, and burned, thus ending balloon racing for the Air Corps.[19]

New Highs

The Air Corps lost a good balloon pilot because he neglected to send his clock to the jeweler to be cleaned before a high-altitude flight. Two Germans, Reinhard Süring and Arthur Berson, had held the altitude record of 35,433 feet for free balloons since 1901. The American record remained 15,997 feet, set by Clifford B. Harmon in 1909. Capt. Hawthorne C. Gray attempted to break both in March 1927, using a silk, rubberized, and aluminum-coated balloon. He carried three cylinders of oxygen, a thermograph, two barographs, an altimeter, a climb indicator, a 50-cent thermometer fastened on the side of the instrument case, and an Atwater Kent radio. Ross Asbill, foreman of the Fabric Department at Scott Field, Illinois, made parachutes so Gray could drop instruments and other equipment without damaging them or the property they fell on. Ballast consisted of 50-pound bags of sand fastened to a rack suspended from the concentric ring. Gray wore a lined flying suit of fawn skin; high, fleece-lined, leather moccasins, laced front and back; wool-knit breeches; and helmet with mask. The Scott Field correspondent for the *Air Corps Newsletter* described his departure:

> On the afternoon of March 9th Captain Gray appeared before the assembled multitude resembling a large, brown bear prepared for winter hibernation; climbed into his wicker basket; glanced at the 80,000 cu. ft. single-ply, hydrogen-filled bag swinging overhead; adjusted the instruments; supervised the distribution of 4,500 lbs. of sand ballast to be expended on ascent; verified the oxygen apparatus; turned on a jazz tune on his radio, and with a cry, "O.K., let her go," slipped up into the sky.[20]

Gray ascended rapidly as he cut sandbags. At 12,000 feet he was still dropping ballast. Radio reception was good. He was cold and numb (the thermometer registered –32 degrees Fahrenheit). The next thing he recalled was regaining consciousness at 17,000 feet with the balloon falling at 1,000 to 1,200 feet a minute. Dizzy and cold, he commenced to drop the remaining ballast. The three knives he brought along to cut the bags had dulled, so he ripped the bags open with his hands. This checked the descent to around 600 feet a minute. Crashing in a ditch near Ashley, Illinois, about 40 miles from Scott Field, Gray sprained an ankle. One of the three planes following the flight picked him up and took him to Scott, where he rested a few hours in the hospital. The radio and some of the instruments shattered in the crash, but the barograph escaped damage. In a flight lasting just 1 hour and 47 minutes, Gray set a new American record with an altitude of 28,510 feet.[21]

Captain Gray flew again on May 4 in the same balloon. He used a different kind of sandbag, however, to ease dumping, and a new oxygen system designed by 2d Lt. Howard H. Couch. Nothing unusual had happened by the time the balloon reached 40,000 feet. Gray felt no dizziness, distress, or lack of oxygen. Frost covered the eyepieces in his mask save for a spot in the center about the size of a 5-cent piece. The rate of ascent decreased. Seeking at least another 1,000 feet, Gray went around the basket testing each sandbag. Finding all empty, he elected to drop the oxygen cylinder he had used up to this time. Switching to another, he let the cylinder go by parachute. The balloon ascended slowly. At about 41,000 feet Gray felt a slight pain in his chest. He remembered the doctor's instructions to descend at the first symptoms of distress. He opened the valve and let out gas for 3 seconds. The balloon dropped rapidly, but Gray did not become concerned until he arrived at around 20,000 feet. His glasses having cleared, he now could see that the gasbag had not taken the parachute-shape expected. At 18,000 feet he removed his oxygen mask and started to throw equipment overboard. The balloon not only was falling too fast for a safe landing but was headed for a swamp. At 13,000 feet a plane piloted by 2d Lt. Leslie P. Holcomb circled the balloon to follow it down, a photographer in the backseat taking pictures with a movie camera.

As Gray neared 8,000 feet, he decided to join the Caterpillar Club. He opened the valve, tied the cord to the concentric ring, climbed to the edge of the basket, and jumped. His parachute opened at once, and he landed safely in a plowed field near Grayville, Illinois, at 1505. Holcomb landed, picked up

Aluminum-coated balloon used in Capt. Hawthorne C. Gray's high-altitude flight.

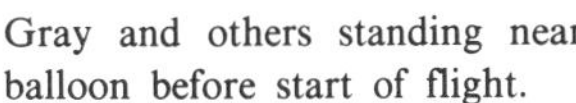

Gray and others standing near balloon before start of flight.

Gray, and flew to Scott Field, about 110 miles away. The balloon, basket, instruments, and most of the equipment Gray had thrown overboard on parachutes were recovered. After the Bureau of Standards calibrated the instruments, Gray received credit for 42,470 feet. That was more than 8 miles and higher than man had ever flown in balloon or airplane. But the Federation Aeronautique Internationale would not make it official because Gray did not come down with his balloon.[22]

On November 4 Captain Gray sought once more an official record, but later the Air Corps stressed the scientific nature of this and his earlier flights. Gray prepared carefully. The failure of the bag to parachute forced him to jump on May 4; a new mechanism enabled him to make the bag parachute. The oxygen from the tanks grew too cold at high altitude; a heater warmed the flow. Instruments sometimes failed because the cold stiffened the oil on the mechanism; Gray sent the 2 barographs and the thermograph to a jeweler to clean and remove all oil. By careful tests he determined how long his oxygen supply would last; the 3 high pressure tanks (each 200 cubic inches) furnished enough for 120 minutes. The valve in the oxygen system was heated so as not to freeze, and was fixed to feed oxygen to revive him should he pass out. His goggles had heating wires to keep them from fogging or icing. A few minutes after Gray began his ascent, Capts. Harry H. Young, pilot, and Harrison H. Fisher, flight surgeon, took off in the PT–1 to follow. The balloon disappeared 35 minutes later in a cloud over McLeansboro, Illinois. Young continued in the southeasterly direction the balloon traveled but did not see it again. Next morning came word that Gray's body had been found in the balloon basket near Sparta, Tennessee.

Gray's log [in part paraphrased], recovered from the basket, told the story of his ascent to 40,000 feet:

2:23 p.m.	Ascent began.
2:40	12,000 feet. "KSD Symptoms of Ricketts" [he was listening to radio station KSD, St. Louis, and low pressure was affecting him]. Oxygen started.
2:50	15,000 feet. "4V" [turned 4 volts on heaters for oxygen and goggles].
3:05	19,000 feet. Temperature zero. Balloon rising. Spilling sand, reading instruments, and tuning the radio—saxophone playing "Traumerei."
3:10	23,000 feet. Snowing. –8 degrees. KMOX [St. Louis] playing "Thinking of You."
3:15	24,000 feet. Still snowing. WLW [Cincinnati] playing "Just Another Day Wasted Away."
3:21	26,000 feet. First oxygen cylinder empty.
3:31	30,000 feet. –35 degrees. Clock stopped.

Gray no longer had any way of knowing how much time he had on oxygen. His log continued:

> 34,000 feet. Cyl. off [cylinder cut loose for ballast], broke antenna, no more music.
>
> 39,000 feet. –28 degrees. [The handwriting had become more and more uncertain since passing 29,000 feet; his mental faculties were being affected by low temperature.]

> 40,000 feet. Sky deep blue, sun very bright, sand all gone. Second cylinder empty, third started.[23]

There the log ended. The instruments recovered from the basket showed that a little after 4:00 p.m. the balloon reached 40,220 feet. The instruments and other evidence revealed that Gray dropped the battery for ballast, sending the balloon up to 42,470 feet. The time was 4:20 p.m. Gray may have been concerned about his oxygen supply at this point. However that may be, he "valved" the balloon at 4:21. Descent to 39,000 feet, achieved at 4:28, was slow. Then he must have "valved" a second time because the descent became abruptly more rapid. By 4:38 his oxygen ran out. He died from lack of oxygen before the balloon came to rest in a tree at 5:20 p.m. The Bureau of Standards verified an altitude of 42,470 feet. The National Aeronautical Association credited Captain Gray with a record. The Federation Aeronautique Internationale did not make it an official world record because the aeronaut, dead before landing, "was not in personal possession of his instruments." The government posthumously awarded Gray the Distinguished Flying Cross for his ascents of March 9, May 4, and November 4, 1927.[24]

Flyers at Wright Field pursued the aircraft altitude work begun years before by the Engineering Division at McCook Field. In 1928, 1st Lt. William H. Bleakley assaulted the record of 38,418 feet set by Navy Lt. C. C. Champion, the past year. During a practice flight on May 31, Bleakley went to 34,000 feet where rough air compelled him to cut the test short. Intent on his work he had forgotten to watch the oxygen gauge. After he commenced to descend, he discovered his oxygen would last just a minute or so.

> I placed myself in a position so that should I "pass out" I would have the control stick set forward. I placed my elbow against my body and my hand on the control stick, leaning forward and dropping my head. At that moment I had my [other] hand on the [radiator] shutter control. As I descended to 32,000 feet I decided to take my

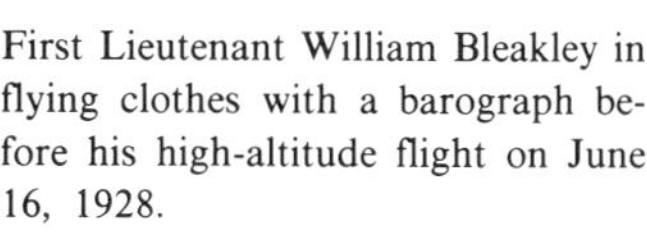

First Lieutenant William Bleakley in flying clothes with a barograph before his high-altitude flight on June 16, 1928.

> hand off the shutter control . . . to throttle down the motor. However, I found I was unable to move my hand and after several efforts gave up. I tried to raise my head . . . but found I was unable. . . . The machine was slowly descending and I kept close watch on the second hand of the clock . . . and tried to count the seconds so as to assure myself as best I could that I was still conscious . . . . I felt as though I were paralyzed. My breathing was very difficult. . . . Knowing that I should not move or exert myself in any manner I tried to keep quiet . . . . My vision was very much blurred. My hearing was very distant and my tongue seemed about four times its size.[25]

Suddenly, at 25,000 feet, Bleakley felt a change. He could move his arm to the throttle and apply his legs and feet to the rudder. As the plane went lower, he grew stronger. But when he landed he could hardly move around. Mechanics helped him out of the aircraft and to the operations office 50 feet away.[26] Nearly two weeks passed before he fully regained his strength.

When Capt. Michael G. Healy, flight surgeon of the Fairfield Air Depot, examined Bleakley on June 15 he found him in fine shape, set for another altitude flight the next day. This time Bleakley went for a record. His plane, the one he used for the practice flight, was 1st Lt. John A. Macready's XCO–5. It was fitted with a new supercharger designed to give sea level pressure up to 30,000 feet, and a duralumin propeller with adjustable pitch. Though having a new oxygen system, Bleakley took along two bottles of gaseous oxygen for an emergency. Dressing for the flight, he put on 2 suits of heavy wool underwear, 5 pair of socks, 2 wool sweaters, a blanket-lined leather flying suit, 2 silk mufflers, 1 pair of silk gloves, 1 pair of heavy fur gloves, 1 pair of moccasins, 1 chamois suit, 1 pair of chamois blanket-lined trousers, and a fur-lined helmet and face mask. He wore experimental goggles with electrically heated lenses. Bleakley took the XCO–5 to 38,000 feet indicated altitude on June 16 and for 18 minutes tried to coax it higher. When the motor overheated and missed, he started to go down. The U.S. Bureau of Standards calculated the altitude at 35,509 feet. He could have used more clothes when the temperature dropped to –52 degrees. The right lens of the goggles fogged up at 28,000 feet on the climb and did not clear up until the plane passed 25,000 feet on the way down. The liquid oxygen system worked perfectly.[27]

On September 23, 1928, 1st Lt. James H. Doolittle flew the XCO–5 for Capt. Albert W. Stevens to test photographic equipment. After climbing for 1 hour and 5 minutes, Doolittle held the plane at 37,200 feet (the ceiling) about 20 minutes while Stevens worked the camera. The oxygen proved insufficient for two men. Stevens passed out; Doolittle became semi-unconscious just as he pointed the nose of the plane downward. Both revived at lower altitude during the 45-minute descent. At the top, the temperature had been –70.6 degrees. The clock, one of the camera's spirit levels, and a spot on Stevens' cheek froze. The camera was ice-covered. Doolittle found the controls stiff and hard to move.[28]

Captain Stevens completed another high-altitude flight in the XCO–5 on October 10, 1928, this time with Capt. St. Clair Streett piloting. Their

Capt. St. Clair Streett, pilot (left) sits in cockpit of plane while Capt. Albert Stevens installs camera in the observer's cockpit before high-altitude photographic flight.

altitude, computed by the U.S. Bureau of Standards, was 37,854 feet. This was 564 feet below the record set by Navy Lieutenant Champion but the highest ever attained by an airplane carrying two men. Since the Federation Aeronautique Internationale had no category for altitude flights in planes carrying more than one man, the record remained unofficial. Stevens and the photographic equipment weighed about 225 pounds which, the Air Corps pointed out, constituted "no mean handicap." The Corps said the real purpose of the flight was not to set a record but to test equipment and experiment with methods for securing altitude measurements from photographs.

Stevens and Streett carried only liquid oxygen on the flight of October 10. They wore electrically heated goggles with a 3/8-inch hole in each lens over the pupil. The hole was not big enough to cause any harm but let the wearer see if frost coated the lens. Without the holes, Streett and Stevens would have been blind; frost forced them to cock their heads this way and that to see. Despite the intense cold (–76 degrees at the top), Stevens had no trouble operating the camera. He used a miniature electric heater a little larger than a rifle cartridge. It was placed in an asbestos pocket across the fingers of his heavy mittens to keep his fingers from becoming numb.

After climbing an hour and a half, Stevens saw the altimeter hand at 40,000, or about 37,500 feet above sea level "yardstick measure." As Streett continued to climb, Stevens took as many pictures as he could during the next 5 minutes. Then he tapped Streett on the shoulder, the signal time was up.

Streett put the nose down. Airspeed rose from 54 to 95 miles an hour. Stevens saw the altimeter hand drop to 37,500 and soon to 34,000. Seven minutes later he was amazed to find they were still at 34,700 feet. He thought: "Streett must like this thin air up here. Why in thunder doesn't he get us out of here? Doesn't he know that at this rate we haven't hardly any gasoline left?"[29] Stevens peered over Streett's shoulder: engine full throttle; supercharger full on; airspeed over 90. In lieu of losing altitude, the plane actually gained a little. Stevens saw Streett pull on the throttle levers from time to time and noticed one twisted out of position. The cold shrank the metal parts of the throttle and supercharger controls, rendering them immovable.

If Streett cut the ignition switches, the water in the radiator and water jackets would freeze. If he turned the switches on and off at intervals to slow the plane and lose altitude, raw gasoline would flow into the supercharger and exhaust manifold, probably causing a fire. With the throttle stuck open, he could not dive sufficiently to overcome the climbing tendency of the aircraft at that altitude. "In other words," Streett said later, "we weren't sure that our wings would stay with us if we dove much in excess of 115 miles per hour because of the very light construction of the plane."[30] The XCO–5 was stuck at 34,000 feet.

The motor sputtered—the gas tank was going dry. Just then, the temperature being relatively warmer (–48 degrees), the controls loosened a little. Streett closed the radiator shutters and pulled back the throttle. The plane began to glide. Over Indiana, 75 miles from Wright Field, Streett looked for a place to land. He picked the largest field in view.

> Now a plane of this type [Stevens explained] has no air feel at all, compared to a regular plane. In other words, it flies like a barn door. Consequently, it was a really remarkable feat when the Captain flew by the end of the field, banked the long wings on the plane at 60 degrees, did a 270 degree turn, landed with a dead stick, rolled to a stop, and still had two-thirds of the field left.[31]

They shed their heavy flying clothes, got 50 gallons of gas at Rushville, Indiana, and flew to Wright Field.[32]

Having snapped some good pictures, Stevens and others studied them to acquire altitude measurements. One photograph showed an altitude of 37,920 feet above ground level, or 38,890 above sea level. Another taken several minutes later showed 38,080 feet above ground level or 39,050 above sea level. The figure 39,050 was 1,190 feet higher than that from the formula of the Federation Aeronautique Internationale, and 556 feet lower than that arrived at by the Bureau of Standards.[33] As Stevens continued his photographic experiments, the Materiel Division was disappointed in its hopes of setting an altitude record. First Lieutenant Harry A. Johnson piloted Stevens on seven flights within a month's time. The last of these was on February 27, 1929. They attained 37,000 feet before being forced to descend when blinded by the sun's reflection on ice particles in the air.[34]

Flight surgeons at Scott Field and Dayton had been keenly interested in the physical condition of Gray, Bleakley, and others going to high altitudes.

At that time, however, the Air Corps had no program for systematic investigation of the psychological and physiological effects. The Air Service had done some research with a low-pressure chamber during and after World War I. But seeing no further need for such work, the School of Aviation Medicine left the chamber behind when it moved from Mitchel Field to Brooks Field, Texas, in mid-1926. The chamber wound up at Dayton, where engineers used it for a time to test equipment and instruments. Then someone pushed it into a corner of a storeroom where it gathered dust. Hence Gray, Bleakley, and other Air Corps flyers were not checked out in a simulated environment of high altitude, low pressure, and cold temperature before their altitude flights.[35]

Blind Flying

Flying in fog or clouds, unable to see the ground and with no horizon, a pilot often became disoriented. His senses contradicted what his instruments told him. He felt he was turning and banking opposite to the true direction. He did not understand this phenomenon and was seldom aware of what was really happening. Having slight faith in instruments, and accustomed to flying by instinct, he tended to obey his senses and rely on his flying ability to pull him through. After all, he learned to fly without instruments. Why did he need them now that he was a qualified pilot?

Capt. David A. Myers, flight surgeon at Crissy Field, California, evinced interest in the problem. He thought a number of young aviators too cocky. Overconfidence in their flying skills compromised safety. When a cocky pilot took a physical examination, Myers extended him an extra session in the Jones-Barany revolving chair. In the regular examination, the doctor turned the flyer in the chair and checked for nystagmus, the involuntary oscillation of the eyeballs that occurs with rapid rotation and dizziness. When the pilot was turned with his eyes open and asked which direction he was going, he invariably gave the right answer. Myers then asked him to close his eyes, gave him a few turns to induce vertigo, gently stopped the chair, and asked the same question. The reply always was that he was going opposite to the way he had started. The aviator became annoyed when he opened his eyes and found the chair not moving. If the flyer still thought he could trust his senses, Myers did it again, turning the chair until the occupant had the sensation of not moving. Opening his eyes, the pilot was chagrined to find himself turning. As Myers said, it was disconcerting for an aviator to discover suddenly he could not tell which way his body turned, or if it turned at all.

Captain Myers tried the induced vertigo test on Capt. William C. Ocker, not because he thought Ocker cocky but to get the reaction of an oldtime pilot. After the test, Ocker left without comment. Later, he returned with a view box and asked Myers to test him again. That was January 1926. Seven years earlier, Ocker had tested one of the first turn indicators made by Elmer A. Sperry and found it useful for flying in bad weather. He thought so much of the instrument that he carried one in his flight bag to put on any plane he flew. The view box he took to the surgeon's office contained a turn and bank indicator so fixed that when he looked into the box he could read the indicator but could not see outside. Seated in the Jones-Barany chair, Ocker started the gyroscope to activate the turn and bank indicator and gazed into the box while Myers turned the chair. Reading the indicator rather than relying on his senses, Ocker answered correctly every time as to the direction

Capt. William C. Ocker

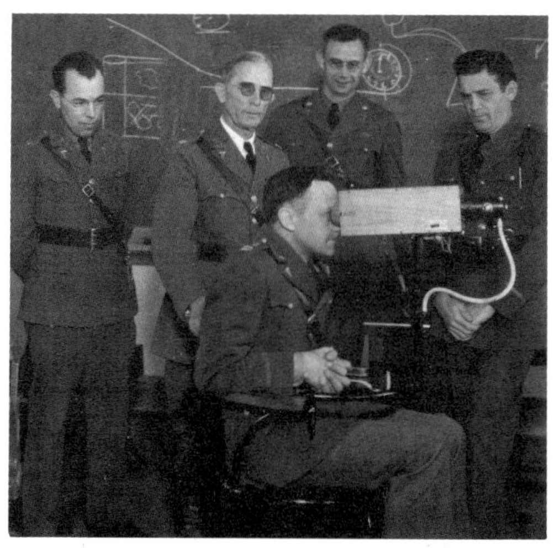

Instruction in use of "Ocker Box," in teaching instrument flying at the School of Aviation Medicine.

of motion, starting, and stopping. Ocker and Myers knew on the spot they had come up with the answer to man's inability to fly without visual reference to the earth. A pilot had to have artificial aids to fly blind. Myers and Ocker spent many hours testing their reactions and those of others, in a revolving chair and in the air. With Ocker at the aircraft's controls, Myers rode behind in a covered cockpit, communicating by pulling strings to tell Ocker what he believed was happening. Demonstrating their "Vertigo Stopper Box," they convinced many pilots of the need for instruments for blind flying. Their work attracted the attention of the School of Aviation Medicine, which tested and confirmed Myers' experiments.

Transferred to San Antonio in 1929, Ocker commanded a school squadron at Brooks Field. In 1930 he was reassigned to Kelly Field, promoted to major, and given $1,000 as partial remuneration for the time and money he spent in developing his Vertigo Stopper Box. Being an Air Corps officer he had to assign his patent on the box to the U.S. Government. As a rule the government did not pay, but in this case Congress voted the money. Ocker continued his experiments assisted by 1st Lt. Carl J. Crane. In 1932 they published a book, *Blind Flying in Theory and Practice.*[36]

Meantime, the Air Corps worked with the Pioneer Instrument Company, Sperry Gyroscope Company, Kollsman Instrument Company, U.S. Bureau of Standards, and others to develop navigation instruments. It also aided the Daniel Guggenheim Fund for Promotion of Aeronautics in studying problems of navigating in fog. The Corps furnished facilities at Mitchel Field, New York, for the Full Flight Laboratory that the fund formed in 1928. It loaned 1st Lt. James H. Doolittle to head the laboratory and do the flying, and further provided a flight assistant, 2d Lt. Benjamin S. Kelsey, and a mechanic, Capt. Jack Dalton. Doolittle used a Consolidated plane, NY–2, to test instruments and practice flying under a hood while Kelsey rode as safety pilot. The instruments Doolittle selected were a Kollsman sensitive altimeter, a Sperry artificial horizon, a Sperry directional gyroscope, and radio equipment with homing-range and marker-beacon indicators.

After testing the instruments on many flights, Doolittle considered everything set for a complete flight, blind from takeoff to landing. It was Tuesday, September 24, 1929. A heavy fog lay over Mitchel Field early that morning. Doolittle took the NY–2 up with the cockpit uncovered, broke through the fog, made a wide circle at about 500 feet then landed. He wanted to go again, alone, under the hood. But the fog was lifting. There might be other aircraft aloft. Harry F. Guggenheim, the fund's president, insisted Kelsey go along. The hood closed, Doolittle warmed up the engine, taxied out, turned, took off on the radio beam, and headed west in a gradual climb. He leveled off at 1,000 feet, turned left 180 degrees, flew 10 minutes, and turned left another 180 degrees. Lined up with the radio beacon on the west side of Mitchel Field, he gradually descended. Leveling off at 200 feet, he held that height until the marker told him he passed the edge of the field. Applying his instruments and procedures previously developed and practiced, he throttled to about 60 miles per hour, went into a glide, and flew the plane "into the ground." He would remember the approach and landing as "sloppy." But he had made a blind flight. "It was," in his words, "the first time an airplane had . . . taken off, flown over a set course and landed on instruments alone."[37]

The work proceeded for a while, but by the end of the year the Guggenheim Fund closed out the project. The Full Flight Laboratory had established the principle of safe fog flying. Commercial and military organizations should perfect the methods for use. The laboratory shut down. The equipment was sent to Wright Field, Ohio, where 1st Lt. Albert F. Hegenberger and the Materiel Division's Fog Flying Unit took up the work. Doolittle likewise transferred to Wright Field. He no longer could support his family on a lieutenant's pay, which was less than what he received from the Guggenheim Fund. On February 15, 1930, he resigned from the Air Corps, took a commission as a major in the Reserve, and went to work for Shell.

After a great deal of experimenting, Hegenberger worked out a variation on Doolittle's landing system. In addition to a Kollsman sensitive altimeter, a directional gyroscope, and an artificial horizon, Hegenberger's plane (a BT–2A) carried a radio compass designed by G. G. Kreusi, a radio engineer at Wright Field. Hegenberger's landing system employed two portable radio transmitters: "A" 1,000 feet and "B" 1.5 miles downwind from the landing place. At the edge of the field stood a radio marker developed by another civilian employee, C. D. Barbulesco. By radio compass Hegenberger brought the BT–2A to transmitter A. He then tuned in B, turned in that direction, and lined the plane up on the two transmitters. Noting the reading on the directional gyro, he continued beyond B, turned, and flew the reciprocal heading. Passing B at 1,000 feet, he lined the plane up on the two transmitters, cut throttle to 20 percent above stalling speed, and descended at 300 or 400 feet per minute. If the plane dropped too low before passing transmitter A, he gave it a bit more throttle to hold altitude. Beyond transmitter A, Hegenberger watched the artificial horizon, directional gyro, and sensitive altimeter. The boundary marker afforded him an added point to check altitude as he glided until the plane landed itself.

At the beginning of May 1932, Lieutenant Hegenberger, 1st Lts. Russel J. Minty and George V. Holloman, and other flyers practiced blind landings at Patterson Field. On May 7, Hegenberger made a complete blind flight, taking off, flying, and landing under a hood, with Barbulesco riding as observer. Hegenberger carried out two more blind flights the following day, Barbulesco observing on one, Kreusi the other. The marker beacon did not work the next day when Hegenberger executed three flights within a space of thirteen minutes with Minty in the open cockpit. Nor was the marker operative when he lifted off for his fourth flight under the hood that day, this time alone. Five minutes later he landed safely. Thus Hegenberger became the first to make a blind solo flight, a feat winning him the Collier Trophy.[38]

William Ocker, David Myers, James Doolittle, Albert Hegenberger, and the many others concerned with instrument flying contributed immeasurably to the advancement of civil as well as military aviation. Ocker and Myers first pinpointed the problem of disorientation occurring when a pilot flew in fog or clouds, unable to see the ground and with no horizon. Ocker's Vertigo Stopper Box was an initial step toward the problem's solution. Then Doolittle, Hegenberger, and others opened the way by testing instruments and experimenting with techniques that would permit flights in atmospheric conditions that previously kept planes grounded. But much work was still needed for perfection of equipment and methods. During this period, Air Corps pilots got some training and practice in instrument flying. However, the business of equipping aircraft with suitable instruments, and of teaching Army flyers to use them, progressed slowly until a desperate need arose during the airmail emergency in 1934. Years passed before instrument landings were common.

The Pan American Flight failed to generate all the goodwill it was expected to. Moreover, the mechanical difficulties which Dargue and his companions encountered, and the flight's slow movement through Central and South America did not portray a bright picture of aviation's future in that part of the world. While it is interesting to observe that Pan American Airways soon began operations in Latin America, it is impossible to determine exactly how, and to what extent, the Air Corps' Pan American Flight influenced that development.

The Maitland-Hegenberger flight from California to Hawaii holds a place in history as a "first" among a series of "firsts." Even so, James Dole's $25,000 may have had as much or more to do in opening the Pacific area to air transportation. Highlighting the need for better and more reliable navigation equipment, the flight owed its success to careful preparation and to Hegenberger's skillful navigation.

The endurance flight of the *Question Mark* drew a great deal of attention at the time and carved a niche in history for Carl Spatz and his crew. The Air Corps (and its successors, the Army Air Forces and the United States Air Force) regarded the flight as a great achievement. Along with several refueling flights about the same time, it held forth the possibilities of extending the range of aircraft through inflight refueling. Nevertheless, the *Question Mark*'s flight did not at once lead to the application of the principle to Air Corps operations.

The flights of Hawthorne Gray, William Bleakley, Albert Stevens, and others engaged in high-altitude work were given less publicity. They have

Chapter XVI

GHQ Air Force Headquarters

The history of General Headquarters Air Force went back at least to 1917 and the men who drew the first plans for the American Expeditionary Force. They conceived aviation as having two principal functions—tactical and strategic. Tactical air units worked with ground units, furnishing direct support. Strategic units operated on their own, beyond the battlezone. William Mitchell, Edgar S. Gorrell, Benjamin D. Foulois, and other AEF members developed plans for an air force to fly strategic missions against Germany. Pershing balked at the idea of independent operations but approved creation of a strategic air force under control of General Headquarters AEF. This air force was never formed, since the war ended before aircraft for strategic operations came on the scene.[1] The postwar organization the U.S. Army adopted in 1920 provided a GHQ Reserve which included aviation to be used as a centrally controlled strike force and for reconnaissance. The Lassiter Board in 1923 suggested an air force for combat operations and special missions under the control of General Headquarters. Army regulations the following year authorized a GHQ Air Force in time of war.

Mobilization plans in the mid-1920s allotted attack, pursuit, and observation aviation to armies, and observation units to army corps, for direct support of ground forces. They also called for a GHQ Air Force of 5,200 officers, 46,000 enlisted men, and 2,300 airplanes. Under the command of an air officer who reported to the commander in chief in the field, the GHQ Air Force would further the commander in chief's strategic and tactical plans. It would operate against enemy air, ground, or naval forces in both direct and indirect support of ground forces.[2] In peacetime, during the 1920s and early 1930s, GHQ Air Force existed solely as a vague "something" that supposedly would come into being when the Army took the field for defense of the United States. Corps area and department commanders controlled the training and operations of Air Corps tactical units designated to form the nucleus of GHQ Air Force in an emergency. The Chief of Air Corps commanded combat units only for some particular demonstration, exercise, or maneuver authorized by the War Department. With War Department approval, the Air Corps temporarily established an air division for maneuvers in 1931, that being the closest it came to exercising GHQ Air Force.

Mitchell and Foulois had long wanted an air force that was separate from and not controlled by the Army. By the early 1930s, however, the attitude of General Foulois and many Air Corps officers was changing. They were discouraged by the repeated failures to secure an independent air force, and seemed inclined to accept, as an interim arrangement, a GHQ Air Force as better than nothing. At the same time, the attitude of the General Staff and several high-ranking officers of other Army branches was gradually growing more favorable toward aviation. A number of these officers felt that creation of a GHQ Air Force might reduce agitation for a separate service. The trend ran toward putting all combat units into GHQ Air Force and assigning just observation aircraft to divisions, corps, and armies. By early 1932, the War Plans Division of the General Staff was describing GHQ Air Force as containing all bombardment, attack, and pursuit, plus some observation.[3] This is what General Patrick had advocated in 1923 but the Lassiter Board had rejected. A series of events in the early 1930s (related in this and later chapters) led to the formation of the headquarters of GHQ Air Force. After assignment of air units to GHQ Air Force, it became a component of the Regular Army.

Coastal Defense

Both the Air Corps and the General Staff regarded coastal defense as one of the chief functions of GHQ Air Force. But for years the Army and Navy quarreled over the role of each in defending the nation's sea frontier. Claiming sole responsibility for operating land-based aircraft for coastal defense, the Air Corps opposed the Navy's attempts to develop air stations and land-based planes for the same purpose. Inclusion of torpedo planes in the Navy's program for Panama and Hawaii violated an Army-Navy agreement limiting the Navy to scouting and patrol planes. From the Navy's disingenuous explanation, the Army understood the planes in the Navy's program were of the scouting or patrol type, or of the so-called "three-purpose" type normally used for scouting and patrol but which might be used incidentally for bombing. Examining the Army's and Navy's five-year programs, the Joint Army and Navy Board reported on August 16, 1928, that it found no duplication.

Late the following year, the General Staff discovered the Navy's program contained torpedo and bombing as well as scouting and patrol aircraft for Panama and Hawaii. The General Staff believed the matter serious enough to require resolution by the President.[4] On February 25, 1930, Secretary of War Patrick J. Hurley requested President Hoover to halt the Navy's procurement of land-based aircraft and development of air stations. The Navy told the President the Joint Board had already settled the matter. Hurley sent a second letter to the White House, and the Navy followed with another in which Secretary Charles Francis Adams said he had met with Hurley but could not resolve the "alleged dispute." Hurley continued to urge the President to restrain the Navy, while Adams asked the President to order the War Department to accept the Joint Board's decision.[5]

The matter still wanted resolution when Gen. Douglas MacArthur became Chief of Staff in November 1930. On January 9, 1931, he and Adm. William V. Pratt, Chief of Naval Operations, reached agreement on employment of aircraft. The general said:

> The naval air forces will be based on the fleet and move with it as an important element in performing the essential missions of the forces afloat. The Army air forces will be land based and employed as an element of the Army in carrying out its mission of defending the coasts, both in the homeland and in overseas possessions.[6]

MacArthur thought this assured the fleet "absolute freedom of action with no responsibility for coast defense." Such division of duties, he continued, "enables the air component of each service to proceed with its own planning, training, and procurement activities with little danger of duplicating those of its sister service." He counted this great progress against seemingly insuperable difficulties.[7]

Preparing for coastal defense, the Air Corps sought permission to set up an Aerial Coast Defense School at Langley Field. This surprised General Staff members for they "supposed that this training was already being accomplished throughout the Air Corps."[8] After MacArthur talked with Foulois, the Air Corps received orders to begin instruction without delay. Turning down the request for a special school, the War Department directed that coastal defense be part of regular Air Corps training. At the same time, the War Department said experimental work and research would be required to develop instruments and training manuals.[9] Taking advantage of the opening, General Foulois formed a Frontier Defense Research Unit at Bolling Field. It consisted of Capt. Lawrence J. Carr, 1st Lt. Glen C. Jamison, 2d Lt. Norris B. Harbold, and ten enlisted men. Although concerned with determining requirements for equipment, personnel, and

Chief of Staff Gen. Douglas MacArthur (left) seeks agreement with Adm. William V. Pratt (right), Chief of Naval Operations, over which service is responsible for coastal defense.

methods, the unit devoted a great deal of effort to problems of instrument flying and navigation. Unless proficient in navigation, Army flyers could not carry out the long overwater flights incident to coastal defense. Expert help appeared in the person of Harold Gatty, the Australian navigator on Wiley Post's around-the-world flight in the *Winnie Mae*. Hired by the Air Corps on the recommendation of 1st Lt. Albert F. Hegenberger, Gatty had been giving a course in dead reckoning in the Office of the Chief of Air Corps. Now he helped the Frontier Defense Research Unit develop and test instruments and methods.[10] As an outgrowth of this, the Air Corps in October 1933 formed units at Langley Field, Virginia, and Rockwell Field, California, to teach navigation and instrument flying.

Planning to create an air force for coastal defense, the Air Corps met rebuff from the General Staff. Since planning affected the Navy as well as other branches of the Army, the staff would handle it, and the Air Corps could collaborate when asked.[11] General Foulois nonetheless went ahead with plans and took them to the Harbor Defense Board, on which he sat.[12] His "Plan for Defense of Our Seacoast Frontiers" conceived a Frontier Air Defense Command for defense against hostile ships and aircraft. Operating under General Headquarters, the command comprised a Frontier Air Patrol and a Frontier Air Force. The plan divided the nation's sea frontier into six frontier air defense zones, some with two or more frontier air defense regions. The patrol operated long-range airplanes, seaplanes, amphibians, and airships from "amphibidromes" furnished with servicing facilities, docks, ramps, and communications. So far as possible, the patrol used commercial facilities. Its aircraft carried radio, navigation equipment, plotting facilities, arms for self defense, and bombs for attacking submarines and destroyers. Sighting the enemy, the patrol craft reported location, time, speed, and course to shore stations for communication to the zone commander. From his command post he communicated with General Headquarters, the zone air force, patrol stations, and Coast Artillery district headquarters.

Under the plan, the air force and each zone concentrated bombardment, pursuit, and attack units at air force stations for administration and training. Auxiliary fields ("protectodromes") protected the air force by dispersal, providing servicing, ammunition storage, and communications. The zone commander issued orders to the zone air force based on information from the Frontier Air Patrol and other sources. General Headquarters shifted units from zone to zone concentrating them where needed.

General Foulois' plan separated air force operations into three phases. In the first, the planes operated under General Headquarters in the area between the limit of the range of frontier air defense forces (estimated at 250 miles) and the line of contact with ground forces. They located, observed, attacked, and destroyed hostile vessels and forces to defend United States territory and afford reasonable assurance against surprise. These actions proceeded in the second phase, the aircraft conducting observation and offensive missions to support coast artillery. This phase, in which the planes were controlled by the frontier commander, lasted from the time the enemy came within range of ground weapons until he was driven off or the operation entered the third phase. The final phase involved all arms on the frontier, with the Air Force in its normal role for land operations.

General Foulois pointed out that the Air Corps lacked units for frontier patrol. The air force might be organized by assigning all existing bombardment, pursuit, and attack units in the United States, but it would fall short of the strength called for by intelligence estimates.[13]

The Harbor Defense Board reviewed the plan and forwarded it to the General Staff. Lt. Col. Leroy P. Collins, handling the case for the War Plans Division, held that the plan "might be approximately 100 percent effective against a surprise attack if we ever got money enough to carry it out," but that kind of money was not to be had. He faulted it for failing to consider the fleet, with its aviation, and naval coastal patrols, which could be looked to for the first information of enemy movements at sea. The plan visualized equal threat from all directions and tied a large part of the air force to a cordon defense that ignored the outstanding characteristics of air forces—mobility. Colonel Collins drafted a directive for his chief, Brig. Gen. Charles E. Kilbourne, to obtain coordination. The General Staff concurred but Foulois did not. Collins revised the directive but Foulois still disagreed. Kilbourne then submitted the case to the Chief of Staff for decision.[14]

Brig. Gen. Charles E. Kilbourne, Chief of the War Plans Division

Adopting Kilbourne's recommendations on January 3, 1933, General MacArthur said the function of the Army's air arm was "to conduct the land-based air operations in defense of the United States and its overseas possessions," using any suitable type of land-based planes. Casting aside the Frontier Air Defense Command, Frontier Air Patrol, and Frontier Air Force, MacArthur identified two distinct classes of aviation concerned with frontier defense. These were corps and army observation units assigned to mobile forces and harbor defenses, and the air force constituting General Headquarters aviation. He planned to concentrate the air force in one or more areas on or before M-day and move all or part of it to threatened areas to operate directly under General Headquarters or under area ground commanders. The air force would cooperate with naval air and surface forces in the same area, adhering to principles set forth by the Joint Board as modified by the agreement between the Chief of Staff and the Chief of Naval Operations "making the Army solely responsible for coast defense."

Retaining the general scheme of three-phase operations as proposed by Foulois, MacArthur underlined the importance of observation, but omitted reference to the seaplanes, amphibians, and amphibidromes in Foulois' plan. MacArthur wanted the observation component of the air force to have planes and airships capable of long-range reconnaissance over land and water beyond the range of corps and army observation planes. Collins explained that to the Army this meant amphibian planes, but the wording was purposely general to cover land-based flying boats should future development justify their use. Having laid down basic principles, MacArthur called for war plans to include employment of GHQ Air Force for coastal defense.[15]

To obtain planes for coastal defense, the Air Corps converted Dolphins (twin-engine amphibious transports built by Douglas) for reconnaissance. In doing so it changed their designations from "cargo" to "frontier patrol," Y1C–21s becoming FP–1s, and Y1C–26s, FP–2s. The new nomenclature lasted just a few weeks before switching to "observation amphibian," the FP–1s now named OA–3s, the FP–2s, OA–4s.[16] When the Air Corps secured public works money for aircraft, it wished to buy long-range amphibians for reconnaissance. However, War Department insistence that the money be spent for combat planes resulted in orders for B–10s and A–17s.[17] The Air Corps hoped to use the long-range reconnaissance mission to procure long-range bombers. It wanted them for strategic operations but could not say so when military policy rested on defense. It saw in the MacArthur-Pratt agreement plausible justification for long-range bombers to defend against sea attack. One of the significant outcomes of the agreement was War Department approval of projects (to be mentioned later) that led to the development of the heavy bombers of World War II.[18]

The agreement between the services did not long survive Pratt's retirement on July 1, 1933. His successor as Chief of Naval Operations, Adm. William H. Standley, repudiated it. The Navy developed land-based planes with the aim of assuming the whole coastal defense responsibility. The Army kept on planning and preparing for a role in coastal defense. The Air Corps pursued its quest for long-range bombers.[19]

Four-Army Plan

In his directive of January 3, 1933, on coastal defense, General MacArthur substituted four coastal frontiers for General Foulois' six air defense zones. This brought coastal defense in line with plans for the four field armies he had created the previous August. As Chief of Staff, MacArthur deemed war planning his most important duty. The General Staff kept plans on file in case of hostilities with various countries; they needed overhauling. Noting that organization of forces for field operations had not progressed much above division level, General MacArthur perceived the need for welding units of the Regular Army, Organized Reserve, and National Guard "into an integrated tactical machine capable of instantaneous response to the orders of the President." The formation of four field armies was a step in forging such a machine. General MacArthur divided the country into four strategic regions—northeast, Great Lakes and northcentral, Gulf and southern, and Pacific Coast—and in each organized a field army commanded by the senior corps area commander. Together the armies comprised a field army group under the War Department Chief of Staff. Besides their normal duties, Regular Army officers were given mobilization assignments at general headquarters of the army group or with field armies. This furnished an organization for firming up war plans, conducting training exercises in peacetime, and mobilizing a field force in an emergency.

MacArthur's four-army plan allotted only observation aviation to ground forces for direct support, and gave General Headquarters all bombardment, pursuit, and attack, plus some observation for long-range reconnaissance. The plan prescribed an air force headquarters, manned by Air Corps officers in their mobilization assignments, to handle war planning and peacetime exercises. Corps area commanders retained responsibility for the training and operations of aviation units in peacetime, but relinquished control to General Headquarters for training exercises and mobilization.[20] The plan for GHQ Air Force was therefore part of the larger one for four field armies, which in turn was embodied in mobilization planning. Every General Staff division, all arms and services, and the corps areas were involved. General Kilbourne's War Plans Division was charged with hammering out the details. The Air Corps' role consisted of supplying information as requested, and eventually in reviewing and concurring, or nonconcurring, with what the War Plans Division produced. Though thus restricted, Foulois managed to make known his position on three matters of utmost importance. He wanted the GHQ Air Force in being in peacetime; the Chief of Air Corps in control of GHQ Air Force units, stations, training, and operations; and GHQ Air Force headquarters in Washington.

General Kilbourne moved slowly and cautiously. Despite the strenuous efforts of aviation enthusiasts to separate aviation from the Army and War Department, the General Staff had retained control over the Air Service during the first half of the 1920s. Having lost part of that power to Assistant Secretary of War for Air Davison in 1926, the General Staff hoped to get it back when he left office. Division chiefs and other General Staff members commonly favored creation of a GHQ Air Force. First, however, they wanted to assess the effect it might have. They felt that once established, GHQ Air Force would claim the complete attention of the airmen to the neglect of aviation for direct support of ground forces. General Foulois' urge to run the GHQ Air Force also influenced the General Staff. A zealous separatist, Foulois had antagonized several of its members by his campaign, particularly his testimony before Congress on behalf of an independent service. The staff suspected him of not always being entirely aboveboard in his dealings with the War Department, and thought he neglected the business of his office. He loved flying and spent many days traveling about the country inspecting Air Corps activities. His reluctance to delegate authority at times delayed the work of the General Staff until he got around to doing his part. Unable to discharge his present duties, how did he expect to shoulder the additional load of commanding GHQ Air Force?[21]

G.H.Q. Air Force (Provisional) Staff, (l. to r.): Maj. Clarence Tinker, Maj. Jacob Rudolph, Brig. Gen. Oscar Westover, Lt. Col. Henry Arnold, Maj. Arnold Krogstad, and Maj. Ralph Royce.

GHQ Air Force (Provisional)

While General Kilbourne and his assistants threshed out details of the four-army organization, the Air Corps secured War Department approval and formed a GHQ Air Force (Provisional) for maneuvers in May 1933. The Air Corps at first planned a swift concentration of units in the Northwest around Puget Sound, Washington, for several weeks of training. With just $19,500 on hand for maneuvers that year, it revised the plan to use facilities at March Field, California. By holding maneuvers there, it would have enough money left for a brief staff exercise later at Puget Sound. Brig. Gen. Oscar Westover, Assistant Chief of Air Corps, became Commanding General, GHQ Air Force (Provisional). His staff, which gathered in Washington on February 28 to ready plans, included Lt. Col. Henry H. Arnold, Chief of Staff; Maj. Arnold N. Krogstad, G–1; Maj. Clarence L. Tinker, G–2; Maj. Ralph Royce, G–3; and Maj. Jacob H. Rudolph, G–4. An antiaircraft exercise at Fort Knox, Kentucky, limited the number of men and planes available for maneuvers in the West. Even so, General Westover formed an air force of 350 officers, 530 enlisted men, and 280 airplanes. He drew on elements of the 1st and 2d Bombardment Wings, 3d Attack Group, 9th and 12th Observation Groups, and transport planes and crews of various stations.

General Westover gave his commanders a general plan for concentration but sent sealed orders for them to open on May 3, the day before the units deployed. Five planes from Mitchel Field, New York, went by way of Salt Lake City. The others took the southern route via El Paso and Tucson. Bad weather prevented some units from moving the first day. A sandstorm caused problems in getting planes into and out of El Paso. Using Army and Commerce Department radio, and commercial telephone and telegraph, General Westover directed unit movements from a temporary command post at El Paso. Save for a few aircraft delayed by motor trouble and the five on the northern route held up by snowstorms, the units completed concentration on May 8, a day late.

With the staff assembled at March Field, Colonel Arnold explained they were trying to "find the right way to handle the GHQ Air Force." He wanted "a fair test."[22] Based at March Field, the units dispersed most days to airports in the Los Angeles-Riverside-San Diego area on alert for the daily operation order. Westover issued orders by radio from a command post or command plane, and now and then by sealed instructions to be opened at certain points during the exercise. From May 12 to 26, GHQ Air Force (Provisional) attacked airdromes, aircraft carriers, and other targets; intercepted and defended against aerial attacks; engaged in aerial combat, pursuit against pursuit, and composite force (bombardment, pursuit, and observation) against composite force. It set aside one day for visitors, featuring control of an air force in the air, an attack on March Field, and a review. When the maneuvers ended, the units flew home under the command of junior officers. Westover, his staff, wing and group commanders and their operations and communications officers, and squadron commanders flew to Seattle for a few days of staff exercises before returning home.

Of the many recommendations resulting from the maneuvers, Westover put in first place immediate organization of Headquarters GHQ Air Force as part of the Regular Army. He advocated making headquarters part of the Office of the Chief of Air Corps, giving the Chief of Air Corps jurisdiction over all air units and stations, and delegating to the Commanding General, GHQ Air Force, responsibility for the development, training, and operation of GHQ Air Force units.[23]

TAN

While Air Corps units returned from the maneuvers, Maj. Gen. James F. McKinley, War Department Adjutant General, asked General Foulois to recommend employment of GHQ Air Force under certain war plans but staying within the statutory authorization of 1,800 serviceable aircraft. The war plans considered were attack by Great Britain (RED), by Great Britain and Japan (RED-ORANGE), and by Mexico (GREEN). On July 13, 1933, the Air Corps recommended a GHQ Air Force of 2,310 planes; distribution of bombardment, attack, and pursuit units among the most critical areas at the beginning of, or just before, war began; an alarm system along the coast to alert GHQ Air Force Headquarters of the enemy's approach by sea or air; concentration of aircraft to meet the threat; coordination of air units by radio; and initial GHQ Air Force defense of the coast to a distance of 200 or 300 miles offshore. On August 11, Secretary of War George H. Dern appointed a committee, headed by Maj. Gen. Hugh A. Drum, Deputy Chief of Staff, to review and revise the Air Corps plan.[24] Events the next day, however, dictated an immediate review of TAN, the war plan for Cuba.

Cuba had been the scene of much unrest and disorder since an unsuccessful revolt against President Gerardo Machado in 1930. The U.S. Government watched the situation very closely, but did not exercise its right under a treaty with Cuba in 1903 to intervene to maintain a government adequate for the protection of life, property, and individual liberty. When the Cuban Army forced Machado out of office on August 12, 1933, a mob seeking vengeance on his followers ran wild in Havana. Reports of shooting, burning, and looting impelled President Roosevelt to send warships. The War Department began reviewing its plans against the possibility the Army might also be called.[25]

General McKinley requested General Foulois on August 12, 1933, to designate an officer to work with the War Plans Division on TAN. Indicating that a directive for the four-army plan would be issued soon, McKinley instructed Foulois to handle the organization of the headquarters of GHQ Air Force until the War Department picked a commander. Headquarters, he said, would be at Langley Field.[26] While trying to figure out the purpose of McKinley's letter, Lt. Col. James E. Chaney, Chief of the Air Corps Plans Division, submitted the name of Maj. Walter H. Frank, the Plans Division member working on TAN. McKinley replied that operational planning should be done by people charged with carrying out the plans. If Frank's mobilization position was with Headquarters GHQ Air Force, his selection to work on TAN was logical; if not, the Air Corps should select someone else.[27]

Colonel Chaney drafted a reply for General Foulois' signature on August 23, basing the letter on a lengthy memorandum he gave the Chief of Air Corps a week earlier. It was not clear what the status of Headquarters GHQ Air Force would be. The Air Corps had neither been given instructions nor asked for recommendations. The little information General McKinley supplied on August 12 did not suffice for sound planning. Chaney asked four questions:

> Is this Headquarters and Staff to be merely a paper organization to be brought into active being only when the War Department Mobilization Plan is put into effect or is it to be an *actual, active peace-time organization supervising, controlling* and

> *operating* the G.H.Q. Air Force in peace-time so that it will be able to carry out its mission in an emergency?
>
> Will the Commander of the G.H.Q. Air Force actually command the Air Force in peace-time?
>
> Will he have charge of its tactical training?
>
> Will he be charged with the tactical inspection of Air Force units to assure efficiency and coordination in tactical training?

Major Frank, he said, should continue to work with the War Plans Division on TAN until clarification of the status of Headquarters GHQ Air Force.[28]

The same day, General Foulois himself wrote The Adjutant General. In lieu of asking questions about the status of GHQ Air Force, he asserted, "G.H.Q. Air Force is an *actual, peace-time organization*" rather than a paper organization to be brought into being on mobilization. As such, he said, "it must be *organized, equipped* and *trained* in *peacetime,* and it must be *administered, supervised, controlled* and *operated* in peacetime by competent military personnel." Believing it unwise and unsound to place Headquarters GHQ Air Force at Langley Field, he asked the General Staff to delay decision on its location until General Drum's committee completed revision of the air plan for defense of the United States.[29] General McKinley responded that GHQ Air Force's status would be cleared up by the forthcoming directive on the four-army organization. He wanted an answer to the question whether Major Frank would be on the staff of the Commanding General, Provisional GHQ Air Force, if TAN became effective. General Westover answered "yes." McKinley then told Foulois he could put off organization of Headquarters GHQ Air Force until General Drum's committee finished its business, but he should complete a provisional headquarters for TAN.[30]

Stepping in to end the paper skirmish, General MacArthur directed that the Assistant Chief of Air Corps in Washington serve ex officio as Commanding General, GHQ Air Force. He also allowed organization of GHQ Air Force headquarters in Washington, D.C., rather than at Langley Field, if that seemed better.[31]

The new President of Cuba, Carlos Manuel de Cespedes, meanwhile made some progress in restoring order, but Fulgencio Batista overthrew him on September 5, 1933. Anti-American feeling flowed strong in Cuba. The United States dispatched nine warships. The Marine Corps mobilized a regiment at Quantico, Virginia. Many Americans favored landing troops at once. Believing "Marines may be landed any day," the *Army and Navy Journal* predicted "the Army will follow."[32]

The War Department checked on the preparedness of the various arms and services to carry out TAN. General Westover, ex officio Commanding General, GHQ Air Force, named his staff: Colonel Chaney, Chief of Staff; Maj. Arnold N. Krogstad, G–1; Maj. Clarence L. Tinker. G–2; Maj. Walter H. Frank, G–3; and Lt. Col. Jacob H. Rudolph, G–4. Capt. Harry H. Young, at Fort Crockett, asked for copies of the Navy's avigation chart of Cuba. Lt.

Col. Albert L. Sneed, at Langley Field, wanted Rand McNally maps, with air trails, for the southeastern states, as well as U.S. Navy avigation charts for the east coast, Cuba, Central America, and the West Indies. Capt. Arthur L. McCullough, Air Reserve, who had resigned from the Regular Air Corps to enter commercial aviation, asked for active duty. He had been operations manager and chief pilot of Compania Nacional Cubana de Aviacion, a subsidiary of Pan American Airways, for nearly 4 years. He believed he knew more about Cuban geography, weather, and aviation facilities than any other person. If needed in the Intelligence Division or any other department, he would gladly serve.[33] The situation in the 8th Pursuit Group was similar to that in most other Air Corps units. The Commander, Maj. Byron Q. Jones, reported the group "ready to take the field on short notice with its present available strength in personnel and planes." But he had only 49 of 115 officers authorized, 536 enlisted men of 915, and 36 planes of 51. He asked that Capt. Harold H. George, 2d Lt. Richard A. Grussendorf, and 15 other officers be returned at once from duty with the Civilian Conservation Corps.[34]

By September the Air Corps had designated units for TAN and arranged to bring them to full peacetime strength. It was holding up tables of movement by rail and air awaiting certain decisions, but could complete the tables within an hour (or, perhaps, several days). The Air Corps had firmed up requirements for gasoline, oil, ammunition, vehicles, and signal equipment, and was finishing plans for air depots at Miami and Havana, and for organizing "the airport of departure (corresponding to a seaport of embarkation)." It would need about a month to prepare an aircraft to carry a five-lens camera to map Cuba. (It ordinarily used a transport for mapping, but for

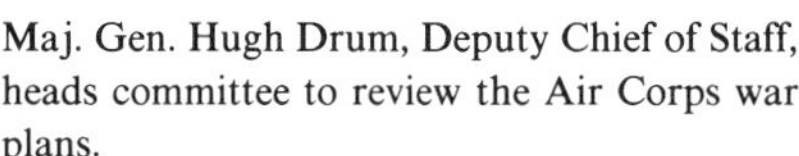

Maj. Gen. Hugh Drum, Deputy Chief of Staff, heads committee to review the Air Corps war plans.

work over Cuba it required an airplane that could defend itself from American-built planes flown by Cuban pilots trained by the U.S. Army Air Corps.) It had to survey facilities at Miami and arrange for gasoline and oil at airfields where units would stop on their way to Cuba. It could do this work better if secrecy surrounding preparations were relaxed. Another handicap was the necessity for working through corps area headquarters, which did not know what was needed, and which would release the units immediately on mobilization.[35]

The disturbances in Cuba went on for several months, but TAN did not go into effect. Dedicated to a Good Neighbor policy, President Roosevelt stood firmly against intervention.

Drum Board

The Drum Board meanwhile completed its review of Air Corps war plans. General Foulois represented the Air Corps, the other members being General Kilbourne; Maj. Gen. George S. Simonds, Commandant, Army War College; and Maj. Gen. John W. Gulick, Chief of Coast Artillery. Considering the stage aviation had thus far reached, the board could see no possibility of land-based bombers flying across the Atlantic or Pacific, attacking the United States, and returning home. Attempts by enemy forces to set up land or floating bases within striking distance would signal the intent to attack. Consequently, the phrase "air defense of the United States" gave a false view of the way aviation should be employed. The board laid down the principle: "Whether operating in close conjunction with the Army or Navy, or at a distance therefrom, all of these agencies must operate in accordance with one general plan of national defense." Identifying seven strategic regions in the United States,[36] the board could not conceive of all being attacked at once. Although air forces alone could not protect any of these areas, the development of aviation rendered overseas invasion more difficult. A "properly constituted GHQ Air Force, a unit heretofore lacking," the board said, could detect the approach of an enemy force, attack it before it reached shore, oppose a landing, and support ground operations against the invader. The board saw a highly mobile air force in a dual role. Strategically, it performed long-range reconnaissance, interdicted enemy reconnaissance, demolished important installations, and interdicted enemy movements. Tactically, it supported ground forces by reconnaissance, demolition, and interdiction during preparation for battle; took part in the battle; and afterwards exploited victory or minimized defeat. The board concluded that GHQ Air Force was needed to discharge war plans and furnish aviation to work with corps, armies, and overseas garrisons.

The Drum Board figured GHQ Air Force required 1,103 aircraft, including 123 in reserve, to carry out RED–ORANGE. The addition of planes for overseas garrisons (578), corps and army observation (332), and training (307) gave the Air Corps a requirement of 2,072 active airplanes and 248 in reserve, a total of 2,320. The War Department, however, had instructed the board to keep within the 1,800 aircraft authorized by law. Making the necessary adjustments, the board called for a GHQ Air Force of 711 active planes. Even this smaller figure far surpassed the number then in combat units earmarked for GHQ Air Force in an emergency. Bombardment units, for instance, possessed 57 of 112 planes in the 711–airplane plan; pursuit, 153 of 271. But the board found ground forces for RED–ORANGE even weaker. Noting that the Army owned 1,600 of the 1,800 aircraft authorized by Congress, the Drum Board urged that the Air Corps be given the full number. Yet at the same time, it insisted this not dilute the strength of the Army's other arms and services. Secretary of War Dern approved the board's report on October 11, 1933.[37]

Headquarters

General Foulois meantime exercised his authority to organize Headquarters GHQ Air Force in Washington on October 1, 1933. General Westover, ex officio Commanding General, named fourteen officers for staff assignments as additional duty, and six others to serve with the staff when the Air Force assembled for training. Except for Major Tinker, General Westover retained the officers he selected earlier for TAN. Stationed at March Field, Tinker could not come in frequently for conferences, so Maj. Follett Bradley, at Mitchel Field, took his place. Five months later, General Westover organized a headquarters squadron at Bolling Field, commanded by Capt. Orlo H. Quinn, to provide administrative assistance and clerical service for Headquarters GHQ Air Force.[38] As Commanding General, GHQ Air Force, Westover in peacetime submitted recommendations and plans on the organization and training of the GHQ Air Force as a unit and its employment in war; commanded the GHQ Air Force when brought together, or when operating as a unit, for any purpose; and suggested ways to enhance its efficiency.[39] Establishment of the GHQ Air Force itself came later, after the Air Corps had been put to the test during an emergency arising from cancellation of government contracts for carrying mail.

During the 1920s, GHQ Air Force existed solely in mobilization plans. In the early 1930s, a stream of events led to the creation of Headquarters GHQ Air Force. The MacArthur-Pratt agreement of January 1931 started the process that turned the Air Corps to planning an air force to operate under the control of General Headquarters for defense of sea frontiers. But the plan required such large outlays of money, men, and materiel that it stood no chance of approval. There followed, however, War Department redefinition of the functions of the Army's air arm and revision of war plans to employ GHQ Air Force for coastal defense. At the same time, the War Department was forming four field armies for mobilization in an emergency. This entailed establishment of headquarters for the armies, the air force, and other units, manned by Regular Army officers. In these mobilization assignments, they engaged in war planning and peacetime exercises. The Air Corps tested the operation of GHQ Air Force by founding a provisional air force for maneuvers in May 1933. That same year, disorders in Cuba, with the possibility that Army aviation units might be thrown into action, disclosed how ill prepared Army aviation was for a military expedition abroad. This prompted the naming of a commanding general and staff for a provisional air force if the war plan for Cuba was invoked. Concurrently, the Drum Board's review of Air Corps war plans focused attention on GHQ Air Force and its needs for men and equipment. This was the sequence of events that culminated in the creation of Headquarters GHQ Air Force on October 1, 1933.

Chapter XVII

Airmail

General Foulois remembered February 9, 1934, as one of the three most significant dates in the history of U.S. air power. On December 17, 1903, the Wright brothers demonstrated powered flight in heavier-than-air craft. On March 19, 1916, U.S. Army airplanes (under Foulois' command) first took the field on tactical operations. And on February 9, 1934, the U.S. Army Air Corps received orders to fly the mail.[1]

Air Corps airmail operations fell into four phases. During the first (February 9–19, 1934), the Air Corps created an organization, assigned men and equipment, prepared planes for the job, and supplied pilots special training so they would be somewhat familiar with the routes they were to fly. A number of serious aircraft accidents occurred that raised questions as to the Air Corps' ability to carry the mail. The second stage, beginning when the Air Corps started airmail operations in unusually foul, cold, weather on February 19, brought more fatal accidents, leading to suspension of flights on March 11. In the next phase, lasting a week, the Air Corps reorganized, overhauled equipment, gave pilots more training, eliminated some routes, and reduced schedules. The last phase, beginning with resumption of operations on March 19, saw the gradual phasing out of Air Corps participation, the renewal of contract flights, and termination of Air Corps airmail activities on June 1, 1934. This in brief is the story of one of the largest projects—and in many ways the most important—undertaken by the Army's air arm in the interwar years.

In his office in the Munitions Building about 1100 that memorable 9th day of February 1934, General Foulois received a telephone call from Harllee Branch, Second Assistant Postmaster General. Branch wanted the general to come to the Post Office Building. Foulois assumed this concerned an interdepartmental aviation committee on which he and Branch sat. However, Branch wanted to discuss the possibility of the Air Corps' taking over airmail operations if the government annulled contracts with the airlines. Asked if the Air Corps could do the job, Foulois said yes. Newspapers had been reporting that the Roosevelt administration and the Democratic Congress were investigating airmail contracts awarded by the previous Republican regime. There were rumors the government might cancel the contracts on grounds of fraud and collusion. General Foulois never thought of the Air Corps' becoming involved. Assuming the mail would go by train, he had made no plans. When Branch broached the subject, Foulois called his office and asked Lt. Col. James E. Chaney and Capt. Edwin J. House to come to the Post Office Building. Others joined the group, including Stephen A. Cisler, who handled airmail matters for the Post Office Department, and Edward Vidal, in charge of the Aeronautics Branch of the Department of Commerce. The conference lasted from noon until three in the afternoon. The men went over the existing mail routes, considering the ones the Post Office thought most essential to maintain. After discussing the suitability of Army aircraft and the training of Air Corps pilots, they decided the Air Corps could handle the airmail under normal winter operating conditions. Asked how much time the Air Corps would need to prepare, the general said "about a week or ten days."

Returning to his office after the conference, General Foulois called in members of his staff and started them on plans. Leaving to report developments to General Drum, Deputy Chief of Staff, he was unaware that President Roosevelt, Postmaster General James A. Farley, and Attorney General Homer S. Cummings had already decided to cancel the airmail contracts. Announcing the decision at a press conference, the President said Secretary of War Dern was placing men and equipment at Farley's disposal to carry the mail. Newsmen gave General Foulois the word while he waited to see General Drum. Foulois told the Deputy Chief of Staff about the conference in the Post Office Building and what he had heard from reporters. Just then General MacArthur walked in and said he had learned the Army would assume airmail operations. Foulois explained again what had happened. Asked by the Chief of Staff if the Air Corps could do the job, Foulois replied, "Yes, sir," but it would be no "picnic." General MacArthur wanted to see a plan of operations as soon as possible. Orders issued by the President and Farley gave Foulois ten days to get ready.[2]

Scheduled airlines had hauled more than seven million pounds of mail in 1933 and had received over $16 million in government payments. The airlines normally carried mail in planes used for passengers and express. But at times they transported it in open-cockpit aircraft, when the weather was below minimums set by the Department of Commerce for passenger service but still good enough for flying. If these airplanes could not operate, the Post Office dispatched the mail by train to its destination or to an airport with better flying conditions. The airlines possessed modern hangars, shops, and offices, and good, well-equipped planes. Their pilots, many of them graduates of the Air Corps' cadet program, had served a year or more as copilots on scheduled operations. They were trained and experienced in using radios and instruments. Their two-way radios let them communicate with ground stations from any point on their route. The airlines operated their own communication networks, but their pilots also used the communications, radio beams, marker beacons, weather broadcasts, and other facilities of the federal airways.

The Air Corps and Post Office Department agreed that initially the Army would fly 18 routes. The route mileage (11,800) amounted to less than half that of the airlines. The Air Corps' schedule called for 62 trips a day, 24 by day and 38 by night. This entailed 40,800 miles of flying, not one-third the distance commercial lines flew with mail each day. In mid-March the Air Corps and Post Office eliminated routes and dropped flights, reducing the daily mileage to 26,100 and night flying from 23,000 to 4,800 miles.[3]

General Foulois gave Maj. Byron Q. Jones command of the Eastern Zone, Lt. Col. Horace M. Hickam the Central, and Lt. Col. Henry H. Arnold the Western. In his first order, Foulois placed airmail operations under Air Corps rules for cross-country flying. Regulations required that each flight be properly authorized, equipment in satisfactory condition, and weather favorable. The flight had to be completed before dark unless the plane carried night-flying equipment, the landing field was lighted, and the pilot experienced in night flying. Regulations contained definitions to assist commanders in judging weather. "Excellent" meant conditions generally "ideal" for flying, the sky clear to partly cloudy, ceiling unlimited, visibility excellent, wind light. Conditions were "good" if the sky was overcast, the ceiling at least 3,000 feet, winds light to moderately strong. The average pilot could take off and get through under "poor" conditions, with a ceiling of 1,000 feet and local showers. Conditions were "dangerous" if there was an extremely low ceiling, poor visibility, fog, high wind, or the field so wet and soft it was unsafe for landing. The regulations directed commanders to relate these definitions to the pilot's skill, the character of the terrain and his familiarity with it, weather changes peculiar to certain localities, and other factors in deciding whether to clear him for a cross-country flight. Before airmail operations commenced, Foulois charged zone commanders with "safeguarding lives and property at all times even at sacrifice of mail service." He ordered them to "inculcate all personnel engaged in air mail operations with the above principle."[4]

Maj. Byron Q. Jones

Lt. Col. Horace M. Hickam

Lt. Col. Henry H. Arnold *William Bruce Arnold collection*

After learning of accidents that took three lives on February 16, General Foulois urged pilots to be careful, especially during the first days after operations began on the 19th. "The safety of pilots, mail and planes," he said, "is of more importance than keeping of mail schedules." He ordered zone commanders to "drill these instructions into your pilots daily until they thoroughly understand the safety first policy of the Air Corps."[5] After two accidents (one fatal) during bad weather in the Eastern Zone on February 22, Major Jones reminded control officers and pilots to take every precaution to prevent flying in unfavorable conditions. Control officers at Newark, Cleveland, Washington, and other terminals bore responsibility for authorizing flights from their stations. Jones told them to study the newest weather map, consult the latest forecast, review the weather sequence over the last several hours, confer with local Weather Bureau personnel, and prepare a weather report for the pilot before clearing a flight. At intermediate stops manned by enlisted men, the senior man prepared a weather report and the pilot cleared himself. Jones ordered no flying when a ceiling under 500 feet was anticipated any place along the route, and he cautioned that was not enough altitude in strong wind over mountains. In addition, he directed pilots on night flights to maintain beacon-to-beacon visibility at all times and urged special care to avoid conditions conducive to icing on airplanes.[6] That night, one of his pilots, a second lieutenant of the Air Reserve, suffered serious injuries when he got lost in bad weather on the Newark-Richmond route and crashed in Maryland.

Following these and several other accidents (including 2 fatal ones about the same time), General Foulois made 2 years service in the Air Corps prerequisite to a pilot's being assigned a night flight unless weather on the route was excellent. He instructed pilots not to commence night flights in unfavorable weather, forbade their continuing at night unless instruments and radio functioned correctly, and told them not to fly when ice might form on the plane, propeller hub, or instrument venturi. Foulois held control officers responsible for seeing that planes passing through their stations received the required inspection, and that radios, instruments, engines, and controls operated satisfactorily. He further limited flight duty to 8 hours in 24, relieved pilots from all duty 1 day out of 4, and prohibited them from making mail or ferrying flights unless experienced in the type of plane assigned for the mission.[7]

When three accidents took the lives of four Air Corps flyers on March 9, President Roosevelt directed Secretary of War Dern to stop the Air Corps from carrying mail except on routes and under circumstances which would "insure against constant recurrences of fatal accidents."[8] General Foulois suspended operations and laid down conditions to be met before the Air Corps resumed carrying mail. To comply with the President's orders "so far as humanly possible," he required zone commanders to give "positive assurances" that: (1) airplanes spotted for mail runs were sufficient for the purpose and well maintained; (2) two-way radio, blind-flying instruments, and night-flying equipment were installed and operating properly on planes used on night runs; (3) airplanes on day runs had flying instruments and radio receivers, with two-way radio installed insofar as possible with sets available in the zone; (4) older and more proficient pilots flew night and hazardous runs; (5) older, seasoned personnel served as control officers and assistants; and (6) experienced Reserve officers called to active duty were assigned to scheduled runs as fast as practicable.[9]

Even a little thing like a thermometer (or absence of it) affected airmail operations and safety. Capt. Caleb V. Haynes, commanding a section of the Eastern Zone, suggested to Major Jones on March 5 that air temperature thermometers be provided for all mail planes. Explaining that pilots needed them to determine when ice might form on their aircraft, he asked for Type C–3 A.C. 33–49. First Lieutenant Uzal G. Ent, adjutant at zone headquarters, wrote to the Materiel Division on March 7 to see if thermometers were available. Brig. Gen. Henry C. Pratt, Chief of the Materiel Division, asked on March 10 for details on requirements, saying the C–3 was standard for photographic planes but procurement would take two months. Major Jones immediately asked the Middletown depot if it had thermometers. When Maj. Lawrence S. Churchill replied no, Jones informed General Foulois he needed thermometers to comply with orders, and they should have lights so pilots could read them at night. His pilots resumed flying the mail, but without thermometers to warn of icing.[10]

Any number of other examples could be cited, the problem of glare from engine exhaust being a case in point. Pilots on night practice runs in O–38s found the glare from the short exhaust stacks of the radial engine so blinding they could not see beacon lights ahead. The glare was especially bad in fog, rain, or snow. Headquarters Eastern Zone received reports of this condition on February 16. The next day, Major Jones asked the Materiel Division to design a collector ring to gather exhaust gases from each cylinder and disperse them through one opening. Calling General Foulois' attention to the problem, and labeling the glare "a serious hazard to younger fliers," Jones recommended immediate action. The reply from Wright Field the same day said the Materiel Division had already furnished collector rings for 0–38s and was testing one for the O–38B, which it expected to release for production the following week. The Materiel Division stated rings for O–38E and F aircraft were available in the depot: "Your requisition on depot will receive normal action if stock is insufficient." Three weeks later, on March 9, Jones received another message from General Pratt: "Purchase order for collector rings will be issued March 14th, calling for complete delivery within 30 days."[11] There the matter stood when General Foulois ordered resumption of operations.

With the notice to begin operations again on March 19, zone commanders were given additional instructions. They were to make clear to their people that, "if on any route, or any duty, the conditions of weather, personnel or equipment are such as to give rise to any doubt as to the safety of moving the mail, that is from the standpoint of human safety, the mail shall not and will not be carried."[12] Major Jones followed with an order not to fly except in "good" weather, that is, with a ceiling of at least 3,000 feet and winds no more than moderately strong. He applied this to Eastern Zone administrative and training flights as well as those carrying mail.[13]

The Weather Bureau and Department of Commerce cooperated to afford the Air Corps the best meteorological data and forecasts available to airmen in the United States. Nevertheless, pilots often encountered unpredictable conditions that forced them to land and sometimes to crash. According to Colonel Arnold, a series of accidents in the Western Department "taught us we could not trust the weather reports because they did not mean anything." Weather shifted swiftly in the mountains. "The only way to find out," Arnold said, "is when you get there, and you have to know by experience whether to keep going or to turn around and go back." Few Air Corps pilots possessed such experience when they began flying mail. Colonel Arnold's opinion of weather services in the Western Zone dipped even lower when he learned how weather observations were made at one station in the

Gen. Foulois stands in front of air mail map designating routes to be covered by Air Corps pilots.

mountains. The observer was a veteran at the business but blind—"his wife goes out, looks up, and [sees] what the weather is and she tells him and he gives it over the telephone."[14]

General Foulois saw that complete compliance with safety orders was hard to achieve. His flyers faced the issue of seeking to adhere to both safety regulations and "the Post Office Department's slogan to put the mail through."[15] Major Jones said his pilots (especially the "youngsters," less so the "older and cooler heads") were inspired with the idea that "by golly, they would carry that air mail, and they carried it," a spirit that at times got them in trouble.[16] Noting that military pilots were trained to "do the job in spite of odds," Capt. Willis R. Taylor, Eastern Zone operations officer, deemed this a dangerous psychology when aggravated by adverse publicity. Pilots responded to newspaper criticism with "we will show them." As a result they flew when they knew they should not.[17] The Chicago papers gave Colonel Hickam, the Central Zone Commander, a bad time. Believing that men who opposed government ownership of transportation and communications set newspaper policy, Hickam knew "we could not be a success in the papers." Newsmen garbled and twisted everything, he said. Realizing "they were trying to jockey me into position to pin something on the Air Corps . . . I shut up."[18] Colonel Arnold also found newspapers "antagonistic." They thought it "their duty to vilify us and call us murderers and everything else," which, of course, "did not have a wholesome effect on the pilots." His young flyers thought they had to put the mail through. They believed signs saying, "Neither rain, snow, hail, or heat stops the mail." They did not realize commercial pilots turned back.[19]

Control officers sometimes succumbed to the pressure and authorized flights in unfavorable weather or with planes and equipment not working properly. The control officer at Cleveland, for example, cleared 2d Lt. Charles P. Hollstein on a night flight to Washington despite a faulty radio and bad weather en route. Hollstein, a Reserve officer who got his wings less than seven months earlier, had little night-flying experience. Mechanics had worked on the radio at Cleveland but could not correct faulty shielding. Soon after Hollstein took off at 0200, the compass light went out so he stopped at Akron to fix it. After leaving there, he passed through snow flurries, the compass light went out again, and the radio made so much noise he could not distinguish signals as he approached Pittsburgh, his first scheduled stop. With visibility so poor he could not see the airport, he decided to try Uniontown, put the Pittsburgh mail on the train, get the radio repaired, secure a flashlight to read the compass, then go on to Washington. He crashed while circling to locate the airport. The plane was a complete wreck, but Hollstein escaped serious injury. He walked into Uniontown, reported, and returned to the plane to get the mail sacks.[20]

Lieutenant Hollstein and other Air Corps pilots carrying mail flew from the same airports the airlines had used for airmail operations and continued to use for passenger service. At Washington, for example, they flew from Washington-Hoover Airport in lieu of Bolling. At certain airports the Air Corps took over National Guard facilities. The Commerce Department and Weather Bureau on occasion let the Air Corps use part of their space. The Air Corps rented whatever hangars, shops, and offices it could get at municipal and commercial airports. At some fields the airlines owned everything, as Eastern did at Raleigh, North Carolina, and Boeing did at Elko, Nevada. Generally, however, the airlines cooperated in making buildings and space available to the Air Corps. Even so, Air Corps operations in many places suffered from lack of adequate facilities. At Chicago, Colonel Hickam ran the mail for a week from a "municipal outhouse" ("It was all boarded up and the plumbing was in bad shape"). He then took over National Guard hangars ("just ruthlessly kicked them out"), and afterwards rented a grocery store across the street.[21] Major Jones planned on directing operations in the Eastern Zone from the airport at Newark, New Jersey, the terminal for all airmail flights into and out of the New York metropolitan area. Unwilling to evict the National Guard to get office space, he sought help from New York City. Mayor Fiorello H. La Guardia gave him several rooms, office furniture, and hangar space for his pursuit plane at Floyd Bennett Field, which Jones used for headquarters until March 13. Failing to obtain a formal commitment from the mayor for the continued use of city facilities and equipment, he moved his headquarters to Mitchel Field.[22] Colonel Arnold made his headquarters at Salt Lake City on the fourth floor of a hotel, affording living as well as office space.[23]

Captain Haynes reported that at Richmond, Virginia, "the Army was forced to establish headquarters in rear of hangars, in tents, sheds, and other places" unsuitable for efficient operations. Servicing, repair, and maintenance of airplanes and engines became "almost an impossible task," Haynes said, when done outside in all kinds of weather.[24] At Atlanta, Georgia, the Air Corps found room under cover for only two planes at a time.[25] At Cheyenne, Wyoming, it rented hangar space for seven planes but, unable to obtain more space, it parked the rest of the planes outside. At Cheyenne, it also rented a supply room from Boeing Air Transport, Inc., for $120 a month and, as Colonel Arnold said, "sponged" on the Commerce Department for additional space. Colonel Hickam thought it "very nice" that Boeing let the Air Corps use offices and washrooms at North Platte, Nebraska. But he said that after the airline lost its contract for gasoline "our men were not welcome there . . . for a week." Getting back into the company's "good graces," the pilots could once again "go in and get warm."[26] At Pasco, Washington, the Weather Bureau spared a small office. At Reno, Nevada, it turned over a good portion of a large room it rented from Boeing. At Seattle, the Air Corps took over a hangar belonging to the Organized Reserve.[27]

When it could, the Air Corps often used its own facilities for administration and maintenance. At Washington, D.C., Bolling Field provided service and repairs for aircraft operating from Washington-Hoover Airport, and for a time served as headquarters for a section of the Eastern Zone. Major Jones sent planes to regular Air Corps stations for major maintenance. Colonel Arnold set up a supply and repair depot at Salt Lake City to avoid sending planes to Rockwell Field, California, for work.[28] The municipal airport of Portland, Oregon, being congested, the Air Corps created airmail headquarters on Pearson Field at Vancouver Barracks in Washington, just across the river from Portland. In taking over 2 hangars and an administrative building, the Corps left Reserve units 1 little office. Twenty-eight enlisted men and 25 civilians maintained planes at Pearson Field. One officer, 3 enlisted men, and 2 civilians worked at the municipal airport, where they rented hangar space for 1 plane at $25 a month. An aircraft completing its flight to Portland unloaded its mail and flew to Pearson Field for servicing. One hopped over from Pearson just before time for a flight to leave Portland.[29]

First Lieutenant George F. Schulgen reported aircraft maintenance arduous and slow at Newark due to bad weather and a dearth of tools and hangar space, the situation being aggravated by the many different kinds of planes and engines requiring work. This was by no means unusual. Capt. Fred C. Nelson, engineering officer at Cleveland, had "too few mechanics and almost no tools."[30] Shortages of mechanics, tools, and shops persisted in many places until the end of mail operations.

The Air Corps acquired most of the men for airmail operations from tactical units, thus enabling the schools and Materiel Division to pursue their normal activities. The older, more experienced men commanded regions and sections, served as control officers, managed servicing points, and filled administrative, supply, communication, and engineering positions. Now and then they flew the mail, but more often the mail runs fell on younger pilots. As a rule, the latter had slight knowledge of instrument flying or the use of the federal airways, and frequently possessed little or no previous flying in the kind of planes given them to carry mail. To procure seasoned pilots, the Air Corps called on Reservists who lost their jobs with the airlines owing to the cancellation of contracts. By the end of March, fifty-one men from the airlines were on active duty.[31]

Many of the pilots flying the mail were Reserve officers on active duty after graduation from Advanced Flying School. Any number of young men must have become airmail pilots in much the same way as 2d Lt. Beirne Lay, Jr. He had decided to become a flyer when he saw the movie *Wings*. Accepted as a flying cadet, he trained at Randolph and Kelly Fields, graduated and received a Reserve commission in July 1933, and went on active duty with the 20th Bombardment Squadron at Langley Field, Virginia. He and his friends ("a lot of eager kids") welcomed a chance to fly the mail. Selected for airmail duty, Lay took an hour of instruction in blind flying, his previous experience being thirteen hours under the hood at Kelly Field. He attended a lecture by 1st Lt. Elwood R. Quesada on airway beams, of which Lay grasped the theory but scarcely the practice. The following morning, after learning he was to fly a night run between Nashville and Chicago, he checked out in a Bellanca (a C–27), got another hour of instrument flying, drew a pistol, ammunition, flashlight, and winter flying clothes, and packed personal belongings—all before 1300. He then left for Chicago, copilot of a Bellanca carrying men to organize and operate a mail station at Sky Harbor, the airport for Nashville. Continuing to Chicago in the Bellanca, he covered in daytime the route he would fly at night.

Before leaving Langley Field, Lay received orders to complete a practice run over his route. At Chicago he learned he would make the trial in a P–12 just in from Selfridge Field. He had checked out in a P–12 while in Advanced Flying School but had never flown one at night. The Commerce Department strip maps for the airway between Chicago and Nashville had not arrived. The Rand McNally state maps he borrowed bore no aeronautical information, so he memorized emergency landing fields, beacons, radio beams, radio frequencies, and identification signals. With the stars shining, Lay taxied the P–12 to the end of the gravel runway, radioed the control tower, secured his clearance, and took off. Tuned to the beacon, he heard dots and dashes marking his route, but almost at once the radio went dead, and the compass started spinning. From the various lines of rotating lights radiating from Chicago, he picked one he thought would lead to Terre Haute, Indiana, the first stop.

The Air Corps had stationed a sergeant and two privates at Terre Haute to handle the mail. When Sergeant Patton heard the plane, he turned on floodlights at the east end of the field. The wind blew strong from the east. Snow covered the ground. Lay circled, blinking his navigation lights, the signal to turn off the floods—but the lights stayed on. He did not want to land with the lights in his eyes, but a crosswind landing was too dangerous. So he faced the lights, which blinded him the last 40 feet to the ground. When the plane stopped bouncing, he headed for the operations office, where he found a redhot stove. The thermometer outside the door showed –8 degrees when Lay reluctantly left the fire. In the air again, he headed for the next stop, Evansville, Indiana, 96 miles, then on to Nashville, 157 more. In clear weather, he saw 2 and sometimes 3 beacons ahead. The return flight to Chicago the next night was uneventful. He arrived at Chicago at daybreak, Monday, February 19, the day the Air Corps commenced flying the mail. Lay did not get much rest, for he had a lot to learn and much to do before departing at 2145 on his first trip with mail.[32]

General Foulois wanted each pilot to have at least one familiarization

Left: Pilot delivers mail at March Field, Calif., from his P–12 and (below) mail is loaded onto B–7 aircraft in Oakland, California.

flight on his route before tackling a mail run. During such flights, planes often hauled sandbags representing a load of mail. On some routes, Air Corps pilots flew as copilots on commercial planes to become familiar with landmarks, radio beams, marker beacons, and landing fields. Employees of the Post Office and the Department of Commerce helped acquaint pilots with mail operations, and several pilots were given a quick refresher in instrument flying at the Air Corps schools at Langley and Rockwell. Then, too, the Air Corps sent instrument training planes (BT–2Bs) to various airports so airmail pilots could practice flying under the hood. (A number of these BT–2Bs were taken from Reserve training centers.)[33]

The inexperience of many Air Corps pilots carrying mail created problems and accounted in part for the high accident rate. Take the case of 2d Lt. William J. McCray, a Reserve officer who had worn wings for just over a year. He had flown less than six hours in A–12s when he crashed one on a mail run from Chicago to Cheyenne. Flying in good weather, he had no trouble until the motor quit. Colonel Hickam said McCray "became engrossed in listening to his radio . . . and let his gasoline tank run dry." McCray actually had plenty of gas but not enough time to switch tanks and restart the motor.[34] Young, inexperienced pilots were not the only ones to make mistakes. Consider what happened to Maj. Charles B. Oldfield, a regional commander in the Western Zone, who got his wings in 1921 and had more than a hundred hours in planes with retractable landing gear. He landed a YB–10 wheels up at Cheyenne after a mail run from Salt Lake City: "I simply forgot I was flying a plane with retractable landing gear."[35] An airline pilot flying for the Air Corps also forgot to lower the wheels on his YB–10 before landing at Toledo with one engine not working.[36] Another airline pilot crashed at Cheyenne while on a training flight two days after coming on active duty with the Air Corps. As Colonel Arnold put it, here was an experienced pilot, one who had been flying commercial planes out of Cheyenne, and "he goes up in an Army plane, and bingo, he goes in a spin and kills himself."[37]

Of 10 pilots flying planes involved in fatal accidents, the oldest had flown for 15 years, 1 had been rated 5 years, another 4, and 2 others 2 years. The other 5 had been out of flying school not more than a year when assigned airmail duty. The one with the least pilot time (364 hours) had graduated from flying school a few months earlier and never before had flown the kind of plane in which he died several minutes after taking off.

Air Corps pilots used sundry types of aircraft for mail. Transports were best for long routes with heavy loads, but the Air Corps owned few of these planes and most of the time needed them to haul people, supplies, and equipment. Bombers served both as transports and mail planes. B–2s, B–4s, and B–6As carried heavy loads but at slow speed. Y1B–7s flew faster, but of the bombers the YB–10s proved best for mail. The Glenn L. Martin Company started delivering YB–10s to the Air Corps as airmail operations got under way. As quickly as possible, the Air Corps assigned the new aircraft to the Eastern and Western Zones for the transcontinental route. The Air Corps initially employed P–12s on shorter routes, such as between Toledo and Detroit, or where the mail was light, as between Atlanta and New Orleans. Though fast, the pursuit ships afforded little room for mail sacks. Further, the load disturbed the plane's balance. The Air Corps soon replaced P–12s with observation planes, which took more cargo and flew at fairly good speed. Observation aircraft transporting mail included O–19Cs, O–25Cs, O–38Bs and O–38Es, and O–39s. Carrying more mail and faster than the observation ships, the A–12 served well in the Central Zone where it was the principal plane in mail service. On the other hand, it was unsatisfactory for operations from fields at high altitudes in the Western Zone. All these various types and models had open cockpits except the transports, YB–10s, O–38Es, and O–39s. One might imagine that a pilot flying the mail during bad weather would welcome a canopy over his cockpit. Some pilots, however, did not like the one on the O–38E because it restricted vision at night and in bad weather.

An O–38E at Cheyenne on March 9 claimed the lives of two men. Right after takeoff, the plane stalled, went into a spin, fell to the ground, and burned.[38] A week later another man died under similar circumstances at Cheyenne.[39] Investigators could not determine the cause in either case. Talking with pilots at Cheyenne, Colonel Arnold learned they did not like to fly the O–38E. Difficult to hold level at high altitude, the aircraft was unsuited for mountainous country and for operations from Cheyenne, where the airport was 6,145 feet above sea level. The O–38E tended to fall off on one wing or to whipstall and go into a spin too quickly. Once it started spinning, a pilot needed 2,000 feet to pull out, and then he had to take great care to prevent spinning again.

Air Corps test pilots suggested several possible causes of these accidents. Colonel Arnold believed the load was improperly distributed. Pilots confirmed that the tendency to spin became more pronounced when the rear cockpit and baggage compartment were loaded with mail or other cargo. Engineers at Douglas Aircraft, the company which built the O–38s, told Arnold that loading the baggage compartment threw too much weight to the back, causing the plane to spin. He should have changed the wing setting. Arnold removed the O–38Es from service.[40] Arnold explained later that airplanes had fixed places for bombs, ammunition, and other things. Normally, one did not put the load anywhere else, but special arrangements were required to put mail in observation and pursuit planes. Speaking of the O–38Es he said:

> Well, we went into that blindly. Unfortunately, we did not know any better. If we had stopped to think a bit, it would have been different. Things moved too fast to give us an opportunity to think much. If we had thought a little, I would know that something like that would happen and so would the rest of us. We did not do much thinking because we did not have a chance to think right; we were just ordered to carry the mail.
>
> Who passed on it—well, I guess I am the fellow who passed on it.[41]

The same kind of thing occurred with navigation instruments. When orders came to carry the mail, the Air Corps owned sufficient directional gyroscopes and artificial horizons for the work. But few were aboard aircraft (the Air Corps was saving them for B–10s and other new ships), and most pilots had never used them. Colonel Hickam equipped the A–12s of the 3d Attack Group before sending them to mail stations in the Central Zone. Major Jones dispatched his mail planes to the depot at Middletown,

Pennsylvania, for installation of instruments. First Lieutenant Henry W. Dorr, engineering officer of the Eastern Zone, said later that the depot was not prepared to do the job properly. The chief object seemed to be to get the instruments into the aircraft one way or the other. Workmen at Middletown mounted instruments on boards and put the boards beneath the regular instrument panel, below eye level, where pilots could not read them accurately. The mounting not being shockproof, the instruments did not work well for very long.[42] Colonel Arnold asserted the installation of directional gyros and artificial horizons was "without rhyme or reason." If an engineer found it "easier to run a cable there . . . he put the instrument there," without any idea of how the flyers would use it. Speaking of the need for logical placement, Arnold said "when a man is flying blind, he has not time to go hunting for instruments."[43]

At least one death could be traced to faulty installation of instruments. First Lieutenant Otto Wienecke had more than 1,000 hours as a pilot when he flew his aircraft into the ground on a mail run from Newark to Cleveland. He was in an O–39, considered by many pilots to be the Air Corps' best mail plane. The directional gyro and artificial horizon were on a board in front of his knees. In heavy snow east of Cleveland and unable to see the ground, he apparently used the gyro to hold the plane on course and watched the artificial horizon to keep the aircraft on level flight. He evidently did not realize that the artificial horizon, meant to be installed at eye level, gave a false reading when viewed from an angle. When he thought the plane perfectly level, it actually was headed down.[44] That accident was one of those that brought airmail operations to a sudden halt in March. Major Jones used the interlude in operations to change the instruments. The need for shockproofing had been recognized earlier, but the Middletown depot did not have people with the expertise and experience to deal with the problem. The Sperry Gyroscope Company sent men to Middletown to assist in designing and building new instrument boards. Nearly all aircraft received the new boards before operations resumed.[45]

For airmail operations the Air Corps had access to all the facilities and services of the federal airways developed and maintained by the Department of Commerce. (*Map 9*) Marked for daylight operations and lighted for night flying, the federal system furnished intermediate fields for emergency landings. Among the flying aids were radio range signals to lead pilots across the country, radio marker beacons that served as milestones and enabled flyers to check their progress, and radio broadcasts reporting weather along the route. In addition, teletypewriter circuits disseminated weather data and maps, and reported the movements of aircraft along the airways.[46]

A number of Air Corps pilots, like Lieutenant Quesada the former flying aide to Secretary Davison and chief pilot in the Air Corps' New York-Chicago mail run, were experienced in flying the airways. The majority were not. Pilots at Langley Field and elsewhere utilized range stations set up by the Air Corps to practice flying on a radio beam. However, the Air Corps had not stressed airways radio and did not plan to employ it in moving units during an emergency. The radios it adopted for tactical operations were unsuited for airways use. Consequently, the Air Corps was unprepared in February 1934 to take full advantage of the facilities and services of the federal airways as airline pilots did in flying mail.

At the outset, General Foulois ordered that each mail plane be fitted with a two-way radio (SCR–183), the best set the Air Corps possessed for the job. In many instances, sets had to be installed before aircraft were ready for airmail duty. Since the Air Corps did not have sufficient SCR–183s to equip all planes, some got only receivers (SCR–192s). The Air Corps ordered more SCR–183s but could not secure delivery for several weeks.[47] Designed for two-way voice communication between airplanes operating together, the SCR–183 was not meant for airways use. A high-frequency set, the SCR–183 transmitted on 6,200 to 7,700 kilocycles and received on 4,000 to 8,000 kilocycles. The Commerce Department's radio beams, marker beacons, and communication stations received both low- and high-frequency transmissions but in the higher frequencies guarded only 3,105 kilocycles for calls from commercial planes. Hence, Air Corps pilots could not adapt the SCR-183 to the airways without its modification or changes in Commerce Department operations. An SCR–183 or SCR–192, however, needed merely a different coil, quickly obtainable and easily inserted, to receive radio beams, marker signals, and weather broadcasts. Several mail planes got these coils before airmail operations commenced, and modification of more sets proceeded rapidly.[48]

No simple modification was attainable to allow the SCR–183 to transmit on lower frequencies. On February 11 the Commerce Department agreed to stand by on 6,385 kilocycles for calls from Air Corps mail planes. After Army pilots set their transmitters to that frequency, the Commerce Department realized it could not monitor 6,385 kilocycles for the Air Corps and at the same time guard 3,105 kilocycles for airlines carrying passengers. The Air Corps therefore procured high-frequency receivers for the Commerce Department to guard 6,385 kilocycles, but it took time for the Air Corps to collect the sets and distribute them. Further, pilots had trouble keeping their transmitters tuned precisely to the frequency.[49] Airlines planes carried powerful (50-watt), long-range (100-mile) transmitters; Air Corps mail planes had weak (7½-watt), short-range (15-mile) sets. With an SCR–183 (or SCR–192) fitted with the correct coil and working properly, a pilot received continuously as he progressed along the airway from one station to the next, but the SCR–183 did not furnish continuous two-way communication.[50] Describing a flight between Salt Lake City and Cheyenne, Colonel Arnold said he had two-way communication a "large part" but not all of the time. He soon lost touch with Salt Lake City. He heard them; they did not hear him. Later, he picked up the signal from the intermediate field at Knight, Wyoming. When he came closer, he called Knight and received a reply. On the next part of the route he heard both Knight and the next station, Rock Springs, Wyoming. Nearer Rock Springs, he talked with the operator there. Arnold found the same thing all the way to Cheyenne,[51] and similar conditions prevailed all along the airways.

The temporary suspension of airmail operations permitted pilots to learn more about flying the airways. Most of the new second lieutenants on airmail duty lacked experience in using radio beams. Like James M. Goodbar of the 3d Attack Group, they perhaps tried it, did not do well, and went back to navigating by compass. Goodbar explained: "The only time I ever flew that [way] or tried to I went at it blind; I did not know exactly what I was trying to do and I just stuck my hand out, tuned in on one beam and it did not occur to me that it would run out some time or other." After Goodbar drew airmail duty, he and other pilots at Fort Crockett, Texas, heard a radio expert explain beam flying. Afterwards, he and the other pilots practiced what they had learned. So, Goodbar said, "we were pretty well up on it when we started flying [airmail]. That is, comparatively speaking." Asked if two-way radio increased his safety and comfort, Goodbar said he relied on it for weather and other information. He added he "made a practice of calling . . . stations that I went by if for no other reason than to check in if they wanted to know where I was." Believing it would be difficult to orient oneself by radio beams if really lost, Goodbar stated he did not think the Air Corps expected him to fly blind by radio.[52]

More training, improvements in equipment and maintenance, schedule changes, reduced mileage, fewer flights at night, more rigid control, greater attention to safety, and better weather helped reduce the number of accidents after operations resumed following the crashes in March. The government soon afterwards began contracting with the airlines to carry the mail, and by June 1 the Air Corps was out of the business.[53]

In the midst of the airmail operations, Secretary of War Dern appointed a committee headed by Newton D. Baker to investigate. Baker's civilian associates were Dr. Karl T. Compton, President, Massachusetts Institute of Technology; Dr. George W. Lewis, Director of Aeronautical Research, National Advisory Committee for Aeronautics, and three ex-members of the Army Air Service: Clarence D. Chamberlin, the transatlantic flyer; Edgar S. Gorrell, former head of the Strategical Aviation Branch of the Air Service, AEF, and now President of the Stutz Motor Car Company; and James H. Doolittle, present manager of the aviation department of the Shell Petroleum Corporation. The military members were those on the Drum Board. Maj. Gen. Hugh A. Drum served as executive vice chairman, with Maj. Albert E. Brown as his assistant and recorder. From the board Secretary of War Dern

Members of the Baker Board, (seated l. to r.): Maj. Gen. Benjamin Foulois, Dr. Karl Compton, Newton Baker, George Dern, Maj. Gen. Hugh Drum, Dr. George Lewis, Maj. Gen. G.S. Simonds; (standing l. to r.): Brig. Gen. Gulik, James Doolittle, Edgar Gorrell, Brig. Gen Kilbourne, and Clarence Chamberlin

wanted a report on the adequacy and efficiency of Air Corps equipment and training for peace and war.

Convening at the Army War College on April 17, 1934, the Baker Board inquired at length into airmail operations to find and recommend ways to improve the Air Corps. It reviewed the work of earlier boards and committees, heard more than a hundred witnesses from the Air Corps, other branches of the Army, other federal departments, business, and industry, and solicited suggestions from Air Corps officers. The board found that Air Corps equipment suitable for combat was not readily adaptable to carrying airmail. Flyers trained for military operations could not be expected in the beginning to carry mail as well as pilots experienced in airmail work. Nonetheless, the board said, Army pilots "met this duty with fidelity which does them great credit as soldiers." The board thought the experience invaluable as a test of equipment, pilots, and readiness.

Interested in everything about the Air Corps, the Baker Board offered comments and recommendations on a wide range of subjects. It regarded 2,320 airplanes, the number suggested by the Drum Board, the minimum necessary for peacetime. Pilots should average three hundred hours of flying a year; receive more training in night, instrument, cross-country, radio-beam, and bad-weather flying; and get more ammunition and live bombs for training. The Air Corps should develop better instruments, communications, and armament; strengthen its meteorological setup; and give tactical units training under various conditions in different parts of the country. The board further urged action on a variety of personnel problems—the officer shortage, overage in grade, stagnation of promotion, and rank commensurate with responsibility. Reference the last, the board proposed temporary promotion of officers as authorized by the Air Corps Act of 1926.[54]

General Foulois considered the board's report, which Dern approved, "the first comprehensive outline of War Department policy with respect to aviation that the Army has ever had."[55] Postmaster General James A. Farley told Foulois the Air Corps undoubtedly would benefit from carrying the mail. The nation and Congress would support the Corps better and see that it secured the best equipment to be had and sufficient money for pilots to get as much flying time as needed.[56] The Chief of the Air Corps felt sure the effects would be beneficial. But progress in carrying out the Baker Board's recommendations came slowly.

In later years, General Foulois vigorously defended his actions and the Air Corps' performance in the airmail episode. As he saw it, time tended to magnify the importance of the event in the history of American military aviation. Reaction to the deaths of Army flyers "forced" the President and Congress to release funds for the Air Corps. Had it not been for those deaths, the nation would have been as unprepared for the Second World War as it was for the first.[57] Although General Foulois thought the airmail performance "magnificent,"[58] others, then and since, have more often termed it a "disaster" or a "fiasco."[59]

Chapter XVIII

GHQ Air Force

Airmail operations temporarily diverted attention from GHQ Air Force. Regular Army officers with mobilization assignments on the staff of Headquarters GHQ Air Force remained too busy with the airmail to perform their assigned Air Force planning function. Nor did the War Department think this the time to do anything about the Air Force that formed the foundation for the report submitted by the Drum Board and approved by the Secretary of War. Showing great interest in the GHQ Air Force, the Baker Board learned that the War Department planned to create the Air Force as a peacetime component of the Regular Army, but had not yet worked out the details or set the time. The Baker Board's endorsement speeded completion of the plan. In September 1934, General Westover commanded GHQ Air Force in a command post exercise pitting a group of field armies against enemy forces that had landed on the east coast of the United States. The following month, Lt. Col. Frank M. Andrews joined the General Staff to firm up the details for GHQ Air Force, which came into being under his command on March 1, 1935. A year of testing confirmed the soundness of the concept but called for organizational changes and adjustments.

Baker Board

When the Baker Board asked all Air Corps, Air Reserve, and Air

National Guard officers for constructive suggestions to improve the Air Corps, most of those responding urged independence from the General Staff and the War Department. Testifying before the board, 1st Lt. Lawrence J. Carr undoubtedly expressed the feelings of many of his fellow airmen when he cried: "Why can't you let us get out and develop what we think we should do and perform our mission?"[1] Denied independence, many were not ready to accept a GHQ Air Force as a temporary alternative.

General Patrick, former Chief of the Air Corps, thought the time for independence had not yet arrived, but told the Baker Board he wanted the Air Force Commander, under the Chief of the Air Corps, responsible for unit training both in the United States and in overseas departments. At the beginning of war, the Chief of the Air Corps could either turn the units over to the Air Force Commander or go into the field himself to command the Air Force, leaving someone else in charge of individual training and materiel. Brig. Gen. Oscar Westover, also convinced GHQ Air Force should be under the Chief of the Air Corps, did not see it mattered whether headquarters was physically part of the Office of the Chief of the Air Corps.[2] Maj. Carl Spatz thought Air Force headquarters should be on an airdrome close to its equipment and to a high-powered radio station. Initially, however, it would be better for the commander to be in the Office of the Chief of the Air Corps to work with that office on organization, doctrine, and policy.[3] Lt. Col. Henry H. Arnold advocated stationing the commander away from Washington to avoid being influenced by little things associated with the routine of the War Department and the Office of the Chief of the Air Corps.[4]

Maj. Walter H. Frank asserted that Air Corps officers agreed the Air Force should include all bombardment, attack, pursuit, and some long-range reconnaissance, with division, corps, and armies having their own observation aviation.[5] Maj. Walter G. Kilner proposed placing the Air Force under a commander who reported through the Chief of the Air Corps to either the Chief of Staff or the Secretary of War.[6] Maj. Leslie MacDill suggested the Air Force be under the Chief of the Air Corps. He said corps area commanders could supply base services but should not control tactical units.[7] Maj. Ralph Royce stressed that "if we have an Air Fleet," it should be under the Chief of the Air Corps. This was essential, he said, because the Chief of the Air Corps supplied tools, machinery, airplanes, maintenance, overhaul, and everything else "which the Air Fleet had to have in order to survive."[8] Capt. Edwin J. House contended the Air Force Commander needed a staff ready "right now" to function on M-day.[9] Maj. John B. Brooks, then on duty with G–3 of the War Department General Staff, concurred with his chief, Brig. Gen. John H. Hughes, that the Air Force must be more than a paper organization. The Air Force commander should be responsible for training and inspection. "If that were done," he said, "I think that would be the greatest single advance we could have for the Air Corps."[10]

Maj. Gen. Fox Conner, Commanding General, First Corps Area, deemed GHQ Air Force necessary. It should consist of all bombardment, attack, and pursuit, everything but observation. In the field it should be under the field commander. He seconded Baker's suggestion that Air Force organization be the same in peace and war "so habit will be fixed."[11] Maj. Gen. Paul B. Malone, Third Corps Area, wanted "to dismiss . . . the hypothesis that we have created in the Air Corps something like a super arm. It is not. Unless it contributes to victory in battle it is not worth the money." He averred the Air Corps, in all of its relationships, should be subordinate in peacetime to the Chief of Staff and in war to the commander in chief in the field, the same as the Infantry, Field Artillery, and Cavalry. In peace, Air Corps units should come under corps area commanders like other branches of the service. Focusing on plans being made for GHQ Air Force, General Kilbourne asked Malone about making the Air Force commander responsible for Air Force training. Malone favored that: "When you get a GHQ Air Force thoroughly coordinated and organized and commanded, then we have a new set-up." Continuing, Malone said he did not want a string tied to Air Corps troops sent to his corps area. He had "a considerable air force at Langley Field" and believed he "should exercise complete control there." Baker asked, "Until the GHQ Air Force is created?" Malone replied, "Until the GHQ Air Force is created and allotted, after which there would be an entirely different set-up."[12]

Maj. Gen. George V. H. Moseley, Fourth Corps Area, harked back to an earlier proposal when he was Deputy Chief of Staff and the four-army plan was being developed. He would organize the Air Corps (minus certain units attached to ground forces) as an air division commanded by a general officer. The headquarters would be in the midwest. The commander would serve as an inspector ten months of the year to see that units followed the proper line of training. During the other two months, the commander would assemble the division, first in one strategic area and then another, to afford training in handling air units and to give pilots flying experience in those areas. Moseley held it was still a good idea.[13] Maj. Gen. Preston Brown, Sixth Corps Area, concluded "the Army Air Corps is about as good an Air Corps as you have ever seen." Having flown with it for several years, he knew it "thoroughly." Save for some observation for army corps and other units, most of the Air Corps could be put into GHQ Air Force. He favored formation of the Air Force in peacetime but wanted to keep training under corps area commanders. Opposed to having the Air Force Commander report directly to the Chief of Staff, he said he did not want to create "any more bureaus or any more separate and distinct organizations that are exempt from somebody's control."[14]

From their questions and comments, Generals Drum and Kilbourne showed they supported establishment of GHQ Air Force in peacetime. They advocated using all bombardment, attack, and pursuit, plus some long-range reconnaissance for the Air Force. It would be under a line officer who had responsibility for training and reported directly to the Chief of Staff in peace and to the field commander in war. Noting the clamor for Air Corps independence, Maj. Gen. Robert E. Callan, G–4, testified that a GHQ Air Force offered all the advantages of a separate air force without violating the principle of unity of command. He suggested the same setup in peace and war.[15] General Hughes, G–3, came before the committee with a plan for GHQ Air Force. It gave command to a line officer, stationed him in the field, and charged him with training in peacetime and combat operations in war. The plan reorganized combat units to give the Air Force greater mobility by turning over the operation of Air Force stations to station complements under corps area commanders.[16] Subsequent to Hughes' testimony, which ensued late in the proceedings, Baker observed that the War Department was making progress in the development of GHQ Air Force. Unless some question remained, or the Air Corps saw cause for complaint, he thought the board could drop the subject.[17]

The Baker Board built its report, dated July 18, 1934, on the U.S. military policy of defense of the homeland and overseas possessions. There was no need for armament beyond the minimum needed for this purpose. Since aviation had bolstered the power of the nation's defense, it was advantageous to national defense policy. Rejecting the creation of an independent air force, the board held that the Air Corps must remain under the General Staff.[18] It then adopted the Drum Board's recommendations to: organize all combat units in an Air Force capable of operating in cooperation with or independent of ground forces; allot observation units to corps and armies for direct support; and leave Air Corps units in overseas possessions under department commanders. The Baker Board further proposed the Air Force in peacetime be commanded by a general officer with broad experience as an airplane pilot, headquarters be in the field away from Washington, and the commander be directly subordinate to the Chief of Staff in peace and to the field commander in war. The Baker Board also suggested the Air Force Commander be responsible for organizing, training, exercising, and inspecting the Air Force. However, corps area commanders would retain jurisdiction over and maintain airfields. This division of functions, the board said, would add to the mobility of the Air Force.[19]

Before the Baker Board finished its work, the President, at the direction of Congress, appointed a Federal Aviation Commission to study aviation in the United States. This commission, headed by Clark Howell, editor of the *Atlanta Constitution*, regarded GHQ Air Force as an experiment and preferred not to comment until after an adequate trial. It did recommend constant study of the Air Force as an independent striking unit and urged

Clark Howell, selected by President Roosevelt to chair the Federal Aviation Commission. *Library of Congress*

development of the force "to its limit by tactical maneuvers and through the procurement of materiel best suited to such independent operations."[20]

Command Post Exercise

While the Howell Commission studied American aviation in the summer of 1934, General Westover and GHQ Air Force fought a soldierless, shotless, bloodless, and painless war.[21] In a game played by staff officers, a coalition of European powers suddenly declared war and landed troops on the New Jersey coast. At this time the United States was already fighting a Pacific power, its fleet was in the Pacific Ocean, and its air force was on the west coast. When Regular Army units available in the East proved no match for the enemy invading New Jersey, General MacArthur mobilized two armies and called on GHQ Air Force to defend the Atlantic frontier.[22]

General Westover's Air Force comprised three wings—pursuit (commanded by Lt. Col. Frank M. Andrews), bombardment (Maj. Willis H. Hale), and attack (Maj. Earl L. Naiden)—and two observation groups (Maj. Walter H. Frank and Maj. Charles B. Oldfield). But only commanders and staffs of armies, corps, divisions, the Air Force, its three wings, and two groups participated. In all, 450 commissioned officers and 550 enlisted men took part; 30 officers and 40 enlisted represented an air force of 1,200 officers and 10,000 men. The defenders (Blue) attached 107 observation aircraft to ground units; the enemy (Black), 153. GHQ Air Force (Blue) consisted of

432 planes against Black's 600. While Black owned more bombers, Blue had faster pursuit craft. Yet the only ships actually flying in the exercise were 10 B–10s Colonel Arnold brought from California to show how fast combat units could move from coast to coast.

The play opened at the Army War College on August 26, 1934. The single Air Force unit set for instant action was Major Frank's 9th Observation Group at Mitchel Field, New York, which at once reconnoitered at sea. The 2d Bombardment and the 8th Pursuit Groups arrived from Langley Field, Virginia, to begin operations the next day. Others joined the defense as they reached the theater. General Headquarters gave Blue air force the missions of locating and attacking Black naval vessels, troop transports, supply ships, and boats landing troops on the coast, and of screening Blue forces from observation and attack by Black aircraft. By continuously patrolling by daylight, Blue pursuit generally succeeded in thwarting enemy observation and attack on Blue troops. Blue planes delayed enemy movements by assaulting marching soldiers. Still the enemy's superior air strength prevented Blue bombers from much success against enemy ships, and hindered operations of attack planes against small boats and the airdromes Black established in New Jersey.

By September 2, 1934, the enemy had pushed inland thirty or thirty-five miles and held a line in New Jersey from Sandy Hook on the north to Salem on the south. Refugees from the war-torn seaboard choked roads. Enemy shells fell on the outskirts of Camden. Black air force, operating from land, attacked cities from New York to Washington, and gathered information about Blue movements and concentrations. All Blue air force units had reached the theater and were engaged. The First Army, commanded by Maj. Gen. Dennis E. Nolan, was ready for combat. General MacArthur shifted General Headquarters from Washington to Raritan Arsenal, near New Brunswick, New Jersey, where General Westover positioned his headquarters. The commander in chief next ordered First Army to seize the offensive before Philadelphia while the Second Army under Maj. Gen. Preston Brown defended New York. General Headquarters commenced calling on Blue air force for direct support of ground forces.

The battle raged. One day Black bombers escorted by pursuit raided Blue communications, bombed Camden and Philadelphia, wrecked one span of a railroad bridge over the Delaware River, blew up the First Army's gasoline reserves, and bombed the Blue airfield at Baltimore. Blue antiaircraft guns downed 6 Black planes near Philadelphia. Blue airplanes caught bombers on the ground at Cape May, Millville, Belmar, and Spring Lake, New Jersey, destroying 36. Another day, Blue bombers hit 6 transports on the Delaware. That afternoon, 108 Blue pursuit ships intercepted 54 bombers and 100 pursuit planes over Philadelphia and Wilmington, Delaware, shooting down 20 planes while losing 18. The same day, Black bombed the

7th Brigade near Lacey, New Jersey, and put 31 armored cars out of action, while Blue bombers aided the 29th Infantry by harassing cavalry.

General Westover considered the enemy's transports the chief objective for Blue air force, but to operate against them successfully demanded neutralization of the enemy's superior air power. Diversion of Blue air force to close support prevented its achieving the requisite superiority to stop the flow of reinforcements and supplies from enemy ships. The exercise clearly revealed to Westover and other airmen the evils resulting from taking an air force from its legitimate business to aid ground forces. Others drew different lessons from the exercise. General Nolan, for example, wanted attack aviation as part of his army. General Malone, commanding III Corps, thought corps aviation should include combat as well as observation planes. General Drum believed the exercise underscored the need to coordinate all means of defense. Someone must "decide what forces are going to be used, at what places they will operate, and what the objectives will be for those forces." All must work together under "some man's conception of what the whole operation will be." There must be unity of command, one man with a fixed determination as to how he is going to win. "I believe," Drum said, "those of us who study the larger problems of National Defense are convinced that in the organization of a GHQ Air Force which is equipped to operate independently, is self-sustained, and capable of being part of a homogeneous team, we have found the solution to the problem."[23]

Organizing GHQ Air Force

There was no question there would be a GHQ Air Force, but many details had to be hammered out before it became reality. Lt. Col. Frank M. Andrews joined the General Staff on October 11, 1934, to assist G–3 in this work. A native of Tennessee and a graduate of the U.S. Military Academy in 1906, Andrews served with the Cavalry in the Philippines, Hawaii, and the United States before transferring in August 1917 to aviation, with promotion to major. During the war he served in the Aviation Division of the Signal Corps, commanded Rockwell and Carlstrom Fields, and supervised the Southeastern District, Department of Military Aeronautics. Among his assignments in the 1920s were duty with American forces in Germany, at Kelly Field (as executive officer, commander of a school group, and Commandant of the Advanced Flying School), and in the Office of the Chief of the Air Corps. Andrews, a military airplane pilot and airplane observer, graduated from the Air Corps Tactical School in 1928, the General Staff School in 1929, and the Army War College in 1933. He commanded the 1st

Brig. Gen. Frank M. Andrews assumes command of G.H.Q. Air Force in March 1935.

Pursuit Group when General MacArthur called him to Washington to assist with Air Force planning.

Most of the details having been settled, the War Department announced on December 27, 1934, that Secretary of War Dern had approved a test of GHQ Air Force. At the same time it announced that Andrews, "a distinguished officer of the Air Corps, an outstanding pilot, with long experience in command of Air Corps tactical organizations," had been designated Air Force Commander.[24] On New Year's Eve, The Adjutant General issued instructions for the establishment of GHQ Air Force on March 1, 1935.[25] So significant an event as the activation of GHQ Air Force seemed to call for an aerial feat of some kind. Perhaps a flight showing off the range of the newest bombing planes was in order, say from Washington to France Field in the Panama Canal Zone by way of Miami. Capt. Harold D. Smith, commanding ten B–12As and crews from Hamilton Field, California, awaited orders at Langley Field for such a flight when Andrews assumed command of the Air Force on March 1. Smith and his men were still waiting on the 5th. Their orders, when they finally arrived, sent them back to California. There was not enough money to go to Panama.[26]

General Andrews (his new job gave him temporary promotion to brigadier general) was short men and equipment for the new Air Force. He required more bombing and gunnery ranges as well as hard-surfaced runways. His men needed better training in night flying, instrument flying, navigation, gunnery, and bombing. He even lacked a definite doctrine for employment of the Air Force.[27] But he was optimistic. The Air Force had the

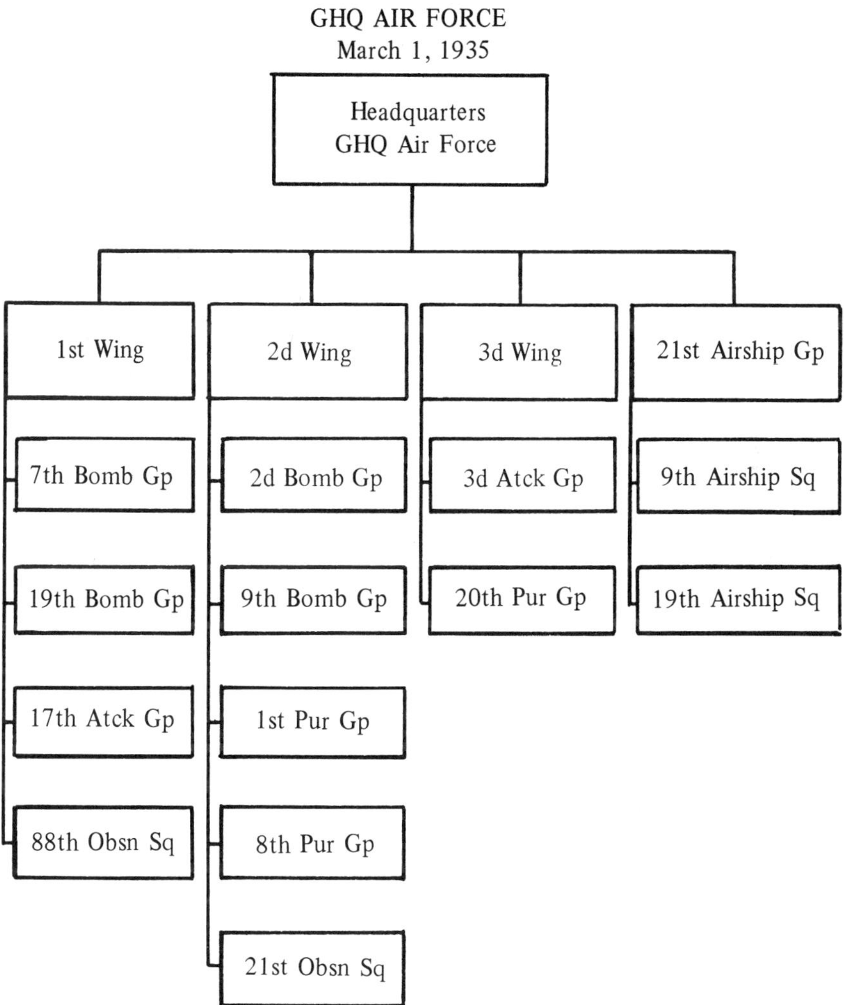

"whole power and authority of the War Department behind it." General MacArthur himself had urged Congress to furnish more money and people: "Congress and the President appear sympathetic and real progress seems in sight."[28] Addressing the troops, General Andrews depicted the Air Force as a "highly mobile force of great striking power" with a war mission of "offensive air operations against enemy air, ground and sea forces." In executing its mission, it would conduct "independent air operations, or air operations in conjunction with friendly ground and naval forces," depending on the circumstances.[29]

For his staff, which he assembled in Washington early in February 1935,

Andrews selected Maj. Hugh J. Knerr to be Chief of Staff, Maj. Harvey S. Burwell, G–1; Maj. Follett Bradley, G–2; Capt. George C. Kenney, G–3; and Maj. Joseph T. McNarney, G–4. All got temporary promotions on March 2, Knerr going to colonel, and the assistant chiefs to lieutenant colonel. They set up headquarters at Langley Field. Although the Baker Board wanted Air Force headquarters away from Washington, it did not specify a place. General Foulois suggested Patterson Field, Ohio. With the contiguous area of Wright Field, it afforded a large tract and extensive facilities for concentrating the Air Force for training and maneuvers. G–3 of the General Staff preferred Langley Field so General Andrews would be near his chief, General MacArthur.[30]

The War Department gave Andrews command of all Air Corps tactical units in the United States except observation squadrons allotted to ground forces. The principal units were three wings, which dropped their pursuit, bombardment, and attack designations and became regional subcommands of GHQ Air Force. Henry H. Arnold commanded the 1st Wing, on the west coast; Henry C. Pratt the 2d, on the east coast; and Gerald C. Brant the 3d, in the central states. On March 2 Arnold and Pratt received temporary promotions from lieutenant colonel to brigadier general, Brant from lieutenant colonel to colonel.

The 3 wings together comprised 9 groups with a total of 30 tactical squadrons (12 bombardment, 6 attack, 10 pursuit, and 2 reconnaissance). General Arnold had his headquarters and an attack group (17th) at March Field, California; one bombardment group (19th) at Rockwell Field, California; and another (7th) at Hamilton Field; plus a reconnaissance squadron (88th) at Brooks Field, Texas. Pratt's wing, headquartered at Langley Field, consisted of the 2d Bombardment and 8th Pursuit Groups at Langley; the 1st Pursuit Group at Selfridge Field, Michigan; and the 9th Bombardment Group at Mitchel Field, with 1 squadron of the 9th and 1 reconnaissance squadron at Bolling Field. All of Colonel Brant's units, including the 3d Attack and 20th Pursuit Groups were at Barksdale Field, Louisiana (wing headquarters and the attack group having joined the pursuit group there at the end of February). The Air Force also encompassed the 21st Airship Group, with headquarters and one airship squadron (9th) at Scott Field, Illinois; and another squadron (19th) attached to the 2d Wing at Langley. Twelve service squadrons supported the tactical units. Additional units of the Air Force were 8 tactical squadrons converted from school groups and detached for duty with Air Corps schools, and 3 tactical squadrons for activation in the future.

As Commanding General, GHQ Air Force, Andrews carried responsibility for the instruction, training, maneuvers, and tactical employment of all elements of his command. Since the Air Corps still handled individual training and materiel, it supplied officers, enlisted men, and equipment to the

Air Force. Department commanders kept jurisdiction over aviation units and activities in overseas areas. Corps area commanders remained in charge of training and operations of observation units allotted to ground forces and continued jurisdiction over stations occupied by Air Force units.[31]

Most Air Force units stayed in place without change of functions, but all reorganized under new tables of organization the War Department approved for service test. The goal was a mobile force divorced to the utmost from functions that would tie it to the ground. Service squadrons became mobile to support tactical units in the field as well as in garrison. Tactical units were "skinned to the bone" so they could move swiftly and sustain themselves in the field for short periods until service squadrons arrived.[32] Reorganization also entailed formation of station complements (fixed units under corps area commanders) to run the Air Force's permanent stations. In some places—Langley Field for one—reorganization involved reassignment of quarters, people, and duties. But most shifted to "the GHQ set up" in short order.[33]

Organization of GHQ Air Force meant temporary promotion to numerous Air Corps officers besides Andrews, his staff, and wing commanders. Army officers advanced slowly under existing laws, and many in the Air Corps held grades one or two below those authorized for the job. The War Department had tried for years to get Congress to change the law. Hoping for a better promotion plan, the Air Corps had not capitalized on the clause in the Air Corps Act of 1926 authorizing temporary promotions. At the urging of the Baker Board, the War Department used the occasion of the Air Force's activation to grant such promotions. Dozens of Air Corps officers in Air Corps units, in overseas departments, and in GHQ Air Force benefited.[34] Majors commanding groups were made lieutenant colonels. Captains and first lieutenants heading squadrons moved up to major. Lieutenants in charge of flights found themselves captains. One list contained names of three hundred officers promoted in a single day. A second announced the promotion of twenty-two instructors, section chiefs, squadron commanders, and staff officers of Maxwell Field, Alabama. A third named sixty officers promoted in Hawaii.[35] So many changes set "the rank and file wondering how it will all end."[36] General Andrews explained that officer rotation in and out of GHQ Air Force, foreign service, Air Corps units, General Staff, corps areas, and other assignments would continue. He told his officers some of them now enjoying temporary rank would lose it when their assignments changed, while others without temporary promotion would receive it.[37]

A law signed by President Roosevelt on July 31, 1935, presented 4,310 Regular Army officers permanent promotions on August 1. The new law specified automatic promotion for second lieutenants after 3 years of commissioned service, and for first lieutenants after 10 years. It also opened promotion to higher rank by authorizing extra majors, lieutenant colonels, and colonels. Over a quarter of the Air Corps' 1,391 officers moved up one step the day the new law took effect.[38] If an Air Corps officer still did not have rank equal to his position, he retained his temporary grade. Promoted to permanent colonel, Andrews continued to wear the insignia of brigadier general as Commanding General, GHQ Air Force.

On March 1, 1935, units of Andrews' command had 60 bombers, 42 attack planes, 146 pursuit ships, and 24 transports, the total being about 40 percent of the number called for by tables of organization. The Air Corps had been equipping one unit at a time with new aircraft coming from the factories. General Andrews concurred in this policy, but under it some Air Force units flew only late model planes while others used only older ones. General Arnold's 7th Bombardment Group, for instance, flew B–12s and B–12As; General Pratt's 2d Bombardment Group, B–6As. The 1st Pursuit Group flew P–26As, the 8th Group P–6s and P–12s.[39] The 9th Group with a long history as an observation unit became a bombardment group after joining the GHQ Air Force on March 1. Still, it continued to fly O–1Gs for another year before securing B–10Bs. The 17th Pursuit Group possessed 52 P–26s when it entered GHQ Air Force (it was soon redesignated an attack group). The men of the 17th rejoiced at the prospect of getting new attack planes. Even so, their enthusiasm was dampened when they learned that Northrop would not have the new aircraft ready for several months.[40] Meantime, the 17th turned over its P–26As to the 1st and 20th Pursuit Groups, enabling the 20th to transfer P–12Es and P–12Fs to the 8th Pursuit Group. The latter then sent P–12Cs and P–12Ds to the 17th for training until new A–17s were available in the spring of 1936.[41]

General Andrews had about 80 percent of the enlisted men but just 40 percent of the Air Corps officers called for in his tables of organization. Strength reports showed, for example, 6 officers instead of 28 in each of the pursuit squadrons at Selfridge Field, 5 of 24 in the 13th Attack Squadron at Barksdale Field, and 5 of 26 in the 31st Bombardment Squadron at Hamilton Field. Cadets alleviated the pilot shortage; about 100 who won their wings flew with tactical squadrons.[42]

Some shifting of units took place later in the year when the Army and Navy swapped stations in California. The Army got Moffett Field, a dirigible base recently constructed at Sunnyvale. Having stopped building dirigibles after losing the *Macon* and the *Akron* (in which Rear Adm. William A. Moffett died), the Navy no longer needed Moffett Field. It wanted, and for many years had been trying to get, Rockwell Field to add to the adjacent naval air station. The Army's struggle to keep Rockwell ended when President Roosevelt ordered the exchange. One of the conditions was that the Army maintain the mammoth dirigible hangar at Sunnyvale in case the Navy required it in the future. The 19th Bombardment Group and 76th Service Squadron moved from Rockwell to March Field on October 25. At the same time, the Commanding General, Ninth Corps Area, disbanded Station Complement, Rockwell Field, using most of the men to form Station Complement, Moffett Field. The Air Corps retained its depot at Rockwell until new facilities became ready at Sacramento in 1939. Having acquired Moffett Field, the War Department transferred the 19th Airship Squadron from Langley Field to Arnold's 1st Wing. About the same time, the 88th Observation Squadron (Long-Range Amphibian) moved from Brooks Field, Texas, to Hamilton Field to be with other units of the wing.[43]

Principles of Employment

When General Andrews took command of GHQ Air Force on March 1, 1935, the War Department was still trying to formulate principles of operations.[44] General MacArthur assigned the work to Brig. Gen. Charles E. Kilbourne's War Plans Division. General Kilbourne circulated a draft late in December 1934, but coordination and revision delayed publication ten months.[45]

Air Force doctrine as enunciated by the Air Corps Tactical School rested on the proposition that

> the principal and all-important mission of air power . . . is the attack of those vital objects of a nation's economic structure which will tend to paralyze the nation's ability to wage war and thus contribute directly to the attainment of the ultimate objective of war, namely, the disintegration of the hostile will to resist.[46]

The Tactical School taught, and Air Corps officers as a rule believed, aviation could be decisive. The General Staff and officers of other arms could not accept this in the absence of demonstration. As the War Plans Division put it: "So far, well-organized nations have surrendered only when occupied by the enemy's army or when such occupation could no longer be opposed."[47] Aviation could assist but could not itself achieve victory. The Tactical School responded by asserting that the advent of air power gave for the first time in history the ability to bring war immediately to the internal structure of an enemy nation. To say, in the absence of demonstration, that air power could not defeat a nation in this manner expressed only opinion. The school opposed inclusion of such contentious matter in a statement of principles. Out of this exchange emerged the following statement for publication:

> The power of air forces has not yet been fully tested. The effect which they are capable of producing and the extent to which they will influence warfare is still undetermined. But it appears certain that skillful use of air forces will greatly affect operations in future wars.[48]

While insisting attack on the enemy's economy was the primary mission of air power, the Tactical School pointed out this depended on the availability of proper equipment. Airplanes did not then have sufficient range to reach the homeland of any major power deemed a possible enemy. Besides, the U.S.

military establishment existed to defend the nation against attack. Under these circumstances, the Tactical School said, the Air Force's chief role was air defense. This involved preventing any foreign power from setting up air bases from which it could threaten the nation's security. It also entailed defeating enemy air forces occupying bases within striking distance of U.S. territory. By concentrating on air defense, GHQ Air Force

> obtains the greatest security from air attack for its own force and other friendly forces, it furnishes genuine support for the operations of other military forces and concurrently establishes the maximum degree of protection from air attack to the civil population and all vital objectives within the nation.[49]

The War Plans Division gave counter-air force operations an important place in the work of GHQ Air Force but not the premier role advocated by the Tactical School.[50]

The principles the War Department published in a training regulation on October 15, 1935, assigned three kinds of operations to GHQ Air Force: Beyond the sphere of influence of the ground forces; in immediate support of ground forces; and in coastal frontier defense and other joint Army and Navy operations. Targets for the first of these were air forces; warships, submarines, supply vessels, and sea transports; munitions factories, refineries, and fuel storage plants; factories producing aircraft and equipment; bridges, railway yards and stations, tunnels, harbor facilities, canal locks, and other critical points in lines of communication; powerplants, powerlines, and other utilities; and troop cantonments and concentrations. While the relative significance of these targets varied with the situation, the enemy air force usually occupied first place. Attacks on such objectives included necessary reconnaissance by long-range observation craft.

Operations in immediate support of ground forces fell into three phases. Before opposing ground forces joined battle, the Air Force interdicted reconnaissance and attack by enemy air forces; reconnoitered to supply security and information; and struck communications, concentrations, moving columns, and ammunition dumps. During the ground battle the Air Force assaulted troops massed for attack or counterattack. Victory having been won, the Air Force harassed and interdicted the enemy's retreat and interdicted enemy aircraft to facilitate pursuit by our ground forces. Following defeat, the Air Force fought enemy planes that tried to prevent withdrawal, and harassed and interdicted enemy forces in pursuit. In coastal defense and other joint operations, the Air Force functioned under procedures agreed to by the War and Navy Departments.

The training regulation reaffirmed the present command structure, with the Air Force under the direct control of the Chief of Staff as the commander in chief of the four armies in peace, and of the commander in chief of field forces in war. The commander in chief directed Air Force operations in several ways. He assigned a broad general mission and permitted the Air Force Commander to select the objectives and conduct the operations. He ordered special missions, designated major objectives for attack, and made the Air Force Commander responsible for carrying them out. He prescribed special missions for execution under his immediate control, or directed the Air Force Commander to support specific army operations under instructions from army commanders.[51]

Testing

General Andrews' chief task during the service test of GHQ Air Force was to determine the organization best suited to a highly mobile Air Force. The 2d Wing at Langley Field had recently learned a lot about mobility during a month of maneuvers in the southeast. On January 9, 1935, 99 officers, 19 flying cadets, and 255 enlisted men took the field under Lt. Col. Byron Q. Jones' command with 29 bombing planes, 44 pursuit planes, an airship, 8 transport planes, 45 trucks, and an ambulance. Their aim was to see if the wing could operate in the field under wartime conditions while constantly changing its theater, taking care of itself with supply by air, and evacuating its sick.

The 2d Wing went first to Miami where Colonel Jones split it into Red and White forces, each consisting of a bombardment and a pursuit squadron. They fought their first battle on the line east and west through Tampa. The commanders of opposing forces could operate from any airports more than 20 miles from the front. Capt. Albert M. Guidera put the Red pursuit squadron (Capt. Harold H. George commanding) at Sarasota and bombardment (Capt. Edward C. Black) at Bradenton. Maj. Willis H. Hale used Auburndale for his pursuit (Capt. Rex K. Stoner) and Lakeland for bombardment (Capt. Robert T. Cronau). The battle began at 1500 on January 15 and went on for 3 days with 2 air forces shooting at each other with camera guns. The wing next moved to Tampa to run a 20-hour check on the planes and to give the men a day of rest. A cold wave made life miserable for the men while they fought in the second theater in northern Florida. When they arrived at Mobile to make a 40-hour aircraft check, cold weather prevented doing the work outdoors and there was no shelter. Moving on, they made the check at New Orleans, where Shushan Airport offered a heated hangar. For the next battle, the opposing forces operated from Montgomery, Alabama, and Columbus, Georgia, after which the wing returned to Langley Field on February 3.

The maneuvers elicited comments to Colonel Jones on tentpoles, bedding, stoves, paper plates, trucks, radios, supplies, distribution points, gasoline and, among other things, per diem funds. Given proper transportation, Jones believed small squadrons of bombardment and pursuit could operate indefinitely from previously prepared fields without the assistance of service squadrons. For maximum power and utmost mobility, he advocated combat crews, airplanes, and administrative overhead, with the service squadrons supplying other personnel. However, he thought it might be more economical and convenient to let combat units be self-sufficient in peacetime. Seeing scant need for separate groups of bombardment, pursuit, and attack, he suggested organization of composite units balanced among the different classes of aviation.[52]

Units of the new GHQ Air Force engaged in exercises and maneuvers to test mobility and gain experience in field operations. The 96th Bombardment Squadron, supported by the 59th Service Squadron, moved from Langley Field to Richmond to operate temporarily from Richard E. Byrd Flying Field.[53] The 20th Pursuit Group and 71st Service Squadron of Barksdale Field went to Tallulah, Louisiana, and the 3d Attack Group and 60th Service Squadron to Tyler, Texas, after which Colonel Brant concentrated his 3d Wing at Shushan Airport.[54]

Units sometimes simulated operations from advance airfields by setting up temporary camps at their home stations. Lt. Col. Howard C. Davidson's 19th Bombardment Group lodged for a day on the south edge of Rockwell

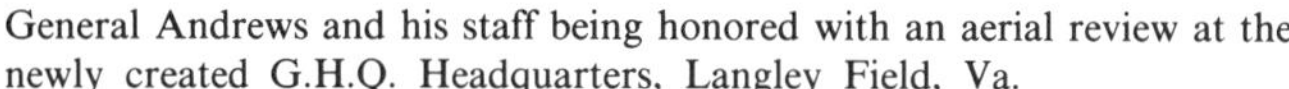
General Andrews and his staff being honored with an aerial review at the newly created G.H.Q. Headquarters, Langley Field, Va.

Field to practice loading ammunition and flying missions under field conditions, with a field kitchen furnishing the noon meal.[55] Preparing for maneuvers in the Imperial Valley of California, Lt. Col. John H. Pirie put his 17th Attack Group in the field for twenty-four hours at March Field. His three squadrons flew their missions from camps they created at widely separated spots on the reservation. Afterwards, 40 officers, 350 enlisted, 30 airplanes, 44 trucks, and 10 motorcycles from the 17th Group and 64th Service Squadron moved to the valley for 10 days.[56] The 34th Attack Squadron from Colonel Pirie's group maneuvered at Big Bear Lake, California, while the 31st Bombardment Squadron from Lt. Col. Clarence L. Tinker's 7th Bombardment Group operated from Medford, Oregon. The entire 7th Group, supported by a portion of the 69th Service Squadron, spent 2 weeks in mock warfare at Merced, California.[57] Other Air Force squadrons and groups took part in comparable activities.

Devising a problem involving bombing planes protected by pursuit, General Arnold concentrated his 1st Wing at Hamilton Field on March 22, 1935. Capt. Harold D. Smith returned that day from Langley Field after waiting in vain for orders to fly to Panama. With Smith's B–12s Arnold mustered 25 bombing planes from the 7th and 19th Groups and 51 P–26As from the 17th Group.[58] The 88th Observation Squadron did not come from Brooks Field, Texas, for the gathering at Hamilton Field but it participated in the next one, at March Field on Saturday, April 13, 1935. About 90 planes joined in the 2-day exercise. Maj. Gen. Paul B. Malone, Commanding General, Ninth Corps Area, was guest of honor of the 1st Wing on Sunday. All officers and cadets on the post turned out in uniform for a formal dinner followed by "hi-jinks, vaudeville, frivolity, and fun."[59] General Arnold apparently had gotten the idea from Canadian officers during a trip to Alaska the previous year. He called such affairs "Wing Dings."

Arnold assembled his wing several more times during 1935, once at Salt Lake City just before officers from Air Force headquarters arrived to inspect his wing. The inspectors—Lt. Col. Walter R. Weaver, Maj. Ennis C. Whitehead, and Maj. Russell L. Maxwell of the Ordnance Department—preceded the Commanding General, GHQ Air Force. When Andrews' bomber touched down at March Field on July 31, 1935, Arnold greeted him with a 11-gun salute, a 32-man honor guard, and a drum and bugle corps playing the "General's March." The ceremony over, Arnold took his chief and long-time friend to his quarters as his guest. When Andrews flew off to inspect Rockwell and Hamilton, Arnold went along in his own bomber.[60]

After returning to Langley Field, Andrews attacked seaplane speed records set by Edwin C. Musick, Capt. Boris Sergievsky, and Charles A. Lindbergh on August 1, 1934, in a Sikorsky S–42 built for Pan American Airways. The general used a B–12A fitted with floats made by the Edo Aircraft Corporation. The 1,000-kilometer course ran from Hampton Roads to New York and back by way of Washington. Carrying two bombs, Andrews went around twice on August 24, 1935. He failed to break the S–42's record for 2,000 kilometers but set a new record for 1,000 kilometers with 2,200 pounds, his speed on the second lap being 165.04 miles per hour against the S–42's 157.58.[61] In October 1935 Andrews obtained a Douglas DC–2 (XC–32), fitted by the Materiel Division as a "flying office."[62]

General Andrews' trip in 1935 included one to Florida on December 1. Assuming a situation in which a coalition of European powers was preparing for military operations in the Caribbean, he concentrated his Air Force for its largest and most important operation of the year. All wings participated (for the first time), but groups lacking modern aircraft—the 2d Bombardment (B–6As), 8th Pursuit (P–12Es and P–12Fs), 9th Bombardment (O–1Gs), and 17th Attack (P–12Cs, P–12Ds, and P–12Es)—did not take part in the fighting. Reorganizing his force, Andrews formed one wing each of bombardment, attack, and pursuit. General Pratt, at Miami Municipal Airport, commanded the pursuit wing of 57 O–26As from Selfridge and Barksdale Fields to defend Miami against the other two wings. General Arnold, commanding bombardment, brought 27 B–10s and B–12As from the west coast to operate from Vero Beach Airport. Colonel Brant, with 28 A–12s from Barksdale commanded the attack wing at Fort Pierce Airport. Operating from Chapman Field were the 88th Observation Squadron, with five O–35s, and a transport squadron made up of crews and aircraft that helped move the tactical units. The field became Andrews' headquarters and the site of an advance depot.

Everyone lived and worked under field conditions. The 59th Service Squadron from Langley Field and the 71st from Barksdale pushed overland to Florida to set up tent cities, establish messes, and prepare airfields for arrival of tactical units by air. It was a big undertaking to collect and move the tents, poles, cots, mattresses, blankets, pillows, mosquito bars, mess equipment, tools, wheels, tires, tubes, spark plugs, cable, gaskets, technical orders, office equipment, supplies, bulletin boards, shelving, latrines, and other things needed to sustain 225 officers, 600 enlisted men, and 125 airplanes in the field for 2 weeks.

Fourteen officers and 190 enlisted men under the command of Maj. Harvey H. Holland deployed with the 59th Squadron. The train consisted of 70 vehicles, mostly trucks, one with a winch and boom to salvage wrecks, and another with a 1,200-gallon gasoline tank. The train also included reconnaissance cars, two Plymouth passenger cars (for General Andrews and Pratt to use in Florida), an ambulance, and two motorcycles. When the squadron left Langley on Monday morning, November 25, 1935, part of the mess section went ahead to have lunch ready when the train arrived at Rocky Mount, North Carolina. The lighter cars, with considerable distance between them, tried to maintain 35 miles an hour. The heavier vehicles followed at not more than 25 miles an hour. Next came the ambulance, wrecking truck, and a party of mechanics in a reconnaissance car to handle repairs and emergencies. A Plymouth with Capt. Charles B. De Shields, the transportation officer en route, and Capt. Edward J. Kendricks, Medical Corps, brought up the rear. Monday night the men slept in hangars at Fort Bragg, North Carolina. On Tuesday the armory at Sumter, South Carolina, could not accommodate everyone; those who slept in tents were uncomfortable on the cold night. Rather than camp again, the squadron drove on to the armory at Jacksonville, Florida, some men not arriving until 0300. After resting Thanksgiving Day, the squadron proceeded to Vero Beach, completing its 1,100-mile trip from Langley on Friday evening. Having set up camp for the bombardment wing, Major Holland took part of the men to Fort Pierce on Sunday to establish camp for the attack wing. That done, he divided his squadron to run the two camps.

Tactical units departed their home stations at 1500 Eastern Standard Time, Sunday, December 1, 1935. All were at their field stations 22 hours and 55 minutes later. The Materiel Command contracted for gasoline and oil at municipal airports where units stopped to refuel. To avoid conflict in the use of airdromes and servicing facilities, Air Force headquarters set aside specific times for different units. The P–26s from Barksdale, for example, had priority at Jacksonville Municipal Airport from 0800 to 1000, those from Selfridge from 1000 to 1210. The west coast bombers, led by Colonel Tinker, completed the flight from March Field to Vero Beach in a little less than 22 hours, despite stops at El Paso, Barksdale Field, and Maxwell Field.

A teletype loop linked the airdromes. Radios furnished air-to-air and air-to-ground communication. Transport planes delivered equipment, supplies, and rations from the depot to the various airports. Mechanics (61 at Vero Beach, 50 at Fort Pierce) worked long hours to keep the planes flying. Operations began Tuesday morning, December 3, 1935, with 18 bombers dropping 36,000 pounds of bombs on the port at Tampa. Colonel Tinker next led the planes south to attack Miami. Observers spotted them over Hollywood. General Pratt ordered Lt. Col. Millard F. Harmon, Jr., to intercept. Harmon put 36 ships in the air within 4 minutes. Pursuit "annihilated" the bombers in a 10-minute battle. While that went on, 18 A–12s led by Maj. William N. Amis bombed an airdrome being built on Virginia Key off Coral Gables, and escaped.

The same pattern prevailed as operations continued. Colonel Brant attacked the city of Miami, its municipal airport, and airdromes in the area. Bombers and observation planes flew many long missions, several spanning 900 or 1,000 miles. General Arnold struck supply depots at Fitzgerald, Georgia, wiped out oil storage tanks at Tampa, and demolished an aircraft carrier off Key West. Returning from these missions, the bombers detoured to hit Miami. In night operations, they dropped flares to simulate bombing.

Upon Foulois' retirement, Maj. Gen. Oscar Westover becomes Chief of the Air Corps.

With good observation and communication, General Pratt repulsed or destroyed the enemy time after time. Even so, he was hard pressed to defend the city when bombers and attack aircraft struck simultaneously. On two occasions the bombing force split, permitting planes to get through to their target. Colonel Brant tried the same tactic without success. One day General Andrews sent a reconnaissance plane to search for an enemy warship 100 miles or so off the east coast of Florida. Finding it, the observer reported latitude, longitude, course, and speed by radio. General Arnold's bombing planes intercepted and attacked.

At war's end on December 10, 1935, the Air Force spent a day working on planes and equipment. Afterwards, the 59th Service Squadron and the bombardment and attack wings moved to Miami. The same day an Air Force crew set a record in an amphibian. Service testing the YOA–5, the five-man crew headed by 1st Lt. Hugh F. McCaffery had flown to San Juan, Puerto Rico, with a stop at Guantanamo, Cuba. They returned nonstop to Chapman Field on December 12, making the flight of just over 1,000 miles in 7 hours and 25 minutes. The next two days the Air Force put on demonstrations for Miami's annual air show, billed as the All-American Air Maneuvers. On December 15 the units started home.[63]

General Andrews now reported to Gen. Malin Craig, who had succeeded MacArthur as Chief of Staff on October 2, 1935. With Foulois retiring, Westover moved up to Chief of the Air Corps on December 22. Arnold then went to Washington to become Assistant Chief of the Air Corps, his post at March Field assumed by Col. Henry B. Clagett. Brant was promoted to brigadier general in his position as Commander of the 3d Wing. Andrews received another temporary promotion, to major general, on December 27, 1935, giving the Commanding General, GHQ Air Force, rank equal to that of the Chiefs of Infantry, Cavalry, Field Artillery, Coast Artillery, and the Air Corps.[64]

Reorganizing

The service test of GHQ Air Force convinced General Andrews it would be sound to form combat units into an Air Force under its own commander and staff. But GHQ Air Force needed reorganization "to permit its development and employment as a strategically and tactically mobile M-day force."[65] Division of authority created serious problems. The Air Force Commander carried responsibility for Air Force training and operations, corps area commanders for the Air Force's permanent stations, and the Chief of the Air Corps for training the people and furnishing the equipment the Air Force needed. On a lower level, wing and station commanders did not share the same interests. One concerned himself with training and operations, the other with such things as discipline, appearance of personnel, and maintenance of buildings and grounds. General Pratt, 2d Wing Commander, had units in three corps areas and dealt with three corps area commanders. This, Andrews said, "is productive of discord and not conducive to a smoothly and expeditiously developed GHQ Air Force." Overlapping jurisdiction interfered with Air Force training and operations. Andrews urged that the stations occupied by his units be designated "air bases," exempted from the control of corps area commanders, and placed under the control of Air Corps officers in the Air Force chain of command.

General Andrews also wanted to reorganize squadrons and station complements. Since skeletonized combat squadrons depended largely on service squadrons for maintenance of personnel and support, they did not have the desired mobility. The exercise in Florida in December 1935 indicated the air force could be "the most mobile and effective of the new engines of war" only if the necessary bases and logistics were prepared in advance. As the Air Force grew, "service, care and supply increased greatly." Tactical use of small formations was fairly simple; employment of large fleets demanded much preparation.[66] Abolish service squadrons, Andrews said, and make combat squadrons self-contained units. Replace station complements with new units to perform services that for reasons of mobility, flexibility, and tactical efficiency, combat squadrons could not do for themselves. He wanted the new service units to concentrate on security, base maintenance, care of the flying field, bombing and gunnery ranges, shops, radio ranges, meteorological service, motor transport, air transport, ordnance, and supply. Besides Air Corps troops, the new service squadrons required signal, quartermaster, ordnance, engineer, and medical personnel.

Among other things, General Andrews desired his headquarters in Washington to facilitate Air Force business. But he did not want members of his staff scattered around the area. They needed to be close together to be ready to function twenty-four hours a day. He suggested either Bolling Field or Fort Hunt, Virginia, eleven miles south of Washington.[67] A board of officers, headed by Col. William S. Browning of the Inspector General's Department, had recently recommended Bolling Field for Air Force headquarters. In addition to Browning, the board consisted of two Air Corps officers, Lt. Col. Follett Bradley and Maj. Rosenham Beam. It had been charged with surveying the personnel situation but had digressed to look into organization.[68] The board also advocated consolidation of the Air Corps and GHQ Air Force under the control of the Chief of the Air Corps. Maj. William E. Lynd, acting chief of the Air Corps War Plans and Training Division, urged approval of this proposal, as well as Andrews' suggestion for making combat squadrons self-contained units capable of operating in the field for short periods without assistance.[69]

General Hughes, G–3, wanted corps area commanders to retain control of Air Force stations. Maj. Gen. George S. Simonds, successor to Maj. Gen. Hugh A. Drum as Deputy Chief of Staff, sided with Andrews. The Adjutant General informed corps area commanders that on July 1, 1936, the Air Force would assume jurisdiction over its permanent, peacetime stations.[70] Three months later the War Department approved reorganization of combat squadrons, abolition of service squadrons, reorganization of group headquarters, and creation of new units in the Air Force chain of command to replace station complements. Other changes on September 1, 1936, attached one reconnaissance squadron (redesignated from long-range observation) to each bombardment group, transferred airship units from GHQ Air Force to corps area control, gave GHQ Air Force control of its photo sections,[71] and inactivated units, among them the bombardment squadron at Bolling Field and the tactical units detached for service at the various schools. The War Department afforded Andrews latitude in reorganizing. For example, he could move personnel among units at various stations so long as he stayed within the station's authorization of grades and ratings.[72]

Shifting some work previously done by service squadrons, Andrews enlarged his combat squadrons to render them more mobile and self-sufficient. The table of organization for a B–10 or B–12 bombardment squadron now called for 36 officers and 167 enlisted men in lieu of 26 officers and 49 enlisted men. The number of aircraft stayed at 13.[73] In the reorganization, group headquarters picked up new names. For instance, Headquarters 7th Bombardment Group became Headquarters and Headquarters Squadron, 7th Bombardment Group, other groups acquiring similar designations. The War Department disapproved Andrews' suggestion to redesignate Air Force stations as bases, but did place them under his control. In doing so, it formed a unit named "base headquarters and air base squadron" at each location to replace the station complement. Base Headquarters and 1st Air Base Squadron, for example, served Langley Field; Base Headquarters and 4th Air Base Squadron, March Field. Owing to this new organizational nomenclature, stations soon became bases in common parlance.[74]

General Andrews touched on problems of coordination between GHQ Air Force and the Air Corps in his report on the service test but made no recommendation. Earlier, though, he proposed an air division be established in the General Staff to act for the Chief of Staff in conducting Air Force and Air Corps activities. In early planning for the Air Force, General Foulois wished to sidestep the problem of coordination by subordinating the Air Force commander to the Chief of the Air Corps. General Westover took the position of his predecessor. The Browning Board recommended the same thing. A board of Air Corps officers studying the Browning Board's work thought the Air Force should be retained as a separate command, but with its commander reporting directly to the Chief of the Air Corps. General Pratt and Colonel Knerr, representing GHQ Air Force, signed the report. Knerr later withdrew his signature because Andrews did not agree with the board's proposal. General Simonds suggested the Chief of the Air Corps be assigned the additional duties of a deputy chief of staff. This would give Westover more authority than Andrews without bringing Andrews under his control. The existing arrangement went on with both Westover and Andrews reporting to Craig.[75] The problem, and differences of opinion, likewise continued.

Westover suggested about a year later that as Chief of the Air Corps he should also be Chief of Aviation, General Headquarters. This would insert him between Craig and Andrews. The latter opposed this but later changed his stand, outlining three possible courses: Designate a chief of aviation with jurisdiction over both Air Force and Air Corps; appoint a deputy chief of staff for aviation; or put the Air Force Commander under the Chief of the Air Corps. While he disliked the last method, it would at least make one person responsible for all aviation activities. Nothing happened until after Arnold succeeded Westover as Chief of the Air Corps in September 1938. He did not get to be Chief of Aviation, General Headquarters, as he desired, but did convince the War Department to do something. On March 1, 1939, the Commanding General, GHQ Air Force, became directly subordinate to the Chief of the Air Corps. Thus Maj. Gen. Delos C. Emmons, who that same day succeeded Andrews as Air Force Commander, reported to Arnold instead of Craig.[76]

The formation of GHQ Air Force in the Regular Army on March 1, 1935, effected a revolutionary change in the organization of U.S. Army aviation. The new unit brought together, under one air commander, active combat units which served as the nucleus of the air force to be mobilized in time of war, but which in peacetime had hitherto been dispersed among various corps areas for administration and training. Hence for the first time, peacetime organization of the air arm paralleled that for war. Further reorganization in 1936 transferred GHQ Air Force bases, formerly under the control of corps area commanders, to the Air Force. Divisions, corps, and

ARMY AVIATION
1935

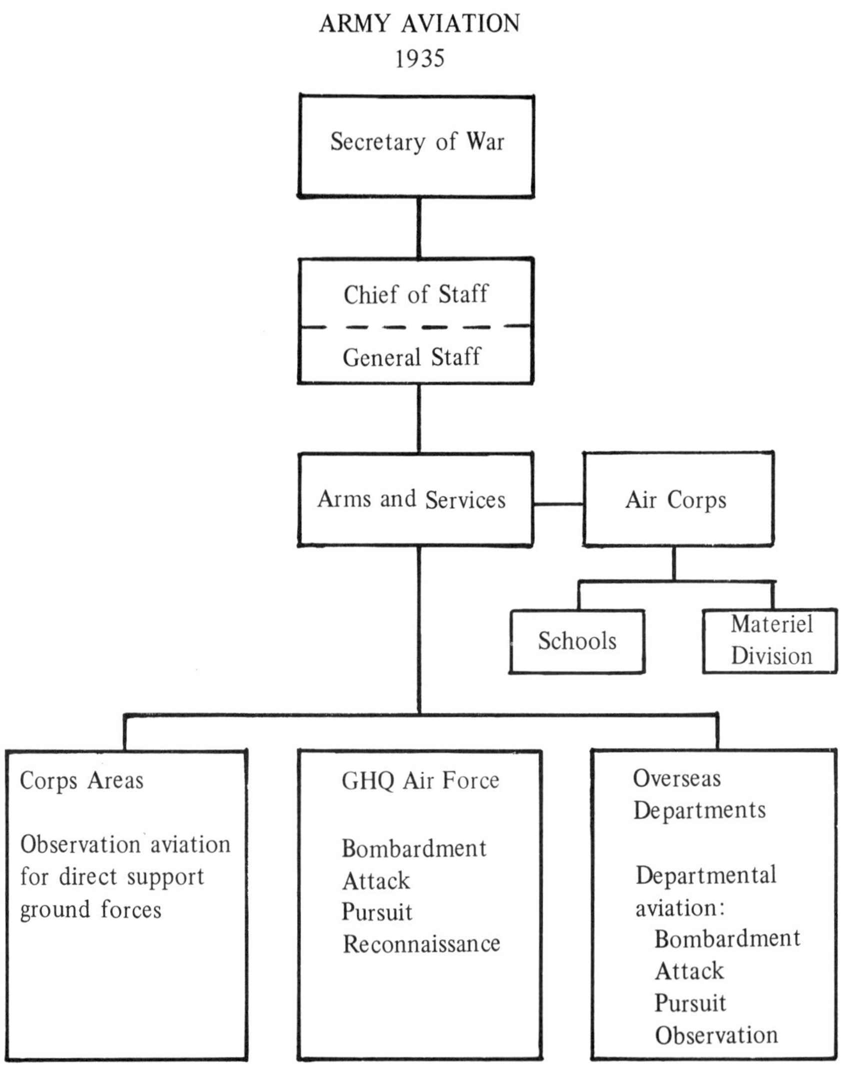

field armies continued to have observation aviation assigned to them and under their control. Nevertheless, field armies lost the pursuit and attack aviation that once had been part of a field army's air force. With this change, all bombardment, attack, and pursuit aviation in the United States, plus some observation aviation for reconnaissance, became part of GHQ Air Force.

In organizing GHQ Air Force in 1935, the War Department placed it under the War Department Chief of Staff in peacetime and in war under the commander in chief in the field. Also under the Chief of Staff, the Chief of the Air Corps retained responsibility for personnel and materiel for the Army's air arm, including the Air Force. This arrangement created problems for both organizations. Subordination of the Air Force commanding general to the Chief of the Air Corps, tried in 1939, failed to produce a satisfactory solution. GHQ Air Force proved a great disappointment to some airmen because it did not bring the independence they had labored for so long. Yet, they found it a vast improvement over the previous arrangement and thought it might be a step—as it turned out to be—toward creation of a separate and independent air force. Furthermore, they used it to try to build a powerful long-range bombing force for the strategic air operations they envisioned as decisive in future wars. The building of this air force was a troublesome and frustrating task until international developments started to clear the way in the late 1930s.

Chapter XIX

Building an Air Force

The formation of GHQ Air Force in 1935, followed by reorganization in 1936, provided a solid framework for building a powerful combat force. However, the further development of air power depended upon the resources available and how they were employed. This chapter tells of the effect the Great Depression had on U.S. Army aviation, and discusses the measures taken to build up personnel strength and achieve technological improvements in aircraft. The disagreement between the War Department and the airmen as to the kind of bombing planes to be bought is treated. So is the conflict within the air arm itself over the vulnerability and protection of bombardment formations. Also included is the improvement of base facilities, especially construction of hard-surfaced aprons and runways. And the chapter considers developments in lighter-than-air aviation, elimination of airships, introduction of motorized observation balloons, and experiments with barrage balloons.

Depression

GHQ Air Force came into being during the Great Depression, when the federal government was cutting military expenses. The number of Regular Army, Organized Reserve, and National Guard airplanes fell after 1932 owing to insufficient appropriations, procurement of large all-metal planes, higher cost due to improved performance and additional equipment, suspension of procurement of aircraft for almost a year, and disposal of those no longer fit for use. The low point came at the outset of 1936, when the Army owned 1,814 planes or about 500 fewer than at the end of the fifth year of the expansion program. Total officers on duty declined from 1,574 (1,254 Regular Army and 320 Air Reserve) in mid-1932 to 1,463 (1,305 Regular and 158 Reserve) 3 years later. Enlisted strength, however, rose gradually from 13,369 to 14,719 during the same period.[1]

Unable to complete the expansion program in the 5 years originally allotted, the Air Corps asked for $34 million to continue the program and meet other expenses during Fiscal Year 1933. The amount requested included $16.9 million for 428 new airplanes, 230 being tactical types. With public revenues falling off and the federal deficit mounting, Hoover asked the Bureau of the Budget to curtail aircraft procurement. The bureau eventually approved $25.4 million for the Air Corps, which Congress appropriated but imposed restrictions. Believing the Air Corps needed new combat planes more than other kinds, it required at least $9 million to be spent for bombardment, pursuit, and attack ships. To assure money for aircraft operation and maintenance, it directed expenditure of not less than $5.9 million for that purpose.[2]

Soon after the onset of the fiscal year, the Democratic National Convention nominated Franklin Delano Roosevelt for President. The party's platform called for a 25-percent reduction in federal spending, a balanced budget and, among other things, relief for the unemployed. Pledged to a "New Deal," Roosevelt carried 42 states in November. Democrats won control of the Senate and House of Representatives. However, a lame duck Congress enacted the appropriation bill for Fiscal Year 1934. The act President Hoover signed just before he turned the government over to Roosevelt on March 4, 1933, contained $26.3 million for the Air Corps. It again specified minimums for combat planes ($7.6 million) and for aircraft maintenance and operation ($9.1 million).[3]

Three weeks into the New Deal, the new Director of the Bureau of the Budget, Lewis W. Douglas, asked cabinet officers to estimate expenditures for Fiscal Year 1934. Suggesting a ceiling for each department, he gave Secretary of War Dern a figure of $196 million for military programs, a cut of $74 million. When General MacArthur protested, Douglas authorized expenditure of $244 million. Due to the curbs on spending, the Air Corps used only $12.6 million of its 1934 appropriation, with $1.4 million going for new planes (26 B–10Bs).[4] Asked by the administration to prepare a study on how Army needs might be met with public works money, the War Department suggested spending $304 million on housing, mechanization, motorization, and modernization. This included $39 million for the Air Corps. While it did not get nearly as much as it wanted, the War Department received $100 million from the Public Works Administration over 2 years, with more to come. Most of it went for construction and other projects providing work for the unemployed, but a portion purchased motor vehicles, ammunition, and aircraft. Allotted $7.5 million the Air Corps acquired 62 B–10Bs and 30 A–17s.[5]

The Air Corps requested $36.5 million for Fiscal Year 1935. The review process, particularly limits set by the Bureau of the Budget, scaled down the amount to $25 million. When the appropriation bill was before Congress early in 1934, the Air Corps commenced carrying airmail. Questions arose as to the quality of Air Corps equipment and training. Sensitive to the political consequences, President Roosevelt asked Congress for an extra $10 million to strengthen the Air Corps. Congress approved half that sum, raising the appropriation for Fiscal Year 1935 to $30 million.[6]

Government economy during the depression took many forms. In June 1932 Congress enacted a bill furloughing federal employees without pay for 1 month during the coming year. This applied to government employees making $1,000 or more a year, including Army officers but not enlisted men. So at times some Air Corps stations seemed almost deserted. Rockwell Field, for instance, showed little activity about the time of Roosevelt's inauguration. Officers who had not gone on furlough earlier were departing now to get back on the job before spring training started. The lame duck Congress extended the law just before adjourning in March 1933, but the new Congress quickly abolished it and authorized the President to reduce federal salaries.[7] Roosevelt acted swiftly, ordering a 15-percent cut, the maximum allowed by law. Air Corps officers gave up nearly twice as much as they had by furlough. Enlisted men no longer enjoyed exemption—the $21-a-month private now got $17.85. Overriding a veto, Congress pared the maximum cut to 10 percent for February 1–June 30, 1934, and to 5 percent for the next year, after which the cuts ceased.[8]

To save transportation costs, Congress changed the law on assignment of officers to overseas posts. For many years, the laws of the United States prohibited assignment of officers and enlisted men for more than 2 years in the Philippines and 3 in the Canal Zone except upon a person's own request.[9] On March 3, 1933, President Hoover signed a law making 3 years the minimum for officer assignments in the Philippines, Canal Zone, and Hawaii.[10] Some men, including 1st Lt. Dale D. Fisher and 2d Lt. Allen R. Springer, got away on a transport that left Manila on March 11, before the new law could be put into effect in the Philippines. Maj. John B. Brooks and Capt. Oliver S. Ferson were among those with enough time overseas to leave in June as planned. But the law added 6 months to Capt. Leland R. Hewitt's stay at Clark Field, and 10 months to 1st Lt. Yantis H. Taylor's.[11] The 73rd Congress changed the law again, in May 1934, limiting assignments of officers and enlisted men to 2 years in the Philippines, Hawaii, and the Canal Zone unless a person asked for a longer term.[12]

The depression affected everything and everyone. It deprived the enlisted men of a bonus for reenlisting.[13] The pursuit pilot who normally flew 200 hours a year now got only 160 to 170.[14] Lacking bombs, the 96th Bombardment Squadron removed bomb racks from its planes and stored them until it could resume bombing practice early in 1935.[15] The Air Corps could not hold its annual maneuvers.[16] The Training Center took fewer cadets.[17] Beginning February 1934, cadets were no longer given Reserve commissions upon completion of advanced flying training. They received their wings but retained cadet status while serving 1 year with tactical units before being commissioned in the Reserve, after which they went on active duty for 1 year.[18] The depression also touched the *Air Corps Newsletter*. It suspended publication in October 1933, not to resume until January 1935.

The Civilian Conservation Corps (CCC) seriously affected the Air Corps and other branches of the Army. Established in April 1933, the CCC furnished temporary employment for 250,000 unmarried men ages 18 to 25, 25,000 World War I veterans (who had failed to obtain the bonus they demanded), and 25,000 experienced woodsmen. The program expanded to a peak of 500,000 men in mid-1935. For "three squares and a buck a day," enrollees worked 5 days a week on reforestation, soil conservation, fire prevention, and similar projects on state and federal lands. The military took no part in supervising the work; military training found no place in the program. But the Army inducted the men, put them through a period of physical conditioning, set up and ran camps, and supplied food, clothing, shelter, medical care, and recreation. It normally assigned 3 officers and a few enlisted men to each camp of 200 men.

Men entering the CCC sometimes went directly to work camps. Usually, however, they passed through an Army post on their way to the woods. Langley Field, Virginia, received its first contingent on May 24, 1933. Twelve hundred men arrived by boat and train from Fort Hunt, where the Army ran a temporary camp for veterans. Divided into companies of 200, the men at first lived in hangars and empty warehouses. After physical examination, each one received trousers, shirt, underwear, socks, shoes, belt, hat, raincoat, and barracks bag. They next established a tent camp on one side of the field and commenced physical training under Air Corps supervision. In about a month they moved to the woods.[19] Mitchel Field, New York, prepared for the arrival of 1,000 CCC members by putting cots in hangars, setting up kitchens, erecting latrines, and fixing showers with hot and cold water.[20] First Lieutenant Robert W. Harper of Chanute Field, Illinois, selected 440 men from 6 counties in Illinois for the CCC. Capt. Edwin F. Carey of Chanute commanded a company sent to Starved Rock, Illinois. A little later, Chanute and Selfridge reported half of their officers absent on CCC duty.[21] Second

Lieutenant Philip B. Foote of Crissy Field served at Camp F–27 in northern California, where the 996th Company made trails to forest lookout stations, built bridges across streams, strung telephone lines, and aided in firefighting.[22] The many Air Corps officers on CCC duty included 1st Lt. John C. Crosthwaite, Commander of Camp Temescal, near Corona, California. The men liked him well enough to offer a "petition" in appreciation for his service.[23]

Crosthwaite's home station, March Field, California, became headquarters for a large district containing 25 CCC camps. Lt. Col. Henry H. Arnold, the Commander of March, also headed the CCC district. In the beginning he heard March Field would receive 500 CCC members. The number climbed to 700, then 1,500, 3,000, and more. Before long, 7,000 men had passed through on their way to the forest. At one point in mid-1933, 29 of the 88 officers assigned to March Field were on duty at camps; 22 others served full-or part-time on Arnold's staff at district headquarters. The 1st Bombardment Wing ceased to exist. Group training stopped. The 7th Bombardment Group consolidated all of its flying in the 11th Squadron; the 17th Pursuit Group did the same in its 95th Squadron.[24] The CCC cut into the training of the National Guard and Organized Reserve, brought Regular Army training to a near standstill and, as General MacArthur reported, "almost destroyed the readiness of units for immediate and effective employment on emergency duty."[25] The CCC seriously impaired the effectiveness of Air Corps units just when it appeared they might be needed in Cuba.

Air Corps pilots in the woods with the CCC flew seldom or not at all. Second Lieutenant George F. Schlatter of the 1st Pursuit group, on duty with the CCC at Roscommon, Michigan, was lucky enough to have one of the group's P–16s to practice flying at Roscommon Airport. Landing hard, the ship nosed over and burned, but Schlatter escaped injury. For Lt. Col. Frank M. Andrews, the group commander, accidents like this raised a question whether pilots on CCC duty for long periods could maintain their proficiency.[26] To get back to its principal business, the Army called Reserve officers for duty with the CCC. By October 1933, March Field had just 13 Regular officers at camps and 18 on full- or part-time duty at district headquarters. Within a few weeks Reservists replaced most of the Air Corps officers in the CCC program.[27]

The Air Corps benefited from some other relief programs. Although Congress in 1935 prohibited use of public works money for aircraft and other military equipment, the War Department continued to receive funds from this source for construction. An allotment of $3.5 million helped build Hamilton Field; $1.7 million replaced wartime buildings at Middletown, Pennsylvania; $265,880 improved Felts Field, Spokane, Washington, for the National Guard. Public works money paved runways at Selfridge Field; enlarged the operations office at Scott Field; repaired a hangar at Fort

Leavenworth; improved Schoen Field at Fort Benjamin Harrison, Indiana, for Reserve training; paved aprons at Marshall Field at Fort Riley, Kansas, Post Field at Fort Sill, Oklahoma, and Mitchel Field; built quarters at Patterson Field and Rockwell Field; put in drainage at Randolph Field; and built a garage for enlisted men's automobiles at March Field.[28]

The Air Corps' financial condition began to mend slowly in the mid-1930s, then faster as the end of the decade drew nearer. The period was marked by progress toward recovery from the depression; failure of arms limitations; war in Ethiopia, Spain, and China; the menace of Hitler; and the clear need for modernizing and strengthening America's defense. These and numerous other developments at home and abroad produced larger appropriations for U.S. military programs. Air Corps expenditures rose from $20.3 million in 1935 to $32 million in Fiscal Year 1936, $41.1 million in 1937, $50.9 million in 1938, and then jumped to $83.1 million in 1939.[29] With more money the Air Corps procured additional pilots and mechanics and bought new and better airplanes and accessories; GHQ Air Force shored up its units, secured extra gasoline for training and operations, and expended more ammunition in bombing and gunnery practice.

People

The Air Corps' enlisted force burgeoned from 14,719 men in June 1935 to 20,838 four years later—an upturn of over 40 percent. Commissioned strength of the Air Corps, Regular Army, grew more slowly (from 1,305 to 1,670), an increase of about 28 percent during the same period. Not until June 1939 did the number of Regular officers reach the 1,650 authorized by law in 1926. Congress meantime raised the authorization, first to 2,092 (1938), then to 3,203 (1939). But these figures were goals for enlarging the peacetime Army over several years, goals soon overtaken by international events.[30] The 5-year plan provided additional pilots by authorizing extended active duty for 550 Air Reserve officers, but the Air Corps did not achieve this objective. A decline during the depression in the number of active duty pilots was offset in part by graduates of the Advanced Flying School who flew with tactical units as cadets. These cadets (about 100 in mid-1934, 140 a year later, and 125 in 1936) were a significant part of the pilot strength of the Air Force. At Selfridge Field, for example, cadets at times accounted for half of the flyers assigned to squadrons of the 1st Pursuit Group.[31]

General Andrews needed more pilots for GHQ Air Force, but assignment of cadets posed problems. A cadet, having been denied a Reserve commission and officer's pay, tended to avoid responsibility instead of seeking it. Since he was enlisted, his commander could not give him officer responsibilities. On the other hand, he had to assume the social obligations of a commissioned officer while drawing less pay than many noncommissioned officers of his squadron. However, being able to accrue flying time helped maintain his morale.[32] After persistent recommendations by General Andrews and others, Congress in 1937 provided money to commission cadets on duty with tactical units. Thus, 97 members of GHQ Air Force became second lieutenants in the Air Reserve on June 20, 1937, and the class graduating from Kelly Field, Texas, that month received Reserve commissions and began active duty on July 1.[33] Under a law to make active service more attractive and expand and strengthen military aviation, these Reservists could stay on active duty from 3 to 5 years, with promotion to first lieutenant after 3 years and a bonus of $500 upon discharge. The Reservist then might try for a job in civil aviation, and perhaps continue his military affiliation through the Organized Reserve or National Guard. Meanwhile, as vacancies arose, he could apply for commissioning in the Air Corps, Regular Army.[34]

The Air Corps also gave extended active duty to Reserve officers who were pilots but not graduates of the Air Corps Advanced Flying School, provided they took a "refresher" course at the Air Corps Training Center. By the end of April 1939, 75 Reservists had completed the refresher.[35]

For 5 years the only officers commissioned in the Air Corps, Regular Army, came by transfer from other branches. Most of them were recent graduates of the U.S. Military Academy. Unable to obtain Regular commissions, and not being permitted to continue on active duty as Reserve officers, many Kelly Field graduates enlisted in the Air Corps until they could return to duty as Reserve officers or secure a commission in the Regular Army. These men and the service benefited from an act of Congress permitting the War Department to call 1,000 Reservists to duty each year, and from this group select 50 for the Regular Army. The other services having no Reservists eligible the first year (1930), the Secretary of War approved 50 Air Reservists for commissioning as second lieutenants in the Regular establishment. The next year, the commissions went to the other branches.[36] As a result of these programs for producing second lieutenants for the Air Corps, the number of officers rose sharply. In mid-1936 the Air Corps counted 1,534 (1,362 Regulars and 172 Reservists) on extended active duty. In 3 years the number grew to 2,516 (1,670 Regulars and 846 Reservists), allowing expansion of GHQ Air Force to more than 1,000 officers (nearly evenly divided between Regulars and Reservists) by June 1939. Nevertheless, most of the increase in the Air Force was in second lieutenants. Noting that two-thirds of his officers were in that grade, General Emmons, commanding GHQ Air Force, pointed out that lack of experience affected both combat effectiveness and the peacetime accident rate.[37]

Aircraft

After falling off during the depression, the number of aircraft started to rise again in 1936. The goal was 2,320 as recommended by the Baker Board in 1934 and authorized by Congress in 1936. MacArthur wanted to buy 800 a year for 3 years to reach the goal as quickly as possible. Afterwards, it would be necessary to buy 500 a year to replace planes that crashed, wore out, or became obsolete, and to keep abreast of other developments. The War Department in the beginning could not muster the money to carry out this program, so it revised the plan in the hopes of attaining 2,320 serviceable aircraft by mid-1940. The number on hand grew from 1,276 in June 1936 to 1,380 in 1937, 1,719 in 1938, and 2,177 in 1939. During this time, the Air Corps replaced many older planes with newer ones of much better performance, thus further strengthening the nation's air power.[38]

Bombers

The YB–10s that the Air Corps first received during airmail operations were a big advance over the older Keystone bombers in performance and striking power. Driven by 2 Wright engines, the B–10 carried a crew of 4, provided internal storage for 2,260 pounds of bombs, and mounted 3 machineguns (in the turret, the rear cockpit, and the floor of the fuselage behind the bomb bay). Its maximum speed was 207 miles per hour, cruising speed 169, service ceiling, 21,000 feet, and range 600 miles (extendable with extra fuel tanks).[39] The Air Corps used 10 of them for a flight to Alaska and back in the summer of 1934. Colonel Arnold planned to go fishing when he finished flying the mail, but General Foulois sent him to Dayton, Ohio, to organize a squadron of B–10s to fly to Alaska. He formed three flights, taking command of one and giving the others to Maj. Hugh J. Knerr and Maj. Ralph Royce. He selected Capt. Westside T. Larson for navigator; Capt. Harold H. McClelland, communications and meteorological officer; Capt. John D. Corkille, engineer; and Maj. Malcolm C. Grow, surgeon. Altogether, 14 officers and 16 enlisted men made the trip. SSgt. Henry Puzenski flew with Arnold as crewchief, as he would for many years. (*Map 6*)

Starting the trip at Bolling Field on July 19, 1934, the B–10 squadron flew in easy stages to Fairbanks, Alaska, in 6 days, covering 4,000 miles in 25 hours and 30 minutes of flying. While in Alaska, the squadron photographed more than 20,000 square miles, mostly along lines from Fairbanks to Nome, Whitehorse (Canada), Anchorage, and Juneau. Colonel Arnold and his men commenced the return trip on Wednesday, August 16, 1934. On Thursday they averaged 175 miles per hour flying 990 miles nonstop from Juneau to

Left: Secretary of War George H. Dern and Lt. Col. Henry Arnold, with totem pole presented to Air Corps from people of Alaska; right (from l. to r.): Glenn Martin, designer of B–10, Ass't. Sec. of War Harry Woodring, Maj. Gen. Benjamin Foulois, and Lt. Col. Henry Arnold; below: B–10s take-off from Nation's capital for Alaska.

Seattle, the longest leg on the entire journey. They stopped at Salt Lake City on Friday night, at Dayton on Saturday, and arrived at Bolling Field on Sunday. "We have proved," Arnold said, "that it is possible to take tactical units of the Air Corps to Alaska quickly and bring them back successfully."[40]

From Bolling Field, Colonel Arnold led the B–10s to his home station at March Field, where a few days later he received orders from General MacArthur to demonstrate how swiftly a squadron of bombers could move from California to New York. Leaving March Field at 0230 on Monday, September 3, 1934, B–10s flew to Amarillo, Texas, and refueled. Poor gasoline caused trouble as the flight continued and bad weather, with worse to the east, forced the squadron to detour to the south and make an unscheduled stop at Shreveport, Louisiana. Weather delayed the planes at Atlanta, and they stopped at Langley Field for an hour to simulate loading bombs before going on to Mitchel Field, New York. While Arnold failed to complete the movement within 24 hours, as he expected, the flying time amounted to just 18 hours and 10 minutes.[41]

Air Corps flyers thought the YB–10 a great plane, far superior to any of their previous bombers. The production model, B–10B, performed even better, and so did the B–12, a Martin plane similar to the B–10 but with Pratt and Whitney engines. The War Department ordered 103 B–10Bs and 32 B–12s for tactical units. Air Corps leaders, however, wanted a bigger plane to carry heavier loads of bombs farther and faster for strategic operations. The War Department, in the spring of 1934, approved an Air Corps project for developing an experimental bomber. Boeing built the plane, the XB–15, which arrived at Wright Field in December 1937 for inspection and testing.

The War Department further approved an Air Corps request for a multiengine bomber to follow the B–10. Three companies sent planes to compete at Dayton in August 1935. Martin offered one resembling a big B–10. Douglas sent a two-engine bomber derived from its DC–2. Boeing entered its Model 299, a "Flying Fortress." Unable to keep its plans secret, Boeing unveiled its "mystery" plane at Seattle in July 1935. On August 20, Boeing test pilot Leslie R. Tower flew the big, four-engine ship nonstop from Seattle to Dayton at an average speed of 252 miles per hour.[42] The Air Corps knew it wanted Fortresses. The Boeing 299 crashed during testing at Wright Field on October 30, 1935. Aboard were Tower and four men from the Materiel Division—Maj. Ployer P. Hill, Chief of the Flying Branch, pilot; 1st Lt. Donald L. Putt, copilot; John B. Cutting, engineer; and Mark H. Koogler, mechanic. Taking off, the plane climbed steeply to 300 feet, stalled, crashed, and caught fire. Tower and Hill died. Investigation disclosed that no one had unlocked the rudder and elevator controls. Consequently, Major Hill could not control the aircraft after it became airborne. That caused the crash, not faulty design, structural failure, or malfunction.[43]

The Air Corps asked for 65 Flying Fortresses. The War Department approved 13 for service testing. Maj. Barney McK. Giles took a crew from the 96th Bombardment Squadron to Seattle in February 1937 to bring back the first one. Designated YB–17, it was the largest plane built for the Army since the Barling Bomber in 1923. Powered by four 930-horsepower engines, it carried a crew of 9, had an automatic pilot, operated at 217 miles per hour (maximum speed 256), and had a service ceiling of 30,600 feet and a range of 2,480 miles. It carried up to 5 tons of bombs, and mounted 5 machineguns (either .30- or .50-caliber) in the nose turret and in blisters on the top, bottom, and both sides of the fuselage. Major Giles arrived at Langley Field with the first YB–17 on March 4, 1937. Maj. Caleb V. Haynes picked up the second one for the 49th Bombardment Squadron; Maj. Vincent J. Meloy the third for the 20th Bombardment Squadron. By August 5, 1937, the 2d Bombardment Group owned 12; the 13th plane went to Wright Field, Ohio, for experiments.[44]

The Air Force naturally wanted to show off its new bombers and what they could do. Lt. Col. Robert Olds, the Group Commander, and Major Giles took the first one to Bolling Field on March 9 to display for 4 days. Newspapers featured stories of the plane's visit. The bomber's "inspiring size, beauty and formidable fighting capabilities" attracted hundreds of people.[45] On May 16, Colonel Olds led 4 of the planes on a cruise from Langley northward to Augusta, Maine, west to Cleveland, and back by way of Pittsburgh and Richmond, passing over 20 cities in 15 states during the 11-hour flight.[46] Six of the B–17s (the Air Corps dropped the prefix after the service test) took part in an aerial review GHQ Air Force put on for an American Legion convention in New York.[47] Later the Air Corps placed one on exhibition at Treasure Island in San Francisco Bay for the Golden Gate Exposition.[48] The Air Force publicized the speed of the B–17s: 5 hours from Miami to Langley Field, 5 from Kelly Field to Langley, and Wright Field to Langley in the "remarkable time" of 1 hour and 45 minutes. Bucking headwinds on January 6, 1938, Colonel Olds flew 2,317 miles from Langley to March in 13 hours and 27 minutes. Returning to Langley 3 days later, he cut his previous time 2 hours and 26 minutes. Olds described the return flight as "routine military training." He said it proved the progress achieved in equipping and training the Air Force and the ease with which reinforcements could be sent to either coast.[49]

An excellent opportunity to demonstrate the long-range capability of the B–17s came to the Air Force in February 1938. The State Department requested a goodwill flight to Buenos Aires for the inauguration of Roberto M. Ortiz as president of Argentina. With Colonel Olds in command, 6 planes, each with its regular crew, took off from Langley Field at 2-minute intervals on Tuesday morning, February 15. Keeping in touch by radio, each flew on its own to Miami. After stopping there for service, Olds hoped to make Lima, Peru, in one hop, but that would depend on the weather south of Panama.

Before departing Miami he told commanders of the other planes to assemble over Colon, Panama. There he decided to continue without landing. All arrived safely at Lima after a nonstop flight of 2,695 miles in 15 hours and 32 minutes. Employees of Pan American Grace Airways (Panagra) helped service the planes during a 7-hour layover. A propeller needing adjustment delayed departure of Maj. Vincent J. Meloy's plane. The other 5 aircraft flew southward to Santiago, Chile, turned eastward across the Andes, assembled near Buenos Aires, and landed at El Palomar Field about 1130 on February 18, 1938. The flying time for 2,200 miles from Lima was a bit over 12 hours. Major Meloy arrived later the same day and that evening Colonel Olds delivered a letter from President Roosevelt to president-elect Ortiz. On Sunday the 6 American planes appeared over the city during the inaugural ceremonies.

The flyers adopted a more leisurely pace on the return trip. Capt. Archibald Y. Smith and his crew could not leave El Palomar with the others on Tuesday due to minor damage to their aircraft when it broke through the concrete while being rolled away from the gas pit. They caught up, however, at Santiago, Chile, where the flight remained overnight. A broken starter delayed Smith on Wednesday, but he rejoined the others at Lima, Peru, on Thursday. After stopping in Panama for a day to service the planes, the men took off Sunday morning with orders to assemble over Norfolk, Virginia. General Andrews, bearing a sheaf of telegrams and messages, greeted the flyers when they stepped out of their B–17s at Langley Field ten hours and forty-five minutes later. Colonel Olds was awarded the Distinguished Flying Cross, and the 2d Bombardment Group the Mackay Trophy, for the goodwill flight.[50]

Three B–17s under Major Meloy's command completed a goodwill flight to Bogota, Colombia, in August 1938.[51] The day after they took off for Bogota, Colonel Olds touched down at Langley Field with the XB–15. Similar in appearance but bigger than the B–17, this "mighty leviathan of the air" (the phrase was the Air Corps') boasted a wing span one-third greater and weighed a third more. There were several new features. Two generators run by auxiliary engines furnished electric power. A separate station in the cockpit let the flight engineer assume many mechanical duties from the pilot and copilot. A passage in the wings behind the engine nacelles enabled the crew to service engine accessories in flight. There were two wheels on each of the main trucks. There were also cooking, sleeping, and toilet facilities. Maj. Stanley M. Umstead and 1st Lt. Leonard F. Harman had begun testing the B–15 at Wright Field in 1938. It could go twice as far as the B–17 but was slower and had a lower service ceiling. Four 1,000-horsepower engines, the most powerful available, did not deliver enough power for the great ship. Even so, the 2d Bombardment Group welcomed it and placed it in service.[52]

The Air Force used the B–15 to haul medical supplies to South America

Above: Six Boeing B–17s en route to Buenos Aires, Argentina; left: Gen. Andrews addresses B–17 crew upon return to Langley Field after goodwill flight to Argentina.

Members of goodwill flight to Bogota, Columbia, (l. to r.): Maj. Harold George, Maj. Vincent Meloy, and Maj. Caleb Haynes.

in February 1939. When an earthquake killed thousands of people and injured many more in Chile on January 24, 1939, General Andrews ordered long-range planes held in readiness for a relief mission. However, he canceled the alert when he learned the Panama Canal Department would send aircraft. Argentina dispatched aid by rail and air and others helped, but Chile needed many more medical supplies quickly. The American Red Cross asked President Roosevelt to send a Flying Fortress. Orders arrived at Langley Field late Wednesday afternoon, February 1, 1939. Given command of the flight, Maj. Caleb V. Haynes selected a crew: Capt. William D. Old, copilot; Capt. John A. Samford, navigator; 1st Lt. Richard S. Freeman, engineer-pilot; 1st Lt. Torgils G. Wold, weather officer; TSgt. Adolph Cattarius, chief aerial engineer; SSgts. William J. Heldt, Harry L. Hines, and David L. Spicer, aerial engineers; Capt. James E. Sands and Pfc. Russell F. Junior, radio operators. They would fly the B–15, but the 2d Bombardment Group also prepared a B–17 to use in case the B–15 could not go. In addition, the group readied a B–17 for Capt. Hilbert M. Wittkop to take in case one plane could not carry all the supplies. By late Thursday all was ready. Friday afternoon Wittkop watched with growing disappointment as soldiers loaded the B–15 with gauze bandages, ether, chloroform, sutures, sulfanilamide tablets, syringes, needles, X-ray plates, surgical gloves, and other items, until all sixty-nine cartons, weighing 3,250 pounds, were aboard.

Taking off at 0635 Saturday morning, Major Haynes climbed above a layer of low clouds and leveled off at 5,000 feet. The clouds dissipated in time for Samford to check drift as the plane crossed the coast just west of Cape Lookout, North Carolina. The flyers identified Palm Beach, Florida, as they passed east of it, well at sea. Five miles east of Miami Beach at 1145, Samford

Field Director of the American Red Cross (left) turns over supplies to Maj. Caleb Haynes, pilot of B–15 on flight to Santiago, Chile, to aid earthquake victims.

set a course straight for Panama. A cartoonist depicted life aboard the "Golden Goose" on the way to Santiago—the B–15 on automatic pilot, Haynes dreaming of decorations, Old napping, Samford relaxing with feet on desk, Junior complaining about Spanish on the radio, Hines working on an engine under Freeman's direction, Wold hanging outside to observe the weather, Heldt cooking, Spicer with coffee and sandwich, Sands resting on a cot, and Cattarius' voice coming from behind a closed door in the rear of the plane. A flash of light at the entrance of the Panama Canal came as a welcome sight at 1855. Ten minutes later the plane landed at France Field. Airborne again at 0400 on Sunday, the flyers reached Lima, Peru, at 1250. The Panagra crew helped service the plane during the 9-hour stop. Lt. Col. Ralph H. Wooten, the U.S. military attache to Chile, along with the head of the Chilean air force and representatives of the Red Cross, met the plane when it landed at Santiago shortly after 0730 Monday morning. In the 49 hours and 18 minutes since leaving Langley Field, the airmen had covered 4,933 miles in a flying time of 29 hours and 53 minutes.

Major Haynes planned to return by easy stages with stops at Lima and Panama. Learning gas would not be available at Lima, he decided to load 4,000 gallons and head for Panama, with Talara, Peru, an alternate airport. Leaving Santiago at 1810 Thursday, the B–15 went all the way, covering 3,200 miles in 19 hours and 55 minutes. At France Field, Haynes received orders to remain in Panama until Monday, fly to Miami, stay there overnight, and arrive at Langley Field at 1100, Tuesday, February 14. Over Langley Field at 1035, Haynes got orders by radio to fly around until 1100. Units of the 2d Wing needed time to get into formation on the concrete ramp along the hangar line. When the time came, Haynes landed, taxied the B–15 to its usual resting place, and stepped out. Andrews went forward to greet the crew informally and tell them what a good job they had done. Then, with the crew lined up on the ramp, he extended a formal welcome. W. D. Millner, Red Cross Director at Langley Field, read a message from Norman H. Davis, Chairman of the American Red Cross. At the close of the ceremony, 52 planes from Langley and Mitchel Fields took off in single file, assembled in formations and passed in review.

General Andrews had planned to present the Distinguished Flying Cross to Major Haynes but Secretary of War Harry H. Woodring decided to make the presentation. At 1245, Major Haynes and his crew left Langley in the B–15 for Washington and the ceremony in the Secretary's office. General Craig read the citation and Secretary Woodring pinned the medal on Haynes' blouse in the presence of General Arnold, officials of the American Red Cross, and representatives of the Chilean government.[53]

Then on June 10, 1939, Major Haynes flew the B–15 to Mexico City to carry home the body of Francisco Sarabia, a famous Mexican flyer who died when his plane crashed in the Potomac River.[54] And on June 30, Haynes and

Captain Old set 2 world records at Wright Field with the B–15 by carrying a payload of 22,046 pounds to 8,228 feet, and 31,164 pounds to 6,561.6 feet.[55]

The bomber competition of August 1935 led to procurement of 13 B–17s and production of the 2-engine plane offered by Douglas. This aircraft, the B–18, employed a crew of six, including three gunners to handle three .30-caliber machineguns firing from nose turret, tail turret, and through a tunnel in the floor of the fuselage. Considerably smaller and lighter than the B–17, the B–18 was slower, had a lower ceiling and shorter range, and carried fewer bombs. Douglas received a contract for 81 of these planes, which the Air Corps labeled medium bombers.

Soon afterwards, in June 1936, Congress increased the authorization of serviceable planes for the Army from the 1,800 provided in the Air Corps Act of 1926 to the 2,320 recommended by the Drum and Baker Boards.[56] The War Department hoped to obtain funds to reach the new goal by June 1940. GHQ Air Force and the Air Corps wanted to buy more heavy bombers. General Andrews considered big bombers—4-engine aircraft that could haul heavy loads long distances—powerful instruments of defense. Easily converted to carry extra fuel in place of explosives, big bombers could also conduct long-range reconnaissance. Andrews believed the nation's strategic position and military policy demanded them. The Army required them in an emergency to reinforce Hawaii, the Canal Zone, and Alaska. The General Staff viewed big, long-range bombers as aggressive weapons not needed under a policy of defense. Smaller bombers possessed greater flexibility. They could support ground forces. More could be obtained for the same money, making it easier to achieve the goal of 2,320 airplanes, and replacements cost less. The Army did not need long-range bombers for defense against hostile fleets because responsibility for defense beyond the coastal area rested with the Navy.[57]

Airmen attempted in various ways to prove the B–18 did not satisfy Air Force needs and that the cost of a fleet of heavy bombers compared favorably with the cost of medium bombers. General Andrews' staff compared the performance of B–18s with B–17s against an enemy fleet approaching San Francisco. The Engineering Section at Wright Field contributed tables 1st Lt. Leonard F. Harman prepared to show speed, endurance, and range for both models with different loads of fuel and bombs.[58] Someone else constructed a table for movement of the fleet from 2,637 miles at sea at 0300 on D-day minus 4 days to landing at 2139 on D-day. To meet the enemy threat, the Air Force planned a series of attacks for B–17s and B–18s. It scheduled the first for 0600 on D-day minus 1 day, the fleet then 912 miles out: Each B–17 carried 2,200 pounds of bombs; the B–18s 1,200 pounds each. If sent on this mission, the B–18s could not take part in the second one that day. If they skipped the first and flew the second, they would convey 3,400 pounds of bombs as compared with the B–17's 4,600 pounds. For the third attack, at

0600 on D-day, with the fleet 350 miles at sea, B–17s hauled 8,000 pounds of bombs, and B–18s 4,400 pounds. For the fourth attack, at 1300, 200 miles out, B–17s transported 8,800 pounds and B–18s 5,300. For the fifth, at 1800, 84 miles at sea, B–17s moved 8,800 pounds and B–18s 6,200 pounds.

From this, the men at Langley Field figured it took 50 squadrons of medium bombers to do the work of 31 squadrons of heavies. With B–17s priced at $176,000 each, and B–18s at $103,000, a fleet of heavies cost $79 million, mediums $77 million. With 2,320 airplanes authorized for the Army, a bomber force of B–17s left 1,857 planes available for pursuit, attack, reconnaissance, observation, transport training, and other purposes, but purchase of B–18s left just 1,569. It required about the same number of enlisted men for both, but nearly 60 percent more officers for a B–18 force.[59] Such arguments failed to persuade the War Department. General Craig directed no more heavy bombers be procured except for experimental purposes until international conditions indicated a need for them. The B–18, he believed, fulfilled reasonable requirements at a justifiable cost.[60] The Army ordered 52 more B–18s making a total of 133, followed by 217 B–18Bs.

Douglas delivered the first B–18 to Wright Field in the fall of 1936. The Air Corps sent the plane to Langley Field later that year for testing. The Air Force commenced receiving B–18s in mid-1937 for its units. The flyers gladly relinquished B–10s and B–12s for the newer planes. The 7th and 19th Bombardment Groups of the 1st Wing converted first, followed by the 2d Bombardment Group then by the 9th Bombardment Group and the 2d Wing. The Hawaiian Department welcomed B–18s when the 31st Bombardment Squadron arrived from Hamilton Field in February 1938. The Panama Canal Zone started getting B–18s late that year, by which time units in the United States had completed conversion.[61]

Pleased with the service test of the B–17s, the Air Force wished to equip its bombardment groups with them. The War Department approved procurement of merely 13 but later upped the number to 39. These planes, with a different nose, and larger rudders, and flaps, constituted the B-model. The Air Corps accepted the first B–17B a month before Hitler invaded Poland.[62] When war broke out in Europe, the B–18 was the standard bomber of GHQ Air Force. Along with B–18s, the 2d Bombardment Group still flew its 12 B–17s, now listed among the "limited standard models" in the Air Corps inventory. The B–18 was likewise standard in Hawaii and the Canal Zone. The Philippine Department possessed but a few B–10s and some B–3As.[63]

Pursuit, Fighter, and Interceptor Planes

The development of bombers with greater speed, range, and defensive fire impacted on pursuit aviation. In the 1920s bombing planes were slower and less maneuverable than pursuit craft. Bombardment formations had to have friendly pursuit for protection against enemy planes. By the early 1930s bombardment and pursuit were more evenly matched, and it appeared bombardment would soon seize the advantage. General Westover called attention to the change in his report on GHQ Air Force (Provisional) in 1933. New bombers flew at speeds above 200 miles per hour. Pursuit needed to be far faster than that to intercept or support bombers. He doubted if pursuit could operate safely and efficiently at such high speeds. Moreover, he thought bombardment aviation had enough speed and firepower to operate without support. He knew nothing that could "frustrate the accomplishment of a bombardment mission."[64]

Questioned about employment of pursuit in defense of bombardment, Maj. Carl Spatz told the Baker Board in April 1934 that until the last two or three years the doctrine "was fairly well fixed in our minds." Pursuit accompanied bombers to afford close protection on the way to and from the objective, or they penetrated ahead of the bombers to assure protection over the target. But bombers now flew faster and farther. Pursuit no longer possessed a great advantage in speed or the ability to go all the way to the target. New methods of protecting bombers needed to be found. Spatz mentioned two: Design a long-range plane for escort work, or give bombers enough guns to defend themselves.[65]

Asked about the future of pursuit operations, Lt. Col. Henry H. Arnold told the Baker Board it "is one of the mysteries that nobody can answer right off hand." He believed the single-seat fighter would be used solely as an interceptor to defend a localized area. To oppose bombers, he visualized a multiseat "fighting" plane faster than bombers, armed with two 37- or 40-millimeter cannon, two .30-caliber guns in front, and flexible guns in back, and carrying fifteen to twenty-five 5- or 10-pound bombs to drop on the enemy's bomber formation. Tests had shown that a plane with 15 bombs covered an area 50 feet wide by 250 feet long so no place in that area was more than 10 yards from a bomb. Any number of fighting planes could be added to widen horizontal coverage with bombs. The fuze setting gave dispersion in depth. "If you put those up top-side," Arnold said, "and sit there out of range of bombardment machine guns, and drop the bombs down, that formation won't stay long."[66]

Later that year Colonel Arnold conducted tests at March Field with the Air Corps' newest planes, B–12s and P–26s. Though he found further study essential to change existing ideas on equipping and training units to attack enemy aircraft, he saw the day of individual combat drawing to a close.

Planes traveled so fast that firing at close range appeared impossible. Armament needed to be changed for work at greater range. Arnold deemed it "extremely doubtful if single-engine Pursuit planes . . . can prevent a formation of modern Bombardment planes from reaching their objective or destroy the planes either en route to or returning from their objective." Present pursuit planes would be of no value at night against camouflaged bombers.[67] General Andrews and Colonel Knerr were among those airmen who thought bombers could get through any opposition offered by pursuit ships or antiaircraft guns.

The information the Air Corps prepared for the Baker Board indicated an average proportion of 58 percent pursuit planes to 42 percent bombardment among the air forces of the leading foreign powers. The ratio in the U.S. Army was then 74 to 26. The Baker Board recommended 52 to 48. Working toward that goal, the Air Corps achieved a ratio of 53 to 47 by the beginning of World War II.[68]

Seeking ways in the early 1930s to protect bombing formations, the Air Corps obtained twenty-five biplane pursuit ships, P–16s, for the 1st Pursuit Group, but pilots did not like them. The rear gun seemed to be of little use. Capt. Earle E. Partridge, who commanded P–16s in the 94th Pursuit Squadron, recalled years later that "the gunner . . . sat looking backward so he could shoot people coming up from the rear. We never did shoot anybody, but the airplane was delightfully built. Put the brakes on too hard, and it would go right over on its back."[69] The group gradually disposed of its P–16s during 1934. Accidents damaged several beyond repair. In other cases, damage the Air Corps usually repaired served as an excuse to drop a plane from the Air Corps inventory.[70]

The usefulness of the rear gunner in a two-seat pursuit ship came into question again during tests of P–30s in 1934. A sharp turn sometimes caused the gunner to black out.[71] Colonel Arnold, one of the many who considered the gunner of no value in fast-moving combat, could not visualize any situation where another type of aircraft could not be better used than biplace pursuit.[72] The Air Corps nonetheless ordered 50 P–30As (redesignated PB–2As). When the Air Corps began accepting the new biplace pursuit (PB) planes in the spring of 1936, it divided them between the 1st and 8th Pursuit Groups. The next year, however, it concentrated them in the 8th Group, where they stayed until removed from tactical use in 1939.[73] In lieu of biplace pursuit, Arnold advocated multiplace fighters. Without such planes for his tests at March Field in 1934, he cast B–12s in the role of multiseat fighters against and in defense of bombardment. Though the B–12s proved unsatisfactory for this work, the tests pointed up the need for multiseat fighting planes. The ideal plane demanded 8 or 10 guns, a crew of 3 or 4 men, 500 pounds of bombs, and a range equal to and a performance superior to bombers. Realizing it would be difficult to secure such an aircraft, Arnold proposed a compromise: provide first for performance and range, then add as much armament as possible. He suggested such a plane be secured without delay for testing. A variety of experiments with bombs dropped on formations of bombers impelled him to urge immediate development of small 2- to 10-pound bombs.[74]

Not everyone believed bombardment aviation to be invincible or in the need for big fighting aircraft to oppose or defend bombers. The loudest protests came from Capt. Claire L. Chennault, instructor in pursuit at the Air Corps Tactical School. Antiaircraft exercises at Fort Knox, Kentucky, in May 1933 convinced him pursuit could intercept bombers before they reached their target if the defensive area had sufficient depth and the defenders received timely warning. He urged development of a superior interceptor, creation of a warning net, and intensive training in interception and attack. Arguing in behalf of singleseat pursuit, Chennault vigorously opposed a multiseat fighter. He thought it impossible to produce the latter with ample speed, range, and firepower. If the plane could be developed, it would cost more than a bomber and would be very expensive to maintain and operate. Its flexible guns, required for the work envisioned, could not fire accurately in high-speed flight. Losing the argument, and being disgruntled with the Air Corps and Army, Chennault retired on disability on April 30, 1937. He left at once for the Far East and got a job with the Chinese government.[75]

Some people, Capt. Ross G. Hoyt for one, saw a need for two kinds of pursuit planes instead of a single all-purpose ship. He suggested an "Interceptor Pursuit" with the requisite speed, rate of climb, and firepower to overtake and deny the operation of hostile aircraft. For the other type he advocated a long-range multiplace plane with performance superior to the best bombers.[76]

The Air Corps ordered an experimental multiplace fighter (XFM–1) from Bell Aircraft Corporation. First Lieutenant Benjamin S. Kelsey took it up on its first flight at Buffalo on September 1, 1937. About the size of a B–10, the low-wing monoplane used 2 Allison 1,150-horsepower engines as pushers and attained a maximum speed of 270 miles per hour. The XFM–1 mounted 2 each .30-caliber, .50-caliber, and 37-millimeter guns and carried twenty 30-pound bombs. It employed a crew of 5—pilot, copilot-navigator, radio operator-gunner, and two gunners stationed in the front of the engine nacelles on each wing. Crewmembers communicated with each other by telephone and could move from one position to another while in flight. The War Department contracted in May 1938 for 12 for service testing. General Arnold said Bell's "Airacuda" caused more comment at home and abroad than any other aircraft that year. With it, he said, "we jumped to an early lead" in attaining a plane capable both of combating enemy bombers and of defending our own bombers on long missions. The Air Corps received none before war broke out in Europe.[77]

The Boeing P–26 became the standard single-seat pursuit plane in the mid-1930s. General Andrews, as previously mentioned, assigned these aircraft to the 1st and 20th Pursuit Groups when he became Commanding General, GHQ Air Force. The Air Corps had already announced competition for designing new planes. Seversky won a contract in May 1936 for 77 P–35s. Shortly afterwards, the Air Corps ordered 3 P–36s from Curtiss for testing and the following year placed an order for 210 P–36As. The P–35 and P–36 were the first single-seat pursuit planes with enclosed cockpits and retractable landing gear available to Air Force units. They cost a lot more but performed much better than the P–26s.[78]

The 1st Pursuit Group got the P–35s, the first arriving at Selfridge Field, Michigan, at the end of December 1937. All three pursuit groups in the United States accepted P–36As after deliveries to Air Force units commenced in September 1938. When the first P–36 reached Barksdale Field, Louisiana, the 20th Group reported it soon would be "flying on silver wings instead of the old faithful blue and yellow."[79] Since adoption of chrome yellow for wings and tails in 1927, fuselage coloring had changed from olive drab to light blue. Besides, squadrons now used identifying coloring on their aircraft. Orange paint on a motor cowling at Selfridge Field, for instance, identified the plane as belonging to the 27th Pursuit Squadron; red indicated the 94th Pursuit Squadron. Upon receiving all-metal aircraft, the Air Corps continued to paint them yellow and blue until 1937. It left the P–35s, P–36s, B–17s, A–17As,

Formation of P–26s, which were the first all-metal monoplane fighters produced for the U.S. Army Air Corps.

and other new aircraft in their natural color. Noting the change, France Field, Panama Canal Zone, reported that the B–17s landing on their way home from their goodwill flight to Argentina were "silver winged." Finding that white did not show up on the silver finish of the new P–36s, the 55th Pursuit Squadron obtained permission to change its identifying color to blue.[80]

With Air Force units getting PB–2s, P–35s, and P–36s, the Army began transferring P–26s from the United States to overseas garrisons. The Philippine Department received the first, early in 1937, followed in turn by the Hawaiian and Panama Canal Departments.[81]

Attack Aircraft

When the 3d Attack Group joined the GHQ Air Force it was flying A–12s, the kind of plane it contributed to airmail operations the past year. The Air Corps, however, had already ordered 110 A–17s from Northrop. Second Lieutenant Edward M. Garvin, engineering officer of the 13th Attack Squadron, went to Inglewood, California, to pick up the first of the group's newer planes at the factory in February 1936. The men at Barksdale Field found both covered cockpits on this model more comfortable than those of the old planes, but with more gadgets. The A–17 flew faster, higher, and farther than the A–12, carried the same number of machineguns, and hauled an equal amount of 100-pound bombs but could tote twice as many 30-pound bombs.[82] The 3d Group also accepted 12 Y1A–18s (two engine, two-seat monoplanes built by Curtiss) in 1937 for service tests. Compared with the A–17s, they were fitted with like armament and flew a little higher and faster but had not the range of the A–17s.[83] Adopting the A–17 as the standard plane for attack aviation, the Air Corps ordered 100 more with retractable landing gear and other improvements (A series) then contracted for another 29.

The 17th Attack Group at March Field, California, received A–17s in the spring of 1936, followed soon afterwards by the 37th Attack Squadron at Langley Field, Virginia. A–17s also went to Panama for the 74th Attack Squadron, redesignated from pursuit on September 1, 1937. Meantime, the 26th Attack Squadron in Hawaii acquired some of the A–12s no longer needed in the United States.[84] Douglas delivered two A–17s as three-seat transports (A–17ASs). General Westover flew one on inspection trips to various parts of the United States and once went to the Canal Zone. SSgt. Samuel Hymes usually went along as mechanic. On one of these trips (September 21, 1938), Westover presented the Daedalian Trophy for safe flying to the 19th Bombardment group at March Field. That same day

Westover and Hymes flew their A–17AS to Burbank, California, to visit the Lockheed plant. A man in the tower watched the aircraft approach from the south, fly over the field, and head out on a wide circle preparatory to landing. Half a mile northwest of the field, at 300 feet, the plane went into a steep bank, whipped over, and nosed down in a power spin. The Army buried Sergeant Hymes with full military honors at Arlington Cemetery the following Monday, General Westover on Tuesday.[85]

Shortly before his death, Westover proposed a bigger, faster, more powerful attack bomber to support ground forces. The General Staff quickly approved. Douglas was already developing such a plane. The Air Corps ordered seventy-seven of them (A–20s) in June 1939 and procured several hundred more for combat in World War II.[86]

Transports

The Air Corps bought an assortment of transports, sometimes just one of a specific model or series. The Air Force needed aerial transportation for mobility. Maj. Hugh J. Knerr, Chief of the Field Service Section of the Materiel Division in 1932, said: "If an Air Force is tied to rail heads and its services of supply dependent upon motor transportation, its mobility is that of the flat car and truck."[87] The Air Service used Martin bombers and DH–4s in the early 1920s to carry passengers and move equipment. In the late 1920s the Air Corps bought cargo aircraft for assignment to the depots and to service squadrons at the various stations. Other planes, particularly bombers, performed as transports as the need arose. The Air Corps employed bombers as well as transports to shift men, equipment, and supplies for airmail operations. Several tactical units carried out experiments in moving by air, for example, Major Knerr's 2d Bombardment Group in 1928. The Air Corps created temporary transport units from time to time for special jobs, the maneuvers in 1931 being a case in point.

With the War Department stressing mobility for both land and air forces, the Air Corps figured an air force associated with a field army of a million men needed a wing of 210 cargo planes, each one capable of carrying three thousand pounds. The skeleton of this organization had to be formed in peacetime for swift expansion in an emergency. As Field Service Section Chief, Major Knerr in 1932 suggested a transport group with headquarters at Wright Field, Ohio, and a squadron at every air depot in the United States. On November 19, 1932, General Foulois approved the creation of a provisional group and four provisional squadrons. As section chief, Major Knerr became group commander; the operations officer at each depot commanded a squadron. With enlisted men as pilots, the squadrons hauled engines, parts, and other equipment to the airfields in their areas, returned items to the depots, and transferred materiel between depots. Then, too, they furnished transportation for maneuvers and for other work beyond the capacity of aircraft assigned to service squadrons.[88]

The squadrons started operations with whatever transports happened to be at hand. Soon, however, they received C–27s. In 1935 the Air Corps bought the Douglas DC–2, a bigger and faster transport than any previously purchased in quantity. The one acquired for testing became General Andrews' command plane. Receiving 20 more (18 C–33s and 2 C–34s) during 1936, the Air Corps distributed them among Air Force stations and Air Corps depots. Early in 1939 Douglas began delivering 35 C–39s (DC–2s with DC–3 tails). In 1939 the Air Corps also procured 13 small passenger planes (C–40s) built by Lockheed Aircraft Corporation.

At a time when the Air Corps did not get enough money to give each station a stock of all items that might be needed, air transportation afforded a fast means for distributing supplies in emergencies. The Materiel Division believed that even with more money for supplies, the Air Corps would find it more economical to keep station inventories low and rely on air transportation to deliver items for depot stocks when called for. Air transportation assured rapid transfer of items among the depots. And it saved time and money in preparing materiel for shipment. Items normally requiring costly wooden crates to go by rail traveled by air in inexpensive cardboard boxes. Engines did not require crating; the transport squadrons hauled them on the dollies used to move engines into and out of storage.[89]

Judging the provisional squadrons a success, the Air Corps, with War Department approval, gave them the status of Regular Army units in mid-1935. But not until May 1937 did it organize a Regular Army group, the 10th Transport Group commanded by Maj. Hugh A. Bivins with headquarters at Patterson Field, Ohio.[90] By that time the squadrons each contained around 50 enlisted pilots and 1 or 2 officers.[91] The group's planes were the new C–33s and the old C–27s, both models being among the 3 aircraft operated by each unit.

The Air Corps justified procurement of transports as "part of our tactical setup," necessary to move "personnel of tactical units at the same rate of speed as the planes with which the tactical units were equipped." When not engaged in tactical movements, transports were of "inestimable value" in distributing supplies.[92] Long-range plans for aerial transportation specified expansion of the 10th Transport Group by assigning more men and aircraft to the squadrons, formation of a GHQ Air Force group of three squadrons with one flight for each Air Force station, and activation of one squadron each for Panama and Hawaii.[93]

Brig. Gen. Augustine W. Robins, Materiel Division chief from January 1935 to January 1939, disliked the present practice and plans that spread the few transports among Air Force and Air Corps stations. Since the 10th Group had to provide transportation for GHQ Air Force and the Office of the Chief of Air Corps upon demand, he wanted all Army transports in the United States assigned to the 10th Group. He especially wanted the GHQ Air Force's C–33s. If he could get these planes, he would make them available on call from Andrews or Westover. At other times, would use them to their utmost capacity to haul supplies. General Westover rejected this scheme. GHQ Air Force kept its transports and called on the 10th Group when in need of additional transportation.[94]

Other Aircraft

Deliveries of new aircraft procured by the Air Corps included PT–13s (beginning in 1936), BT–8s (1936), and BT–9s (1936) for the Training Center, as well as BT–9s for Reserve training; O–46s (1936) and O–47s (1938) for observation units of both the Regular Army and the National Guard; and Y10A–8s (1937) and OA–9s (1938) for overseas garrisons. The Air Corps also ordered 180 basic combat planes (BC–1s), first delivered by North American Aviation in the spring of 1938. The Air Corps and Air Force used them for courier service and general flying. Navigators, bombers, and copilots, together with Air Corps officers detached for duty at West Point, New York, Fort Leavenworth, Kansas, and elsewhere, flew them to maintain proficiency. Assignment of BCs to the Training Center took place in mid-1939.[95]

Fields and Runways

Changes in the construction and performance of aircraft dictated changes in airfields. During the early 1930s, most military and civilian pilots in the United States still flew from grass fields or sod landing strips. In fact sod was better than a hard surface when a pilot relied on a tailskid for braking. But under heavy use and without good drainage and maintenance, a grass field grew soft, muddy, rutted, or dusty (as Langley Field did), depending on the season. Brakes and tailwheels on airplanes eliminated the need for soft ground for landing. Heavier and speedier planes had to have a longer, harder, smoother surface for taking off and landing.[96]

To improve fields for Army operations, the Air Corps decided in 1934 to build "all-weather landing mats or runways."[97] The program would naturally take considerable time to complete. General Andrews listed absence of runways among Air Force deficiencies in 1935. He needed hard-surfaced ones, usable in all kinds of weather, for his B–10s and B–12s. He would need them even more when he got the bigger, heavier planes he wanted.[98] At many stations (Mitchel and Langley Fields for instance), paving began with "warming-up aprons" in front of the hangars. Using methods similar to those in building highways, workmen at Scott Field laid 25-foot sections of concrete, 6 to 8 inches thick, for an apron 100 feet wide by 1,000 feet long. Men working around aircraft no longer waded through mud in wet weather; propellers on engines being tested or warmed up no longer threw up huge clouds of dust during a dry summer. When the Hawaiian Department built a new area at Wheeler Field for the 18th Pursuit Group, it planted grass on the field but put in a concrete apron.[99]

Concrete containing portland cement produced the best runways, landing mats, and aprons but it cost a lot. The Army tried less expensive materials. The cheapest was cinders compacted by rolling (used at Bellows Field, Hawaii), but such runways could not carry heavy weights and did not stand up well. Oiled earth, just a bit more expensive, afforded a more stable surface. The Army adopted this method to construct runways at Langley Field in 1937. The contractor dug up the ground to a depth of five inches, mixed in a bituminous compound, and compacted it with a roller. Having trouble stabilizing the earth and sealing the surface against water, he switched procedures several times and redid some of the work. Even so, an inch of asphaltic concrete had be be added a year later to seal the surface.[100]

At Luke Field, Hawaii, workmen spread a six-inch layer of crushed rock, rolled it, oiled it, and covered it with fine stone. Reports from Hawaii within a few months told of weeds poking through the "new and widely-advertised landing mat."[101] At Moffett Field, California, putting together a mat entailed hauling in rock, stabilizing it with emulsified asphalt, and finishing it with a coat of asphalt and sand.[102] An asphalt runway built in 1938 permitted flying at Nichols Field, Philippine Islands, regardless of the condition of the rest of the field.[103]

Seeking more room and higher ground, the Air Corps started work in 1932 on a new field just south of Bolling Field, between the Potomac River and the Anacostia hills. With living quarters ready in the spring of 1936, officers and enlisted men moved to "new" Bolling. Flying continued at "old" Bolling while hundreds of workers (furnished by the Work Projects Administration) graded and leveled ground for the new flying field. The Air Corps opened the new runways, made of asphalt-choked slag, in 1938.[104]

How well a runway stood up depended upon many things, including the kind of materials, thickness, quality of construction, drainage, use, and maintenance. Hamilton Field reported late in 1937 that heavy bombers were beating the landing mat to pieces, but the mat was being reinforced.[105] B–17s at Langley Field broke up the runways so as to require repaving with six inches of concrete in 1940. Inside a year after completion of runways at "new" Bolling, parts were no longer safe for heavy planes. There were many holes; one section had sunk about a foot. Engineers traced the trouble to poor drainage and the porosity of runway surfaces. They improved drainage, repaired the runways, and resurfaced them with two inches of asphaltic concrete.[106]

The Navy got "old" Bolling Field as part of a deal whereby it acquired Rockwell Field, California, and Luke Field, Hawaii. To replace Luke, the Army selected 2,500 acres near Fort Kamehameha for a new air station named in honor of Lt. Col. Horace M. Hickam, killed in an aircraft accident at Fort Crockett, Texas, on November 5, 1934. Brig. Gen. Barton K. Yount, commanding the 18th Composite Wing, moved his headquarters from Fort Shafter to Hickam Field in September 1937, long before completion of construction. The 31st Bombardment Squadron arrived from the United States the following February, but the 5th Bombardment Group and the air depot did not finish transferring from Luke to Hickam until early in 1939.[107]

The 91st Observation Squadron, after nearly fifteen years at Crissy Field, California, moved in June 1936 to Fort Lewis, Washington. Crissy's location at the Presidio, within the city of San Francisco (on the bay just inside the Golden Gate), was not an ideal place for a flying field. Fog frequently rendered flying hazardous and often halted operations. But construction of Golden Gate Bridge drove away the 91st Squadron.[108] Political opposition blocked attempts in the early 1930s to move the Technical School from Chanute Field, Illinois (a location the Air Corps regarded as unsatisfactory). Compromise enabled the school to establish a branch at Denver in 1937 under the command of Lt. Col. Junius W. Jones. The War Department named the place in honor of 1st Lt. Francis B. Lowry, a Denver resident killed in action in France. Two departments, photography and armament, arrived in February 1938.[109] Construction began in 1938 on a base for GHQ Air Force at Tacoma, Washington. The War Department named it for Col. William C. McChord, killed in an airplane accident on August 18, 1937. It did not receive its first combat units until June 1940.[110]

The Last of the Airships

The Air Corps' lighter-than-air branch came near extinction. In 1939 the Corps converted 2 of its 4 airship companies to balloon units, which the War Department insisted on having to keep alive balloon observation for artillery adjustment. That left 1 airship company (later designated squadron) at Langley Field for coastal defense and another airship squadron (a service unit) and a group headquarters at Scott Field, Illinois, for training. The lighter-than-air branch in early 1933 consisted of 20 officers and 650 enlisted men; 4 nonrigid airships, of which only 1, the new TC–13 (350,000 cubic feet), was suitable for tactical work; 2 observation balloons in service and 4 in storage; and about a dozen free balloons for training and racing. Goodyear held a contract for a new ship, the TC–14. But Congress prohibited procurement of any additional ships beginning in July 1933, and in 1935 forbade reconditioning ships that became unsafe for operation.[111] Having more airship pilots and balloon observers than jobs, the Air Corps kept the Balloon and Airship School closed. Men with lighter-than-air ratings worked in engineering, supply, procurement, and other jobs. Most of those who also held airplane pilot ratings (about one-third of the 73 airship pilots in 1933) served in the heavier-than-air branch. The other airship pilots went to flying school at San Antonio, Texas, 1 or 2 to a class, to get new ratings, after which they seldom returned to lighter-than-air work. Lt. Col. Arnold N. Krogstad, Chief of the Air Corps Personnel Division, observed in 1935 that many airship pilots had not seen a ship in 5 to 10 years, and that balloon squadrons got little official work (one of them just 20 hours in years).[112]

The Navy's loss of the dirigibles *Akron* in 1933 and *Macon* in 1935 affected not only naval aviation but the Army Air Corps as well. In the wake of the *Akron*, William Mitchell urged the government to stop wasting money on battleships and build a fleet of giant dirigibles. Testifying before a congressional committee investigating dirigible disasters, he asserted: "In a great war, if you had 50 airships properly equipped, that is, 20,000,000-foot [gas capacity] airships, you could sweep the seas with them."[113] They could remain aloft long periods, were "not particularly vulnerable to attack," could carry immense loads of weapons, "and sink anything on top of the water or under the water."[114]

General Foulois, called to testify about the loss of the *Roma*, showed no such enthusiasm for airships. Questioned by Representative John J. McSwain of South Carolina on the functions of airships in combat, he could not "visualize any particular tactical use to which they would be put that could not be performed better by an airplane."[115] Asked on another occasion why the Air Corps did not abandon airships, Foulois said when it secured long-range observation and bombing planes it would no longer need airships. Helicopters, and autogiros, then being investigated by the Air Corps, might substitute for balloons in artillery adjustment. Meantime, the Air Corps needed to keep abreast of developments and maintain a nucleus of officers trained in lighter-than-air operations.[116]

General Westover, an airship pilot and balloon observer before he became an airplane pilot, held a more favorable opinion of lighter-than-air aviation than Foulois. He saw the two branches supplementing each other rather than competing. Aircraft could not entirely replace observation balloons for artillery adjustment. Coastal defense required airships for observation and reconnaissance, especially in areas with poor visibility such as in the northeastern and northwestern areas of the United States. Airships could stay up much longer than airplanes, move slower, and hover. This made them valuable for spotting mines and submarines and for protecting convoys.[117]

The few officers still in lighter-than-air aviation, together with those who held balloon and airship ratings but now served in other jobs, did not abandon all hope of reviving the dying service. But they got little support for their plans and proposals.[118] General Andrews found the airship units of scant use to GHQ Air Force.[119] Since he did not want them, the War Department transferred the units to corps area control in July 1936. Lacking money for new ships and prohibited from reconstructing old ones, the War Department disbanded the airship organization in June 1937, transferred the people to other activities, and gave the airships to the Navy. The Air Corps' lighter-than-air branch now comprised Headquarters 21st Balloon (formerly Airship) Group and a balloon depot at Scott Field; the 1st and 2d Balloon Squadrons at Fort Sill, Oklahoma, and Fort Bragg, North Carolina; and a new balloon squadron, the 3d (formed from the 19th Airship Squadron) at Fort Lewis, Washington.[120]

With the 3d Balloon Squadron established at Fort Lewis, the Air Corps named the field there in honor of Capt. Hawthorne C. Gray. Under the command of Maj. Michael E. McHugo, the squadron began operations with a C–3, the standard captive observation balloon. Soon, however, the squadron received a C–6, a motorized balloon the Air Corps was buying for all of the squadrons. It used helium instead of hydrogen.[121] Like the C–3, it served as a captive balloon for observation. But when the time came to move, the squadron hauled it down, removed the basket, attached a small car with an 85-horsepower engine and two cockpits, and flew it to the next observation post. In this way the squadron avoided congested roads, trees, overhead wires, and other obstructions usually encountered in towing an inflated balloon at the end of a cable.[122]

When it appeared in 1938 that Scott Field would become headquarters for GHQ Air Force (a plan soon abandoned), the Air Corps, with War Department approval, moved the balloon depot and supplies to Duncan Field, Texas, sold the big airship hangar and a small balloon hangar for junk, and disposed of the balloon equipment at Scott Field.[123]

The Air Corps wanted to move Headquarters 21st Balloon Group to Brooks Field, Texas, to organize a balloon school. Ten years had passed since it had trained observers at Scott. It needed 300 more observers for balloon units in mobilization plans. It wanted to run 2 sessions a year at Brooks, with 5 officers of the Regular Army and 15 from the National Guard in each class. The War Department thought the proposal significant enough to merit further study. Two months later, on September 21, 1938, the small band of lighter-than-air enthusiasts in the Air Corps lost their leader (General Westover) in an aircraft crash. Instead of letting the Air Corps open a school, the War Department inactivated the 21st Balloon Group at Scott Field on June 1, 1939.[124] Thus, the lighter-than-air branch consisted of 3 balloon squadrons, plus 10 officers and 350 enlisted men, when war broke out in Europe.[125]

Barrage Balloons

The Air Corps received little support for barrage balloons for defense against hostile aircraft. The French, British, and Germans used barrage balloons during the 1914–1918 war. When the General staff studied antiaircraft defenses in the early 1920s, General Patrick advocated balloons as an effective and relatively inexpensive defense for the Panama Canal, the Capitol in Washington, the White House, and important dry docks, wharves, factories, and bridges. The War Department made the Air Service responsible for development and use of barrage balloons, but Maj. Gen. Frank W. Coe, Chief of Coast Artillery, objected. Since the Coast Artillery controlled antiaircraft batteries, it should also control the balloon barrage. The solution: Divide the responsibility, the Air Service developing equipment, the Coast Artillery controlling use.

Following a number of experiments and tests, the Engineering Division at McCook Field, Ohio, gave Goodyear a contract for 3 balloons, which the Air Corps and Coast Artillery began testing at Langley Field in 1927. The War Department considered barrage balloons in defense plans. The antiaircraft defenses of the Panama Canal, for instance, called for 74 balloons in 2 barrages, 1 at either end of the canal. With barrage balloons competing with other activities for money, interest waned. The Air Corps did not continue its project after it used up its balloons in tests and experiments. The balloon barrage dropped from plans for defense of the Panama Canal.[126] After the British and French undertook sizable barrage balloon programs in 1937, the Air Corps decided to use $5,000 of its money to buy 1 balloon for experiments. It sent the balloon to Fort Sill in March 1939 for the 1st Balloon Squadron to test and to Panama to obtain data on deterioration in storage.[127] But it took war in Europe to move the U.S. War Department to further development of barrage balloons.[128]

At the beginning of the Second World War on September 1, 1939, GHQ Air Force possessed merely 48 percent of the number of officers, 39 percent of the enlisted men, and 53 percent of the aircraft then authorized for peacetime. Two-thirds of the officers were inexperienced second lieutenants assigned to combat units. Further, the aircraft assigned included just 12 heavy bombers, these being the B–17s acquired in 1937 for service testing. Nonetheless, the Air Force of September 1, 1939, reflected a vast improvement of combat capabilities since establishment of the GHQ Air Force on March 1, 1935.

During the 4½-year period, the officer strength of GHQ Air Force grew by 82 percent, while the number of airplanes nearly doubled. Performance improved markedly in all types of aircraft. The B–10, standard for GHQ bombardment units in 1935, had been replaced by the B–18, extending the range of the bombing force and permitting bigger bombloads. Progress from P–26s to P–35s and P–36s meant faster climb, greater speed, higher ceiling, and longer range for pursuit operations. Replacement of A–12s with A–17s afforded attack units aircraft with greater speed, range, and bombload. The future promised even greater improvements with the delivery of B–17Bs, P–40s, and A–20s already on order. The next chapter explores other factors affecting the combat capabilities of GHQ Air Force.

Chapter XX

Crew Training

Short of money, people, and materiel, lacking adequate facilities, and unable to obtain the kind of airplanes he thought necessary for GHQ Air Force, General Andrews stressed training to attain the most air power from the resources available. This chapter tells of measures taken to enhance the proficiency of Air Force personnel—individually and as members of organized combat crews—in piloting, navigating, bombing, and shooting. It also treats factors that strengthened combat crew performance: development of new bombing and gunnery ranges, improvements in bomb handling, the introduction of the Norden bombsight, and experimentation with different bombing methods.

Pilots, Navigators, Bombers, and Gunners

For a strong, effective air force, General Andrews required men skilled in flying, navigating, bombing, and shooting. They needed lots of training, more than they had been getting. Many of the flyers in tactical units were young men not long out of flying school and without the experience to make them really useful to their units. Among the older more experienced men, many loved to fly and did so at every opportunity. Others, by inclination or the nature of their assignments, flew just enough to stay eligible for flying pay. Noting a shortage of money had curtailed flying, the Baker Board urged that pilots be given more time in the air. It further recommended, in the light of deficiencies disclosed by the airmail project, they get more training in cross-country, instrument, and night flying. And the board suggested the War Department provide a means for determining whether the Army's pilots actually qualified for flying duty.[1]

Setting standards for flying proficiency, the War Department required airplane pilots to fly at least 100 hours a year, including 35 hours in navigation, 10 on instruments, 10 at night, and 5 in formation. The 35 hours in navigation entailed one 500-mile flight with two intermediate stops and, for pilots in the United States, two 200-mile flights (that could be combined with the 500-mile flight) over Commerce Department airways, using airways communications and flying aids. Pilots in tactical squadrons surpassed War Department standards. For instance, the 2d Bombardment Group's training program for Fiscal Year 1936 called for 45 hours of navigation and 30 hours on instruments for each pilot. The men averaged 91.5 hours in the former and 58 in the latter.[2]

Taking seriously lessons of the airmail project and recommendations of the Baker Board, the Air Corps paid more attention to training for blind flying. As circumstances permitted, it procured more and better instruments and equipment for use on airways and for flying at night and in bad weather. Even so, a person might suffer some bad moments before mastering the technique of flying blind, as did a cadet with the unlikely name of Josephus Jones. Failing to stay on a radio beam, he got lost. Unable to see anything outside the cockpit, and forgetting what the textbook said, he cut wide arcs as he tried desperately to find the course. Suddenly, the ship fell off on its right wing, went into a spin, and Jones went over the side. Landing on the hangar floor, he timidly approached the instructor: "Please, Lieutenant, couldn't we start the problem over again; I've never been in a Link Trainer before."[3]

Edwin A. Link, Jr., invented his instrument trainer in 1928, but the Air Corps showed little interest until its pilots ran into trouble trying to fly blind while carrying mail. Then the Air Corps quickly ordered 6. The next year it bought 15 of an improved model, and 2 years later it ordered 24.[4]

Link's trainer resembled a little airplane. Sitting on a fixed base, it could assume the positions of an aircraft in flight. The hooded cockpit contained stick and rudder, flight instruments, and earphones and microphone to receive signals and communicate with the ground. Since the trainer was unstable, the pilot needed to fly it constantly. A person experienced in using instruments had no trouble controlling the machine, but a novice might, as Josephus Jones discovered.[5] Cadets became acquainted with the trainer in flying school. Rated pilots used it to increase and maintain skill in blind flying. The Air Corps furnished one or more of the devices for each of the principal flying fields in the United States and overseas departments. With one installed in Hangar 4 at Hamilton Field, pilots of the 7th Bombardment

Link Trainer

Group took a 15-hour course. Near its end, each flew the Link on a long cross-country trip confronting all kinds of conditions, from bumpy air and ice to split beams and false cones of silence.[6] Each airplane pilot proved his proficiency in instrument flying once a year by taking a test in either an aircraft or a Link trainer. General Westover qualified when checked by Maj. William E. Farthing in 1936.[7]

General Andrews, who set high standards for pilots of GHQ Air Force, found that newly rated men fresh out of flying school had not enough experience with multiengine planes to go straight into bombardment squadrons. He therefore sent them to pursuit and attack units, and for a year confined their flying to the vicinity of the station, unless in formation under direct control of their flight or element commander. He made 750 hours of flying as a rated pilot prerequisite for assignment to a bombardment unit. There the pilot underwent air and ground instruction on the type of plane his unit used. If it had B–10s, and he passed the test, he qualified as a pilot. If it flew B–17s or B–18s, he became a copilot for 50 hours, which qualified him to pilot. After serving 7 years and flying 2,000 hours as a rated airplane pilot, he qualified as airplane commander.[8]

In 1937 the Air Corps provided a new rating (military airplane pilot) for men who held the airplane pilot rating for 12 years and logged 2,000 hours.[9] Those immediately eligible included some captains (Earle E. Partridge, for one) as well as older members of the service with higher rank. Two years later the Air Corps revised its ratings and began reclassifying its flyers. Graduation from the Air Corps Advanced Flying School made a man a pilot. After 10 years service with the Air Corps or in U.S. naval aviation, and with an Army record of 1,800 hours as pilot, he became a senior pilot. With 15 years service and 3,000 hours of military flying, or 20 years and 2,000 hours (pilot and copilot counted at 100 percent, other flying at 50), he advanced to command pilot.[10]

The commanding general of GHQ Air Force required pilots of bombing planes to learn navigation. Flyers on long trips to sea or over land obscured by clouds or darkness needed to be skillful in calculating position and course by dead reckoning and celestial observation. Schools created at Langley Field and Rockwell Field in 1933 taught navigation (or avigation as the Air Corps then preferred to call it)[11] and instrument flying until closed during the airmail project. The school at Rockwell reopened in October 1934 but shut down again the following June when General Andrews ordered each of his bombardment groups to train navigators.[12]

Intent upon qualifying all of his bombardment and reconnaissance pilots as navigators, Andrews told group commanders to give 4 men at a time 12 weeks of instruction under a graduate of the school at Rockwell. The course, stressing dead reckoning and celestial navigation, included 180 hours of study on the ground and 66 hours of navigation in the air. Capt. Archibald Y. Smith ran the school for the 2d Bombardment Group at Langley Field with 1st Lt. Joseph A. Miller and 2 enlisted men (a clerk and a draftsman) to assist him. Four officers made up the first class, but Smith soon dropped 2 because he lacked equipment. Smith kept ahead of schedule with the ground courses, but equipment shortages, mechanical trouble, bad weather, and other duties put the students behind in their flying. First Lieutenant David H. Kennedy of the 49th Squadron and William O. Senter of the 20th completed the course after Smith got the time extended from the end of October to the 1st of December. For their graduation test Kennedy and Senter flew to Jacksonville and back by dead reckoning and celestial navigation. Four officers, including Capts. Richard E. Nugent and Robert F. Travis, comprised the new class beginning January 6, 1936.[13] The bombardment groups at Mitchel Field, March Field, and Hamilton Field set up similar programs. So did the overseas departments, with 1st Lts. John W. Egan and Curtis E. LeMay running the school at Luke Field, Hawaii.[14] Later Brig. Gen. Henry C. Pratt, Lt. Col. Robert Olds, Lieutenant Egan (who had become Navigation Officer, 2d Bombardment Group), and others suggested the Air Force establish a central school to train navigators, but General Andrews rejected the idea.[15]

Besides becoming navigators, pilots assigned to bombardment units also had to be bombers (or bombardiers, to use a term gradually coming into general use in the late 1930s).[16] To appreciate the problem of sighting, a pilot needed to be a bombardier; to understand the pilot's problem during bombing, the bombardier needed to be a pilot. Bombing accuracy hinged on coordination between the two.[17] Pilots of bombing crews likewise learned to shoot. On Keystone bombers, one man often acted as copilot, navigator, bombardier, and front gunner. On B–10s and B–12s, the navigator-bombardier also manned a gun. B–17s carried both navigator and bombardier, but one man handled both jobs on the B–18s. To be eligible as a B–18 commander, a 19th Bombardment Group member was required to qualify as a B–18 pilot, a celestial navigator, an expert bombardier, and an expert aerial gunner. Short of men with such qualifications, the group's commander, Lt. Col. Harvey S. Burwell, permitted squadron commanders to select acting aircraft commanders.

Commanders formed their flyers into permanent crews assigned to specific airplanes. The aircraft commander supervised the crew's training, measuring efficiency by ability to fly, bomb, and shoot. A combat crew of the 19th Group, for example, became fully trained when every officer qualified as pilot or copilot, celestial navigator, expert bombardier, expert gunner, and camera gun operator. At the same time, all enlisted men—flight engineers, radio operators, and armorers—had not only to master their individual specialties, but become expert gunners and camera gun operators.[18]

Ranges

Men in tactical units accomplished part of their training with bombardment trainers, camera obscura, and camera guns. Honing real skill required lots of practice dropping real bombs and shooting live ammunition. The 8th Pursuit Group at Langley Field used 303,082 rounds of .30-caliber and 11,167 of .50-caliber ammunition in Fiscal Year 1936 firing at aerial and ground targets. That same year, the group's attached 37th Attack Squadron expended 75,000 rounds of ammunition and 2,063 bombs in training. Commissioned pilots and enlisted gunners of the 8th Attack Squadron at Barksdale Field, Louisiana, spent 2 weeks in the spring of 1936 practicing aerial gunnery and firing "for the record" to earn their gunnery qualifications. During that time, the 8th Squadron spent 60,244 rounds of ammunition and wore out 8 gun barrels. Of the 51 officers and men who fired, 34 qualified, 26 rated "Expert."[19]

The 19th Bombardment Group dropped nearly 8,000 bombs during one 11-month period. It used some for demonstrations and maneuvers, and a few for tests, but most went for what men called "440–40." The reference was to the training regulation of that number, which required bombardiers to drop 5 bombs from each of 4 altitudes (4,000, 7,000, 10,000, and 15,000 feet), set forth the method for scoring, and prescribed the degree of accuracy needed to qualify. Colonel Burwell, the 19th's Commander, went all out to increase the number of expert crews so his group could carry out its mission of getting to its target and bombing it accurately.[20]

Having a good bombing range within reasonable distance of its home station, the 19th Group could practice whenever it wanted to. On the other hand, the 8th Attack Squadron had to travel from Barksdale Field to Galveston, Texas, or Miami to fire machineguns at aerial targets. Many other units also lacked proper bombing and gunnery facilities. GHQ Air Force and the Air Corps tried to improve existing ranges and develop new ones, but the work proceeded without benefit of uniform criteria or specifications. The initiative usually came from the local commander. He sought a range within reasonable distance, available whenever he wanted to use it, a place large enough and so situated that bombs and machineguns would not endanger life and property, an area with light air traffic, where the neighbors would not complain, and where land could be obtained at a reasonable price. The 2d Bombardment Group, for example, figured that for practice bombs the target should be at least 3,000 yards from any habitation and in a sparsely settled area so the bombing run did not pass over towns or villages. For demolition bombs the target needed to be even farther away, at least 6,000 yards.[21] Such places were not always easy to find, but Lt. Col. Henry H. Arnold discovered one in the Mojave Desert, California, and was developing it when his wing became part of GHQ Air Force.

March Field had no bombing and gunnery facilities when Arnold became commander in 1931. His men flew to San Diego to practice, using an area of Rockwell Field to drop dummy and small demolition bombs, and going out over the water to release larger bombs and fire their guns. Arnold found this unsatisfactory because too many other planes occupied the same airspace, and clouds and fogs often interfered with practice. Then, too, fishermen complained that bombs killed fish. Seeking a better place, Arnold selected Muroc Dry Lake, about 80 miles north of March Field. He described the lake bed, an area of about 6 square miles, as "level as a billiard table." Adjacent lay 75 square miles of wasteland. The federal government owned about half of the tract, but 34 square miles belonged to the railroads and 4 were school lands. Arnold believed the government could get title in exchange for other public land. The dry lake made an ideal landing field, and the tract supplied sufficient space for the whole wing to bomb and shoot at the same time without endangering anyone. President Roosevelt furnished public land for a range in February 1934. Arnold began improving and using it while the government worked to acquire title to additional land.[22]

The 19th Bombardment and 17th Attack Groups at March Field used Muroc regularly for bombing and gunnery. The other group in General Arnold's wing (the 7th Bombardment) was at Hamilton Field, too far away to use Muroc very often. The Air Corps selected a site at Honey Lake in

Seventeenth Attack Group bombing on Muroc Dry Lake, Calif.

northeast California, but Arnold thought this site (about 200 miles from Hamilton Field) too distant, and General Andrews agreed. Lt. Col. Clarence L. Tinker, 7th Group Commander, also agreed, but knew no place nearer Hamilton Field where his group could drop live bombs and fire at aerial targets. The state of California opposed having a range at Honey Lake. Carl B. Sturzenacker, Chief of the Division of State Lands, said "the troops have to have some place to play," but they did not need one of the richest grazing districts in the state. He suggested they find another spot to throw bombs and shoot. The War Department persisted for a while but eventually dropped the project.[23]

Hamilton Field afforded suitable space for dropping practice bombs, and Colonel Tinker marked a danger zone in San Pablo Bay adjacent to the field so gunners could fire at targets on the seawall. He also put a target on the bay for them to bomb, but fog often prevented its use, especially in winter. After the Honey Lake project fell through, the 7th Group stationed a detachment at Mather Field (then inactive) and conducted bombing and gunnery there. To shoot at aerial targets, however, the group flew over the Pacific between Bolinas Point and Bodega Head. Every aircraft carried an inflatable life raft, and crewmembers wore life vests under their parachutes. As an added precaution, Tinker sent four enlisted men (including a radio operator and a cook) with a boat to a rescue post on Point Reyes whenever his group engaged in aerial firing.[24]

Plum Tree and Mulberry Islands plus areas in Chesapeake Bay and the Atlantic gave units at Langley Field facilities for all kinds of gunnery and bombing.[25] For many years, units of Mitchel Field went to Aberdeen, Maryland, to do their bombing and gunnery. They further used facilities at Langley Field and Chapman Field, Florida. In 1937 Lt. Col. Carl W. Connell, commanding the 9th Bombardment Group, secured permission to drop practice bombs at Camp Upton, an inactive post in the middle of Long Island. The group also obtained a range for aerial gunnery in the Atlantic off Fire Island between Babylon and Eastport. Firing out to sea, gunners kept a sharp watch for ships and stopped shooting when one appeared.[26]

The 1st Pursuit Group continued to use Camp Skeel, Michigan, along with an area of Lake Huron for aerial gunnery.[27] In the winter the men practiced in the south, at Chapman, Eglin, or Barksdale Fields. Barksdale, an area of over 22,000 acres (including lakes, bayous, and woods) afforded suitable space for the 3d Wing to practice bombing and firing on ground targets from the air but not for gunnery against towed targets.[28] Men from Barksdale sometimes practiced gunnery at Chapman,[29] but more often went to Galveston, Texas. In 1937, for instance, pilots of the 20th Pursuit Group, flying P–26s from the airfield at Fort Crockett, Texas, fired .30- and .50-caliber ammunition at a target towed by a P–26. For night work they flew from the municipal airport, larger and better lighted than the field at the fort.[30] The Air Corps tried to procure land for a bombing and gunnery range near Rockport, Port Lavaca, Matagorda Island, or elsewhere along the Gulf of Mexico between Galveston and Corpus Christi. It was to be used by the Training Center as well as by units from Barksdale Field. The effort failed due in part to the high prices people wanted for their land. So to practice gunnery over the gulf, units at Barksdale proceeded to go to Fort Crockett or Galveston Municipal Airport.[31]

The range known as Chapman Field, situated a few miles south of Miami, dated from World War I when the government bought 850 acres for an aerial gunnery school. Declaring the field surplus, the War Department offered it for sale in 1921, but after bids came in a question arose concerning the government's title to part of the tract. While the government tried to clear the title, the Air Corps reopened the field late in 1931 for bombing and gunnery. Units from Mitchel, Langley, Selfridge, Scott, and Barksdale practiced at Chapman. The 2d and 3d Wings used the area for exercises, and General Andrews established headquarters there for Air Force maneuvers in 1935. However, Chapman became less important after the Air Corps developed a new range at Valparaiso, Florida.[32]

The Air Corps Tactical School became interested in Valparaiso (about 130 miles from Montgomery), after failing to obtain land nearer Maxwell Field for bombing and gunnery. The Valparaiso Realty Company in October 1934 offered to give the government 1,460 acres of land, including an airdrome. The tract bordered on Choctawhatchee National Forest and Choctawhatchee Bay, the latter separated from the Gulf of Mexico by a narrow strip of land. Col. John F. Curry, Tactical School Commandant, urged the War Department to accept. While the government worked out details, 45 officers, 110 enlisted men, and 36 airplanes from Maxwell occupied Valparaiso's municipal airport on the weekend of March 28–29, 1935, to practice aerial gunnery over the gulf. The next weekend, 30 students and 28 enlisted men from Maxwell again used the airport while practicing gunnery.

Pending acceptance of the realty company's offer, the government leased the land. The Tactical School placed a detachment there to develop and run the place. The realty company helped by putting up small buildings and preparing a recreation area. Permanent construction by mid-1936 including barracks, mess hall, latrine, headquarters building, storage buildings, radio shack, and a boathouse for two rescue craft. Enlisted men lived in tents. Men from Maxwell Field went to Valparaiso frequently to practice gunnery and bombing for a few days or a week at a time. The 15th Observation Squadron came from Scott Field for gunnery, and soon did units from Selfridge and Langley. After accepting donation of the land in March 1937, the War Department named the place Eglin Field in honor of Lt. Col. Frederick I. Eglin, who died in an aircraft accident near Fort McClellan, Alabama, on January 1, 1937.[33]

Overseas units at Albrook Field, Panama Canal Zone, practiced bombing and aerial gunnery on Panama Bay and fired at ground targets on a range the Panama Canal Department developed at Rio Hato.[34] The 18th Pursuit Group at Wheeler Field in Hawaii shot at ground targets on the combat range at Schofield Barracks, practiced aerial gunnery off the north shore of Oahu, and went to Bellows Field at Waimanalo for field exercises and gunnery.[35]

The 5th Bombardment Group at Luke Field found no place in the islands or along the many miles of beaches that it could call its own bombing target. It therefore constantly experimented to find suitable targets to place at sea.[36] Like the 30th Bombardment Squadron in the United States and the 74th Attack Squadron in Panama, the group tried aluminum slicks. A plane dropped a glass bottle containing aluminum powder which spread on the surface of the water for bombers and gunners to aim at. The 5th Group thought the material rather expensive. Flyers could usually see it easily from 7,000 or 8,000 feet but sometimes had trouble spotting it when approaching the sun. The slick disappeared swiftly, lasting about a hour in a calm sea, less in rough water. But aluminum powder often enabled the bombers and gunners to practice when they lacked other targets.[37]

Bombs

The Air Force secured guns, ammunition, and bombs from the Ordnance Department. Ordnance service companies of the corps areas stationed men at airfields to prepare, maintain, and repair ordnance and deliver it to combat units. The squadrons put munitions aboard their planes and cleaned and maintained their own guns. After transfer of Air Force stations to General Andrews' control in 1936, ordnance personnel at the fields became part of the 10th Ordnance Service Company. The company was commanded by Lt. Col. Russell L. Maxwell, the ordnance officer on Andrews' staff. It consisted of six platoons, one at each Air Force station.[38]

Bombardment squadrons used 100-, 300-, 1,100-, and 2,000-pound bombs, which the Ordnance Department shipped to the airfields. Demolition bombs arrived filled with TNT, but practice bombs (mostly old demolition bombs) came empty. The local ordnance platoon as a rule put a little black powder in these practice bombs for a "spotting charge" and filled the steel cases with sand to the weight of service bombs. The platoon also attached fins, installed suspension lugs, and inserted fuzes. Attack and pursuit units used small bombs of 17, 30, or 100 pounds designed for fragmentation, demolition, smokescreens, mustard gas, or other chemicals. The Air Force used Browning machineguns, mostly .30-caliber but some .50-caliber on the newer pursuit ships. Receiving cartridges in wooden boxes, ordnance platoons loaded the ammunition into links on belts for delivery to the squadrons.

Capt. Charles B. De Shields, armament officer of the 59th Service Squadron at Langley Field, moved bombs and ammunition to Richmond for a mobility test with the 96th Bombardment Squadron in July 1935. His stock for the exercise consisted of nine 2,000-pound bombs, eighteen 1,100-pounders, and twenty-seven 600-pounders, plus seven 140-pound boxes of machinegun ammunition. He and his section (4 noncommissioned officers and 20 privates) hauled the sand-filled bombs to Richard E. Byrd Flying Field by truck, set up a bomb dump, and moved the bombs to the flight line as needed by the 96th Squadron. The squadron loaded 9 planes for 14 missions from Byrd Field. Since it was testing mobility—not practicing bombing—the bombs were removed from the planes before takeoff. De Shields and his men took them back to the dump to use another day. Thus the 59th's armament section handled each bomb a number of times, and at the end hauled them back to Langley Field.

The 96th used slings and hoists to lift bombs onto aircraft. But De Shields possessed no special tools, equipment, or mechanical devices of any kind. He relied on manpower to lift bombs into and out of 1½- and 2½-ton trucks to move them from place to place. To raise a bomb from the ground to the truckbed, his men leaned two channel irons (9 feet long and 6 inches wide) against the end of the truck and slid the bomb up. Some men pulled on a ¾-inch manila rope from the truckbed, others pushed from the ground. The trucks had no bracing or other means to keep the bombs from the rolling and shifting that often damaged fins. To unload the truck, the men used the rope to let the bombs slide down the channel iron. They moved the bomb into position on the flight line or in the dump by fastening a rope to it and dragging it along the ground. After the exercise, De Shields asked for trailers to carry bombs, and a truck with a boom and hoist.[39]

Capt. Edward P. Mechling, Ordnance Department, was somewhat better prepared for handling bombs during an exercise at Langley Field in December 1938. A squadron of 9 B–17s and 2 squadrons of B–18s, commanded by Colonel Connell from Mitchel Field, flew 2 missions with demolition bombs to targets on Plum Tree Island. Brig. Gen. Arnold N. Krogstad, 2d Wing Commander, decided to use the entire bomb allotment then available. The Ordnance Department shipped more than 300,000 pounds of bombs of various sizes (100 to 2,000 pounds) to Langley Field. Captain Mechling and his men removed the bombs from the railroad cars, took them to storage areas, prepared them for use, and delivered them to the planes. Preparations kept 20 men busy for a month. Just before the exercise, 1 officer and 16 men arrived from Mitchel and Selfridge Fields to help.[40]

Several of the men under Mechling's command were graduates of the School of Aviation Ordnance which Colonel Maxwell had established at Langley Field in 1937. This school, directed by Mechling, gave students from all over the Air Force training in supply and maintenance of ammunition, assembly and delivery of ordnance material to combat units, and inspection and repair of armament. Each class enrolled about a dozen men and lasted about 2 months.[41] Throughout the exercise, Captain Mechling's men had 34 trailers, furnished by the Ordnance Department and modified at Langley Field. They needed them all to haul the 378 bombs scheduled for 1 mission. "Prime movers" (trucks and a few light tractors) pulled trailers 2 at a time. A tracklaying tractor with a crane proved a good, safe means for lifting bombs on and off trailers and moving them at storage sites. However, having but 1 crane, the men still did much of the lifting by hand. They were duly grateful for the low beds on some trailers that made for easier loading and unloading by hand.[42]

Sights

A bombing crew needed a good sight to achieve any reasonable degree of success in hitting a ship or a factory. The Air Corps eventually found such a sight in the Navy Department. In the early 1920s the Army Air Service used various models of a World War I sight designed by Lt. Comdr. Harry E. Wimperis of the Royal Naval Air Service. General Mitchell put some of these (Mark IIIs) on planes to bomb the ships in 1921. The Air Corps, after much experimentation, adapted a new sight (the D–4) in 1926. When in perfect repair and under ideal conditions, a D–4 gave good results in bombing from heights up to 8,000 feet, but at higher altitudes errors often became excessive. In the early 1930s the Air Corps acquired C–4s. Large and cumbersome, the C–4 did better at high altitude but had no advantage over the D–4 under 8,000 feet. Luckily, something better was on the way.[43]

In October 1931, Army officers witnessed tests the Navy made against the USS *Pittsburgh* with a bombsight invented by Carl L. Norden. This sight solved the problem of ascertaining the angle at which a bomb should be released to achieve a direct hit on the target. Gyroscopic stabilization kept the telescopic sight on the target despite the plane's roll, pitch, or turn. A clockwork-like mechanism synchronized data inserted by the bombardier on ballistics, altitude, speed, and drift, and calculated the precise moment for bomb release. Other persons used gyroscopic stabilization and mechanical synchronization for bombsights, but no one had produced an instrument of manageable size and weight which did the job so well as Norden's.

Impressed by the tests on the *Pittsburgh*, General Foulois wanted Norden sights for the Air Corps. The Navy, however, had classified the device Secret and would not permit the Army to talk to Norden about a contract. Even so, Foulois arranged for the Navy to buy twenty-three for the Air Corps. When Norden delivered the first one at the beginning of April 1933, the Navy invited the Air Corps to send two representatives to its proving ground at Dahlgren, Virginia, to witness acceptance tests. General Foulois sent Maj. James L. Hatcher, Ordnance Department, and 1st Lt. Clarence S. Thorpe, Air Corps, from Wright Field, and afterwards the two men took the sight there for further testing. The results proved so favorable that on April 17, 1933, the Air Corps asked the Navy to buy seventy-eight more.[44]

To prevent disclosure of secret information, General Foulois kept to a minimum the people involved until he could issue detailed instructions to govern shipment, storage, and use. The Materiel Division established a course at Wright Field to instruct a few officers in care, maintenance, and operation so eventually they could go out and teach others. The first course, in July 1933, showed the need for additional ballistic data for proper calibration of sights. After the necessary data had been collected and the course revised, another class was scheduled for February 1934, but the airmail project forced cancellation.[45]

Norden delivered some sights in February 1934, more in April, and the last of the first lot in May. The Navy secretly notified the Air Corps of each delivery, and the Materiel Division sent a plane to Dahlgren to bring the sights to Wright Field to be locked in a vault. Eventually the property officer picked them up on his account with a notation, "found on Post." The Materiel Division could not distribute the sights to depots and airfields because it wanted instructions. Reserving allocation to himself, Foulois had not yet made his desires known.[46] Meanwhile, the second class at Wright Field finished the course on September 1, 1934. On October 3, General Foulois instructed commanders at the stations to take special steps for storing and safeguarding the new sights. He said distribution would begin as soon as commanders completed preparations.[47]

While the Materiel Division accumulated a stock of Norden sights, the Air Corps ran low on D–4 and C–4 ones. The 7th Bombardment Group, for instance, could not get sights for its new B–10s. Lt. Col. Henry H. Arnold, the wing commander, mentioned the problem in talking with Maj. Gen. Hugh A. Drum in Washington early in November 1934. General Foulois soon received a note saying the Deputy Chief of Staff wanted to know the reason for the delay and when sights would be furnished. Foulois explained that the Air Corps was buying a new sight under conditions of the strictest secrecy. Station and unit commanders required thorough indoctrination in his policy concerning the sight. They needed men properly trained for care and operation and they had to provide adequate safeguards for storage and use. "Then and only then," Foulois said, "will they be intrusted with the new sights."[48]

A little later, after assuming command of the Hawaiian Department, General Drum observed the Navy bombing very accurately at medium altitudes using Mark XV (Norden) sights. He borrowed several from the Navy for Brig. Gen. Delos C. Emmons, 18th Wing Commander, to try with the department's Keystone bombers. After brief but intensive training, Air Corps officers secured excellent results. Judging the Mark XV far superior to the C and D sights then available in Hawaii, General Drum wanted sufficient Mark XVs, or sights equally good, to equip all of his bombers without delay.[49] Under instructions from General Westover, the Materiel Division chief had just allocated twelve M–1 sights (the Air Corps' designation for the Navy's Mark XV) for shipment to Hawaii during 1935, with enough more in 1936 to equip every bomber in the department.[50]

In May, General Westover and Rear Adm. Harold R. Stark cleared the way for distribution of Norden sights to Army stations. Stark, Chief of the Navy's Bureau of Ordnance, approved Army regulations relating to security of M–1 sights and eased restrictions by reclassifying the Norden sights Confidential.[51] Westover told Brig. Gen. Augustine W. Robins, chief of the Materiel Division, that since the Navy inspected and calibrated new sights at Dahlgren, a second inspection need not be made at Wright Field. Approving distribution of the sights by Fairfield Air Depot, Westover directed Robins to turn the sights over to the supply officer as soon as the depot could provide the proper security.[52]

When Arnold received M–1s for the 1st Wing, he found they surpassed other sights but had some objectionable features.[53] For one thing, Norden designed the sight to meet Navy requirements for use in planes of moderate speed, bombing from medium altitude. The Army needed a sight for use at fast as well as moderate speed, and at high and low as well as medium altitude. Then, too, production of M–1 sights lagged behind schedule, the Air Corps receiving new bombing planes faster than it received sights. Air Corps officers could not discuss such matters with Norden without first securing permission from the Chief of the Air Corps and having a Navy representative (usually a junior officer) present.[54] Nonetheless, the Air Corps over a period of time obtained modifications and quicker delivery.

On a bombing mission with an M–1, the pilot flew to the place selected to begin the bombing run (initial point), leveled the plane, and put it on the proper heading to take it to the target. The straight-and-level flight required for sighting rendered the plane vulnerable to antiaircraft fire, so the pilot-bombardier team kept the bombing run as short as possible. The bombardier, in command of the aircraft during the run, lined up the target in the M–1's telescope. Using data from his instruments, tables, and calculators, he set the M–1 for the ballistics of the bomb to be dropped, altitude, air and ground speeds of the plane, and drift caused by crosswind, making subsequent corrections as necessary. The M–1 synchronized all this information bearing on the path the bomb would take after it left the plane.

By means of a pilot direction indicator (PDI), the M–1 automatically signaled the pilot which way to turn to stay on course for the target. In bombing systems lacking a PDI, the bombardier called instructions to the pilot by interphone: "Steady on course . . . turn right . . . steady . . . stop . . . right again . . . stop . . . turn left . . . stop."[55] With a PDI, a needle on a dial in the cockpit told the pilot which way and how much to turn. In communication with the pilot by interphone, the bombardier might call for a zig-zag course to avoid antiaircraft fire, but the pilot could not make any large or sudden turn or change altitude. As the plane neared the target, the bombardier lifted the trigger on the sight and held it up.

When a bombardier released bombs by mechanical methods, some time elapsed between his perception of the time to drop the bombs and pulling of the trigger, and again between the pulling of the trigger and when the bombs came out of the racks. This was not long, but long enough to make quite a difference where the bombs hit. Development of an electrically operated release mechanism had reduced this delay. Eliminating the time it took the bombardier to react cut it more. The M–1 automatically determined the precise moment of bomb release and instantaneously pulled the trigger operating the electrical release. With bombs away, the bombardier turned off the sight and the PDI. The pilot took charge and headed for home (or perhaps for another run down of the range to get more practice). The bombardier covered the sight so after the plane landed no one outside could see it.[56]

Norden improved the performance of his bombsight by developing an automatic pilot to use with it. This let the bombardier, using servo controls, to direct the course of the plane during the bombing run, changing direction as need be to keep the sight on target. Norden's automatic pilot, which the Air Corps called "automatic flight control equipment,"[57] erased the time lag inherent in signaling course corrections to a human pilot. With the sight disengaged, Norden's pilot served the same purpose as the automatic pilots the Air Corps obtained on its B–17s and B–18s and bought for installation on B–10Bs. However, the Air Corps' automatic pilots could not be coupled to the Norden sight to permit the bombardier to steer the plane.

The Air Corps learned of Norden's new device in 1935 but could not get it prior to Navy testing and approval for procurement. As the tests neared their end in mid-1937, GHQ Air Force asked the Air Corps to acquire sufficient equipment for its B–17s and for a squadron of B–18s for service tests. Col. Hugh J. Knerr, Chief of Staff, GHQ Air Force, felt sure all bombardment units would eventually secure the new device.[58] Obtaining one set of equipment, the Materiel Division put it on a B–18, tested it, and turned the plane over to the 2d Bombardment Group on May 27, 1938. After seven weeks of tests, Lt. Col. Robert Olds, the group commander, recommended: Expedite procurement and installation of automatic flight control equipment on all bombing and reconnaissance aircraft, giving B–17s priority.[59] The Air Corps had begun work months earlier to procure Norden's equipment to replace automatic pilots already on the Army's B–15s and B–18s and for Douglas to put on new B–18s. Nevertheless, equipment for the change did not begin to become available until early 1939.[60]

Methods

Combat crews learned to work in flights, squadrons, groups, and wings. Most unit commanders (General Arnold being a notable exception) were graduates of the Air Corps Tactical School. In some cases, they had served on the faculty and helped to write the manual that formed a foundation for unit operations. The pursuit text covered objectives, equipment, formations, tactics, and employment.[61] Others dealt with bombardment and attack in similar fashion. But books did not tell all. Changing conditions confronted commanders with new problems. Units needed to work out many things for themselves. Training still entailed much experimentation with equipment,

dropped from 15,000 feet, they put 32 in an area representing an airdrome. They did better on the next mission when they scored 77 hits with 101 bombs dropped on the area occupied by a munitions factory and bomb dumps.[65]

Beginning bombing from high altitude (above 10,000 feet) in the summer of 1936, the 9th Bombardment Squadron made the important discovery that "any errors in sighting or releasing were considerably magnified by the time the bombs reached the ground."[66] The 19th Bombardment Group attained poor results 2 years later when it bombed a target the size of an aircraft carrier from 20,000 feet. Reporting errors of 400 to 1,600 feet, the group explained the bombs had been released by students or average bombardiers. Also, the windows on some of the B–18s frosted so neither pilot nor bombardier could see the target or horizon.[67]

While commanding the 1st Wing at March Field in 1936, General Emmons assigned a number of experiments to his units. The Tactical School's bombardment manual discussed various formations (javelin-up and javelin-down, stagger, wedge, column, dispersed-column) and ways to release bombs (singly, by salvo, and in trail).[68] Emmons wanted to determine the bomb patterns produced under different conditions so the proper ones could be selected for each bombing mission. The 9th Bombardment Squadron dropped inert bombs from several altitudes, flying different formations, altering distances and intervals between planes, and employing various combinations of altitude, speed, and wind direction. After a formation completed its drop, men went out to record the location of each bomb and measure the distance from the target.[69]

Lt. Col. George C. Stratemeyer seemed never to want for ideas for the 7th Bombardment Group, which he commanded from 1936 to 1938. One Saturday he assigned separate missions to his three squadrons, followed by assembly of the group at a designated point 80 minutes after taking off. The following Saturday he sent the planes out individually with instructions to form squadrons at three widely separated points, assemble in group formation, fly to the assigned target, take bombing positions (individual planes echeloned to the right from 8,000 to 10,000 feet with 500 yards between planes), attack, rally, and return to Hamilton Field. Nor was the colonel at a loss for a plan for the subsequent Saturday, or the next, although on the latter occasion at least he received some help from his operations officer, 1st Lt. Frederick L. Anderson, Jr.[70]

Units practiced gunnery and bombing at night as well as by day. The 35th Pursuit Squadron, for example, fired tracer ammunition at ground targets illumined by flares.[71] The 18th Pursuit Group fired live ammunition at a floating target at night.[72] An A–17 illumined by Coast Artillery searchlights served as a target for the 20th Pursuit Group, attacking with empty guns.[73] Experimenting with night bombing, the 7th Bombardment group used a bonfire to represent incendiary bombs dropped by an observation plane to provide an aiming point for bombardiers.[74] The 19th Bombardment Group dug a trench in the shape of a cross, filled it with oil, and set it on fire for a target. However, pilots could not see it very well from a distance.[75]

In California, Capt. Kenneth N. Walker, 9th Bombardment Squadron Commander, worked out a way to take a formation through fog from one clear area to another. Approaching a fog bank, he announced by radio the magnetic heading and speed to be flown and assigned different altitudes to various flights. Circling to the left, the planes left the formation at one-minute intervals. On the other side, Walker announced the altitude for assembly and circled until all the planes arrived. The flight leaders pulled out, and after being joined by their wingmen took their places in javelin formation.[76]

Units in the West sometimes found haze from forest fires almost as bad as fog. It forced pilots to fly on instruments, and bombardiers could not pick up their target until within two or three miles of it. Hence there was little time for the pilot to get on course and for the bombardier to figure drift and rate of travel.[77] Army flyers always sought ways to overcome atmospheric conditions hampering their operations. After Capt. Albert F. Hegenberger's famous blind flight of 1932, the Air Corps bought equipment and trained some pilots to land on instruments. Still, perfection and routine use of blind landings remained for the future. A formation of bombers could not land at a field covered with low fog, but with the landing area itself clear it could land if it could get down through the clouds. The 2d Bombardment Group devised a method to avoid collision during the descent.[78]

To take advantage of clouds for protection against hostile airplanes and antiaircraft fire while bombing, units framed methods for bombing from or through an overcast. First Lieutenant Douglas M. Kilpatrick, Jr., and other members of the 2d Bombardment Group developed a "100% instrument approach." The bombing run began at a point visible to the crew, such as a lighthouse or a railway junction, and as near to the target as possible but at least 5 minutes distant. From this initial point, the pilot took the plane into the overcast on a course to the target. The bombardier released the bombs at the time and place determined by dead reckoning. From its tests, the group calculated probable error (that error which was just as likely to be exceeded as not). With the initial point 30 miles from the target, it expected probable errors of 1,200 to 1,800 feet long or short in range and 500 to 700 feet deflection to either side. This bombing procedure was practicable solely for very large targets, though in all bombing the chances of getting hits rose when bombs were released in train. Another method employed the same procedure except that the plane descended out of the overcast just in time for the bombardier to use his sight. Both methods required highly trained navigators and bombardiers.[79]

The 19th Bombardment Group tried bombing by dead reckoning after a flight of fifty miles. Simulating a release, the group figured the bomb would hit one-quarter of a mile from the point selected. This showed bombing by dead reckoning to be possible—"for a very large target."[80] The 9th Bombardment Group experimented with bombing through the overcast by simulating overcast conditions and using Mitchel Field for a target. Working from dead reckoning, the bombardier sighted the vertical position of the aircraft at time of release, marked it on a diagram of Mitchel Field, and plotted an imaginary fall of bombs in train. Lt. Col. Carl W. Connell, the group commander, reported that such tests seemed to show that nine or more bombing planes dispatched at intervals from an initial point within twenty-five miles of the target could successfully bomb an area the size of Mitchel Field. The 9th Group, like others, also practiced bombing from under an overcast by using a low-altitude attachment to the Norden bombsight.[81]

General Emmons, commanding GHQ Air Force, asked commanders in May 1939 to conduct as much training as possible under conditions of poor visibility or low overcast. He wanted to uncover difficulties that would prevent successful bombing and to discover what equipment and techniques would better chances of success.[82]

Airways and Weather Service

Accurate, up-to-date weather information and good communications were essential for movement of planes and units for both peacetime training and wartime operations. Federal airways, though useful, did not always furnish service when and where the Air Force needed it. The Army's communications system, run by the Signal Corps, gave priority to administrative messages and did not function fast enough for aircraft movements and operations. Aircraft sometimes reached their destination before the message heralding their arrival. However, for the flight of B–10s to Alaska in 1934, Colonel Arnold and his communications officer, Capt. Harold M. McClelland, got the Signal Corps to alert its ground stations to give messages concerning the flight precedence over other military traffic.

McClelland and other Air Corps officers worked with the Department of Commerce on further development of airways. But as time went by, they became convinced that the solution to the Army's problem lay in establishing special facilities to handle military flights. In 1937 a committee of Air Corps officers headed by Lt. Col. Robert Olds proposed an "Army Airways Control System" under Air Corps jurisdiction, for communicating with planes in flight, dispatching and reporting arrival of aircraft, disseminating meteorological information, and controlling aircraft during bad weather.

Late the next year, the War Department created the Army Airways

Communications System under control of the Chief of the Air Corps to operate radio stations to facilitate air traffic between Army airfields in the United States. With the country divided into three communications regions, the Air Corps on November 15, 1938, activated three communications squadrons, the 1st at March Field, 2d at Langley, and 3d at Barksdale, each with detachments at various airfields in its area. Squadron Commanders, Capt. Russell A. Wilson in the 1st Communications Region and 1st Lt. Lloyd H. Watnee in the 3d, also served as regional control officers. They took over radio stations the Signal Corps had operated for the Air Corps at the airfields. Because they needed more equipment, several months elapsed before all stations were functioning.[83]

The Air Corps was also dissatisfied with the Army Meteorological Service run by the Signal Corps. Weather stations at the airfields were supervised by Air Corps officers, usually graduates of a course in meteorology at Massachusetts Institute of Technology or California Institute of Technology. The Signal Corps supplied equipment and enlisted observers and forecasters. The Air Corps tried a number of times to set up an independent service. Success came in 1937, when the War Department split the meteorological service among the branches of the Army. The Signal Corps continued to develop, buy, and distribute equipment. At the same time, the Field Artillery, Chemical Warfare Service, and other branches ran their own weather services to meet their own peculiar needs. On July 1 the Air Corps organized three weather squadrons, the 1st at March Field, 2d at Langley Field, and 3d at Barksdale Field. Each squadron, responsible for weather service in its region, consisted of detachments operating weather stations at the airfields.[84]

The Air Corps continued to send a few officers (1st Lts. Don Z. Zimmerman, Leo P. Dahl, and Thomas S. Moorman, Jr., among others) to civilian institutions to study meteorology. To train enlisted men to forecast, it established the Air Corps Weather School at Patterson Field, Ohio. Lt. Col. Junius H. Houghton became commandant; Capt. Don McNeal, Signal Corps, who had run the Signal Corps school at Fort Monmouth, served as senior instructor. Twenty-five students from many different stations began the 5-month course on September 1, 1937; 20 graduated on January 28, 1938; 26 more entered on February 1. Observers received training on the job at Air Corps weather stations until creation of a school for observers at Scott Field, Illinois, in September 1939.[85]

The training of combat units improved markedly during the last half of the 1930s, after the airmail operation of 1934 had exposed training deficiencies. The Baker Board urged more and better flying training. The War Department and the Air Corps placed far greater emphasis on flying proficiency. More money became available for training. And the formation of

GHQ Air Force in 1935 brought the training of all combat units in United States under the direction of one man, a flying officer.

General Andrews shaped the training program of GHQ Air Forc make his men skilled in flying, navigating, bombing, and shooting. organized men of his bombardment squadrons into crews for training operations. He required officers to become proficient in navigation, bomb ment, and gunnery as well as piloting. And he called upon enlisted me qualify as gunners in addition to their individual specialties such as fli engineer, radio operator, or armorer. His men used Link trainers and ot devices for practice on the ground, but they spent more time than ever be in the air. Development of large ranges at Muroc Dry Lake, California, Valparaiso, Florida, afforded the best facilities thus far available for bomb and gunnery practice. Acquisition of the Norden bombsight refined bomb accuracy and moved the Air Force much closer to the long-time goal precision bombardment. At the same time, the Air Force experimented w mass, or area, bombing, and worked on techniques for bombing dist targets overlaid with clouds.

Chapter XXI

Operations

Exercises and maneuvers gave General Andrews, as well as field army, corps area, and department commanders, opportunities to test aviation organization, personnel, equipment, and methods. The largest and most significant of these operations during the 1930s dealt with coastal defense, in which Army aviation searched for and attacked ships approaching U.S. territory; and with air defense, employing aircraft to protect specific places or localities from attack by hostile aviation. The latter entailed the use of civilian observers, the development of warning nets, and the testing of air raid blackout procedures. Also, the Air Corps put on acrobatic performances as part of its public relations program, and did civic work.

Exercises and Maneuvers

Being responsible for direct support of ground forces and for independent operations under the commander in chief, Air Force units often played roles in army exercises and maneuvers. Maj. Gen. Frank Parker, for instance, used attack and observation aviation in a Third Army command post exercise in 1936. Maj. Gen. George S. Simonds employed the 1st Wing as an army air force during a Fourth Army command post exercise in 1937. Maj. Gen. Albert J. Bowley used the 17th Attack Group in Fourth Army maneuvers in 1939.[1]

Maj. Gen. Charles E. Kilbourne drew attack, pursuit, and bombardment from 2d and 3d Wings of GHQ Air Force for Second Army maneuvers in the summer of 1936. On this occasion, troops of two corps areas worked separately. Those of V Corps operated near Fort Knox, Kentucky, those of VI Corps northwest of Camp Custer (near Battle Creek), Michigan. In the latter region, two National Guard divisions fought a Regular Army brigade and a mechanized force. Operations in the other area were on a similar scale. Pursuit planes from Selfridge Field, Michigan, helped defend Fort Knox, Kentucky, against bombers from Langley Field, Virginia, and attack ships from Barksdale Field, Louisiana. Units from Selfridge, Barksdale, and Langley took part in the maneuvers in Michigan. So did the observation squadron and photo section of each National Guard division, an observation squadron and photo section from Scott Field, Illinois, and the 1st Balloon Squadron from Fort Sill, Oklahoma.

For the Air Force, the high point of the Second Army maneuvers came with a mission for which the 96th Bombardment Squadron won the Mackay Trophy. On August 13, 1936, Capt. Richard E. Nugent led 3 B–10Bs to attack ground forces in Michigan. With more than 600 miles to fly and the attack scheduled for 2200, Nugent took off from Langley at 1630. Soon the planes ran into dense haze with solid overcast. Nugent went on instruments, 1st Lt. Joseph A. Miller navigating. The pilots on either wing, 1st Lt. Edwin G. Simenson and 2d Lt. Burton W. Armstrong, Jr., flew tight formation until they encountered thunderstorms and heavy fog. When they no longer could see the navigating lights on Nugent's ship, they separated. Weather was better near Camp Custer, and Nugent circled just below the clouds for 15 minutes until the other 2 planes got in formation. He led them up through the clouds and toward their objective. Gliding down through the clouds, the flyers released flares 8,000 feet over the target at 2158. The B–10s next maneuvered

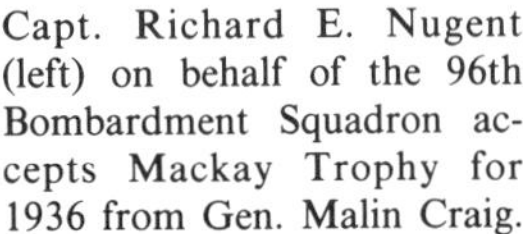

Capt. Richard E. Nugent (left) on behalf of the 96th Bombardment Squadron accepts Mackay Trophy for 1936 from Gen. Malin Craig.

in the area and worked with searchlights before ending the flight at Selfridge Field.[2]

General Andrews concentrated GHQ Air Force in California for maneuvers in May 1937. Brig. Gen. Delos C. Emmons, 1st Wing Commander headquartered at March Field, commanded the defense of Los Angeles, represented by factories, docks, warehouses, and other targets outlined in lime on the ground at Muroc Dry Lake. The defenders included 2 squadrons of P–26s and 1 of PB–2s at Muroc. The 63d Coast Artillery (Antiaircraft), under Emmons' command for the maneuvers, deployed its guns, sound locators, and searchlights around the "city." Brig. Gen. Gerald C. Brant, 2d Wing Commander with headquarters at Hamilton Field, commanded the attacking force—3 squadrons of B–10s operating from Fresno, Stockton, and Merced; and 3 squadrons of A–17s at Bakersfield, Visalia, and Delano. Concentration began on May 1 with transport planes, trains, and trucks moving men, equipment, and supplies to the various locations occupied by the Air Force. Andrews outlined the general scheme of maneuvers, which lasted from May 10 to May 23, 1937, with both day and night missions. Operations envisioned B–10s bombing the city and A–17s using tear gas against the airdrome and antiaircraft emplacements. A–17s would lay smokescreens to reduce the accuracy of antiaircraft fire against bombing planes. Pursuit ships would intercept and attack bombers with camera guns.

In these maneuvers, pursuit planes (P–26s) were but a shade faster than the bombers (B–10s), the latter having the advantage at higher altitudes. Bombardment's best defense against pursuit in the target area was to fly at top speed. Pursuit then found interception difficult and, where it succeeded, could make just one pass during the period of bombing, this being particularly true at high altitudes. Pursuit tried several tactics against bombardment. An attack by individual aircraft from the rear and above on bombers in javelin-down formation made pursuit vulnerable to the bombers' rear guns. Attacking string formation from above and to the side, pursuit passed so quickly that bombardment had little time to fire. At the same time, this gave pursuit but an instant to bring its guns to bear. Attack from above and ahead, with pursuit diving and pulling up under the bomber formation to fire from a stalled position, bared pursuit to fire from the bombers' forward and tunnel guns. The best tactic seemed to be an attack from the same level at long range, where single pursuit ships presented illusive targets for the flexible guns on the bombers. This promised considerable success against bombers in close formation if pursuit employed large guns calibrated for long-range firing.

In bombardment operations, close formations were good for bombing and for defense against enemy aircraft, but increased the chances of being downed by antiaircraft fire. This danger could be sharply reduced by flying in columns of single bombers separated by 10 to 15 seconds and staggered laterally and by at least 1,000 feet in altitude, by quickly losing altitude and turning 180 degrees upon release of bombs, and by departing in the direction from which they attacked. The best time for bombing (that with the least interference from antiaircraft fire) was during the 15 minutes just after dawn and the 15 minutes right before dark. Then there was too much light for effective illumination by searchlight and too little light to render planes visible to ground observers. The use of black paint on bombers made illumination by searchlight more difficult at night, but gave the planes a clearer outline by day.

The maneuvers afforded many other lessons. A complete folder was needed for each contemplated target, since mission success depended upon recognition of the objective. Night precision bombing appeared possible if proper flares or long-burning incendiary bombs could be developed and dropped by attack, reconnaissance, or bombardment aircraft to mark the aiming point. Planes outlasted personnel, suggesting a need for alternate crews. Bombing planes required extra bomb-bay fuel tanks for long cross-country deployments. More aerial transportation was desired to give combat units greater mobility. The supply system was good. Bombardment units had to have more clerks. The wing message center wanted fifty-percent additional men. And among other things, tents were undesirable in hot places.[3]

The following May, General Andrews assembled some 450 officers, 2,300 enlisted men, and 130 airplanes from his 3 wings for maneuvers in the Northeast. These maneuvers included 2 operations of especial interest and importance—interception of the Italian liner *Rex*, and the first blackout in the United States—both will be discussed later.

Coastal Defense

Bombardment and reconnaissance aircraft often flew far to sea to intercept and attack ships. To this end, officers studied navigation, dead reckoning and celestial, and practiced both. Those of the 31st Bombardment Squadron, for example, heard so many lectures and worked out so many problems that they dreamed of rhumb lines and logarithms. The student's immediate goal was to plot his position within 1 mile in 100 and estimate time of arrival within 1 minute of each hour out. He then knew sufficient aerial navigation to tackle problems of coastal defense.[4] One came when an enemy fleet represented by the U.S. Army Transport *Meigs* threatened the west coast in 1936. The 7th Bombardment Group plotted the ship's course from her weather reports and from information supplied by other ships. At 0800 on July 17, Lt. Col. Clarence L. Tinker (7th Group Commander) sent planes to reconnoiter. He ordered the bombers off an hour later, the *Meigs* then 186 miles off the coast. The bombers intercepted 1 hour and 7 minutes after leaving the Golden Gate. Colonel Tinker dropped aluminum powder a mile behind the ship to make a slick for the planes to attack with 100-pound bombs.[5]

U.S. Army transports often became targets for interception and attack. Maj. Idwal H. Edwards, for instance, led 8 B–12s of the 5th Bombardment Group from Luke Field, Hawaii, to intercept the *Republic*. With 1st Lt. John W. Egan navigating, the group missed the estimated time of interception by 15 seconds. After circling 1,000 feet over the transport, 175 miles off Makapuu Point, the planes returned to Luke Field.[6]

The commander with 1 or more amphibians often used them as guard and rescue ships for land-based planes. The Y10A–8 of the 75th Service Squadron, for example, patrolled while P–26s from Wheeler Field, Hawaii, operated offshore.[7] Amphibians from Hamilton Field, California, often accompanied the 7th Bombardment Group's B–10s and B–12s to sea. Before dispatching bombers to intercept the *Meigs* on July 17, 1936, Colonel Tinker sent an amphibian 100 miles out on course. When the bombers reported interception, 2 more amphibians flew out from Hamilton Field. Tinker and the amphibian commanders reported every 15 minutes. Plotters at headquarters tracked the planes so rescue could begin at once if a plane was forced down.[8]

An amphibian ("duck") served as a rescue ship for a squadron of land-based bombers on a flight from Langley Field to the Panama Canal Zone and back in February 1937. Nine B–10Bs of the 2d Bombardment Group made the trip under the command of Maj. Jasper K. McDuffie. The flight supplied training in navigating over water and simulated reinforcement of canal defense. The squadron stopped at the municipal airport at Miami for service both going and coming. The duck, an OA–5 commanded by Capt. Archibald Y. Smith, could not fly as fast as the bombers. It therefore left Miami an hour early, to be overtaken and passed by the squadron half way across the Caribbean.

The squadron made the 1,100-mile hop from Miami to Albrook Field, Panama, in 8 hours on Saturday, February 6. Returning the following Tuesday, the squadron had just passed the duck when No. 90's left engine quit. The bombers were then at about 1,800 feet, with clouds above, and waves below stirred up by a 37-mile-per-hour crosswind. Capt. Cornelius W. Cousland, commanding No. 90, radioed he could not hold altitude with only one motor. The odds were against a safe landing in the rough sea. Even if Cousland succeeded, Smith might not be able to pick up the 4 men from the disabled plane. McDuffie notified Smith and shore stations of the emergency, and ordered 2 bombers, 41 and 43, to stay with the plane until the duck arrived.

take station near San Francisco, where General Emmons' men would use *Utah* for bombing practice. Brown changed course. At 1128 another message came, asking when he would arrive on station. He replied that he would be there by daylight the next morning. He took the orders for altering course and breaking radio silence to mean the exercise was over for the day. Thus, he was not prepared for attack when 15 minutes later a lookout reported 4-engine bombers on the port beam at low altitude. The crew immediately went to bombing stations.

Maj. Caleb V. Haynes piloted the lead B–17, his crew including 1st Lt. Curtis E. LeMay, navigator, and 1st Lt. Douglas M. Kilpatrick, Jr., bombardier; his passengers, General Andrews and Lt. Col. Robert Olds. After getting the scouting aircraft's report of *Utah's* position, heading, and speed, LeMay had laid a course to intercept. Approaching the area, the formation spread out to search. Seeing only fog below, Haynes went down through the overcast. There she was.[20] The story continues from *Utah's* diary:

> At 1147 planes started bombing from about 600 feet. Planes well within gun range during approach, and could have been subjected to heavy Anti-Aircraft fire. Ship started to make radical and irregular changes in course.
>
> At 1159 last plane dropped bombs. Total number of splashes observed was 50. Total number of hits on ship 3.[21]

The B–10s arrived at 1203, 3 minutes too late. At 1210 the fog cleared. Haynes accused the Navy of hiding *Utah* in fog to prevent the Army from bombing. Andrews thought a real enemy would use clouds and fog to conceal its ships.[22]

At 0940 Saturday morning, B–10s found *Utah* 55 miles southwest of the Farallon Islands (or about 90 miles from the Golden Gate). With ceiling and visibility unlimited, they commenced bombing from about 12,000 feet at 0957. Members of *Utah's* crew observed hits from gun turrets and other protected spots. They secured from bombing quarters at 1108, after being told bombing was over for the morning and would begin again at 1245. Men went on deck to score the hits, but scurried for cover when B–17s started bombing at 1114. The crew returned to bombing quarters, and the ship made radical changes in course. Although word came at 1345 that bombing had been completed, *Utah's* crew stayed at bombing quarters another 40 minutes.

A few months later, in May 1938, General Andrews launched Air Force maneuvers by sending three B–17s to find an enemy fleet (represented by the Italian liner *Rex*) threatening the Northeast. He wanted to show the Air Force could intercept any enemy fleet long before the ships could come near our shore. This required newspaper and radio coverage and photographs to prove interception and get publicity. Lt. Col. Ira C. Eaker, head of the Air Corps Information Division, took charge of public relations.[23] The steamship line agreed to cooperate. When the B–17s left Mitchel Field at 0830 on May 12, 1938, *Rex* was 725 miles out, headed for New York City. Major Haynes piloted the lead plane, No. 80. It carried Maj. Vincent J. Meloy (flight commander), an announcer and two engineers of the National Broadcasting

Company, and an NBC transmitter for a network broadcast. Capts. Cornelius W. Cousland and Archibald Y. Smith piloted planes 81 and 82 carrying newsmen, including C. B. Allen of the *New York Herald Tribune* on 81, and Hanson W. Baldwin, military and naval correspondent of *The New York Times*, on 82. Maj. George W. Goddard, the Air Corps' ace photographer, rode with Cousland.

Swinging over Sandy Hook, Haynes flew out to sea at 170 miles an hour. He knew from a report radioed by *Rex* during the night where the ship expected to be at noon. The job of finding that spot in the middle of the ocean fell upon his navigator, Lieutenant LeMay. Heavy overcast kept the planes low until around 1000, when a break in the weather let LeMay get a good check of speed and drift. The aircraft separated on entering a cloud front about 1100 but reassembled 10 or 15 minutes later on the other side. After taking drift again, LeMay estimated interception of *Rex* at 1225. At noon the

Left: Maj. Gen. Frank Andrews outlines plans for the May 1938 G.H.Q. Air Force maneuvers to Brig. Gens. Arnold N. Krogstad and Delos C. Emmons; below: B–17 "Flying Fortresses" intercept liner *Rex*.

flight flew through an area of scattered rain squalls, and the flyers feared they might miss *Rex* in one of the storms. Cousland spotted the ship dead ahead, "There she is. There she is." Then on radio he told Haynes where to look: "Eighty one to eighty. Twelve o'clock."[24] Two minutes later the 3 B–17s passed *Rex*. The time was 1225! The aircraft circled, passengers on deck waved, Goddard shot with his 4x5 Graflex, NBC went on the air and Meloy talked with *Rex's* captain. The planes then headed home and, after a miserable flight in bad weather, landed safely at Mitchel Field about 1630. Newspapers nationwide gave the story the publicity Andrews and Eaker sought. Many used Goddard's shot of 2 B–17s flying alongside *Rex* at smokestack height.[25]

One-Hundred Miles

Air Corps officers remembered *Rex* for an order limiting their flights offshore to 100 miles. The way General Arnold told it:

> Somebody in the Navy apparently got in quick touch with somebody on the General Staff, and in less time than it takes to tell about it, the War Department sent down an order limiting all activities of the Army Air Corps to within 100 miles from the Shoreline of the United States.[26]

Ira Eaker, Robert Olds, and Carl Spaatz (he changed the spelling of his name in 1938) were among those who blamed the Navy for this constraint on Army aviation, but they never saw the order. Arnold said he tried several times to get a copy from the War Department but without success.[27]

Time made the "mystery" of the order[28] all the more mysterious. Eaker said some years later that the Secretary of the Navy protested to the Secretary of War. On another occasion he said General Andrews called him shortly after completion of the mission and told him Gen. Malin Craig, Army Chief of Staff, was very angry about the flight because it encroached on the Navy's mission. Later Eaker recalled he was in Andrews' office when Craig gave Andrews the order by phone.[29] Spaatz, then executive officer of the 2d Wing, said later he received a telephone call from the War Department imposing the 100-mile limit.[30] Lauris Norstad, assigned to the 9th Bombardment Group at Mitchel Field in 1938, remembered the restriction as being "very real at the time," but found his memory "hazy on the legal background."[31] In retrospect, St. Clair Streett, a member of the War Department General Staff, and Joseph T. McNarney, on Andrews' staff in 1938, thought the Army imposed the 100-mile limitation but did not agree on when. Streett believed it followed the interception of *Rex*, but McNarney felt positive it came earlier. Stanley D. Embick, Deputy Chief of the War Department General Staff from May 1936 until September 1938, held that the Army imposed the limit as a safety measure.[32]

Restrictions as to the distance Army airplanes flew to sea existed long before the *Rex* affair. For instance, orders for an Army-Navy exercise in the Philippines in 1929 limited land planes to 25 miles seaward because of high winds and heavy sea.[33] At least twice in 1936, the War Department instructed Andrews to conduct upcoming Army-Navy exercises so that his land-based planes would not fly more than 100 miles offshore. This, The Adjutant General explained, was to "minimize danger to personnel and materiel."[34] A joint exercise later in 1936 took place within 50 miles of the entrance to Pensacola Bay. Afterwards, Capt. John P. Doyle, Jr., commanding 18th Reconnaissance Squadron, suggested that future exercises be held farther from shore.[35] The presidential directive for the exercise against *Utah* in August 1937 specified that the action be confined to within 500 miles of the shoreline. The Army, however, insisted on limiting the distance to 300 miles.[36]

During an exercise in the Chesapeake sector in November 1937, B–17s of the 2d Wing operated as far as 200 miles to sea. However, an order signed by Lt. Col. Carl Spaatz, the wing's executive officer, prohibited B–10Bs from going more than 100 miles offshore. General Brant, the Wing Commander, said the B–10s "did not have the necessary range of action or engine safety factor to warrant dispatching these planes so far to sea."[37] Events immediately after interception of *Rex* on May 12, 1938, gave no hint of new constraints on Army aviation. Spaatz signed an order, dated May 17, 1938, for 2d Wing participation in an Army-Navy exercise in an area not to exceed 300 miles seaward.[38] On June 12, 1938, three B–17s on a training flight from Langley Field intercepted and exchanged greetings with a steamer, *Queen of Bermuda*, about 300 miles at sea, an event the Air Corps reported in its *Newsletter*.[39] The 5th Bombardment group intercepted the U.S. Army Transport *Republic* off Hawaii on July 18, 1938, and again on September 9, the first time 285 miles and the second 329 miles at sea.[40] On August 11, 1938, the 23d Bombardment Squadron at Hickam Field, Hawaii, made what it called "the longest over-water mass flight of Army planes ever attempted in the Hawaiian Department." Maj. Harold W. Beaton commanded 5 B–18s that flew to Frigate Shoals (a distance of some 550 miles) and back.[41]

General Craig meantime asked his staff to prepare a directive limiting operation of Army aircraft to no more than 100 miles from land. Exceptions were made for interisland flights and for some other situations, such as ferrying aircraft to Panama.[42] A reporter covering an air defense exercise at Fort Bragg, North Carolina, early in October 1938 heard of a secret order setting a 100-mile limit on Air Corps operations offshore except with special permission of the Secretary of War.[43] A week later, Brig. Gen. Arnold N. Krogstad, 2d Wing Commander, issued instructions calling for 2d Wing operations up to 200 miles offshore during an Army-Navy exercise at the beginning of November. When General Arnold learned about it, he asked

AIR CORPS–GHQ AIR FORCE RELATIONSHIPS

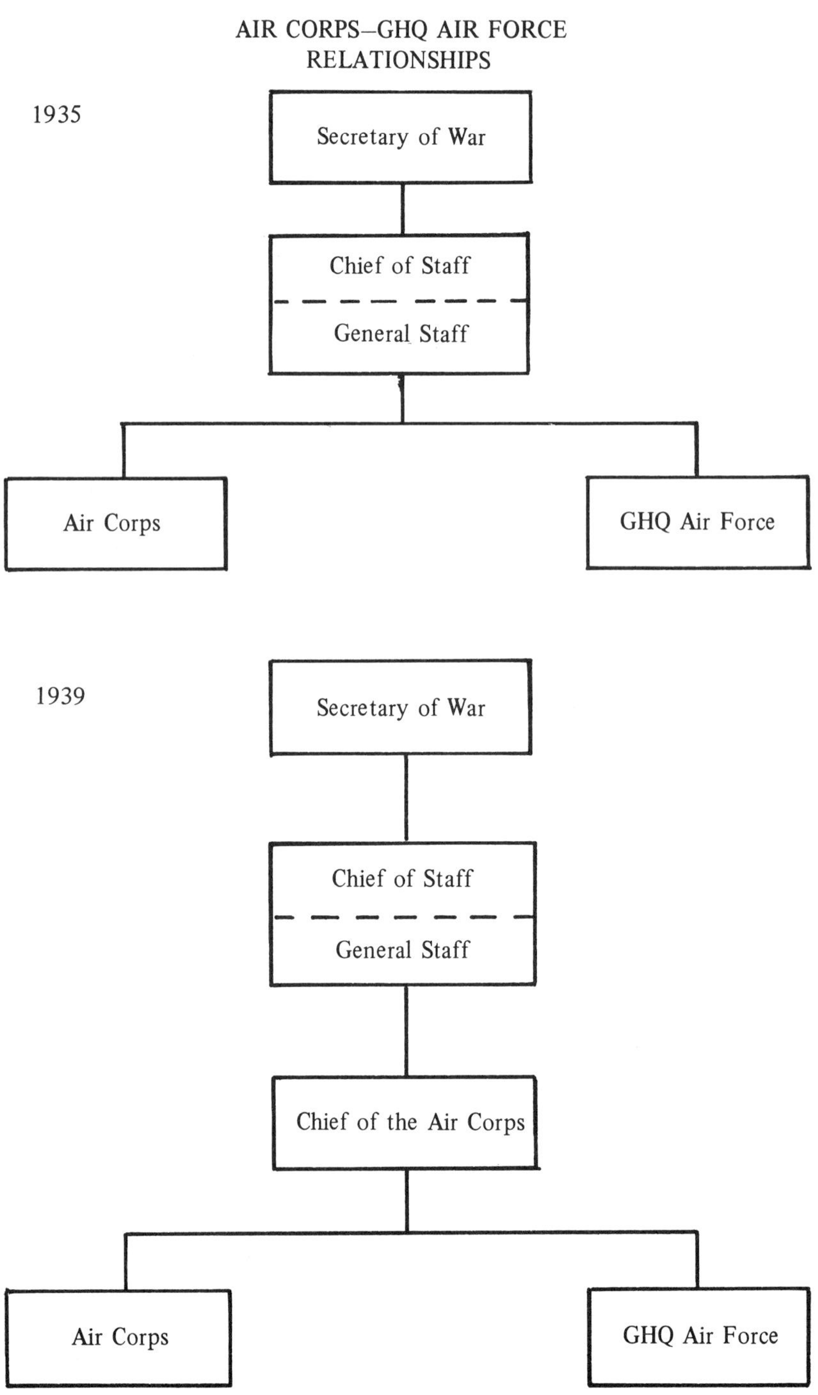

General Andrews whether this conformed to War Department instructions. Andrews said he thought so, but the War Department restricted the Air Force to 100 miles. Reporting the exercise, Krogstad recommended that future operations not be limited to 100 miles.[44]

In December 1938 the War Department asked General Andrews to comment on a Navy plan for a joint exercise in New England the next spring. The inclusion of a 100-mile limit in the plan evidently reflected the curb the War Department had imposed for the November exercise. General Andrews wanted to know if this was what General Craig desired. To find out he sent Col. George Brett, his Chief of Staff, to ask Brig. Gen. George C. Marshall, Deputy Chief of the War Department General Staff. Brett told Marshall that Andrews objected to the restriction because "it took a 1000-mile weapon and reduced its operating range to 100 miles."[45] Marshall informed Brett later that Craig did not object to maneuvers more than 100 miles offshore, with or without the Navy, if Andrews requested authority well in advance. At the War Department's request, the Navy revised the plan to delete the 100-mile limit.[46]

General Andrews quickly requested permission for each of his bombardment and reconnaissance squadrons to make 6 flights beyond 100 miles to sea for navigation training. He said the planes would search solely for Navy, Coast Guard, Army, and other government vessels,[47] and the flights would be given no publicity. Then the General Staff asked General Arnold to comment on whether he thought "the 100-mile limit, as at present governing, is too restrictive for adequate training in aerial navigation." Arnold recommended removal of all restrictions on flights to sea by planes under the control of the Commanding General, GHQ Air Force.[48]

The War Department rendered a decision 3 weeks later, on March 16, 1939. During that time, General Emmons succeeded General Andrews as Air Force Commander, and the chain of command changed. Now the Commanding General, GHQ Air Force, reported to the Chief of the Air Corps in lieu of the Chief of Staff. While retaining the 100-mile constraint on GHQ Air Force, the War Department allowed Arnold to authorize longer flights requested by Emmons if the Air Force used only government vessels and planes as targets, the flights and policy governing them were given no publicity, and the Air Corps reported each authorization to The Adjutant General.[49] The 2d Wing at once asked for authority to fly beyond 100 miles in the Army-Navy exercise scheduled off New England in April. Arnold approved and notified The Adjutant General.[50] During this exercise, held in better-than-average weather, the enemy fleet succeeded in coming within easy striking distance of the coast. This, according to Rear Adm. William T. Tarrant, who commanded the Navy's coastal defense forces, underscored the need to extend the defensive area outward "to the utmost limit of the capabilities of defending aircraft or even beyond."[51]

Maj. Gen. Delos C. Emmons (left) succeeds Gen. Frank Andrews as Commanding General, G.H.Q. Air Force in 1939, and Gen. George C. Marshall (right) becomes Chief of Staff.

General Marshall, preparing for his new assignment as Chief of Staff when General Craig left on July 1, 1939,[52] devoted some time while Deputy Chief of Staff to learning about aviation. General Arnold played a significant role in his education. So did General Andrews, who took Marshall on a tour of airfields and aircraft factories in the summer of 1938. Making plans to assume the duties of Chief of Staff, Marshall sought a senior airman to represent aviation on the General Staff. He chose Andrews, who in his permanent rank of colonel was at San Antonio as Air Officer, VIII Corps Area. Promoted to brigadier general of the line, Andrews joined Marshall on August 4, 1939, as Assistant Chief of Staff, G–3.[53]

Twenty days later, General Arnold issued new instructions for overwater flights. He limited single-engine planes, as well as multiengine ones incapable of flying on half of their engines, to thirty miles from land except under three conditions: An airplane capable of operating from water accompanied the flight; surface vessels had been placed on known stations as a safety precaution; or he specifically authorized an exception. He let multiengine aircraft capable of flying on half of their engines to go a total distance, from takeoff to destination, equal to fifty percent of the range specified for the aircraft. These curbs did not prevent local commanders from imposing tighter ones if needed for safety.[54]

Air Defense

Aerial operations defending the nation against hostile ships went under the name "air defense" in the early 1930s, but people also used the term in several other ways. Representative John J. McSwain of South Carolina, Chairman of the House Military Affairs Committee, had something broader in mind when he spoke of air power being "the first line of defense."[55] Air Corps officers, including General Foulois, employed it in referring to aerial operations defending the nation, a frontier, or a specific place against a hostile force of any kind—land, sea, or air. To General Westover and others it meant all aerial operations against a hostile air force, including attacks on the enemy's airfields and on supply lines supporting his air power, as well as against his aircraft in the air.[56] At times people used "air defense" and "antiaircraft defense" as synonymous for employment of all available means—antiaircraft guns and passive measures as well as airplanes—to prevent attack by air.[57] There were also those who, like Claire L. Chennault, thought of air defense as defending a place or area by intercepting aircraft and frustrating their attack.

McSwain, like others, watched the earth shrink and saw oceans grow narrower as the range of aircraft increased. Addressing the House of Representatives in 1935, he asked: "Today who will dare say . . . that America is secure from invasion and attack by air power, even with 3,000 miles of water on her east and 8,000 miles of water on her west?"[58] An enemy could launch planes from ships at sea or from bases within flying distance

Congressman John J. McSwain *Library of Congress*

from the United States. Defense demanded means for detecting, intercepting, and destroying aircraft bent upon attacking American territory.

The Air Corps and Coast Artillery tested a warning system, interceptors, and antiaircraft guns in an exercise at Fort Knox, Kentucky, in May 1933. A line from Indianapolis to Cincinnati divided two warring states, Blue to the north, Red to the south. Fort Knox represented a rail and supply center to be defended against the Blue air force based at Patterson Field, Ohio, 160 miles away. Blue included a bombardment group of B–2s, B–7s, and B–9s from March and Langley Fields; an attack group of A–8s from Fort Crockett, Texas; and P–16s (acting as attack) from Selfridge Field. The 1st Pursuit Group, commanded by Maj. George H. Brett, constituted the chief component of Red air force. Besides the 17th Pursuit Squadron with P–6Es and the 27th Pursuit Squadron with P–12Es for interception, the group employed some P–16s and an Organized Reserve squadron for observation. Three regiments of antiaircraft artillery from Fort Totten, New York, Fort McClellan, Alabama, and Fort Sheridan, Illinois, placed guns, searchlights, and listening devices at Fort Knox, which served as defense headquarters. The 1st Pursuit Group operated from Bowman Field, Kentucky, 26 miles away, in the direction of Dayton, Ohio.

The exercise tested a warning system that Capt. Claire L. Chennault, pursuit instructor at the Air Corps Tactical School, devised after studying British methods for employing ground observers to detect and report hostile aircraft. During exercises in Ohio in 1931, the Air Corps had achieved scant success in intercepting planes reported by ground observers. Chennault's system, much more elaborate and more highly organized, provided clearer, more precise, and faster reporting. It afforded better methods for plotting and tracking hostile planes, greatly improving the chances of intercepting them before they reached their objective. The defensive sector for the exercise in 1933 was 120 degrees wide, extending toward Dayton from the center at Fort Knox. It comprised some 16,000 square miles, split into 12 subsectors (named) of 10 degrees each, and five sections (numbered) each 25 miles deep. Soldiers from ground branches manned 69 observation posts (named) situated at 7- to 8-mile intervals in 3 bands on arcs of circles about 50, 75, 100 miles from Fort Knox. The 2 soldiers assigned to each post received a little training in aircraft identification.

When an observer at "Emily" (the post at Rising Sun, Indiana) in "Funny 4" sighted 3 planes, high and to the left, flying southwest, he noted the time and quickly jotted down the information on a printed form. He then picked up a phone: "Flash, Louisville 125." Giving the call priority, the operator put it through to the intelligence center at Fort Knox. The soldier receiving it entered the items on the same kind of form used by the observer: Emily, 3 bombers, 1:44, high, left, southwest. Another soldier, wearing a telephone headset connected with the operations office at Bowman Field, read the data into the phone as fast as it appeared on the form. With this system, Bowman Field received some reports within 1 minute of sighting, the average time during the exercise being 2.65 minutes. As fast as the information came in, men in the operations office plotted it on a large map.

Besides commercial telephone, observers could use two other channels to report. They telephoned Postal Telegraph, which sent the message by Morse code to Louisville for relay on a Morse circuit to the intelligence center. Or they phoned one of the radio stations the Signal Corps set up at Batesville, Indiana, and Owenton, Kentucky, for rebroadcast to the intelligence center. Telegraphers and radio operators used the standard form to record the information. The average time for nearly 1,000 messages transmitted by the 3 methods was 2.7 minutes, well within the 4 minutes the 1st Pursuit Group figured it needed for interception. Defense forces likewise secured information about the enemy from 1st Pursuit Group observation planes that maintained surveillance of Patterson Field, which lacked defense. A transport plane with powerful radio circled near Cincinnati, relaying the observation aircraft's reports of bombers taking off.

On most interception missions, Capt. Ross G. Hoyt, the 1st Group's operations officer, led the planes in the air while Major Brett directed operations from Bowman Field. During periods scheduled for defensive operations, Brett held his flyers on alert at the squadrons' operations tents, on the field at the line of ships. When the "flash" reached the Bowman Field operations office, he issued orders by a public-address system. The men on the line acknowledged orders and answered simple questions with code sent by pushing a button that worked a signal light in the group's operations office. The men already knew this system, for it resembled the one they had at Selfridge Field.

Each squadron possessed 6 SCR–183 2-way radios and 5 SCR–192 receivers for 22 planes. Watching the map as flash reports came in, Brett radioed the information to his flyers. The 1st Group employed 3 methods of interception. In the first, Hoyt headed the planes for a point on a line between the reporting post and Fort Knox. As he acquired more radio data from Bowman Field, he adjusted his ships as necessary to keep them between the enemy and the target. If apprised of fresh sightings, he deployed his aircraft on a 25-mile front, 1 squadron (normally 18 planes) on each side of him, and headed toward the last sighting. Under the second method the squadrons operated separately, proceeding to "standby" areas to await new information and instructions by radio. The third method employed squadrons, under their own commanders, in 6-plane flights in patrol areas extending across the entire defensive area about 50 miles out. When a flight reported the enemy in sight, Hoyt assembled part or all of the group upon the flight in contact with the enemy.

Rain hampered activities on May 15, the first day of the exercise. The next morning the 1st Pursuit Group took off at 0853 to intercept 9 bombers coming from the north. The 17th Pursuit Squadron intercepted west of Borden, Indiana (about 40 miles from Fort Knox) at 0931. The 27th Pursuit Squadron contacted the same bombers north of Louisville at 0943, and found 9 attack planes 10 or 12 miles west of Fort Knox at 1005. Three formations of bombers got past and attacked Fort Knox. Three P–12s went up to drive off a B–7 that tried to jam the radio at Bowman Field. When they dove to attack, the pilot of the B–7 opened his throttles wide and, the historian of the 1st Pursuit Group wrote, "literally walked away."

With the warning system functioning faster and more accurately, just one formation of enemy planes filtered through to attack on May 17. On the 18th, the pursuit pilots got off in less than 1½ minutes and intercepted a formation of bombers within 12 minutes after takeoff orders. During the exercise, which ended on May 24, defenders identified 27 enemy formations in daylight; the 1st Pursuit Group intercepted 17. Most contacts between opposing planes came in Section 2 (25–50 miles out), and a number took place in Section 3 (50–75 miles). Others occurred in Section 1 (0–25 miles) but, under the rules of the exercise, interception that near Fort Knox did not count. In addition the defenders identified 19 formations at night. The 1st Pursuit Group plotted movements of hostile planes but, because of inadequate night-flying facilities, attempted only 1 interception at night. On the evening of May 17, a P–6E on observation saw 8 B–2s 5 miles north of Madison, Indiana; a P–12E picked up the same bombers 5 minutes later. When the pursuit pilots opened their radio transmitters, the bombers switched off their lights, dove into clouds, and disappeared. Night operations brought searchlights into play. On the evening of May 21, six P–12s worked with the searchlights against B–9s.[59]

The Joint Antiaircraft-Air Corps Exercise of May 1933 provided arguments for Captain Chennault and others campaigning for a warning net and interceptors. Operations by GHQ Air Force (Provisional) in California, held at the same time, led General Westover to assert that nothing could stop bombers. Colonel Arnold said the same thing after running tests with the 1st Wing the following year. In various training exercises conducted by the 1st and 2d Wings of GHQ Air Force and by overseas departments, pursuit opposed bombardment. During Air Force maneuvers in Florida in December 1935, bombers reported their position by radio to simulate reporting by a warning system. However, the next big exercise involving ground observers took place in California in 1937. The Southern California Edison Company furnished a warning service, without cost to the government, for the 1937 maneuvers carried out in May by the Air Force and Coast Artillery at Muroc. The initiative for setting up this service came from F. L. Eley, an Edison engineer who held a commission in the Naval Reserve, and Lt. Col. Claude M. Thiele, commanding 63d Coast Artillery. Employees at eleven of

Edison's powerhouses and substations served as observers, using the company's telephone system to report sightings.[60]

Eley and Thiele also assisted in organizing an exercise wherein 4 public utilities, a railroad, and 3 government agencies cooperated with the Army in April 1938. Observers manned 85 posts in Southern California. Their sole guidance was a letter that included silhouettes of planes and instructions for filling in message forms and telephoning flash reports. The 1st Wing flew missions designed solely to test the reporting by civilians, who ignored all aircraft except those in formations of 3. The communication systems of the organizations being tied together, flash messages went to Southern Edison at Alhambra (near Los Angeles), which relayed them to March Field. Due to delays in the relay, the time between observation post and March Field averaged 5 minutes, not nearly so good as at Fort Knox in 1933.[61]

Eley and Thiele suggested the Army form a warning service from people and facilities of power companies. Brig. Gen. Delos C. Emmons, 1st Wing Commander, believed such an organization would supply an efficient system at small cost to the government. Andrews and Westover supported the idea in general, as did Brig. Gen. Joseph P. Tracy, commanding 9th Coast Artillery District, Maj. Gen. George S. Simonds, Commander of Fourth Army, and Maj. Gen. Archibald H. Sunderland, Chief of Coast Artillery. But they did not agree on how to organize and develop a warning service. Who should control it—the Air Corps, GHQ Air Force, Coast Artillery, Signal Corps, or the local sector or army commander? The War Department lacked policy and doctrine. Much study would be needed.[62]

GHQ Air Force once more simulated a warning service during maneuvers in the Northeast in May 1938 by requiring hostile planes to report their own positions. The following October, an active network of 329 observation posts functioned during an exercise in North Carolina. Brig. Gen. Fulton Q. C. Gardner, Commander of 4th Coast Artillery District, directed defense of Airdrome A at Fort Bragg against Black air force, commanded by General Krogstad. Black consisted of three bombardment squadrons, one each of B–17s, B–18s, and B–10Bs, a reconnaissance squadron of B–18s, and an attack squadron of A–16s. Based at Langley Field, Black flew out over the sea and came back over the coast between Wilmington and the North Carolina-Virginia line to represent hostile planes from aircraft carriers.

Blue forces defending the airdrome included antiaircraft batteries with a total of twenty-four 3-inch guns, twenty-four searchlights, and one-hundred twenty .50-caliber machineguns. The interceptor force, under Lt. Col. William E. Kepner, comprised a squadron of P–35s and two squadrons of PB–2As operating from Pope Field at Fort Bragg, Knollwood Airport at Pinehurst, and the Commerce Department's intermediate field at Maxton. Col. Sanderford Jarman of the Coast Artillery commanded Aircraft Warning Service with a warning net extending all the way to the Outer Banks and employing 1,800 observers. Reserve officers, enlisted men, and Coast Guardsmen manned some observation posts, but civilian volunteers (tradesmen, housewives, lawyers, farmers—people from many walks of life) served two-thirds of them. Most observers reported by commercial telephone to the nearest of three control centers (New Bern, Wilson, Fayetteville), which relayed reports to General Gardner's headquarters. Radios operated by the Signal Corps, the Army's radio amateurs, or the Coast Guard handled reports from areas not having commercial telephones.

Brig. Gen. William Bryden, commanding Fort Bragg and directing the exercise, scheduled operations for 8 hours a day, 4 by day and 4 at night, for 6 days starting at 0400 on October 10. Hanson W. Baldwin and other newsmen followed the action on an 8- by 16-foot map. They listened to loudspeakers as sighting reports came in, and saw red lights appear to show the location of the posts reporting. They heard Colonel Kepner alert his pilots and saw green lights come on to represent friendly airplanes and airfields. And they witnessed interception, appearing on the map as a rapidly blinking light.

With the warning net functioning, Kepner watched the boards as plotters recorded sightings and tracked the enemy's course. Kepner's squadrons on ground alert (planes in formation for takeoff, pilots nearby, engines warmed up frequently) could become airborne in seven minutes. When the situation warranted, he reduced that time by ordering pilots of one squadron to their planes, with engines off but warm. Having precise information of the approach of hostile planes and a good idea of their number, he sent his ships—a flight, a squadron, or more—into the air and directed them to interception by radio. To avoid collision, pursuit did not go closer than a thousand feet to the enemy.

Bad weather kept pursuit grounded on the 13th, but bombers flew all 6 days of the exercise. On the 11th, for instance, General Krogstad sent 1 B–17 and 2 B–18s ahead of an attacking force to reconnoiter, report on weather en route and at the target, and take photographs during the attack. Colonel Kepner's planes intercepted the reconnaissance ships on the way in. Bombing and attack aircraft followed reconnaissance by about 2 hours. Twelve A–17s, skimming treetops on the way to the target, arrived at 1014 without being intercepted, and laid a smokescreen to blind the antiaircraft gunners when the bombers attacked at 1020. To Maj. Hugh N. Herrick, a Coast Artillery officer riding in the lead A–17, the smoke looked like a cloud made by a brushfire. It appeared to be thin and to cover only a small area. He noted how little smoke 12 planes carried, and thought the ships good targets for machineguns while putting it down. Eight minutes later, on the way out, the A–17s came under attack by pursuit.

The bombers flew a different route. Coming in high over the coast, B–17s entered the defensive net at 0915. Pursuit intercepted at 1001, but the bombers pressed on to the target. Three B–17s, arriving a little early, came under antiaircraft guns before they released their bombs from 25,000 feet at 1014. The rest of the B–17s attacked 5 minutes later. The B–10s, also intercepted, attacked from 12,000 feet at 1021. The smoke laid by the A–17s screened some of the bombers from antiaircraft fire but did not interfere with bombing. Pursuit attacked again as the bombers withdrew.

Artillery defended Airdrome A against night attacks by Black air force while Colonel Kepner's plotting section tracked hostile aircraft and simulated pursuit operations. One night, however, pursuit experimented in working with searchlights against enemy planes. To permit sound locators and listening posts to detect the enemy, pursuit stayed as far away as possible until searchlights illumined the bombers. To avoid collision in the dark, the opposing air forces agreed beforehand on maneuvers to be executed.[63]

The exercises at Fort Bragg included an air raid drill like one held during Air Force maneuvers in the Northeast during May 1938. On the earlier occasion, General Andrews chose Farmingdale, New York, for blackout because the Seversky and Grumman factories made it a typical target for bombardment. Europeans knew about drills, but Americans had no experience with such defensive measures. Someone heard, however, that a surgeon in Europe was operating when the main switch was pulled for a blackout. To avoid anything like that in Farmingdale, the Air Force secured the assistance of civilian authorities in asking people to turn off lights in their own homes and places of business. The sound of aircraft approaching was followed by sirens and radio warnings, then darkness. Pilots reported they could not see a thing: "It might just as well have been a wheat field in the middle of Texas." Still, enough lights burned in the surrounding area to let bomber crews detect the blacked-out target and put their sights on it.[64]

The blackout during the exercise at Fort Bragg in October 1938 lasted 45 minutes and covered about a fourth of the state of North Carolina. Six B–17s, commanded by Lt. Col. Robert Olds picked up newsmen at Pope Field, flew eastward, circled, and came back as the enemy. Receiving a report of sighting of the raiders near New Bern, North Carolina, at 1900, Colonel Jarman ordered: "Blackout immediately." Warned by phone, the guard at the firehouse at Goldsboro sounded the firebell 7 times, notified factories, and informed the Carolina Power and Light Company. In Goldsboro, as in 65 other cities and towns in 21 counties, public officials, civic leaders, and newspaper publishers had helped prepare the people for the raid, so everyone would know what to do. Atlas Plywood at Goldsboro gave 7 shrieks on its whistle, Borden Manufacturing 7 blasts on its siren. Carolina Power blinked lights in homes and other buildings. If anyone missed all those signals, he may have heard the warning broadcast by WPTF at Raleigh. Thus alerted, people turned out lights. A man on the Wayne Bank Building, the tallest building in town, watched for lights and reported them by telephone. Police, Boy Scouts, and members of the American Legion told motorists to switch off their lights. From 12,000 feet in the air, Hanson W. Baldwin saw towns and cities plunged into darkness. Even so, automobile headlights outlined rural roads sufficiently so flyers could identify communities as if there was no blackout. Colonel Olds and his men easily found the darkened target. At 1945, defense headquarters ordered "lights on." Searchlights at Fort Bragg probed the sky, but clouds between 6,000 and 8,000 feet kept them from spotting B–17s bombing from 10,000 feet.[65]

The Hawaiian Department held its first air raid in May 1939,[66] and the Canal Zone conducted a similar drill five months later.[67] But as yet neither the War Department nor the U.S. Government had a passive defense program, and no steps had been taken to organize defense of the civilian population.

Acrobatics

Air raid drills required public support, as did military aviation in general. Army pilots continued to give exhibitions at air races and air shows to gain publicity and goodwill.[68] Seventy-five Army planes performed at the All-American Air Races at Miami, Florida, in January 1935. Pilots of the 1st Pursuit Group, led by Capt. George P. Tourtellot, put on an act including some "bomb bursts" in which the planes dived toward the stands. This "thrilling but dangerous formation" violated safety regulations of the Department of Commerce. Eugene L. Vidal, the department's director of aviation, witnessed the stunt and at once grounded Captain Tourtellot and Maj. Ralph Royce, the group's commander, for a day. The pursuiters went on to give a good show without their leaders, even executing the "bomb burst," but higher and farther out.[69]

The entertainment at Miami featured performances by Three Men on a Flying Trapeze, an acrobatic team that Capt. Claire L. Chennault had formed at Maxwell Field to work out and demonstrate pursuit tactics and maneuvers. The other members were 1st Lt. Haywood S. Hansell, Jr., and SSgt. John H. Williamson, with Sgt. William C. McDonald an alternate. Both Williamson and McDonald were graduates of the Air Corps Advanced Flying School and first lieutenants in the Air Reserve. The Three Men on a Flying Trapeze put on their first big show in September 1934 at the National Air Races in Cleveland. Setting safety standards surpassing those of the Commerce Department, they built their show on precision, accuracy, and speed of execution. By staying away from the stands and maintaining a safe altitude, they gave spectators a perfect view of all their maneuvers. Flying P–12Ds, Chennault, Hansell, and Williamson commenced the performance with 3 consecutive loops in close vee formation, followed by Immelmann turns, slow

Left: "Three Men on Flying Trapeze," perform at races in Miami (l. to r.): Sgt. William C. McDonald, Capt. Claire L. Chennault, and SSgt. John H. Williamson; below: aerial demonstration by Skylarks in P–12s at Cleveland Air Races in 1937.

rolls, snap rolls, and other maneuvers. The team ended their 13-minute performance by flying in review the full length of the field in a vee banked up about 60 degrees.[70]

Lieutenant Hansell left the team to concentrate on his studies as a member of the class of 1934–35 at the Air Corps Tactical School. Chennault, Williamson, and McDonald performed at races in Miami in January 1935, Cleveland in September, and Miami again in December.[71] The 2 enlisted men, who appeared in public as Reserve officers, applied for Regular Army commissions in 1936, competing with 473 other Reservists for 52 vacancies. Not selected, Williamson and McDonald purchased their discharges and became aviation instructors in China, where Chennault would also find employment before many months.[72]

Capt. Charles D. McAllister formed another acrobatic team, the Skylarks, at Maxwell Field, the other members being 1st Lts. Carl R. Storrie and Clayton E. Hughes, with 2d Lt. Wilbur W. Aring alternate. They made a good impression at the 1937 air races at Cleveland, where Storrie announced over the public-address system while the others performed. One observer deemed them worthy successors to the Three Men on a Flying Trapeze. Another, likening their precision to that of the ballet corps which Samuel L. (Roxy) Rothafel had organized for Radio City Music Hall, referred to the flyers as the Roxyettes of the Air. But reassignments soon broke up the team.[73]

Public Affairs

Participation in civil affairs likewise created goodwill for Army aviation. Maj. Caleb V. Haynes' flight to aid victims of the earthquake in Chile (mentioned earlier) was one of many mercy and rescue missions undertaken by Army and National Guard flyers. Another was the bombing of Mauna Loa on the island of Hawaii in 1935, when an eruption of the volcano sent a stream of molten lava down the side of the mountain toward Hilo. Col. Delos C. Emmons, 18th Composite Wing Commander, flew over the lava flow with Dr. Thomas A. Jaggar, volcanologist, the day before Christmas. They opted for bombs to stop or divert the lava, which was moving down the mountain toward the town at the rate of 800 feet an hour. Colonel Emmons placed Lt. Col. Asa N. Duncan, Commander of the 5th Composite Group, in charge of the operation. Colonel Duncan took 10 bombers, 2 observation planes, and 2 amphibians to Hilo, and set up headquarters at the airport there the day after Christmas. As soon as one of the amphibians could be refueled, it took off for Mauna Loa with Dr. Jaggar accompanied by the bomber pilots. The volcanologist pointed out 2 places to be bombed. The next morning, Duncan sent off 5 bombers at 20-minute intervals with two 300-pound practice bombs each for sighting and two 600-pound demolition bombs with ⅒-second fuzes. Having dropped their bombs, the pilots returned to Hilo, got another load, and bombed a second time. After flying over, Jaggar said: "The hits were remarkably accurate, and exactly where I wanted them distributed."[74] The bombs slowed and stopped the lava flow, saving the town.

Army flyers participated in a number of scientific projects. For instance, they assisted Dr. Robert A. Millikan, of the California Institute of Technology, by carrying instruments to 25,000 feet to record cosmic ray activity. They flew for Johns Hopkins University to determine the presence of bacteria in the upper air. And they collected weather data at high altitude for a study of a new air-mass method of forecasting.[75] After an initial failure, 2 Air Corps officers won fame and honor on a scientific expedition into the stratosphere. Their ascension to an altitude of 13.7 miles grew out of Capt. Albert W. Stevens' long-standing interest in aerial photography and scientific observation at high altitude. Early in 1933 he submitted a plan for a balloon flight to gather data on composition of air, wind direction and velocity, temperature, pressure, cosmic rays, solar spectrum, and effects of altitude on radio transmission. With a big balloon and an airtight gondola, he also hoped to break records set by Auguste Piccard, a Swiss physicist, who ascended to 51,775 feet in May 1931 and to 53,152 feet in August 1932. Foulois approved the project, provided someone besides the Air Corps paid for the balloon, gondola, hydrogen, and lead ballast. Stevens interested the National Geographic Society in a joint project, the society paying the expenses and the Air Corps supplying the people. Meanwhile, Navy Lt. Comdr. Thomas G. W. Settle and Marine Maj. Chester L. Fordney ascended to 61,236 feet in a 600,000-cubic-foot hydrogen balloon built by Goodyear.[76]

Goodyear made a 3-million-cubic-foot hydrogen balloon, the *Explorer*, for the National Geographic Society-U.S. Army Air Corps Stratosphere Flight. The Dow Chemical Company constructed the gondola of Dowmetal,

Goodyear's hydrogen balloon, the *Explorer*, prepares for the National Geographic Society-U.S. Army Air Corps Stratosphere Flight, under the command of Maj. William E. Kepner.

a light-weight alloy. Captain Stevens, who became scientific observer for the flight, obtained assignment of Maj. William E. Kepner as the flight commander, and Capt. Orvil A. Anderson as officer-in-charge of ground operations and alternate pilot. Kepner and Anderson selected a spot 12 miles southwest of Rapid City, South Dakota, for stratocamp. With the aid of many individuals and organizations, they had everything ready by July 9, 1935. After a delay due to weather, Kepner, Stevens, and Anderson lifted off early on the morning of July 28. Discovering a rip in the balloon bag at 60,000 feet, Kepner started down immediately. At 18,000 feet the crew opened a door in the gondola. After inspecting the damage, Kepner believed he could land and save the instruments. But the bottom of the balloon dropped out, making it nothing more than a parachute that might fall at any moment. Kepner ordered the others to prepare to jump. The balloon exploded but all three men parachuted safely to earth, landing near Holdrege, Nebraska, just a short distance from where the gondola crashed. Although the crash damaged or destroyed some instruments, much important data survived.[77]

The National Geographic Society decided to try again, using helium instead of hydrogen. Goodyear built *Explorer II* with a volume of 3.7 million cubic feet, and Dow built a larger, lighter gondola. Both Kepner and Anderson were selected for the Air Corps Tactical School, but that meant passing up the stratosphere flight. Kepner went to school and Anderson became pilot of *Explorer II*. Being senior, Stevens commanded the two-man crew. The site and general arrangements stayed the same as before. Stevens expected to make the flight in June, but got no suitable weather until July 11. During inflation the balloon suddenly collapsed and fell on the gondola, trapping three men on top. Rescued quickly from under the mass of fabric, none sustained injury other than a few scratches. After studying the fabric at Akron, Goodyear altered the design, made a new top for the balloon, and sent *Explorer II* to stratocamp.

Stevens waited more than a month for good weather. During inflation on the night of November 10, the fabric ripped 17 feet. Repair delayed takeoff from 0530 to 0700 the next morning. After inspection and tests at 16,500 feet, the crew started to discharge ballast at 0840. The balloon ascended until above 72,000 feet at 1050. To go higher would require dumping more ballast, which might not leave enough to land safely. After the crew carried out observations and tests for 1 hour and 30 minutes, Anderson valved the balloon to start it down. Opening the ports at 16,000 feet the men prepared to land. Anderson and Stevens pulled the ripcord with the gondola 2 feet or less over a field near White Lake, South Dakota, and the bag deflated at once. When the gondola touched ground and rolled on its side, Anderson and Stevens crawled out. They set an altitude record of 72,395 feet. More important, they brought back valuable data about cosmic rays, electrical conductivity of the atmosphere, vertical distribution of atmospheric ozone, composition of the stratosphere, microorganisms in the stratosphere, and brightness of sun, earth, and sky.[78]

U.S. Army aviation made considerable progress during the 1930s in both coastal and aerial defense. Improvement in coastal defense was illustrated by differences between the "flop" of 1931, when the 2d Bombardment Group failed to find the *Shasta* sixty miles off the North Carolina coast, and the successful interception of the *Rex* in 1938, when planes of the same group flew straight to their objective, more than seven hundred miles at sea. During the interval, acquisition of larger planes with longer range enabled GHQ Air Force to extend coastal defense operations farther out to sea. Then, too, better training and more experience greatly enhanced the airmen's ability to find a hostile fleet reported to be approaching the United States.

All Army flyers learned something about navigation, but with the exception of Albert F. Hegenberger and a few others, they generally received little instruction or practice in dead reckoning and celestial navigation until the Air Corps in 1933 established two small navigation schools, at Langley

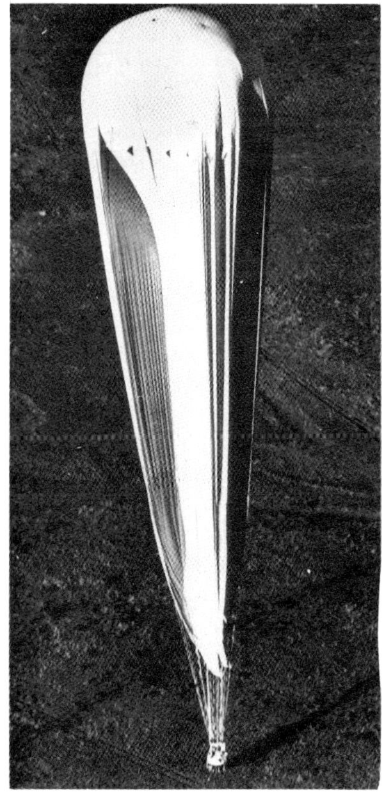

Above: Capt. Albert Stevens (left) and Capt. Orvil Anderson, two-man crew of *Explorer II*; right: bag of *Explorer II* deflates and falls to Earth.

and Rockwell Fields. Additional Army airmen underwent navigation training and skill levels rose after bombardment groups set up schools in 1935 to qualify all of their pilots as navigators. Those who showed exceptional aptitude naturally secured the most significant missions, as 1st Lt. Curtis E. LeMay did to find the *Utah* and intercept the *Rex*. It was not absence of navigational expertise but the Navy scouting force's incorrect position report that prevented Army planes from finding the *Utah* on August 12, 1937. The error in the position report was evidently an honest mistake. It was not a deliberate attempt by the Navy to cause the Army mission to fail, as believed by Army airmen who blindly assumed the anti-Navy attitude William Mitchell had taken from the early 1920s. Nor is there any sound evidence that after the *Rex* affair the Navy was responsible for imposition of a 100-mile limit on Air Force operations at sea, as Army airmen generally believed. Rather, the 100-mile restriction was the Army Chief of Staff's reaction to Air Force violation of War Department policy on publicity concerning such operations. It should also be noted the constraint was tempered by exceptions, the granting of which soon became the prerogative of the Chief of the Air Corps.

While improving its capabilities in coastal defense, GHQ Air Force worked on problems of aerial defense against enemy bombers attacking industrial areas, harbors, cities, important military installations, and other strategic targets in the United States. The air defense system of the 1930s employed pursuit airplanes working in cooperation with antiaircraft guns. And it entailed establishing around the target a defensive area with an aircraft warning net of observation posts and communications to detect and report the presence and location of hostile bombers. Maintaining pursuit planes on alert on the ground, the air force commander dispatched them against enemy planes reported by the warning net. Afterwards he radioed further information coming in from observation posts, so pursuit could intercept and keep the enemy from reaching and attacking the target. Late in the 1930s, blackout of the target area was added to conceal the target from enemy bombers at night. However, at the beginning of the war in Europe in 1939, air defense of the United States still suffered. This stemmed from disagreements among the many Army elements involved in one way or another, and from lack of War Department policy and doctrine on organization and control. Moreover, no civil defense program had yet been developed.

Chapter XXII

Mobilization

When he became Secretary of War in 1936, Harry H. Woodring found the General Staff revising mobilization plans. Believing the nation incapable of recruiting, training, organizing, and equipping forces as rapidly as called for in existing plans, he asked for a "completely adequate and thoroughly practicable" program. The result was the Protective Mobilization Plan of 1937, the title of which, Woodring said, "is fully indicative of our intent." He explained: "In our mobilization planning and our military preparations we contemplate no aggression against any power on earth; we visualize only the possible necessity for armed defense of our own domains."[1]

Under the new plan, active Regular Army units became available to General Headquarters on mobilization day, and active units of the National Guard by M-day plus 30. These, with some fillers, formed an initial force of 400,000 officers and men—165,000 Regular Army, 235,000 National Guard. Thus protected, the nation gained time to create such additional forces as might be required for defense.[2] The initial protective force consisted of units of GHQ Air Force, ready to move on M-day plus 1 to stations assigned by the color plan in effect, [3] plus the 19 active National Guard observation squadrons and their photo sections, to be mobilized by M-day plus 30. In addition, the Air Corps figured it would need 3,000 Reserve officers on M-day. The paragraphs that follow trace developments in National Guard and Reserve aviation in the 1930s and measures taken at the end of the period to strengthen Regular Army aviation.

Secretary of War Harry H. Woodring

National Guard

The National Guard squadrons had disposed of their JN–4 Jennies and received new planes in 1927. At that time, each squadron was authorized 5 trainers (PT–1s) and 3 standard observation planes (O–2s), but units often did not have the full quota. In 1929 the authorization changed to 2 primary trainers (PT–1s), 1 advanced trainer (BT–1), and 5 standard observation ships, with O–11s. O–17s, O–2Hs, O–38s, and O–38Bs being the principal kinds assigned during the next few years. Another change in 1933 allotted the squadrons 8 standard observation aircraft which came to include O–38Es (beginning in 1933), O–43s (1934), and O–46s (1936), all of these being still in use in the fall of 1939. By that time, however, nearly half of the Guard's ships were new O–47As, and more of these were on order. The shift over the years from training to standard service planes permitted Guard squadrons to develop as operational units supporting National Guard divisions. Assignment of late-model planes gave the Guard an aircraft capability comparable to that of observation squadrons of the Regular Army.[4]

Air Guard units trained to support ground forces with artillery adjustment, infantry contact and liaison, and aerial photography. Most National Guard pilots flew frequently, averaging about a hundred hours a year in the mid-1930s, but pilots of some units flew two or three times that many hours. Training in night and blind flying commenced in the early 1930s but received more emphasis (as it did in Regular Army units) after the Air Corps' airmail operation. Since observation planes usually operated individually in division aviation, Guard pilots devoted little time to formation flying. Only on occasion did they undertake long cross-country flights, and then usually to attend air races or National Guard conventions. Their gunnery improved during the late 1930s, when they spent more time on the range as the Air Corps upgraded its gunnery facilities and let the Guard use them.[5]

Air Guard units trained with ground troops whenever possible during summer camp. To work with a division became a momentous event, one first experienced in 1935 by the 41st Division Aviation of the Washington National Guard. The 41st, a "split" division, consisted of units from Oregon, Idaho, Montana, and Washington. The commander was Maj. Gen. George A. White, Adjutant General of Oregon. Reviewing plans for camp at Fort Lewis, Washington, Maj. Robin A. Day, instructor for the 116th Observation Squadron, Washington National Guard, concluded that the division's training officers did not understand aviation. General White, who had no air officer on his staff, needed someone to advise on employment of aviation. So Col. Roy C. Kirtland, Air Officer, Ninth Corps Area, joined the division at Fort Lewis for two weeks.

General White commanded some 7,600 Guardsmen at camp. His aviation included the 116th Photo Section as well as the 116th Observation Squadron. Major Day and 12 other pilots moved with these units from Felts Field, Spokane, Washington, to Fort Lewis. They brought with them all 6 of the squadron's airplanes (1 O–38, 2 O–38Bs, and 3 O–38Es) and borrowed 1 from the California National Guard. Six planes carried radios for working with infantry and artillery. To avoid interfering with daytime operations, the squadron performed 20- and 40-hour checks on its aircraft at night. Since Fort Lewis lacked lighting for night flying, soldiers set highway pots to outline the field one night so the flyers could participate in operations with infantry. Some days 5 planes operated with infantry or artillery while the other 2 flew 5 miles out over the Pacific, 1 towing a target for the pilot and observer of the other ship to shoot. The squadron took vertical and oblique photographs of airports in western Washington for the War Department. The photo section made a mosaic for General White from photographs of the area around Fort Lewis.[6]

The 108th Observation Squadron, Illinois National Guard, usually trained for 2 weeks during the summer at Camp Grant, near Rockford, Illinois. However, in 1936 it went to Michigan for Second Army maneuvers. The pilots flew missions for the National Guard division (33d) as well as for VI Corps and Second Army. First Lieutenant Monro MacCloskey, Adjutant, 33d Division Aviation, thought observation of mechanized cavalry one of the most interesting parts of the exercise. Airborne before daylight, pilots and observers found the cavalry moving to attack the division. Headlights on the vehicles enabled observers to radio reports to the division's command post on the cavalry's strength, movement, and disposition. Lt. Roscoe Burley and the men of the 108th Photo Section used a trailer fitted with equipment for developing and printing pictures. Parking it just outside the division's command post, they tested it for the Materiel Division. In one trial they turned out 316 prints in 1 hour. They were proud of the part they played in another project. The 108th Observation Squadron received a call from the command post to photograph terrain in front of the 33d Division. A pilot and observer on alert took off at once, and 8 minutes later the photo section delivered a wet print to the command post.[7]

During summer camp in 1937, the 115th Observation Squadron and 115th Photo Section of the California National Guard joined their division (40th) in Fourth Army maneuvers.[8] Squadrons and Reservists of Seventh and Ninth Corps Areas participated in 1939 in a Fourth Army command post exercise for units of the Regular Army, National Guard, and Organized Reserve.[9] Meanwhile, in the 1930s, several people proposed changes in National Guard aviation. Lt. Col. Sumpter Smith of the Alabama National Guard, an instructor in pursuit and gunnery during World War I, suggested each squadron be given a pursuit plane. Many of the Guard flyers were pursuit pilots and graduates of the Air Corps Training Center who had served for a year or longer in the Regular Army. Smith thought the Guard would lose them to the Regular Army in an emergency. If they kept up on pursuit flying, they could "step right in and go to work."[10] Discussing this with National Guard officers, the Chief of the Air Corps, General Foulois, discovered some who agreed with Smith, but others wanted to stay with observation. The general himself thought the Guard should have other types of planes besides observation, but units never received them.[11]

While Chief of the Air Corps, General Fechet considered turning some of the Guard's observation squadrons into air force units.[12] On the other hand, the National Guard Bureau favored expansion of observation. Maj. Gen. George E. Leach, chief of the bureau, suggested the Regular Army's corps and army observation be transferred from the Regular Army to the National Guard. The Air Corps Plans Section concluded the Regular Army needed all ten observation squadrons then assigned to the corps areas and departments, and the effectiveness of the Regular Army as an M-day force would be impaired without constant training with observation aviation. The National Guard could not provide proper training and service to the Regular Army.[13] Nothing came of these and similar proposals made later.[14] When the time arrived in the autumn of 1939 to expand National Guard aviation, the bureau was given more observation squadrons. It was to furnish observation for ground forces that the federal government commenced to call National Guard squadrons to active duty in September 1940.

Air Reserve

The Air Corps relied on the Air Reserve for the many officers for mobilization, including 3,000 officers for the initial protective force on M-day. GHQ Air Force, for instance, needed 914, including 540 lieutenants ready as combat pilots. Observation squadrons attached to ground forces required 299; the Air Corps Training Center and technical schools, 581; and the Materiel Division, 422. When the Air Corps figured those requirements in 1937, the Air Reserve total was about 2,900 officers, having fallen from 6,000 over the past 10 years. Death, discharge, transfer, and resignation took many, but the Air Corps eliminated some who were unfit, and other officers did not seek reappointment when their 5-year commissions expired. Further, the Reserve Officers' Training Corps failed to produce large numbers of pilots to replenish the Reserve.

General Patrick's efforts in the 1920s to enlarge the number of ROTC units from 6 to 32 resulted in the addition of 1, at New York University. Enrolling about 1,000 students, the 7 units graduated about 100 a year. Since the Air Corps was furnished no money to send these young men to flying school, the program turned out nonrated second lieutenants. After General Fechet became Chief of the Air Corps, he wanted to abolish ROTC because he thought the graduates of "no more use to us than if they had never been in school." However, he would make a temporary exception in the case of the Massachusetts Institute of Technology, keeping that unit to get engineers. Some colleges objected, but the Air Corps dropped 3 units in 1929 and 3 more in 1932. Students already enrolled at Massachusetts Institute of Technology continued their course until the last class of 11 students graduated in 1935.[15]

General Fechet proposed that pilots be obtained for the Reserve by taking in licensed transport pilots. The Air Corps conducted an experiment to see if graduates of flying schools approved by the Department of Commerce for transport pilots could begin training as Air Corps pilots with the advanced course at Kelly Field. Of 12 who applied, the Air Corps found 6 to be physically, mentally, and morally eligible for appointment as flying cadets. When none of the 6 proved ready for advanced training, the Air Corps sent them to Brooks Field for training, with each to advance as rapidly as his ability warranted. Further study of the matter showed that more than 15,000 civilian pilots were licensed by the Aeronautics Branch of the Department of Commerce. Only about 6,000, those licensed as transport pilots, seemed worth considering. The others had to have as much training in an emergency as persons with no flying experience. Many transport pilots were already affiliated with the Army or Navy Reserve, and most of these would need further training of 1 to 3 months before being ready for the front. And few would be available for combat because most would be wanted either in the aeronautical industry or in the air transportation system.[16]

The Air Corps nevertheless provided a way for an airline pilot or other civilian to secure an Army rating of airplane pilot and a second lieutenancy in the Air Reserve without going near Randolph or Kelly Fields. In fact, if he had been there and washed out, he could not qualify for rating and commissioning without returning and completing the course. On the other hand, he could obtain Air Corps wings and gold bars by showing he had flown 400 hours as pilot, passing a physical examination, demonstrating his ability as a flyer, and completing a written examination.[17] To identify flyers who might be drawn into service in an emergency, the Air Corps Reserve Division commenced registering civilian pilots in 1937. By mid-1939 its list contained the names of about 4,000 men, 1,478 of whom held commercial licenses, 290 limited commercial licenses, and 2,235 private licenses.[18]

While trying to acquire more men for the Reserve, the Air Corps also attempted to rid the Reserve of officers not fully qualified. The seriousness of the problem was evident in statistics for the Reserve in 1930. Of a total 5,700 Reservists, only about 200 were combat-ready pilots. There were 975 others who could fly tactical planes but needed gunnery, bombing, communications, and other training before flying with tactical units in wartime. And 769 more could handle training planes but lacked tactical training of any kind. More than a third of the Reserve officers were without ratings. Over 70 percent of the rated pilots had completed just primary training and thus held solely the junior rating. The Reserve comprised old pilots who had lost flying proficiency; young men who could not attain it; those who no longer met physical standards for pilot duty; and others who, because of the pressure of business or for sundry reasons, had lost interest. The Air Corps could use a number of these in administrative or technical work, and assign others as navigators or observers during an emergency. But what it needed most was pilots ready for combat.[19]

The Air Corps attacked the problem by offering the rating airplane pilot to Reserve junior pilots who had 200 hours of piloting (including 100 hours in service planes, 75 alone, and 20 during the past year), and who passed physical, written, and flying tests.[20] Having provided a way for Reservists to upgrade their ratings, the Air Corps began dropping those who did not do so. It also eliminated many nonrated officers. As a result, the number of Reserve officers fell by mid-1937 to 2,900. About 70 percent of that number were airplane pilots and 13 percent junior pilots; 5 percent held observer, balloon, or airship ratings; and 12 percent no rating at all. These actions, spread over several years, diminished the size but increased the quality of the Air Reserve.[21]

Air Corps efforts to perfect Reserve training were hampered by a shortage of funds. When the Corps withdrew Jennies from Reserve training centers in 1927, it could supply but a few PT–1s to replace them. A bit later, however, it contributed O–2s. Besides these, the centers in the first half of the 1930s flew BT–2Bs, O–1s, and O–19s.[22] At times the Air Corps let the Reservists use a few of its service aircraft for inactive duty training.[23] Like the Air Reserve Association, founded in 1934, the Air Corps wanted more and better equipment for the Reserve, including up-to-date service planes,[24] but could not spare the money. The first significant advance came in August 1936, when the Corps began buying BT–9s for the Reserve and for the flying school at Randolph Field. Built by North American Aviation at Inglewood, California, the BT–9s resembled tactical aircraft more than any other training plane. Later, the Reserve received BT–9As and BT–9Cs each capable of mounting a forward-firing machinegun with camera and a flexible gun in the rear cockpit. These gradually replaced observation aircraft for Reserve training so by August 1939 the Reserve owned 31 BT–9As, 61 BT–9Cs, and 6 O–46As.[25]

The training program called for each Reserve pilot to fly 48 hours a year, encompassing time in both training and service planes and during both inactive and active (14-day) duty. The Air Reserve Association, terming this the "irreducible minimum," wanted 72 hours.[26] General Foulois thought the time should be expanded—"if you have only forty-eight hours you have no business flying."[27] Colonel Arnold doubted the value of Reservists coming back on duty with tactical units after being out a while, even though they got their 4 hours a month.[28] Actually, few Reservists flew 48 hours a year, the average being 20 hours in 1921 and 17 in 1933. Shortage of money and withdrawal of planes from Reserve centers for airmail operations decreased flying again the following year. In 1935, however, Reservists flew 36,000 hours, more than twice the number of the previous year. Flying time climbed to over 46,000 hours in 1938, when 1,037 pilots averaged 40.9 hours, 318 of them in excess of 48.[29]

Until the mid-1930s, Reservists did most of their inactive duty flying in the vicinity of the field, by daylight, and in good weather. The Air Corps restricted the flying of individual Reservists to 30 miles from the airdrome until 1930, when it extended the distance to 100 miles. With more flying time available in 1935, the Air Corps stretched the cross-country limit for Reservists to 250 miles. Extra flying time and additional and better equipment allowed Reservists to get more practice at night, on instruments, with radio, in formation, and in gunnery.[30]

Members of the Air Reserve, like Reservists of other Army branches, seldom secured 14 days of active duty in summer camps more often than once in 3 years. The work of one group at Mitchel Field in 1935 typified the training at Air Reserve camps in the mid-1930s. The pilots averaged 18 hours in the air, making reconnaissance, photographic, cross-country, and night flights, and using camera guns on towed targets. They also attended lectures on navigation, meteorology, radio communications, and chemical warfare; took part in a gas drill; and fired pistols and flexible machineguns on the ground-target range.[31] In 1936 another group flew 2 O–1Gs, 1 O–1E, 1 O–25A, and 4 PT–3As in summer camp at Camp Ripley, Minnesota. Each pilot fired 75 rounds at a sleeve and 75 at a ground target; flew the tow plane (the 0–25); dropped sixteen 17-pound dummy bombs at a 100-foot circle; performed at least one adjustment for 75-millimeter guns; flew infantry contact missions; dropped and picked up messages; took photographs; practiced spot landings from 90, 180, and 360 degrees; and completed a cross-country flight at night.[32]

Reserve units occasionally exercised with ground units. For instance, the 325th Observation Squadron at Bowman Field, Kentucky, became part of the defending force in an air defense exercise at nearby Fort Knox in 1933. But its operations did not favorably impress Capt. Claire L. Chennault, who wrote: "This squadron is equipped with obsolete airplanes and, despite the enthusiasm and ability of its personnel, could do little except provide a ferry and messenger service between Fort Knox and Bowman Field."[33] Other examples were the 403th Pursuit Squadron at Kansas City, Missouri, which worked with the Iowa National Guard in a Minnesota camp during 1935,[34] and the 376th Observation Squadron, which joined Fourth Army maneuvers while encamped in the summer of 1937.[35]

Corps areas sometimes called units of the Organized Reserve for summer camp, and in other cases it was individuals. Either way, the emphasis fell on individual training. Bombardment and pursuit squadrons received no preparation for tactical operations, and little or no instruction in unit administration, the plan being to have individual Reservists fill vacancies in Air Force and Air Corps units during an emergency.[36] In 1937, however, the Air Corps reorganized the Reserve into training squadrons, and revised the program to give unit as well as individual training.[37] The Chief of the Air Corps, General Westover, called the formation of training squadrons "a far-reaching step in the right direction."[38] Reservists put the scheme to test at Schoen Field, Indiana, in the summer of 1937. On active duty two weeks, they formed themselves into a squadron wherein they filled all staff and command positions, prepared flying schedules, and conducted operations. Afterwards, the Air Corps went ahead with a plan specifying 131 training squadrons. The new units, like the old ones, fell under the jurisdiction of corps area commanders, but each was associated with a Regular Army unit. An officer's assignment to a training squadron was taken as a mobilization assignment to a Regular Army unit.[39] Even so, the Air Reserve actually remained as it had been, a pool from which assignments would be made during an emergency.

While trying to build up the Reserve for mobilization, the Air Corps was drawing more and more men from the Reserve for service in peacetime. In mid-1939, it counted 846 on active duty, and 228 others said they would be available for active duty before M-day. Of the 228, 143 were active pilots, 52 had been disqualified for flying, 14 were nonrated, 6 ineligible for assignment because of failure to meet training requirements, and 13 unclassified. Many of them were between 35 and 50 years of age. About half had been on duty with the Civilian Conservation Corps; the rest had little active service, and that mostly at summer camp once in 3 years.[40]

The Air Corps' policy was to take only graduates of its training center, under 35 years of age, for extended active duty. The turnover among these men was rapid. Since they could secure release from active service anytime they wished, commanders were reluctant to give them much responsibility. Many took slight interest in work on the ground; they just wanted to fly. Some tried to build up as much time in the air as possible, hoping to get jobs with the airlines. Others found other work more to their liking than Army service.[41] In June 1939 the Air Corps was short 633 of the 3,000 Reserve officers needed on M-day. Lt. Col. Harry H. Young, Chief of the Air Corps Reserve Division, estimated that but half of the men then enrolled in the Reserve would be available. The remainder—including airline pilots, inspectors for the Civil Aeronautics Administration, and executives in the aircraft industry—could not be spared from their civil jobs.[42]

Toward War

In June 1939, the Fourth Army conducted a command post exercise assuming a situation similar to that General MacArthur used for his command post exercise in 1933: War broke out in Europe; the United States proclaimed neutrality; an Asiatic ally of one of the belligerents struck the west coast without warning, his flyers destroying airplanes at Hamilton and March Fields and crippling the air depot at Sacramento, his troops landing on the California coast and destroying powerplants, oil refineries, and aircraft factories. The U.S. Army went into action against the invaders; the United States declared war; Guardsmen and Reservists mobilized to defend the nation.[43] But real events unfolded differently. Mobilization commenced without awaiting M-day and proceeded piecemeal over many months before the Japanese attack on Pearl Harbor plunged the United States into war.

When Britain and France averted war in the Munich crisis of 1938, America's military leaders placed no faith in Hitler's pledge to demand no more territory. Nor did they share British Prime Minister Neville Chamberlain's belief he had won "peace for our time."[44] War in Europe seemed imminent. The military might of Germany, Italy, and Japan threatened the security of the United States. German and Italian infiltration of Central and South America raised fears of hostile planes at bases within striking distance of our shores. All of this, but chiefly the growth of the German air force, convinced President Roosevelt the U.S. Army needed more aircraft to defend the United States and the Western Hemisphere.

The Air Corps then owned 2,100 serviceable planes, 220 below the number authorized by Congress. President Roosevelt believed the Air Corps needed 20,000 but did not think Congress would approve so many. He called for production of 10,000 aircraft for the Army over 2 years. Generals Arnold, Craig, and Marshall were among those who heard the President outline his program at the White House on November 14, 1938. Afterwards, they tried to balance airplane production with provisions for pilots, maintenance, supplies, and facilities. The War Department also sought to balance the Army as a whole. The Navy asked for more money. This was not what the President had in mind, but he adjusted his program and on January 12, 1939, asked Congress for $300 million to produce at least 3,000 aircraft. Congress responded by raising the Army's airplane authorization from 2,320 to 5,500, approving procurement of 3,251 planes, appropriating money to start the program, and raising the officer authorization to 3,203 and the enlisted to 45,000.[45]

The Air Corps plan approved by the War Department cut the number of observation squadrons for corps and division aviation from 14 to 10, replaced attack aviation with light bombardment, and upped the number of combat groups from 14 to 24. The combat groups consisted of 5 heavy bombardment, 6 medium bombardment, 2 light bombardment, 2 pursuit fighter, 7 pursuit interceptor, and 2 composite. Each heavy and medium bombardment group included a long- or medium-range reconnaissance squadron with the same kind of planes as the bombardment squadrons. Thirteen groups went to GHQ Air Force, 11 to overseas departments.[46]

With an objective of 5,500 aircraft, the Air Corps planned for 3,300 of combat types, 2,073 trainers, and 127 miscellaneous (cargo, photo, and amphibian). It set aside 40 percent as a revolving reserve to keep 3,337 planes operating in peacetime and to replace initial losses in wartime. The Air Corps had 39 B–17Bs on contract, which with the 13 B–17s on hand made a total of 52 heavy bombers. Contracts for 155 B–18s boosted the number of medium bombers to 377. The Air Corps was also buying more P–36s for interceptors and YFM–1s for fighters. When the President signed the appropriations act on April 26, 1939, $50 million was immediately available. The Air Corps quickly ordered 524 P–40s from Curtiss, 13 YP–38s from Lockheed, 12 YP–39s from Bell, 13 YP–43s from Republic, 14 F–2s from Beech, 7 B–24s from Consolidated, and 186 A–20s from Douglas—all of them new models. The P–40 had first flown in October 1938, Douglas' light (attack) bomber in December, and Lockheed's twin-engine, twin-boom interceptor in January 1939. Consolidated had just begun construction of the prototype of its heavy bomber.[47]

The program increased the number of officers by 90 percent, enlisted men by 140 percent. Secretary of War Harry H. Woodring announced on June 1, 1939, the Air Corps would recruit 23,644 enlisted men during the next year. The Air Corps called second lieutenants of the Reserve to expand the officer corps. But as General Arnold pointed out, it could not commission all of them in the Regular Army at one time without creating future promotion problems. It therefore decided to spread the additional Regular Army commissions over 10 years. Reservists would outnumber Regulars at first, but this condition would reverse as the program progressed.[48] Some of the newly recruited enlisted men filled vacancies but most went to school before being assigned to units for duty. The Air Corps opened a school at Scott Field, Illinois, to give enlisted men a basic course before they began technical courses at Chanute Field, Illinois, or Lowry Field, Colorado. In addition it contracted with seven civilian schools to train aircraft mechanics.[49]

The number of cadets entering flying training had been growing, from 340 in Fiscal Year 1937 to 659 in 1938 and 872 in 1939. To procure enough pilots for the 24-group program, the Air Corps planned to start new classes every 6 weeks beginning July 1, 1939, each with about 400 students. It expected nearly half to wash out, mostly during the first 3 months of the course.[50] To accommodate so many students, the Air Corps contracted with 9 civilian flying schools for primary training and transferred specialized training from the advanced course to tactical units. After 3 months at a civilian school, cadets moved to Randolph Field for 3 months of basic training, then to Kelly Field for 3 months of advanced training. Those who completed the 9-month course received wings, the rating airplane pilot, commissions as second lieutenants in the Air Reserve, orders for 3 years of active duty, and assignments to tactical squadrons for further training.[51]

To cope with the huge buildup in personnel and equipment, the Air Corps adopted a policy of using temporary construction for housing and permanent structures for technical buildings. It put up tents at Kelly and Brooks Fields, erected barracks, mess halls, and recreational buildings at Langley, March, Scott, and elsewhere; expanded facilities for technical training at Scott and Lowry; and improved utilities at several stations, among them Bolling Field, Duncan Field at San Antonio, and Fairfield depot in Ohio. Other work encompassed a warehouse at Kelly, a radio building at March, and an addition to the hospital at Schofield Barracks, Hawaii.[52] (*Map 10*)

Construction had already begun on McChord Field, Washington. Congress furnished money to construct a number of other bases planned by the Air Corps for some time. GHQ Air Force got two, one in the northeast at Chicopee Falls, Massachusetts (Westover Field), the other in the southeast, at Tampa, Florida (MacDill). The program also afforded an additional base at

Bruja Point (Howard Field) in the Panama Canal Zone, one in Puerto Rico (Borinquen Field) to defend the Caribbean approach to the canal, and one at Anchorage, Alaska (Elmendorf). And it provided for depots at Mobile, Alabama, and Ogden, Utah.[53]

Thus, as the threat of war grew ever more ominous, the United States hurried to strengthen its defenses. During an extensive expansion program during the summer of 1939, the Air Corps called Reservists to active duty, recruited larger numbers of flying cadets, contracted with civilian schools for help in training pilots and mechanics, began to construct new air bases and to expand facilities at older ones, and placed big orders for aircraft to get more and better equipment for training and operations and to build an operational reserve. The Air Corps then aimed at twenty-four combat groups ready in two years, but Hitler's invasion of Poland on September 1, 1939, and subsequent events rendered that objective obsolete long before it could be achieved.[54]

MAP 10

Chapter XXIII

Summing Up

At the beginning of the Second World War, an independent air force still lay a number of years in the future. Men bent on building an armada of long-range heavy bombers for a strategic offensive found themselves commanding medium bombers for defense. To set U.S. Army aviation of 1939 against the airmen's hopes and desires is to conclude that little had been achieved during the previous twenty years. However, to compare conditions with what they had been in 1919, reveals significant changes and notable progress.

Consider first the matter of organization which aroused so much controversy and consumed so much effort in the years between wars. World War I established aviation as a combatant arm of the U.S Army, integrated it with other arms under division, corps, and army commanders, and gave it the mission of supporting ground forces. Aerial operations during the war convinced Army commanders of the value of aviation, and of the need to r tain it as an integral part of their forces. This carried over into the postwar organization of the Army Air Service. But in the absence of active corps and armies after demobilization, aviation units, along with other field forces, were controlled by corps area and department commanders under the supervision of the War Department General Staff. The Air Service furnished trained men and aeronautical equipment for field forces. Mobilization planning in the 1920s envisioned formation in an emergency of an air force under the control of General Headquarters. It would be employed by the commander in chief in the field to support ground operations or to work separately in carrying out the commander in chief's strategic plans, as the situation might dictate at any specific time.

Many, but not all, Army airmen took exception to both the peacetime and wartime organization of the air arm. These airmen wanted aviation free from Army control, believing that only when divorced from ground forces and given an independent mission would aviation realize its full war-making potential. Then, they held, air power would be decisive in warfare. Army commanders clung to their control over aviation, regarding the air arm as just one of several arms comprising a single force to work under a single head toward a single objective. Hence, in the 1920s the War Department vigorously defended the status quo against attacks from airmen and their allies in Congress and among the general public.

Persons seeking to change aviation organization put forth a variety of plans differing greatly in detail but generally following one of four general forms: (1) Create a new executive department of air with jurisdiction over all aviation in the federal government; (2) form a department of defense with an air force coequal to the Army and Navy; (3) retain Army aviation under the War Department but free it from General Staff supervision by placing it directly under the Secretary of War; and (4) establish an air force, consisting of all Army combat aviation, commanded by an airman under General Staff supervision. The first of these plans, favored by Assistant Secretary of War Benedict Crowell and Brig. Gen. William Mitchell, got its big play in 1919 and early 1920. The second, that for a department of defense, then came to the fore, along with a movement to deal separately with civil aviation which in 1926 produced a civil aeronautics bureau in the Commerce Department. General Patrick was among those who thought the time for independence from the Army had not yet arrived. He therefore proposed the third plan, which would create an air corps in a position similar to the one occupied by the Marine Corps in the Navy Department. General Staff recognition of the growing importance of air power, coupled with the War Department's desire to quiet agitation for an independent air force, led in 1935 to adoption of the fourth scheme—creation of GHQ Air Force under General Staff control. This solution was the least acceptable to airpower enthusiasts.

Chief among the many factors preventing airmen from attaining the type of organization they sought was their inability to marshal convincing evidence supporting their views. Neither the World War I experience nor the existing state of technology bore out their assertions concerning air power. So the airmen based their arguments on hopes and wishes, and took their stand on conjecture as to the future capabilities of aeronautical equipment and techniques. Time and again, the War Department General Staff turned aside the airmen's claims as to the effectiveness of air power by pointing out that "this had not yet been demonstrated." The airmen's efforts to win overwhelming public support proved futile. Americans for the most part were greatly interested in what went on in the air, but they did not rise to assist the flyers' struggle for independence. They wanted to avoid the larger defense forces and greater expenditures of the taxpayers' money that seemed inevitable if the airmen got their way.

After several years of controversy and numerous studies and investigations, Congress in 1926 replaced the Air Service with the Army Air Corps, a change more of semantics than substance. The Air Corps assumed from the Air Service responsibilities to furnish trained men and aeronautical equipment for field forces commanded by corps area and department commanders. At the same time, Congress created the new position of Assistant Secretary of War for Air to which F. Trubee Davison was appointed to assist the secretary in aeronautical affairs. Since this did not remove the Air Corps from General Staff supervision, it added an extra line of authority and gave the Chief of the Air Corps two bosses whose responsibilities were not clearly defined and whose views sometimes conflicted. Denied independence, airmen wanted the power of the Assistant Secretary strengthened. The General Staff wished to abolish the position that cut into its authority. President Franklin D. Roosevelt resolved the matter when, more as an economy measure than anything else, he left the post vacant after Davison departed in 1932.

More significant than establishment of the Air Corps were changes in the composition of the Army's air arm. The interim organization in 1919 contained nearly equal numbers of lighter- and heavier-than-air units. The balloon and airship men held great hopes for the future of lighter-than-air aviation for both military and civil uses. Dirigibles looked promising for reconnaissance, bombardment, and transportation. But the dirigible's slow speed, poor maneuverability, and high vulnerability to both enemy action and natural elements offset advantages of range and carrying capacity. The lighter-than-air enthusiasts found little support from other Air Service members, from elsewhere in the Army (except in the artillery which wanted observation balloons), from members of Congress, or from the general public. Officers of the heavier-than-air branch occupied key positions in the air arm. And with the money, men, and materiel available for Army aviation severely limited, the tendency was toward building up the heavier-than-air branch. A number of dirigible disasters, including the Navy's loss of the *Shenandoah*, *Akron*, and *Macon*, retarded and then halted airship programs in both the Army and the Navy. So the Army's lighter-than-air branch declined and virtually disappeared, eclipsed and consumed by heavier-than-air aviation.

The composition of the heavier-than-air branch itself changed significantly as emphasis shifted from "air service" (auxiliary to ground forces) to "air force" (a separate element with a separate mission). No one in a position of responsibility in the twenties and thirties—not even Mitchell—challenged the proposition that military aviation consisted of both forms of aviation and that both were essential. But no consensus existed as to the mission of each, the types of aviation each needed, how total resources should be apportioned between the two, and what form organization should take. In the Army reorganization of 1920, the War Department assigned observation, pursuit, and attack to ground forces for "air service" in direct support of ground operations. It allocated bombardment aviation to "air force" to work with ground forces or operate as a separate striking force, as directed by the commander in chief in the field. Airmen insisted only observation be designated "air service," and all combat aviation (bombardment, pursuit, and attack) be designated "air force" for independent operations. As a corollary, they would cut back on observation aviation and enlarge the combat air force. By the beginning of World War II, the airmen had gained all these objectives save complete independence.

During the period between wars, observation strength diminished while combat strength increased. Whereas the air arm of 1920 had 14 observation and 13 combat squadrons, observation squadrons dropped to 10 in 1939 while combat squadrons rose to 45. Over that same span, combat aviation was withdrawn from assignment to ground units, leaving just observation for air service supporting ground forces. All combat aviation in the United States came together in GHQ Air Force, formed on March 1, 1935, under the command of an airman, to support ground operations or act as a separate arm as the commander in chief in the field might deem necessary from time to time. Thus, beginning on March 1, 1935, the Army's air arm consisted of four elements: the Air Corps, which supplied men and materiel; observation units assigned to corps areas for direct support of ground forces; aviation units comprising the air forces of three overseas departments (Philippine, Hawaiian, and Panama Canal); and combat units in the United States making up GHQ Air Force.

Several things contributed to these changes. The airmen's insistent demands undoubtedly played a part, though how much would be difficult to determine. Two other interrelated factors may have exerted more influence. Improvements in aeronautical equipment and techniques during the twenties and thirties made the airplane (particularly the bomber) a powerful weapon, and greatly increased the combat capabilities of the Army's air arm. Army officers in key positions saw these changes taking place. They began to realize air power might be better employed if massed under a single commander rather than dissipated among ground forces. Loath to relinquish control over such a powerful and promising weapon, they created GHQ Air Force as an active component of the Regular Army. This, as the General Staff hoped, quieted agitation for an independent air force.

Unsuccessful in achieving complete independence, the airmen further failed to secure a separate budget. With aviation appropriations separated from the Army's, the airmen thought they could obtain more money, sufficient to build the large and powerful air force they envisioned. While it is possible they might have gained additional funds, it seems extremely doubtful they would have received all they wanted. They would still be laboring against all the forces that had kept military expenditures—indeed government expenditures in general—at a relatively low level.

Direct appropriations for Army aviation in the early 1920s averaged about $13 million a year, much less than the Air Service requested and far below what the airmen thought they deserved. This level of appropriations held the number of men in the air arm to around 10,000, well below the 17,500 Congress had authorized in 1920. It counted little with the airmen that similar conditions prevailed throughout the Army. Still, one thing working against the airmen was their inability to produce convincing facts to support their views fully. Another was the American age-old antipathy against a large standing army. In an atmosphere of pacifism and isolationism, peace pacts and disarmament conferences, economy in government, a military policy of minimum defense, with no enemy in sight, and needing proof of the efficacy of air power—the nation would not support a powerful air force.

Nonetheless, Congress in 1926 authorized modernization and expansion of the Army's air arm. Maj. Gen. Mason M. Patrick, Chief of the Air Service, had worked hard to build up Army aviation, and the General Staff long urged the air arm be strengthened. Various studies and investigations pointed in the same direction. Approving a 5-year expansion program, Congress supplied over $147 million in direct appropriations for aviation during the 5-year period (1927–32). The total of men and aircraft grew about 50 percent. But the airmen complained because the 5-year program did not give them more—much more. Further, they lost ground in the Great Depression, even after the need for better equipment and training became apparent in the 1934 airmail operations. Improvement was slow until international conditions hastened the enlargement of the Army's air arm in the late 1930s. All told, the 20 years between wars witnessed substantial growth in the Army's air arm—twice as many men, two and a half times the planes, and a third more tactical units.

Routine training characterized the Army air arm's work in the years between world wars. Exercises and maneuvers at times provided variety, but rarely did the air arm receive orders for military operations. Aviation units served in Germany with the army of occupation immediately after the Armistice. Other units patrolled the southern border between mid-1919 and mid-1921 to prevent raids from Mexico into the United States and to stop smuggling. On another occasion, a squadron joined Army forces the federal government sent to West Virginia to assist civil authorities during a coal miners' strike.

More often, the men of the Army's air arm engaged in peaceful activities of a civil nature. They flew missions to: aid victims of floods, earthquakes, and other natural disasters; search for lost persons; provide aerial ambulance service for people sick or injured; patrol national forests to report fires; and take aerial photographs for the U.S. Geological Survey. Army airmen refined techniques for aerial crop dusting, helped eradicate mosquitoes and other pests, and bombed lava from an erupting volcano to divert the flow from a town that lay in its path. They contributed greatly to the establishment and development of a nationwide system of airports, landing fields, and airways—benefiting civil aviation as much, if not more, than military aviation. Now and then the Army furnished men and equipment for scientific studies and experiments. In fact, one of the great flights in the annals of aviation between the wars was made by two Army airmen, Maj. Albert W. Stevens and Capt. Orvil A. Anderson. On November 11, 1935, they ascended by balloon to 72,395 feet on a scientific expedition sponsored jointly by the National Geographic Society and the Army Air Corps.

The principal purpose of the National Geographic-Army Air Corps flight on November 11, 1935, was to study the stratosphere, but it also aimed at setting a new altitude record, which it did. Competition to see who could fly highest, fastest, and farthest added a great deal to the advancement of aviation in America through the 1920s and 1930s. Initially, Army flyers held an advantage in this game, for at the Armistice in 1918 the Army Air Service occupied the premier place in American aviation. The first aircraft many Americans saw were those of Army men touring the country for the Victory Loan, going around the rim of the United States, racing coast to coast, recruiting for the peacetime air arm, locating landing fields, laying out airways, and looking for forest fires. Competing with them for public attention were ex-Army pilots who bought surplus Army planes and went barnstorming.

Enthusiastic aviators themselves, Army flyers encouraged others to join the flying game. They promoted aviation for transportation, postal service, firefighting, crop-dusting, surveying, exploring, scientific research, and sport. However the peacetime air arm, small in size and short of money, had to curtail activities. Army flyers came under stricter discipline, their flying under greater restraints. Civil aviation grew and advanced. Civilians commenced winning more of the races, setting more records, and capturing more headlines. Still, they did not entirely shut out the Army flyers.

During the twenties and thirties, a never-ending series of pioneering and record-setting flights by aviators of many nations marked aviation progress. U.S. Army flyers played a prominent part. Among their achievements were records in balloon altitude, aircraft endurance with aerial refueling, and aircraft speed. They were the first to make these flights—nonstop transcontinental, around-the-world, and blind solo. Such flights in part reflected technological progress but at the same time assisted in further advancement. Metal replaced wood, wire, and doped fabric in aircraft construction. Performance, reliability, and safety improved. Airplane engines grew more powerful. Enhancing aircraft performance were streamlining, superchargers, adjustable metal propellers, tailwheels, brakes, retractable landing gear, flaps, automatic oxygen systems, and other innovations. Helping to bolster the aircraft's combat capabilities were directional gyroscopes, artificial horizons, and lighted instrument panels. There were also landing lights, voice radios for air-to-air and air-to-ground communication, Norden bombsights, automatic pilots, electrically controlled bomb releases, and other novelties. Greater aircrew comfort and safety were achieved through free-fall parachutes, enclosed cockpits, inflatable life rafts, and survival kits. These and newer operational techniques, together with more and better training in flying, navigating, bombing, and shooting, raised the efficiency of the Army's air arm as an instrument of national defense.

The Army's air arm was smaller but more powerful and effective at the start of the Second World War than at the end of the First. Nothing, however, was ready for a war of the scope and character the Army Air Forces would fight from 1941 to 1945. More—much more—of everything would be needed: men, money, facilities, and equipment, especially heavy bombers. The deficiencies seemed to be in quality as well as numbers. None of the aircraft series in the inventory on September 1, 1939, would play a major combat role during World War II. But planes already on order, such as the A–20, P–40, B–17, and B–24 would see service in the war. Others, including the P–38 and B–29, were in various stages of development. And there were planes not yet conceived, the P–51 being a case in point. Though the airmen spent much time between wars theorizing and experimenting with employment of air power, combat would disclose tactical shortcomings. For instance, a great deal more experimentation would be needed in actual operations to find the best way for executing a raid using hundreds of bombers, and for protecting them from destruction by enemy planes. Further, war revealed that the airmen of the 1930s, obsessed with strategic bombardment, had neglected tactical air power, the air support of ground forces that would be vital to the campaigns of American and Allied armies in North Africa and Europe.

As for leadership, the Air Corps roster of September 1939 listed men like Henry H. Arnold, Carl Spaatz, and Ira C. Eaker. They had the vast experience in aviation, staff work, and command to guide the wartime activities of the Army Air Forces. Yet the exigencies of war would swiftly advance men of limited background into positions of greater responsibility, which they discharged with varying degrees of success.

In the matter of organization, the GHQ Air Force of the late 1930s came to be the prototype of the Army's numbered air forces of World War II. Still, significant changes and extensive elaboration became necessary due to the great size of the numbered air forces and the diversity of their functions. Organizational developments included the creation of subordinate commands for bombardment, fighter, air defense, troop carrier, and tactical air operations. Ultimately, the division of air force functions along strategic and tactical lines led to assignment of strategic and tactical operations to separate air forces. Examples were the Eighth and Ninth in the European theater of operations, and the Fifteenth and Twelfth in the Mediterranean.

But all that lay in the unforseeable future. While the developments of the twenties and thirties did not prepare the Army's air arm for the war it would fight, they laid a foundation upon which to build the Army Air Forces of World War II. Consider, then, the status of Army aviation at the time of the Nazi invasion of Poland. On September 1, 1939, Maj. Gen. Henry H. Arnold presided over an Air Corps of some 26,500 men and 2,200 planes. Under him, Brig. Gen. Barton K. Yount supervised training programs, Brig. Gen. George H. Brett managed materiel activities, and Maj. Gen. Delos C. Emmons commanded GHQ Air Force. High priority went to a gigantic effort begun earlier in the year to expand the air arm by adding personnel, equipment, facilities, and units.

GHQ Air Force, then 4½ years old, still consisted of the original 3 wings headquartered at March, Langley, and Barksdale Fields. These fields and three others (Mitchel, Selfridge, and Hamilton) were all under Air Force control. They served as permanent stations for 4 bombardment, 3 pursuit,

Seven of 10 Air Corps generals, Sept. 1939, (from l. to r.): Brig. Gen. Arnold N. Krogstad, Brig. Gen. Frederick L. Martin, Maj. Gen. Delos C. Emmons, Maj. Gen. Henry H. Arnold, Brig. Gen. Barton K. Yount, Brig. Gen. George H. Brett, and Brig. Gen. Jacob E. Fickel

and 2 attack groups, to which were assigned 10 bombardment, 4 reconnaissance, 9 pursuit, and 6 attack squadrons. The squadrons were usually kept below authorized strength in accord with a long-time Air Corps policy. It prescribed the use of skeletonized units in peacetime and counted on expanding them in an emergency, rather than maintaining a smaller number of units at full strength and adding new ones as resources became available. The proficiency of the older Air Force members stood higher than ever before—in piloting, navigation, bombardment, gunnery, communications, engineering, command, staff, administration, and other skills. But many of the younger men filling a high percentage of the Air Force flying positions were sorely deficient in combat unit experience.

In September 1939 the B–18 medium bomber served as the standard aircraft for GHQ Air Force bombardment units. The 2d Bombardment Group, however, also flew the B–17 heavy bombers it had acquired in 1937. Pursuit units flew P–35s and P–36s. The standard for attack was the A–17, though both attack groups (then on the verge of being redesignated "light bombardment") used B–18s besides. All of these Air Force planes were fairly new and for the most part fully equipped and well maintained. At the beginning of September 1939, Air Corps orders with industry included many new pursuit and attack planes (P–40s and A–20s) and a number of heavy bombers (B–17Bs and B–24s) for the Air Force.

The equipment situation was somewhat different in overseas garrisons, where a little over one-third of the Army's combat squadrons were situated. Of the 3 overseas departments, the Hawaiian had the strongest air component. Its 18th Wing comprised 2 groups with a total of 3 bombardment, 2 reconnaissance, 1 attack, and 2 pursuit squadrons. More than two-thirds of the 18th Wing's bombers were B–18s, the rest older B–12s. Pursuit and attack squadrons flew P–26s and A–12s, models no longer used in GHQ Air Force. The Panama Canal Department had 1 bombardment, 2 reconnaissance, 1 attack, and 2 pursuit squadrons. Its aircraft consisted of B–18s, P–26s, and A–12s. The Philippine Department's air arm was made up of 1 squadron each of bombardment, pursuit, and observation. Army airmen there flew B–10s, P–26s, and 3 series of observation planes (O–19Cs and O–19Es and O–46As). In addition, Army aviation included 9 observation and 3 balloon squadrons assigned to work with ground forces in the United States.

GHQ Air Force was equipped and trained to defend the United States against attack from land or sea. This accorded with national policy and with War Department plans and directives governing the conduct of national defense. At no time during the past 20 years had either the Army's airmen or the War Department thought the air arm strong enough to discharge its defense responsibilities. But neither could persuade Congress and the American people to provide the means for creating and maintaining adequate air power. The air arm was not unique in this respect. Other branches of the Army had suffered the same way, sometimes even more. In recent months, however, the situation had changed due to developments in international affairs. Out of the Munich crisis came more money for military aviation, new authorizations of men and planes for the Army's air arm, and a goal of 24 groups. The Air Corps quickly ordered more airplanes, contracted for training pilots and mechanics, called more Reservists to active duty, expanded facilities, and took other actions to give the Army 24 combat-ready groups by mid-1941. This program barely started when the blow fell on Poland on September 1, 1939.

The fall of France in May 1940 and later world events rendered the 24-group program obsolete. In the 27 months from the German invasion of Poland to the Japanese attack on Hawaii on December 7, 1941, the goal rose from 24 groups to 41, to 54, and then to 84. The story of U.S. Army aviation in that period centers on preparations made as the United States moved from neutrality to war.

Appendices

Appendix 1

U.S. Army Aviation
Principal Facilities: 1919, 1939

The list below shows the principal U.S. Army aviation facilities at the beginning of 1919 and in September 1939 by name and location, with the major agency having primary jurisdiction over aviation activities. See glossary for explanations of abbreviations.

Facility	*Location*	*Jurisdiction*	
		1919	*1939*
Albrook Field	Corozal, C.Z.		PCD
Baker's Field	Rochester, N.Y.	DMA	
Barksdale Field	Shreveport, La.		GHQAF
Barron Field	Everman, Tex.	DMA	
Bolling Field	Anacostia, D.C.	DMA	AC
Brindley Field	Commack, L.I., N.Y.	DMA	
Brooks Field	San Antonio, Tex.	DMA	AC
Call Field	Wichita Falls, Tex.	DMA	
Carlstrom Field	Arcadia, Fla.	DMA	AC(I)
Carruthers Field	Benbrook, Tex.	DMA	
Chanute Field	Rantoul, Ill.	DMA	AC
Chapman Field	Miami, Fla.	DMA	Fourth CA
Clark Field	Fort Stotsenburg, P.I.		PD
Crissy Field	San Francisco, Calif.		Ninth CA(I)
Henry J. Damm Field	Babylon, L.I., N.Y.	DMA	
Dayton supply depot and acceptance park	Dayton, Ohio	DMA	
Camp Dick	Dallas, Tex.	DMA	
Dorr Field	Arcadia, Fla.	DMA	AC(I)
Duncan Field	San Antonio Air Depot, Tex.		AC

Facility	*Location*	*Jurisdiction*	
East Field	Otay, Calif.	DMA	
Eberts Field	Lonoke, Ark.	DMA	
Eglin Field	Valparaiso, Fla.		AC
Elizabeth supply depot and acceptance park	Elizabeth, N.J.	DMA	
Ellington Field	Houston, Tex.	DMA	Eighth CA(I)
Emerson Field	Camp Jackson, S.C.	FA	
Fairfield Air Depot	Patterson Field, Ohio		AC
Florence Field	Fort Omaha, Nebr.	DMA	
France Field	Fort Randolph, C.Z.	PCD	PCD
Garden City supply depot	Garden City, L.I., N.Y.	DMA	
Gerstner Field	Lake Charles, La.	DMA	
Godman Field	Fort Knox, Ky.	FA	Fifth CA
Gray Field	Fort Lewis, Wash.		Ninth CA
Camp Greene	Charlotte, N.C.	DMA	
Hamilton Field	San Rafael, Calif.		GHQAF
Hazelhurst Field	Mineola, L.I., N.Y.	DMA	
Hickam Field	Oahu, T.H.		HD
Hiestand Field	Fort Myers, Fla.	DMA	
Houston supply depot	Houston, Tex.	DMA	
Indianapolis repair depot	Indianpolis, Ind.	DMA	
Kelly Field	San Antonio, Tex.	DMA	AC
Langley Field	Hampton, Va.	DMA	GHQAF
Lawson Field	Fort Benning, Ga.		Inf
Lee Hall balloon school	Lee Hall, Va.	DMA	
Little Rock supply depot	Little Rock, Ark.	DMA	
Los Angeles supply depot	Los Angeles, Calif.	DMA	
Love Field	Dallas, Tex.	DMA	
Lowry Field	Denver, Colo.		AC
Lufbery Field	Wantagh, L.I., N.Y.	DMA	
Luke Field	Ford's Island, T.H.	HD	HD
McCook Field	Dayton, Ohio	BAP	
March Field	Riverside, Calif.	DMA	GHQAF
Marshall Field	Fort Riley, Kans.		Cav
Mather Field	Sacramento, Calif.	DMA	GHQAF
Maxwell Field	Montgomery, Ala.		AC
Middletown Air Depot	Middletown, Pa.	DMA	AC
Mitchel Field	Garden City, L.I., N.Y.	DMA	GHQAF
Moffett Field	Sunnyvale, Calif.		Ninth CA
Fort Monroe balloon station	Fort Monroe, Va.	CA	
Montgomery repair depot	Montgomery, Ala.	DMA	
Muroc Bombing Field	Muroc, Calif.		GHQAF

Facility	*Location*	*Jurisdiction*	
Camp Morrison	Newport News, Va.	DMA	
Morrow Field	Detroit, Mich.	DMA	
Nichols Field	Rizal, P.I.		PD
Olmsted Field	Middletown, Pa.		AC
Park Field	Millington, Tenn.	DMA	
Park Place	Houston, Tex.	DMA	
Patterson Field	Fairfield, Ohio		AC
Payne Field	West Point, Miss.	DMA	
Penn Field	Austin, Tex.	DMA	
Pope Field	Fort Bragg, N.C.	FA	Fourth CA
Post Field	Fort Sill, Okla.	DMA	FA
Randolph Field	San Antonio, Tex.		AC
Ream Field	Oneonta, Calif.	DMA	
Reilly Field	Camp McClellan, Ala.	FA	
Rich Field	Waco, Tex.	DMA	
Richmond supply depot	Richmond, Va.	DMA	
Rockwell Field	San Diego, Calif.	DMA	
Roosevelt Field	Mineola, L.I., N.Y.	DMA	
Ross Field	Arcadia, Calif.	DMA	
Sacramento Air Depot	Sacramento, Calif.		AC
Sacramento supply depot	Sacramento, Calif.	DMA	
San Antonio Air Depot	San Antonio, Tex.		AC
San Francisco supply depot	San Francisco, Calif.	DMA	
Scott Field	Belleville, Ill.	DMA	AC
Selfridge Field	Mount Clemens, Mich.	DMA	GHQAF
Sherman Field	Fort Leavenworth, Kans.		C&GSS
Souther Field	Americus, Ga.	DMA	
Taliaferro Field	Hicks, Tex.	DMA	
Taylor Field	Montgomery, Ala.	DMA	
Valentine Field	LaBelle, Fla.	DMA	
Waco Camp	Waco, Tex.	DMA	
Fort Wayne	Detroit, Mich.	DMA	
Wheeler Field	Schofield Barracks, T.H.		HD
Camp John Wise	San Antonio, Tex.	DMA	
Wright Field	Dayton, Ohio		AC
Wilbur Wright Field	Fairfield, Ohio	DMA	

Appendix 2

U.S. Army Aviation
Field Forces
1920

Second Corps Area

1st Observation Squadron	Mitchel Field
5th Observation Squadron	Mitchel Field
14th Photo Section	Mitchel Field

Third Corps Area

HQ 2d Wing	
HQ 1st Army Observation Group	Langley Field
50th Observation Squadron	Langley Field
88th Observation Squadron	Langley Field
7th Photo Section	Langley Field
19th Airship Company	Langley Field
10th Corps Observation Squadron	Bolling Field
99th Observation Squadron	Bolling Field

Eighth Corps Area

HQ 1st Wing	
HQ 1st Day Bombardment Group	Kelly Field
11th Bombardment Squadron	Kelly Field
20th Bombardment Squadron	Kelly Field
96th Bombardment Squadron	Kelly Field
166th Bombardment Squadron	Kelly Field
2d Photo Section	Kelly Field
HQ 1st Pursuit Group	Kelly Field
27th Pursuit Squadron	Kelly Field
94th Pursuit Squadron	Kelly Field
95th Pursuit Squadron	Kelly Field
147th Pursuit Squadron	Kelly Field
HQ 1st Surveillance Group	Fort Bliss
8th Surveillance Squadron	McAllen, Tex.
Flight	Laredo, Tex.

12th Surveillance Squadron	Nogales, Ariz.
Flight	Douglas, Ariz.
90th Surveillance Squadron	Sanderson, Tex.
Flight	Eagle Pass, Tex.
104th Surveillance Squadron	Fort Bliss
Flight	Marfa, Tex.
1st Photo Section	Fort Bliss
8th Airship Company	Camp Bierne

Ninth Corps Area

9th Corps Observation Squadron	Mather Field
91st Corps Observation Squadron	Rockwell Field
14th Balloon Company	Camp Lewis
24th Balloon Company	Camp Lewis
15th Photo Section	Crissy Field

Field Artillery School — Fort Sill

135th Observation Squadron
23d Balloon Company

Field Artillery Firing Center — Camp Knox

31st Balloon Company
4th Photo Section

Infantry School — Camp Benning

32d Balloon Company

Ordnance Proving Ground — Aberdeen, Md.

258th Heavy Bombardment Squadron
18th Balloon Company

Philippine Department

HQ 1st Observation Group
2d Observation Squadron
3d Observation Squadron
6th Photo Section
17th Balloon Company
27th Balloon Company

Hawaiian Department

HQ 2d Observation Group
4th Observation Squadron
6th Observation Squadron
11th Photo Section
3d Balloon Company
21st Balloon Company

Panama Canal Department

HQ 3d Observation Group
7th Observation Squadron
12th Photo Section

Summary

Wing Headquarters		2
Group Headquarters		7
Squadrons		27
Bombardment	5	
Pursuit	4	
Surveillance	4	
Observation	14	
Airship Companies		2
Balloon Companies		10
Photo Sections		9

Appendix 3

U.S. Army Aviation
Field Forces
1926

Second Corps Area

HQ 9th Observation Group	Mitchel Field
61st Service Squadron	Mitchel Field
1st Division Air Service	
5th Observation Squadron	Mitchel Field
14th Photo Section	Mitchel Field
9th Division Air Service	
1st Observation Squadron	Mitchel Field
8th Photo Section	Mitchel Field

Third Corps Area

HQ 2d Wing	Langley Field
HQ 2d Bombardment Group	Langley Field
11th Bombardment Squadron	Langley Field
20th Bombardment Squadron	Langley Field
49th Bombardment Squadron	Phillips Field
96th Bombardment Squadron	Langley Field
2d Photo Section	Langley Field
59th Service Squadron	Langley Field
19th Airship Company	Langley Field
8th Division Air Service	
99th Observation Squadron	Bolling Field
3d Photo Section	Bolling Field

Fourth Corps Area

4th Division Air Service	
22d Observation Squadron (less det)	Maxwell Field
4th Photo Section	Maxwell Field
Det, 22d Observation Squadron	Pope Field

Fifth Corps Area

5th Division Air Service	
88th Observation Squadron	Wright Field
7th Photo Section	Wright Field

Sixth Corps Area

6th Division Air Service	
15th Observation Squadron	Chanute Field
5th Photo Section	Chanute Field
HQ 1st Pursuit Group	Selfridge Field
17th Pursuit Squadron	Selfridge Field
27th Pursuit Squadron	Selfridge Field
94th Pursuit Squadron	Selfridge Field
95th Pursuit Squadron	Selfridge Field
57th Service Squadron	Selfridge Field

Seventh Corps Area

7th Division Air Service	
16th Observation Squadron	Marshall Field
9th Photo Squadron	Marshall Field

Eighth Corps Area

2d Division Air Service	
12th Observation Squadron	Fort Sam Houston
1st Photo Section	Fort Sam Houston
HQ 3d Attack Group	Fort Crockett
90th Attack Squadron	Fort Crockett
8th Attack Squadron	Fort Crockett
60th Service Squadron	Fort Crockett

Ninth Corps Area

3d Division Air Service	
91st Observation Squadron	Crissy Field
15th Photo Section	Crissy Field

Field Artillery School

44th Observation Squadron	Post Field
23d Photo Section	Post Field

Air Corps Tactical School

50th Observation Squadron	Langley Field
20th Photo Section	Langley Field

Panama Canal Department

HQ 6th Composite Group	France Field
7th Observation Squadron	France Field
24th Pursuit Squadron	France Field
25th Bombardment Squadron	France Field
12th Photo Section	France Field
63d Service Squadron	France Field

Hawaiian Department

HQ 5th Composite Group	Luke Field
6th Pursuit Squadron	Luke Field
19th Pursuit Squadron	Luke Field
23d Bombardment Squadron	Luke Field
72d Bombardment Squadron	Luke Field
4th Observation Squadron	Wheeler Field
11th Photo Section	Luke Field
65th Service Squadron	Luke Field

Philippine Department

HQ 4th Composite Group	Camp Nichols
28th Bombardment Squadron	Camp Nichols
3d Pursuit Squadron	Clark Field
2d Observation Squadron	Kindley Field
6th Photo Section	Camp Nichols
66th Service Squadron	Camp Nichols

Summary

Wing Headquarters		1
Group headquarters		7
Squadrons		32
Bombardment	8	
Pursuit	8	
Attack	2	
Observation	14	
Airship Companies		1
Photo Sections		13
Service Squadrons		7

Appendix 4

Air Units in the Five-Year Plan
May 1926

AIR COMBAT FORCES, U.S

GHQ Air Force
- 1 Air Brigade
  - 1 Communications Squadron
  - 2 Bombardment Wings
    - 2 Bombardment Groups
      - 6 Bombardment Squadrons
      - 2 Service Squadrons
      - 2 Photo Sections
    - 3 Pursuit Groups
      - 9 Pursuit Squadrons
      - 3 Service Squadrons
    - 1 Airship Company

Army Air Units
- 1 Attack Wing
  - 1 Attack Group
    - 3 Attack Squadrons
    - 1 Service Squadron
    - 1 Photo Section
  - 1 Pursuit Group
    - 3 Pursuit Squadrons
    - 1 Service Squadron

Corps and Division Air Services
- 3 Observation Groups
  - 9 Observation Squadrons
  - 3 Service Squadrons
  - 6 Photo Sections

ZONE OF INTERIOR

Special Service Schools
- 1 Training Center
  - 1 Wing
    - 3 School Groups
      - 12 School Squadrons

            3 Service Squadrons
            3 Photo Sections
    1 Tactical School
    1 Technical School
    1 Balloon and Airship School
        1 Airship Group
            4 Airship Companies
            1 Service Squadron
            1 Photo Section

Miscellaneous
- 1 Office, Chief of Air Service
- 1 Engineering Division
  - 1 Engineering School
- 4 Depots
  - 3 Heavier-than-air
  - 1 Lighter-than-air

OVERSEAS

Hawaii
- 1 Wing
  - 1 Pursuit Group
    - 4 Pursuit Squadrons
    - 1 Service Squadron
  - 1 Bombardment Group
    - 4 Bombardment Squadrons
    - 1 Service Squadron
  - 1 Observation Group
    - 2 Observation Squadrons
    - 1 Attack Squadron
    - 1 Service Squadron
    - 1 Photo Section
- 1 Air Depot

Panama
- 1 Wing
  - 1 Pursuit Group
    - 4 Pursuit Squadrons
    - 1 Service Squadron
  - 1 Observation Group
    - 2 Observation Squadrons
    - 1 Bombardment Squadron
    - 1 Service Squadron
    - 1 Photo Section
- 1 Air Depot

Philippines
- 1 Observation Group
  - 1 Observation Squadron
  - 1 Photo Section
  - 1 Bombardment Squadron
  - 1 Pursuit Squadron
  - 1 Service Squadron
- 1 Air Depot

Alaska
- 1 Composite Squadron

APPENDIX 4

Summary
(Tactical Units)

Brigade Headquarters		1
Wing Headquarters		5
Group Headquarters		16
Squadrons		52
Bombardment	12	
Pursuit	21	
Attack	4	
Observation	14	
Composite	1	
Airship Companies		1

Appendix 5

GHQ Air Force Unit List
March 1, 1935

HQ and HQ Squadron, GHQ Air Force	Langley Field
HQ 1st Wing	March Field
HQ 7th Bombardment Group	Hamilton Field
9th Bombardment Squadron	Hamilton Field
11th Bombardment Squadron	Hamilton Field
31st Bombardment Squadron	Hamilton Field
HQ 19th Bombardment Group	Rockwell Field
30th Bombardment Squadron	Rockwell Field
32d Bombardment Squadron	Rockwell Field
93d Bombardment Squadron (Inactive)	(Rockwell Field)
HQ 17th Attack Group	March Field
34th Attack Squadron	March Field
73d Attack Squadron	March Field
95th Attack Squadron	March Field
88th Observation Squadron, LR Amph	Brooks Field
38th Observation Squadron, LRLB (Inactive)	(Rockwell Field)
89th Observation Squadron, LRLB (Inactive)	(March Field)
HQ 2d Wing	Langley Field
HQ 2d Bombardment Group	Langley Field
20th Bombardment Squadron	Langley Field
49th Bombardment Squadron	Langley Field
96th Bombardment Squadron	Langley Field
54th Bombardment Squadron (Detached)	Maxwell Field
HQ 8th Pursuit Group	Langley Field
33d Pursuit Squadron	Langley Field
35th Pursuit Squadron	Langley Field
36th Pursuit Squadron	Langley Field
37th Attack Squadron (Attached)	Langley Field
HQ 9th Bombardment Group	Mitchel Field
1st Bombardment Squadron	Mitchel Field
5th Bombardment Squadron	Mitchel Field
99th Bombardment Squadron	Mitchel Field
14th Bombardment Squadron	Bolling Field
HQ 1st Pursuit Group	Selfridge Field
17th Pursuit Squadron	Selfridge Field
27th Pursuit Squadron	Selfridge Field
94th Pursuit Squadron	Selfridge Field

21st Observation Squadron, LR Amph	Bolling Field
18th Observation Squadron, LRLB (Inactive)	(Mitchel Field)
41st Observation Squadron, LR Amph (Detached)	Kelly Field
HQ 3d Wing	Barksdale Field
HQ 3d Attack Group	Barksdale Field
8th Attack Squadron	Barksdale Field
13th Attack Squadron	Barksdale Field
90th Attack Squadron	Barksdale Field
51st Attack Squadron (Detached)	Maxwell Field
HQ 20th Pursuit Group	Barksdale Field
55th Pursuit Squadron	Barksdale Field
77th Pursuit Squadron	Barksdale Field
79th Pursuit Squadron	Barksdale Field
87th Pursuit Squadron (Detached)	Maxwell Field
42d Bombardment Squadron (Detached)	Kelly Field
40th Attack Squadron (Detached)	Kelly Field
43d Pursuit Squadron (Detached)	Kelly Field
48th Pursuit Squadron (Detached)	Chanute Field
HQ 21st Airship Group	Scott Field
9th Airship Squadron (Detached)	Scott Field
19th Airship Squadron (Attached to 2d Wing)	Langley Field
Service Squadrons	
56th	Selfridge Field
57th	Selfridge Field
58th	Langley Field
59th	Langley Field
60th	Barksdale Field
61st	Mitchel Field
64th	March Field
69th	Hamilton Field
70th	Hamilton Field
71st	Barksdale Field
76th	Rockwell Field
100th	Bolling Field

Appendix 6

GHQ Air Force Unit List
September 1, 1936

HQ & HQ Squadron, GHQ Air Force	Langley Field
HQ & HQ Squadron, 1st Wing	March Field
Base HQ and 4th Air Base Squadron	March Field
HQ & HQ Squadron, 17th Attack Group	March Field
34th Attack Squadron	March Field
73d Attack Squadron	March Field
95th Attack Squadron	March Field
HQ & HQ Squadron, 19th Bombardment Group	March Field
30th Bombardment Squadron	March Field
32d Bombardment Squadron	March Field
38th Reconnaissance Squadron	March Field
23d Photo Section	March Field
Base HQ and 5th Air Base Squadron	Hamilton Field
HQ & HQ Squadron, 7th Bombardment Group	Hamilton Field
9th Bombardment Squadron	Hamilton Field
11th Bombardment Squadron	Hamilton Field
31st Bombardment Squadron	Hamilton Field
88th Reconnaissance Squadron	Hamilton Field
HQ & HQ Squadron, 2d Wing	Langley Field
Base HQ and 1st Air Base Squadron	Langley Field
HQ & HQ Squadron, 2d Bombardment Group	Langley Field
20th Bombardment Squadron	Langley Field
49th Bombardment Squadron	Langley Field
96th Bombardment Squadron	Langley Field
21st Reconnaissance Squadron	Langley Field
2d Photo Section	Langley Field
HQ & HQ Squadron, 8th Pursuit Group	Langley Field
33d Pursuit Squadron	Langley Field
35th Pursuit Squadron	Langley Field
36th Pursuit Squadron	Langley Field
37th Attack Squadron (Attached)	Langley Field
Base HQ and 2d Air Base Squadron	Mitchel Field
HQ & HQ Squadron, 9th Bombardment Group	Mitchel Field
1st Bombardment Squadron	Mitchel Field
5th Bombardment Squadron	Mitchel Field
99th Bombardment Squadron	Mitchel Field
18th Reconnaissance Squadron	Mitchel Field

8th Photo Section	Mitchel Field
Base HQ and 3d Air Base Squadron	Selfridge Field
HQ & HQ Squadron, 1st Pursuit Group	Selfridge Field
17th Pursuit Squadron	Selfridge Field
27th Pursuit Squadron	Selfridge Field
94th Pursuit Squadron	Selfridge Field
HQ and HQ Squadron, 3d Wing	Barksdale Field
Base HQ and 6th Air Base Squadron	Barksdale Field
HQ and HQ Squadron, 3d Attack Group	Barksdale Field
8th Attack Squadron	Barksdale Field
13th Attack Squadron	Barksdale Field
90th Attack Squadron	Barksdale Field
5th Photo Section	Barksdale Field
HQ & HQ Squadron, 20th Pursuit Group	Barksdale Field
55th Pursuit Squadron	Barksdale Field
77th Pursuit Squadron	Barksdale Field
79th Pursuit Squadron	Barksdale Field

Appendix 7

U.S. Army Aviation Field Forces September 1, 1939

HQ & HQ Squadron, GHQ Air Force	Langley Field
HQ & HQ Squadron, 1st Wing	March Field
Base HQ and 4th Air Base Squadron	March Field
HQ & HQ Squadron, 17th Attack Group	March Field
34th Attack Squadron	March Field
73d Attack Squadron	March Field
95th Attack Squadron	March Field
HQ & HQ Squadron, 19th Bombardment Group	March Field
30th Bombardment Squadron	March Field
32d Bombardment Squadron	March Field
38th Reconnaissance Squadron	March Field
Base HQ and 5th Air Base Squadron	Hamilton Field
HQ & HQ Squadron, 7th Bombardment Group	Hamilton Field
9th Bombardment Squadron	Hamilton Field
11th Bombardment Squadron	Hamilton Field
88th Reconnaissance Squadron	Hamilton Field
HQ & HQ Squadron, 2d Wing	Langley Field
Base HQ and 1st Air Base Squadron	Langley Field
HQ & HQ Squadron, 2d Bombardment Group	Langley Field
20th Bombardment Squadron	Langley Field
49th Bombardment Squadron	Langley Field
96th Bombardment Squadron	Langley Field
21st Reconnaissance Squadron	Langley Field
HQ & HQ Squadron, 8th Pursuit Group	Langley Field
33d Pursuit Squadron	Langley Field
35th Pursuit Squadron	Langley Field
36th Pursuit Squadron	Langley Field
Base HQ and 2d Air Base Squadron	Mitchel Field
HQ & HQ Squadron, 9th Bombardment Group	Mitchel Field
1st Bombardment Squadron	Mitchel Field
5th Bombardment Squadron	Mitchel Field
99th Bombardment Squadron	Mitchel Field
18th Reconnaissance Squadron	Mitchel Field
Base HQ and 3d Air Base Squadron	Selfridge Field
HQ & HQ Squadron, 1st Pursuit Group	Selfridge Field

17th Pursuit Squadron	Selfridge Field
27th Pursuit Squadron	Selfridge Field
94th Pursuit Squadron	Selfridge Field
HQ & HQ Squadron, 3d Wing	Barksdale Field
Base HQ and 6th Air Base Squadron	Barksdale Field
HQ & HQ Squadron, 3d Attack Group	Barksdale Field
8th Attack Squadron	Barksdale Field
13th Attack Squadron	Barksdale Field
90th Attack Squadron	Barksdale Field
HQ & HQ Squadron, 20th Pursuit Group	Barksdale Field
55th Pursuit Squadron	Barksdale Field
77th Pursuit Squadron	Barksdale Field
79th Pursuit Squadron	Barksdale Field

Hawaiian Department

HQ & HQ Squadron, 18th Wing	Hickam Field
Base HQ & 17th Air Base Squadron	Hickam Field
HQ & HQ Squadron, 5th Bombardment Group	Hickam Field
23d Bombardment Squadron	Hickam Field
31st Bombardment Squadron	Hickam Field
72d Bombardment Squadron	Hickam Field
4th Reconnaissance Squadron	Hickam Field
50th Reconnaissance Squadron	Hickam Field
Base HQ and 18th Air Base Squadron	Wheeler Field
HQ & HQ Squadron, 18th Pursuit Group	Wheeler Field
6th Pursuit Squadron	Wheeler Field
19th Pursuit Squadron	Wheeler Field
26th Pursuit Squadron	Wheeler Field

Panama Canal Department

HQ & HQ Squadron, 19th Wing	Albrook Field
Base HQ and 15th Air Base Squadron	Albrook Field
HQ & HQ Squadron, 16th Pursuit Group	Albrook Field
24th Pursuit Squadron	Albrook Field
29th Pursuit Squadron	Albrook Field
44th Reconnaissance Squadron	Albrook Field
74th Attack Squadron	Albrook Field
Base HQ and 16th Air Base Squadron	France Field
HQ & HQ Squadron, 6th Bombardment Group	France Field
25th Bombardment Squadron	France Field
7th Reconnaissance Squadron	France Field

Philippine Department

Base HQ and 20th Air Base Squadron	Nichols Field
HQ & HQ Squadron, 4th Composite Group	Nichols Field
2d Observation Squadron	Nichols Field
3d Pursuit Squadron	Nichols Field
28th Bombardment Squadron	Clark Field

Assigned to Corps Areas

First Corps Area	
97th Observation Squadron	Mitchel Field
Third Corps Area	
3d Observation Squadron	Langley Field
Fourth Corps Area	
16th Observation Squadron (Less Flight B)	Pope Field
2d Balloon Squadron	Pope Field
Fifth Corps Area	
12th Observation Squadron (Less Flight C)	Godman Field
Sixth Corps Area	
15th Observation Squadron	Scott Field
Eighth Corps Area	
22d Observation Squadron	Brooks Field
Ninth Corps Area	
91st Observation Squadron	Gray Field
82d Observation Squadron	Moffett Field
3d Balloon Squadron	Gray Field

Assigned to War Department Activities

Field Artillery School	
12th Observation Squadron, Flight C	Post Field
1st Balloon Squadron	Post Field
Infantry School	
16th Observation Squadron, Flight B	Lawson Field
Cavalry School	
1st Observation Squadron	Marshall Field

Summary

Air Force Headquarters		1
Wing Headquarters		5
Group Headquarters		14
Squadron (Heavier-Than-Air)		55
Bombardment	15	
Reconnaissance	8	
Pursuit	15	
Attack	7	
Observation	10	
Balloon Squadrons		3

Appendix 8

Cost of U.S. Army Aviation Fiscal Years 1919–1939

Direct Appropriations

Each fiscal year (July 1 to June 30) Congress appropriated money for the Air Service/Air Corps to pay for new aircraft and other aeronautical equipment, research and experimental work, flying fields and ground equipment, improvement of stations, gasoline and oil, helium for dirigibles and balloons, transportation of materiel, travel expenses of officers, nonpersonal services, salvage of wrecked aircraft, damaged caused by aircraft, and various miscellaneous items. These direct appropriations, which ranged between a low of about $12.6 million (1924) and a high of $70.6 million (1939) are often taken as the measure of government financial support of Army aviation. The figures below are for direct appropriations for the Air Service/Air Corps for Fiscal Years 1919 through 1939 and show these appropriations as a percentage of the Army appropriations for all military purposes (i.e., total Army appropriations less funds for civil projects). The figures are rounded to the nearest tenth of a million dollars and to the nearest tenth of one percent.

Fiscal Year	*Direct Appropriations (millions)*	*Percent of Total Army Military Appropriations*
1919	$925.3	9.3
1920	25.0	3.2
1921	33.0	8.4
1922	19.2	5.9
1923	12.9	4.8
1924	12.6	4.9
1925	12.8	5.0
1926	14.9	5.7
1927	15.3	5.7

Year	*Appropriations*	*Percent of Total*
1928	20.6	7.3
1929	24.6	7.9
1930	34.7	10.4
1931	35.8	10.5
1932	31.5	9.4
1933	25.4	8.8
1934	23.3	8.4
1935	27.4	10.7
1936	45.4	13.3
1937	59.4	15.5
1938	58.6	14.1
1939	70.6	15.7

The above figures are derived from a compilation made in November 1939 by Thomas R. Baldwin of the Legislative Reference Service, Library of Congress, and published in the *Congressional Record*, 76th Congress, 3d session, Volume 86, part 1, pages 556–57. Several other compilations exist for the same period but the figures do not always agree, and no one can be taken with any reasonable assurance as to its completeness and accuracy. Mr. Baldwin's has been used because it appears at least as good as and perhaps better than any of the others.

Indirect Appropriations

Direct appropriations accounted for only part of the money the U.S. Government spent on Army aviation. Other funds came from appropriations to other branches of the Army and other funding categories. These indirect aviation appropriations provided pay for officers and enlisted men. They also paid for radios, medical services, bombs and ammunition, food and other Quartermaster supplies, tear gas and other chemicals, construction, Organized Reserves, Reserve Officers' Training Corps, printing and binding, and among numerous other things, contingent expenses. The War Department Special Committee on Army Air Corps (Baker Board) reported that indirect expenditures exceeded direct expenditures in the fifteen-year period from July 1, 1919, to June 30, 1934 (*Report*, page 53).

In addition, the cost of aviation included the value of materiel issued to the Air Service/Air Corps from the war reserve. This accounted for nearly seventeen percent of the total cost of Army aviation for Fiscal Years 1919–1939.

The following figures for expenditures for Army aviation are derived from computations the War Department made for the Baker Board and printed as Exhibit 4 in the Board's *Report*. For the present purpose, the figures have been rounded to the nearest tenth of a million dollars.

Expenditures

Fiscal Year	*From Appropriations* Direct (millions)	Indirect (millions)	*From War Reserve (millions)*	*Total (millions)*
1920	$25.9	$19.5	$15.0	$60.4
1921	32.3	14.8	15.0	62.1
1922	18.1	14.5	15.0	47.6
1923	12.9	13.1	10.4	36.4
1924	13.3	14.6	12.8	40.7
1925	14.5	15.6	10.0	40.1
1926	17.0	16.9	12.1	46.0
1927	17.0	18.8	9.4	45.2
1928	21.5	24.3	8.0	53.8
1929	27.4	25.4	6.5	59.3
1930	32.0	28.0	10.7	70.7
1931	36.6	35.6	6.3	78.5
1932	29.8	42.3	4.8	76.9
1933	24.7	32.3	2.1	59.1
1934	19.3	32.9	--	52.2
Cumulative total				829.0

In that fifteen-year period, the total cost of aviation ranged between 13.1 and 22.7 percent of total military expeditures, with an average of 18.2 percent.

Appendix 9

Government Officials

with dates of service

Presidents of the United States

Woodrow Wilson	March 1913	March 1921
Warren G. Harding	March 1921	August 1923
Calvin Coolidge	August 1923	March 1929
Herbert Hoover	March 1929	March 1933
Franklin D. Roosevelt	March 1933	April 1945

Secretaries of War

Newton D. Baker	March 1916	March 1921
John W. Weeks	March 1921	October 1925
Dwight F. Davis	October 1925	March 1929
James W. Good	March 1929	November 1929
Patrick J. Hurley	December 1929	March 1933
George H. Dern	March 1933	August 1936
Harry H. Woodring	September 1936	June 1940

Chiefs of Staff

Maj Gen Peyton C. March	May 1918	June 1921
General of the Armies John J. Pershing	July 1921	September 1924
Maj Gen John L. Hines	September 1924	November 1926
Gen Charles P. Summerall	November 1926	November 1930
Gen Douglas MacArthur	November 1930	October 1935
Gen Malin Craig	October 1935	August 1939
General of the Army George C. Marshall	September 1939	November 1945

Heads of Army Aviation

Directors of Air Service		
John D. Ryan	August 1918	November 1918
Maj Gen Charles T. Menoher	January 1919	June 1920
Chiefs of the Air Service		
Maj Gen Charles T. Menoher	June 1920	October 1921
Maj Gen Mason M. Patrick	October 1921	July 1926

AVIATION IN THE U.S. ARMY

Chiefs of the Air Corps		
Maj Gen Mason M. Patrick	July 1926	December 1927
Maj Gen James E. Fechet	December 1927	December 1931
Maj Gen Benjamin D. Foulois	December 1931	December 1935
Maj Gen Oscar Westover	December 1935	September 1938
Maj Gen Henry H. Arnold	September 1938	June 1941
Commanding Generals, GHQ Air Force		
Maj Gen Frank M. Andrews	March 1935	March 1939
Maj Gen Delos C. Emmons	March 1939	June 1941

Assistants to the Chiefs of the Air Service/Air Corps (temporary brigadier generals)

In the Office of the Chief		
William Mitchell	July 1920	April 1925
James E. Fechet	April 1925	December 1927
Benjamin D. Foulois	December 1927	June 1929
William E. Gillmore	June 1929	June 1930
Benjamin D. Foulois	July 1930	December 1931
Oscar Westover	December 1931	December 1935
James E. Chaney	July 1934	May 1935
Henry H. Arnold	December 1935	September 1938
Walter G. Kilner	September 1938	November 1939
Barton K. Yount	February 1939	October 1940
At the Air Corps Training Center		
Frank P. Lahm	July 1926	July 1930
Charles H. Danforth	July 1930	June 1934
James E. Chaney	May 1935	July 1938
Barton K. Yount	July 1938	January 1939
With the Materiel Division		
William E. Gillmore	July 1926	June 1929
Benjamin D. Foulois	June 1929	July 1930
Henry C. Pratt	July 1930	July 1934
Augustine W. Robins	January 1935	January 1939
George W. Brett	February 1939	October 1940

Commandants

Tactical School		
Maj Thomas DeW. Milling	1920	1921

Commandants

Tactical School		
Lt Col Charles H. Danforth	1921	1924
Maj Oscar Westover	1924	1926
Lt Col Clarence C. Culver	1926	1929
Lt Col Jacob W. S. Wuest	1929	1930
Lt Col Roy C. Kirtland	1930	1931
Lt Col John F. Curry	1931	1935
Col Arthur G. Fisher	1935	1937
Brig Gen Henry C. Pratt	1937	1938
Col Millard F. Harmon, Jr.	1938	1939
Col Walter R. Weaver	1939	1940

Technical School		
Maj Frederick L. Martin	1922	1924
Maj William C. McChord	1924	1928
Maj Leo G. Heffernan	1928	1928
Lt Col Seth W. Cook	1928	1929
Maj Leo G. Heffernan	1929	1930
Lt Col Jacob W. S. Wuest	1930	1931
Lt Col James A. Mars	1931	1934
Col Junius W. Jones	1934	1938
Col Gerald C. Brant	1938	1940

Appendix 10

Aircraft in Use in the U.S. Army 1919–1939

This special appendix pictures the principal aircraft mentioned within the narrative of this volume. Although over one hundred aircraft that were in the inventory of the U.S. Army from 1919 to 1939 are illustrated, this appendix is not an encyclopedic reference nor an exhaustive list of all the planes serving the Army during those twenty years. Photographs of specific aircraft models cited in the text were not always available. Therefore, only one model of a particular aircraft is shown throughout as the closet representative of that design. The editors have relied extensively on the standard work by Gordon Swanborough and Peter M. Bowers, *United States Military Aircraft Since 1908*, revised edition, London: Putnam & Co., Ltd., 1971. Readers may refer to this book as a reliable compendium of technical data on the aircraft shown here.

For ease of reference, the aircraft illustrations are alphabetized according to their official military designations (where applicable) or the manufacturers' designations. The names of aircraft manufacturers are supplied, but count for the purposes of alphabetization in this listing only where there is no numerical designation, such as in the case of the Laird Swallow, which is carried under "L". The "X" (experimental status) and "Y" (service test status) models are shown under their designations as on–line aircraft: the XP–38 is thus shown under P–38. The addition of the numeral "1" to the "Y" prefix (as in the Y1O–26) between 1931 and 1936 indicated that the aircraft had been procured with "F–1" funds rather than the regular Army Air Corps construction funds. The official symbols typically in use during the period covered in this book are listed below. (Symbols in the appendix not appearing below are generally manufacturers' designations, both domestic and foreign.)

A	Attack
AT	Advanced Trainer
B	Bomber
BC	Basic Combat
BT	Basic Trainer
C	Cargo
CO	Corps Observation
F	Photographic
FM	Fighter, Multiplace
GA	Ground Attack
LB	Light Bomber
NBL	Night Bombardment, Long Distance
NBS	Night Bombardment, Short Distance
O	Observation
OA	Observation Amphibian
P	Pursuit
PN	Naval Aircraft Factory (Pursuit)
PT	Primary Trainer
PW	Pursuit, Water Cooled
R	Racer
TA	Training, Air Cooled
TW	Training, Water Cooled

Curtiss A–3 (A–3B shown)

Curtiss A–8 (XA–8 shown)

Consolidated A–11

Curtiss A–12

Northrop A–16

Northrop A–17

Curtiss A–18 (Y1A–18 shown)

Douglas A–20

Fokker (Atlantic) AO–1

Curtiss AT–4

Curtiss AT–5 (AT–5A shown) *Denver Public Library*

Keystone (Huff-Daland) B–1 (XB–1 shown)

Curtiss B–2

Keystone B–3 (B–3A shown)

Keystone B–4

Keystone B–5 (B–5A shown)

Keystone B–6 (B–6A shown)

Douglas B–7 (Y1B–7 shown)

Boeing B–9 (YB–9 shown)

Martin B–10

Martin B–12

Boeing B–15

Boeing Y1B–17

Douglas B–18

Consolidated B–24 (YB–24 shown)

Boeing B–29

North American BC–1

Breguet 14 (14A–2 shown)

Douglas BT–1

Seversky BT–8

North American BT–9

Standard Caproni

Douglas C–1

Fokker (Atlantic) C–2

Ford C–3

Ford C–4 (C–4A shown)

Sikorsky C–6

Fokker (Atlantic) C–7 (C–7A shown)

Lockheed C–12 (Y1C–12 shown)

Fokker C–14 (YIC–14 shown)

Bellanca C–27 (Y1C–27 shown)

Douglas C–33

Douglas C–34

Douglas C–39

Lockheed C–40

Engineering Division CO–5 (XCO–5 shown)

De Havilland DH–4 *Denver Public Library*

De Havilland DH–9 (RAF D.H.9B shown)

Douglas DWC

Fairchild F–1 (variant of C–8)

Beech F–2

Bristol F.2B

Fokker D–VII

Bell FM–l (XFM–1 shown)

Boeing GA–1 (Engineering Division GAX shown)

Curtiss H–4 (built for British)

Standard SJ–1

Curtiss JN–4 *Denver Public Library*

Laird Swallow

Keystone (Huff-Daland) LB–1

Keystone (Huff-Daland) LB–5

Keystone LB–6

Keystone LB–7

Keystone LB–10

Packard-Le Pere LUSAC–11

Martin MB–1 *Glenn Martin Co.*

Curtiss-Martin MB–2 (redesignated NBS–1) *Denver Public Library*

Thomas-Morse MB–3

Thomas-Morse MB–6

Engineering Division (Whiteman-Lewis) NBL–1 "Barling Bomber"

Curtiss NC–4

Nieuport 28

Consolidated NY–2

Curtiss O–1

Douglas O–2 (O–2J shown)

Douglas O–7

Curtiss O–11

Consolidated O–17 (XO–17 shown)

Thomas-Morse O–19

Douglas O–25 (O–25C shown)

Curtiss O–26 (Y1O–26 shown)

Douglas O–32

Douglas O–35 (XO–35 shown)

Douglas O–38 (O–38B shown)

Curtiss O–39

Douglas O–43 (O–43A shown)

Douglas O–46 (O–46A shown)

North American O–47 (O–47B shown)

Handley Page O/400

Loening OA–1 (OA–1A shown)

Loening (Keystone) OA–2

Douglas OA–3

Douglas OA–4

Douglas OA–5 (YOA–5 shown)

Grumman OA–9

Orenco (Curtiss) D

APPENDIX 10

Curtiss P–1

Curtiss P–3 (XP–3A shown)

Curtiss P–6 (P–6E shown)

Boeing P–12 (P–12E shown)

Berliner-Joyce P–16 (XP–16 shown; became PB–1)

Consolidated P–25 (Y1P–25 shown)

Boeing P–26

Consolidated P–30 (became PB–2)

Seversky (Republic) P–35

Curtiss P–36 (Y1P–36 shown)

Lockheed P–38 (YP–38 shown)

Bell P–39 (YP–39 shown)

Curtiss P–40

Republic P–43

North American P–51 (**XP**–51 shown)

Naval Aircraft Factory PN–9

Consolidated PT–1

Consolidated PT–3

Consolidated PT–11D

Consolidated PT–12

Stearman PT–13

Loening PW–2

Curtiss PW–8

Boeing PW–9 (XPW–9 shown)

Verville R–1

Verville-Sperry R–3

Curtiss R3C–1

Loening R–4

Thomas-Morse R–5

Curtiss R–6

Curtiss R–8

Salmson 2A–2

Thomas-Morse S–4 *National Archives*

Sikorsky S–42

Royal Aircraft Factory SE–5

S.P.A.D. XIII

Ansaldo SVA–5

Fokker T–2

Dayton-Wright TA–3

Travel Air Model R

Dayton-Wright TW-3

Engineering Division VCP-1

Engineering Division VCP–R

Vought VE–7

Notes

Notes

Chapter I

Demobilization

1. Benedict Crowell and Robert F. Wilson, *Demobilization* (Vol IV of *How America Went to War*) (New Haven, 1921), 1–2; *Order of Battle of the United States Land Forces in the World War (1917–19), Zone of the Interior*, Vol III, pt 1 (Washington: U.S. Army Historical Division, 1949), p 72.

2. Progress Report, Division of Military Aeronautics, Nov 13, 1918, quoted in Ernest L. Jones, "Chronology of Aviation" (MS in AFHRC 168.6501–37).

3. Maurer Maurer, ed, *Air Force Combat Units of World War II* (Maxwell AFB, Ala., 1961); Maurer Maurer, ed, *Combat Squadrons of the Air Force, World War II* (Maxwell AFB, Ala., 1969).

4. In some instances such enlisted detachments were broken up at an embarkation camp in France and reorganized into companies of about 250 men and 2 officers for shipment.

5. Crowell and Wilson, *Demobilization*, Chaps II and III.

6. *Ibid*, Chap V.

7. *Division of Military Aeronautics Weekly News Letter*, Jan 4, 1919, p 7.

8. Crowell and Wilson, *Demobilization*, 105–108.

9. DMAWNL, Nov 23, 1918, p 3 (citing War Department Circular 75, Nov 20, 1918); Jones Chronology, Dec 28, 1918, citing DMAWNL.

10. DMAWNL, Nov 23, 1918, p 3.

11. USA Hist Div, *Order of Battle*, 1949, III, pt 1, 72–73.

12. DMAWNL, Jan 18, 1919, p 6.

13. 41 U.S. *Statutes at Large* (hereafter *Stat*) 286; *Air Service News Letter*, Sep 30, 1919, p 1.

14. ASNL, Sep 3, 1919, p 4.

15. DMAWNLs, Feb 1, 1919, pp 15–16, Mar 29, 1919, p 16; ASNL, Apr 26, 1919, p 2.

16. ASNLs Apr 12, 1919, p 1, May 24, 1919, p 5, Aug 27, 1919, pp 1–2, Sep 3, 1919, p 7, Sep 9, 1919, p 5, Sep 18, 1919, p 10, Sep 23, 1919, p 3, Sep 30, 1919, p 1, Oct 8, 1919, p 5; Jones Chronology, Sep 14, 1919.

17. ASNL, Dec 13, 1919, p 3. The nine were Field E. Kindley, John O. Donaldson, Reed M. Chambers, Harvey W. Cook, Martinus Stenseth, James A. Healy, Clayton L. Bissell, Arthur R. Brooks, and Harold H. George.

18. Jones Chronology, Dec 1, 1919.

19. ASNLs Apr 20, 1920, p 13, May 26, 1920, pp 17, 20, Jun 2, 1920, p 15, Jun 10, 1920, pp 14–15, 20–23, Jun 18, 1920, pp 10, 12, Jun 25, 1920, p 10, Jul 10, 1920, p 15.

20. Jones Chronology, Jun 30, 1920; 41 *Stat* 759.

21. USA Hist Div, *Order of Battle*, III, pt 1, 107–115.

22. United States Spruce Production Corporation, *History of the Spruce Production Division, United States Army and United States Spruce Production Corporation* (Portland, Oregon, n.d.), pp 93–105; *Report on Audit of the United States Spruce Production Corporation*, H. Doc. 235, 80th Cong, 1st sess, 1947.

23. Crowell and Wilson, *Demobilization*, 206–7.

24. *Ibid*, 135.

25. *Annual Report of the Director of Air Service*, 1919, pp 15–17.

26. Crowell and Wilson, *Demobilization*, 288.

27. *Ibid*, 212–13.

28. *Annual Report of the Director of Air Service*, 1919, pp 27–32; Crowell and Wilson, *Demobilization*, 281.

29. DMAWNL, Dec 7, 1918, p 11.

30. ASNL, Jun 6, 1919, p 7; *Annual Report of the Director of Air Service*, 1920, pp 40–41; Crowell and Wilson, *Demobilization*, 272–73.

31. Crowell and Wilson, *Demobilization*, 280; ASNLs, Apr 5, 1919, p 3, May 4, 1920, p 1.

32. ASNL, May 4, 1920, p 17.

33. ASNLs, Jul 10, 1920, p 13, Aug 27, 1920, p 13, Nov 20, 1920, p 14.

34. ASNLs, May 15, 1923, p 6, Jun 4, 1923, p 6, Oct 27, 1923, p 19, Jan 7, 1924, p 11, Jun 30, 1924, p 9.

35. Crowell and Wilson, *Demobilization*, 208; ASNL, Jul 19, 1923, pp 5–6.

36. *Annual Report of the Director of Air Service*, 1919, pp 27–32, 1920, pp 40–41.

37. *Annual Report of the Chief of Air Service*, 1922, p 20, 1923, pp 40–41.

38. ASNLs, Jun 30, 1922, p 4, Sep 6, 1922, p 14, Feb 20, 1923, p 3, May 2, 1923, p 14, May 15, 1923, p 18.

39. *Annual Report of the Chief of Air Service*, 1923, pp 40–41.

Chapter II

The Flying Game

1. Jones Chronology, Nov 14, 1918; DMAWNLs, Feb 1, 1919, p 3, Feb 22, 1919, p 9.

2. ASNL, Apr 12, 1919, p 4.

3. Jones Chronology, Jan 6, 1919, citing DMAWNL, Jan 11, 1919, p 1.

4. DMAWNL, Mar 1, 1919, p 15.

5. DMAWNLs, Feb 1, 1919, p 1, Mar 8, 1919, p 4, Mar 15, 1919, pp 4–5, 15–16; *Aircraft Year Book* (New York, 1920), pp 140–46.

6. DMAWNL, Feb 8, 1919, p 2.

7. DMAWNL, Mar 22, 1919, p 7.

8. ASNLs, Apr 5, 1919, pp 5–7, Apr 12, 1919, p 3, May 3, 1919, pp 2, 12, May 10, 1919, pp 2–3, May 24, 1919, pp 1–2.

9. Jones Chronology, Apr 14, 1919.

10. ASNL, Apr 19, 1919, p 1.

11. Not the first transcontinental flight, but the first by the U.S. Air Service.

12. DMAWNLs, Dec 7, 1918, p 17, Dec 14, 1918, p 3, Dec 21, 1918, p 18, Jan 4, 1919, p 1, Feb 8, 1919, pp 4–5, Feb 15, 1919, p 12; *Aircraft Year Book* (New York, 1919), pp 336, 343; Jones Chronology, Dec 4, 1918.

13. ASNL, Jun 10, 1920, p 16.

14. ASNL, Nov 23, 1920, p 9.

15. ASNL, Oct 31, 1924, p 11. Henry H. Arnold and Ira C. Eaker used the phrase "This Flying Game" as the title for a book.

16. DMAWNL, Dec 14, 1918, p 10.

17. Between September 1, 1917, and November 11, 1918, when most of the flying had been in training, 264 Air Service men had lost their lives in fatal accidents in the United States. The flying time in that period was 811,072 hours, the fatality rate being 1 death in each 3,072 hours of flying. From November 11, 1918, to March 13, 1919, there were 57 fatalities in 125,864 flying hours, or 2,208 hours per fatality. The higher rate, the Air Service concluded, was the result of the kind of flying being done (ASNL, Apr 12, 1919, p 11).

18. ASNLs, Aug 21, 1920, p 10, Jun 2, 1920, p 9.

19. ASNL, May 4, 1920, p 11.

20. DMAWNL, Jan 4, 1919, p 16; ASNL, Dec 19, 1919, p 3.

21. Lowell Thomas and Edward Jablonski, *Doolittle* (Garden City, N.Y., 1976), pp 43–45.

22. Jones Chronology, January 21, 1919, citing various sources, including Macauley's report and the *Saturday Evening Post*, May 10, 1919.

23. Jones Chronology, April 14, 1919, citing Macauley's report and other sources.

24. ASNL, Sep 17, 1920, pp 7–9.

25. Jones Chronology, Jul 25, 1919; "Coastal and Trans-Continental Flights," *U.S. Air Service*, II (Sep 1919), 21.

26. Rutherford S. Hartz, "Report on Round the Rim Flight," in ASNL, Nov 18, 1919, pp 1–8; *Aircraft Year Book* (New York, 1920), pp 269–270; "Coastal and Trans-Continental Flights," *U.S. Air Service* II (Sep 1919), 21; Rutherford S. Hartz, "Round the Rim of the United States," *U.S. Air Service* II (Dec 1919), 32.

27. ASNLs, Aug 22, 1919, p 12, Sep 1, 1919, p 3, Sep 9, 1919, p 5, Sep 18, 1919, pp 9–10, Sep 23, 1919, p 1, Sep 30, 1919, p 5, Oct 8, 1919, p 5, Oct 18, 1919, p 8, Oct 31, 1919,

pp 4–5; "Coastal and Trans-Continental Flights," *U.S. Air Service* II (Sep 1919), 21–22.

28. Jones Chronology, Sep 2, 1919.

29. ASNL, Apr 19, 1919, p 1; Jones Chronology, Apr 14, 1919.

30. ASNL, Aug 22, 1919, pp 1–2; *Aircraft Year Book* (New York, 1920), pp 270, 272; Ray Landis Bowers, "The Transcontinental Reliability Test: Aviation After World War I" (M.A. thesis, University of Wisconsin, 1960), pp 32–35.

31. ASNLs, Aug 27, 1919, p 1, Oct 8, 1919, pp 1–2 (including Burt's report), Nov 7, 1919, p 1; Jones Chronology, Sep 26, 1919.

32. Air Service, *Report on First Transcontinental Reliability and Endurance Test*, in *Air Service Information Circular* (Heavier-than-Air), I (Feb 5, 1920), Preface (signed by Menoher); ASNL, Dec 13, 1919, p 1 (statement by Mitchell). The paragraphs which follow are based principally on (1) the *Report* cited above, (2) Bowers, "The Transcontinental Reliability Test," and (3) items in these ASNLs: Sep 23, 1919, p 2, Oct 8, 1919, pp 2–4, Oct 18, 1919, pp 1–3, Oct 25, 1919, pp 1–4, Oct 31, 1919, pp 1–4, Nov 7, 1919, pp 2–4 (Hartney's account), Nov 15, 1919, pp 1–2 (Gish's report), Nov 18, 1919, pp 9–12 (list of winners), Nov 25, 1919, pp 14–15 (more winners), Dec 13, 1919, p 1 (statement by Mitchell), Dec 26, 1919, pp 9–10 (relative standing of contestants). Also, see the following articles in *U.S. Air Service* II (Nov 1919): As related by Maj John C. P. Barthoff, "From Pacific to Atlantic in an SE–5," pp 14–17; Capt John O. Donaldson, "Twice Across the Continent in a Single-Seater," p 25; and 1st Lt Belvin W. Maynard, "Most Dramatic Incident in My Flight," p 26.

33. ASNL, Nov 7, 1919, p 4.

34. ASNL, Oct 31, 1919, pp 1–4; Glenn L. Martin, "Lessons of the Transcontinental Race," *U.S. Air Service* II (Nov 1919), 12–13.

35. Interview with Mitchell, Oct 18, 1919, in Air Service, *Report on First Transcontinental Reliability and Endurance Test*, pp 33–35.

36. ASNLs, Apr 27, 1920, p 4, May 4, 1920, pp 12–13. Maynard soon left the service (see Bowers, "The Transcontinental Reliability Test," p 217). There were reports that the Rev. Belvin W. Maynard would return to the little church in North Carolina where he preached before the war. Later he was one of the incorporators of the Pioneer Aviation Company, which in the spring of 1921 was planning to establish passenger service between New York and Cleveland (*U.S. Air Service* V (Feb 1921), 30). He died when his plane crashed during an exhibition at the fairgrounds at Rutland, Vermont, on September 8, 1922 (ASNL, Nov 16, 1922, p 6; *U.S. Air Service* VII (Oct 1922), 36).

Chapter III

Reorganization

1. *Annual Report of the Director of Air Service*, 1919, pp 2–7 and chart.

2. Northeastern, Eastern, Southeastern, Central, Southern, Western, Panama Canal, Hawaiian, and Philippine Departments.

3. The command structure was prescribed by War Department General Orders 132, December 19, 1919. See also, Notes on the Application of G.O. 132 to Air Service Units, in AFHRC 167.4–2, January 1920; Army Regulations 1583 (Changes to Army Regulations 96), March 20, 1919; Army Regulations 170–10, May 1, 1924, December 15, 1924, and September 27, 1926; and *Annual Report of the Director of Air Service*, 1920, Chart 1 (facing page 6), page 32, and *passim*.

4. Chase C. Mooney and Martha E. Layman, *Organization of Military Aeronautics, 1907–1935* (*Congressional and War Department Action*) (AAF Hist Study 25, Maxwell AFB, Ala., 1944), pp 37–48.

5. *Ibid*, pp 38–40.

6. H.R. 7925, 66th Cong, 1st sess. Proposals to separate the Air Service from the Army had been offered in Congress before the Armistice, but the submission of the Curry bill marked the beginning of a major campaign toward that end.

7. S. 2693, 66th Cong, 1st sess, Jul 31, 1919.

8. S. 3348, 66th Cong, 1st sess, Oct 30, 1919.

9. Mooney and Layman, *Organization, 1907–1935*, pp 36–47; R. Earl McClendon, *The Question of Autonomy for the United States Air Arm, 1907–1945* (Maxwell AFB, Ala.: Air University, 1950), pp 72–100; Thomas H. Greer, *The Development of Air*

Doctrine in the Army Air Arm, 1917–1941 (USAF Hist Study 89, Maxwell AFB, Ala., 1953), pp 14–25; Robert F. Futrell, *Ideas, Concepts, Doctrine: A History of Basic Thinking in the United States Air Force, 1907–1964*, 2 vols (Maxwell AFB, Ala.: Air University, 1971), I, 15–20.

10. 41 *Stat* 759, 768.

11. 41 *Stat* 104; *Annual Report of the Director of Air Service*, 1920, pp 12–13.

12. 41 *Stat* 953; *Annual Report of the Chief of Air Service*, 1921, p 15.

13. 42 *Stat* 72; *Annual Report of the Chief of Air Service*, 1921, p 16, 1922, pp 10–11.

14. 42 *Stat* 736; *Annual Report of the Chief of Air Service*, 1923, pp 23–24, 78.

15. 42 *Stat* 1396; *Annual Report of the Chief of Air Service*, 1924, pp 11–18.

16. 43 *Stat* 491, 704; *Annual Report of the Chief of Air Service*, 1925, pp 4–6, 92.

17. 43 *Stat* 906; *Annual Report of the Chief of Air Service*, 1926, pp 7–8.

18. 41 *Stat* 768.

19. *Ibid*

20. 39 *Stat* 175; 40 *Stat* 244; DMAWNLs, Dec 21, 1919, p 10, Jan 4, 1919, p 11; ASNL, Oct 31, 1919, p 7; Jones Chronology, Oct 16, 1919 (citing AR 1584 1/2, Oct 16, 1919), Jun 4, 1920 (citing WD GO 49, Aug 14, 1920); Robert D. Ewin, "Silver Wings," in Gilbert Grosvenor, *et al*, *Insignia and Decorations of the U.S. Armed Forces*, rev ed (Washington: National Geographic Society, 1944), pp 119–202.

21. 39 *Stat* 175; 40 *Stat* 244; 41 *Stat* 768.

22. DMAWNL, Nov 30, 1918, p 2; 41 *Stat* 109, 769.

23. Jones Chronology, Jun 30, 1920; ASNL, Sep 21, 1920, p 9; *Army Register*, 1921, pp 1084, 1092.

24. ASNLs, Jun 18, 1920, p 11, Jul 10, 1920, p 14, Jul 29, 1920, p 17, Aug 21, 1920, p 16.

25. ASNL, Sep 17, 1920, pp 20–23; *Annual Report of the Chief of Air Service*, 1921, p 5.

26. The total number of vacancies in the Army was about 4,000, but not all were to be filled at that time.

27. *Annual Report of the Chief of Air Service*, 1921, p 5, 1922, pp 4–5, 1923, p 6, 1924, p 29, 1925, p 14; *Annual Report of the Chief of Air Corps*, 1926, p 36; ASNLs, Mar 15, 1921, pp 14–15, Jun 10, 1921, p 1.

28. *Annual Report of the Chief of Air Service*, 1923, p 6; *Annual Report of the Chief of Air Corps*, 1926, p 36.

29. *Annual Report of the Chief of Air Service*, 1923, pp 6–8; *Annual Report of the Chief of Air Corps*, pp 36, 41–42; ASNLs, Apr 2, 1925, p 11, Mar 2, 1926, pp 12–13, Mar 19, 1926, p 12, May 8, 1926, pp 8, 15, Jun 8, 1926, p 9.

30. *Annual Report of the Chief of Air Service*, 1924, pp 31–35; *Annual Report of the Chief of Air Corps*, 1926, pp 41–42.

31. 42 *Stat* 721, 840; *Army Register*, 1927, p 568, 1932, p 451.

32. *Annual Report of the Chief of Air Service*, 1921, p 6; Jones Chronology, Nov 30, 1921, citing WD GO 57, 1921.

33. 41 *Stat* 768; 44 *Stat* 179, 1902; *Annual Report of the Chief of Air Service*, 1922, p 5.

34. *Annual Report of the Chief of Air Corps*, 1926, pp 36, 41.

35. ASNLs, Jul 16, 1920, p 16, Jul 29, 1920, pp 13, 17, Sep 21, 1920, p 12, Sep 17, 1920, p 13, Nov 20, 1920, p 16.

36. 41 *Stat* 759, 1098; *Annual Report of the Chief of Air Service*, 1921, p 8.

37. Memo, War Plans Division for War Department Chief of Staff, Jul 6, 1921, in AFHRC 145.93–101.

38. 42 *Stat* 74; *Annual Report of the Chief of Air Service*, 1921, p 8; ASNLs, Aug 19, 1921, p 11, Aug 24, 1921, p 14, Sep 9, 1921, pp 9, 13, Oct 14, 1921, p 13, Oct 19, 1921, p 12.

39. *Annual Report of the Chief of Air Service*, 1922, p 6.

40. 26 *Stat* 158; 44 *Stat* 651.

41. ASNLs, Jun 20, 1923, pp 9–10, Jul 10, 1923, p 22, Jun 12, 1924, p 22, May 24, 1926, p 16.

42. *Annual Report of the Chief of Air Service*, 1923, pp 8–9, 1925, p 18; *Annual Report of the Chief of Air Corps*, 1926, pp 43–44.

Chapter IV

Training

1. *Annual Report of the Director of Air Service*, 1919, pp 6, 45–49, 1920, pp 29–30; ASNL, Oct 18, 1919. Cadets who had begun training before the Armistice were permitted to complete it. Some officers and enlisted men received flying instruction at various active airfields while the new training program was being developed.

2. 41 *Stat* 109.

3. ASNLs, Sep 17, 1920, p 15, Nov 9, 1920, pp 8, 14, Dec 29, 1920, p 4, Jan 19, 1921, p 1, Feb 25, 1921, p 6, Jun 16, 1921, p 18, Aug 27, 1921, p 3, Sep 9, 1921, p 9, Feb 28, 1922, p 3, Mar 9, 1922, p 3, June 30, 1924, p 10, Apr 2, 1925, p 5, Aug 22, 1925, p 14, Apr 5, 1926, p 11; *Annual Report of the Chief of Air Service*, 1925, p 46b.

4. 41 *Stat* 109, 768; 42 *Stat* 74, 724; *Annual Report of the Chief of Air Service*, 1923, p 8, 1925, p 18, 1926, p 44. The number on duty fluctuated greatly during a year's time as some cadets finished training and new ones entered. In Fiscal Year 1925, for example, the number varied from lows of 89 and 75 in August and February, during periods between classes, to highs of 180 and 182 in September and March when new classes began. The average strength in Fiscal Year 1925 was 129; the next year it was 156.

5. *Annual Report of the Chief of Air Service*, 1921, p 8, 1924, p 37, 1926, p 47.

6. *Annual Report of the Director of Air Service*, 1920, pp 29–30; ASNL, Sep 9, 1921, p 9.

7. Charles A. Lindbergh, *We* (New York, 1927), pp 105–6, 113–155; ASNLs, Jun 30, 1924, p 10, Apr 2, 1925, p 5, Aug 6, 1925, p 20; Kenneth S. Davis, *The Hero: Charles A. Lindbergh and the American Dream* (Garden City, N.Y., 1959), p 111.

8. ASNL, Jun 5, 1925, p 12; *Annual Report of the Chief of Air Service*, 1924, pp 86–87, 1925, p 46a, 1926, p 48.

9. ASNL, Nov 18, 1921, pp 6–7.

10. ASNLs, May 3, 1919, pp 5–8, May 4, 1920, pp 4–5, Aug 21, 1920, p 13, Oct 22, 1920, p 7, Nov 9, 1920, p 9, Jan 6, 1922, p 14, Mar 29, 1922, p 5, May 18, 1922, p 13.

11. *Annual Report of the Chief of Air Service*, 1921, p 18.

12. ASNLs, Aug 21, 1920, p 13, Sep 24, 1920, pp 5–6, Feb 26, 1921, p 4, Mar 15, 1921, p 19.

13. ASNLs, Oct 8, 1920, pp 1–2, Jan 4, 1921, p 16, Jan 28, 1921, pp 8, 14.

14. ASNLs, May 4, 1920, pp 4–5, Jul 29, 1920, pp 4–5, Aug 13, 1920, p 16, Nov 9, 1920, p 18, Mar 21, 1922, pp 3–4.

15. ASNLs, Feb 20, 1922, pp 3–4, Jan 21, 1923, p 8, Apr 16, 1926, p 17.

16. Jones Chronology, Feb 8, 1922, citing AR 95–15 and Air Service Circular 15.

17. ASNL, Jul 29, 1920, pp 1–2; Jones Chronology, Feb 19, 1920.

18. Jones Chronology, May 5, 1925, citing AS Cir 50–7; *Annual Report of the Chief of Air Service*, 1924, p 84, 1926, p 131; *Army Register*, 1927, p 269.

19. *Annual Report of the Chief of Air Service*, 1922, p 13, 1923, p 43.

20. ASNLs, Aug 16, 1922, p 2, Dec 6, 1922, p 12, Jan 23, 1923, p 3.

21. ASNL, Jan 23, 1923, p 3.

22. ASNL, Jul 10, 1923, pp 3–4; Mason M. Patrick, *The United States in the Air* (Garden City, N.Y., 1928), pp 111–13; Maj Herbert A. Dargue, "Training a Major-General to Fly," *U.S. Air Service* Vol VIII (Sep 1923), 18–19, 21–22.

23. *Annual Report of the Chief of Air Service*, 1922, p 13, 1923, p 43; ASNLs, Aug 16, 1922, p 2, Apr 2, 1925, p 5.

24. ASNLs, Jan 23, 1923, pp 3, 11, Mar 19, 1926, pp 4, 10, Apr 16, 1926, p 7, May 24, 1926, p 10.

25. *Annual Report of the Director of Air Service*, 1919, pp 50–51, 1920, pp 33–34; *Annual Report of the Chief of Air Service*, 1921, p 18; ASNLs, Sep 24, 1920, p 12, Nov 20, 1920, p 3, Dec 14, 1920, p 11; Jones Chronology, Mar 22, 1920, citing ASNL, Jun 10, 1920.

26. Jones Chronology, Jun 11, 1919, Jul 23, 1919, Nov 6, 1919, Jan 31, 1920, Feb 15, 1920.

27. Air Service, A History of [a] U.S. Aerostation, n.d. [*ca.* 1920], in AFHRC 248.211–83H; *Annual Report of the Director of Air Service*, 1919, p 7, 1920, p 35; ASNLs, May 16, 1919, p 7, Sep 9, 1919, pp 2–3, Sep 23, 1920, p 1, Oct 22, 1920, p 7.

28. ASNLs, Dec 2, 1921, p 6, Jan 20, 1922, pp 4–5; Proceedings of a Board of Officers in Regard to the Accident to Airship Roma, in AFHRC 248.211–83G; John H.

Reardon, The Semi-Rigid Airship "Roma," MS in AFHRC K146.01–123.

29. Memo, Maj Oscar Westover, Nov 23, 1922, in AFHRC 168.65481–2; Air Service Estimates for FY 1922 (2 Dec 1920), in AFHRC 167.40112 (1921); Balloon and Airship Div, Financial Program for FY 1921 (11 Nov 1920), in *ibid*; Air Service Estimates for FY 1923 (3 Oct and 20 Dec 1920), in AFHRC 167.40112 (1923); Comparison of Estimate of Funds for FY 1924, in AFHRC 167.40112 (1924).

30. *Hearings before House of Representatives, Committee on Public Funds on H.R. 11548*, 67th Cong, 1st sess, Dec 5–14, 1922, pp 65–66, 72–73, Testimony of Dr. Richard Moore, Chief Chemist, Bureau of Mines, and of Maj Oscar Westover.

31. ASNLs, May 3, 1919, pp 12–14, Oct 31, 1919, p 6, May 10, 1920, p 2; memo, W. J. Barry for Walsh, Jan 16, 1926, in AFHRC 168.65418–2; 43 *Stat* 1110; memo, Maj Oscar Westover for Chief of Air Service, Dec 14, 1921, in AFHRC 167.40112 (1923); *Aircraft Year Book* (New York, 1921), pp 219, 228–29, 1922, p 177, 1923, p 298, 1924, p 304, 1925, p 266, 1926, p 283.

32. *Annual Report of the Chief of Air Service*, 1922, p 13, 1923, p 43, 1924, p 84, 1925, p 46a, 1926, p 131; ASNLs, Jul 26, 1922, pp 16, 18, Jan 23, 1923, p 7, Dec 29, 1923, pp 4–5, Dec 8, 1925, p 10.

33. ASNLs, Jun 4, 1922, pp 13–14, Jan 6, 1923, p 2, Feb 11, 1926, p 4; 1st Lt Orvil A. Anderson, Report on Pony Blimp Crash (1922), in AFHRC 168.7006–35; John Henry Scrivner, Jr., "Pioneer into Space: A Biography of Major General Orvil Arson Anderson" (Ph.D. dissertation, University of Oklahoma, 1971), pp 28–30.

34. Memo, Chief, Sup Div, AS, to Exec, AS, Jun 19, 1925, in AFHRC 168.65418–2; ASNLs, Jun 20, 1923, p 18, Jul 10, 1923, p 3.

35. ASNLs, Apr 2, 1925, p 11, Aug 6, 1925, pp 11–12, Jones Chronology, Jan 9, 1926; Scrivner, "Anderson," pp 35–37; "The New RS-1 Army Semi-Rigid," *Slipstream* VII (Mar 1926), 24–25 .

36. *Annual Report of the Director of Air Service*, 1920, p 30; *Annual Report of the Chief of Air Service*, 1922, p 13; ASNLs, Jul 29, 1920, pp 11–12, Mar 17, 1921, pp 4–5; Jones Chronology, Feb 25, 1921.

37. *Annual Report of the Director of Air Service*, 1920, p 30; *Annual Report of the Chief of Air Service*, 1921, p 19.

38. ASNLs, Jan 6, 1922, p 19, Feb 7, 1922, pp 2–3, Feb 17, 1922, p 13.

39. ASNLs, Feb 20, 1923, pp 10–11, Dec 4, 1924, p 11; *Annual Report of the Chief of Air Service*, 1923, p 45.

40. *Annual Report of the Chief of Air Service*, 1922, p 18, 1923, p 47, 1925, p 46a; 1926, p 130; ASNL, Sep 27, 1922, p 4.

41. *Annual Report of the Chief of Air Service*, 1925, p 46a, 1926, p 131.

42. Robert T. Finney, *History of the Air Corps Tactical School, 1920–1940* (USAF Hist Study 100, Maxwell AFB, Ala., 1955), pp 4–6, 53, 66; *Annual Report of the Chief of Air Service*, 1921, p 17.

43. Finney, *History*, pp 7–9, 11–13. Major Sherman's longhand draft of *Air Tactics* is in AFHRC 248.101–4A, Draft, as is the published version, Training Regulation 440–15 (248.101–4A, 1922). *Air Tactics* was the first of the many texts, manuals, and other documents which record the development of air doctrine at the Tactical School during the 1920s and 1930s.

44. Finney, *History*, pp 8–9, 13.

45. John J. Powers, "Founding of the Air Force Institute of Technology," in *Air University Review* Vol XV (Sep-Oct 1964), 36–50.

46. *Annual Report of the Chief of Air Service*, 1921, p 19, 1922, p 13, 1923, p 46, 1924, pp 85–86, 1925, p 46a, 1926, pp 130–31, 136–37; ASNLs, Oct 8, 1920, p 2, Mar 15, 1921, p 1, Oct 1, 1923, p 23; Thomas and Jablonski, *Doolittle*, pp 69–73.

47. *Annual Report of the Chief of Air Service*, 1923, pp 46–47, 1924, p 88, 1925, p 46a, 1926, p 131.

48. *Annual Report of the Chief of Air Service*, 1921, pp 22–23, 1922, p 16; ASNL, Dec 14, 1920, p 6; Green Peyton, *50 Years of Aerospace Medicine* (Brooks AFB, Tex., 1968), pp 27–53.

49. *Annual Report of the Chief of Air Service*, 1921, p 23, 1922, p 16.

50. ASNL, Jun 12, 1924, pp 11–12.

51. *Annual Report of the Chief of Air Service*, 1925, p 47.

52. *Annual Report of the Chief of Air Service*, 1921, p 25, 1922, p 16; Jones Chronology, Jul 26, 1921, Jan 24, 1922, Jul 1, 1922; 42 *Stat* 724.

53. Peyton, *50 Years*, pp 260–61.

54. *Ibid*, pp 57–58.

Chapter V

Tactical Units

1. Reorganized in the case of those squadrons which remained active after returning from France.

2. *Annual Report of the Chief of Air Service*, 1920, pp 31–37; *Aircraft Year Book* (New York, 1921), pp 204–6.

3. *Report of a Committee of Officers Appointed by the Secretary of War to Consider in All Details a Plan of War Organization for the Air Service* (*27 Mar 1923*), p 2 [hereafter cited as *Lassiter Bd Rprt*], in AFHRC 145.93–253.

4. *Annual Report of the War Department*, 1921, p 24.

5. Maj Gen Mason M. Patrick, General Statement to the Lassiter Board, Mar 17, 1923, in *Lassiter Bd Rprt*, App II, p 2.

6. 41 *Stat* 1098; 42 *Stat* 68; *Annual Report of the Chief of Air Service*, 1921, p 10.

7. 42 *Stat* 724; *Annual Report of the Chief of Air Service*, 1922, p 8.

8. ASNLs, Jul 26, 1922, pp 12–14, Aug 10, 1922, p 15, Aug 16, 1922, pp 8, 14–16, Sep 6, 1922, p 10; *Annual Report of the Chief of Air Service*, 1923, p 53.

9. Info Div, OCAS, USAF Activities, May 31, 1923, and monthly issues through June 30, 1926, in AFHRC 167.4114.l.

10. *Annual Report of the Chief of Air Service*, 1921, pp 10–11.

11. AR 95–10, Nov 17, 1921.

12. *Annual Report of the Chief of Air Service*, 1922, pp 8–9; ltr, The Adjutant General to Chief of Air Service, Dec 18, 1922, 1st Indorsement, CAS to TAG, Jan 19, 1923, 2d Ind, TAG to CAS, Jan 26, 1923, 3d Ind, CAS to TAG, Feb 7, 1923, in *Lassiter Bd Rprt*, App I.

13. Ltr, TAG to Maj Gen William Lassiter, ACS, G–3 (Operations and Training), Mar 17, 1923, in *Lassiter Bd Rprt*, p 2. General Lassiter (USMA, 1889) had been a field artillery and division commander in the AEF and a member of the Dickman Board. Other members of the Lassiter Board were Brig Gen Briant H. Wells, Chief of the War Plans Division; Brig Gen Hugh A. Drum, who had been Chief of Staff, First Army, AEF; Brig Gen Stuart Heintzelman (USMA, 1899), who had been Chief of Staff, IV Corps, and later of Second Army, AEF, and now was Assistant Chief of Staff, G–2; Col Irvin L. Hunt (USMA, 1899), QMC; Lt Col John W. Gulick, a coast artillery officer detailed to the General Staff; Lt Col Frank P. Lahm, veteran airman on duty in G–3, WDGS; and Maj Herbert A. Dargue, who served the committee as recorder. Dargue's notes taken during the meetings of the board are in AFHRC 145.93–101. The Center also has a folder (145.93–102) which Dargue identified as containing "much of the working data used in preparation of Lassiter Board Report."

14. *Lassiter Bd Rprt*, pp 3–6. The board made some changes and adjustments in numbers but in general accepted the plans presented by General Patrick, Chief of Air Service.

15. *Lassiter Bd Rprt*, p 6.

16. AR 95–10, Apr 20, 1923, Apr 16, 1924.

17. Chronological Record of Action Taken on the War Department Committee Report on the Organization of the Air Service [prepared in OCAS sometime in 1925 and extended later through June 24, 1936]. The earlier version is in the file with Dargue's Notes (AFHRC 145.93–101). The later one is with Dargue's copy of the *Lassiter Board Report* preserved in the war plans office of the Air Service/Air Corps/Army Air Forces and now in AFHRC 145.93–253. Also see JB Report 349, Sep 11, 1923, with indorsement by the Secretary of War, in AFHRC 145.93–101, and *Annual Report of the Chief of Air Service*, 1924, pages 140–44.

18. House of Representatives, Select Committee of Inquiry into Operations of the United States Air Services, *Hearings on Matters Relating to the Operations of the United States Air Services*, 68th Cong, 1st sess, 1925, pp 8–9 and App C [hereafter cited as *Lampert Cmte Rprt*]; Futrell, *Ideas, Concepts, Doctrine*, I, 25–26.

19. *Report of the President's Aircraft Board* (Washington: Government Printing Office, 1925), pp 1, 3, 6–7, 10–12, 15–21, A3–A5 [hereafter cited as *Morrow Bd Rprt*]; Futrell, *Ideas, Concepts, Doctrine*, I, 26–29.

20. Futrell, *Ideas, Concepts, Doctrine*, I, 25–28.

21. McClendon, *Question of Autonomy*, pp 124–133; Martha E. Layman, *Legislation Relating to the Air Corps Personnel and Training Programs, 1907–1939* (AAF Hist Study 39, Maxwell AFB, Ala., 1945), pp 28–31; Guido R. Perera, *A Legislative History of*

Aviation in the United States and Abroad for the Period 1907–1940 (Washington, 1941), pp 42–54.

22. 44 *Stat* 780.

23. *Annual Report of the Chief of Air Service*, 1926, p 218.

24. ASNLs, Aug 21, 1920, pp 6–7, 13, Oct 8, 1920, p 11, Oct 29, 1920, pp 8–9, Nov 9, 1920, p 7, Apr 15, 1921, p 8, May 6, 1921, p 12, May 23, 1921, p 13, Feb 11, 1922, p 3.

25. *Annual Report of the Director of Air Service*, 1920, p 29.

26. *Air Service Information Circular*, Jun 12, 1920; ASIC, Jun 30, 1920. Sherman's manual is included in Maurer Maurer, ed, *The U.S. Air Service in World War I*, Vol II: *Early Concepts of Military Aviation* (Washington: Office of Air Force History, 1978), 313–402.

27. Ltr, TAG to Corps Area and Department Commanders, Jul 2, 1925, in AFHRC 167.44–1; *Annual Report of the Chief of Air Service*, 1923, p 44, 1924, p 89, 1925, pp 48–63 (which contains the program for 1925), 1926, p 126; AR 95–10, Apr 16, 1924.

28. *Annual Report of the Chief of Air Service*, 1925, pp 49–63.

29. ASNLs, Jul 10, 1920, p 12, Dec 14, 1920, p 6, Jan 12, 1921, p 4, Jun 30, 1922, p 4, Jul 19, 1923, p 18, Oct 12, 1923, pp 3–4, Feb 14, 1925, p 18, Apr 2, 1925, pp 22–23; *Annual Report of the Chief of Air Service*, 1922, p 14, 1923, p 52, 1925, pp 50–52.

30. For example, see ASNLs, Oct 29, 1920, pp 19–20, Nov 20, 1920, p 13.

31. ASNLs, Apr 20, 1920, pp 2–3, Apr 27, 1920, pp 6–8, Aug 21, 1920, p 1, Dec 29, 1920, pp 3–4, Jan 4, 1921, pp 9, 19, Apr 14, 1923, pp 10–11. See also the many notes on free ballooning in Jones Chronology, especially for the year 1920.

32. ASNL, Sep 14, 1923, pp 6–7.

33. *Annual Report of the Chief of Air Service*, 1925, p 74.

34. ASNLs, Oct 20, 1924, p 19, Oct 31, 1924, p 7, Dec 4, 1924, pp 7–8.

35. ASNL, Apr 2, 1925, pp 1–3 (Lanphier's report); *Annual Report of the Chief of Air Service*, 1925, pp 74–75.

36. ASNLs, Oct 20, 1924, pp 18–19, Oct 31, 1924, pp 19–20, Nov 17, 1924, pp 21–22, May 26, 1926, pp 5–6, Jun 8, 1926, p 13; *Annual Report of the Chief of Air Service*, 1926, pp 158–160.

37. ASNL, Dec 8, 1925, pp 15–16; *Annual Report of the Chief of Air Service*, 1926, p 153; *The New York Times*, Oct 8, 1925, p 1.

38. ASNLs, May 8, 1926, pp 1–2, May 24, 1926, p 11; hist, 1st Pursuit Gp, 1918–1941, p 171, in AFHRC GP–1–HI (FTR).

39. *Annual Report of the Chief of Air Service*, 1925, p 49; ASNL, Apr 30, 1924, p 6.

40. One squadron, the 13th, soon received Dayton-Wright XB–1As designed for night observation.

41. Tng Div, AC Adv Fly Sch, Training Manual 2, Air Forces, pt III, Attack Aviation (1926), pp 1–5, in AFHRC 248.222–55/1926; ASNLs, May 26, 1920, p 8, Oct 19, 1921, pp 3–4, Nov 9, 1921, p 9, Dec 9, 1921, p 7, Feb 17, 1922, p 4, Apr 7, 1922, p 3, Jul 12, 1922, p 4, Feb 20, 1923, p 3. For data on aircraft characteristics and performance the author has used charts compiled periodically by the Technical Data Section of the Engineering Division, "Characteristics and Performance of United States Airplanes" [hereafter cited as Airplane or Aircraft Characteristics], in AFHRC 203–17, supplemented by various published works, principally the following: Martin C. Winrow, ed, *Aircraft in Profile* (Garden City, N.Y., 1965–); Frederick G. Swanborough, *United States Military Aircraft Since 1909* (London and New York, 1963); James C. Fahey, *U.S. Army Aircraft (Heavier-than-Air), 1908–1946* (New York, 1946); Historical Office of the Army Air Forces, *The Official Pictorial History of the AAF* (New York, 1947), pp 163–213.

42. The Development and Use of Attack Aviation Equipment as Used by the Third Attack Group, Kelly Field, Texas (1925), in AFHRC 284.222–34c; *Annual Report of the Chief of Air Service*, 1925, p 65; ASNL, Mar 5, 1923, p 12.

43. Tng Div, AC Adv Fly Sch, Attack Aviation (1926), pp 28–30; ASNL, May 31, 1924, p 5.

44. Swanborough, *U.S. Military Aircraft*, pp 199–200, 552, 572.

45. ASNLs, Oct 27, 1923, p 9, Dec 7, 1923, pp 6, 11, Jul 8, 1925, p 16; Aircraft Record Card, AC 64216, in AFHRC ACR–1; Henry H. Arnold, *Global Mission* (New York, 1949), pp 110, 128–29.

46. ASNL, Aug 12, 1920, pp 6–7.

47. ASNLs, Apr 12, 1921, p 13, May 6, 1921, p 12, May 23, 1921, p 13, Jun 30, 1921, p 12, Feb 7, 1922, p 6, Feb 20, 1922, p 3, Apr 26, 1922, p 5; Doris A. Canham, *Development and Production of Fighter Aircraft for the United States Air Force, 1911–1949* (Wright-Patterson AFB, Ohio, 1949), pp 17–21.

48. Canham, *Fighter Aircraft*, pp 20–30.

Chapter VI

Reserves

1. *Annual Report of the Chief of Air Service*, 1921, p 17,

2. DMAWNL, Dec 14, 1918, p 2.

3. DMAWNL, Feb 1, 1919, p 11; ASNL, Aug 22, 1919, p 6.

4. *Annual Report of the Director of Air Service*, 1920, p 23.

5. ASNLs, Apr 20, 1920, pp 8–9, May 26, 1920, p 20, Jul 20, 1920, p 9, Aug 5, 1920, p 11.

6. ASNLs, Jul 10, 1920, pp 17, 19, Jul 16, 1920, p 12.

7. ASNL, Jul 10, 1920, pp 4, 17.

8. *Annual Report of the Chief of Air Service*, 1921, p 41.

9. 39 *Stat* 187.

10. 41 *Stat* 759, 775.

11. *Annual Report of the Chief of Air Service*, 1922, pp 30–31, 1923, pp 72–73, 1924, p 132, 1925, p 80; *Annual Report of the War Department*, 1923, pp 139–141, 1924, pp 144–46, 1925, pp 133–35, 1926, pp 157–160.

12. 41 *Stat* 780.

13. *Annual Report of the War Department*, 1923, pp 141–42, 1926, p 160.

14. Ltr, CAS to TAG, Mar 3, 1921; ltr, TAG to CGs of CAs, Mar 15, 1921, and attached charts listing RA and NG as well as OR units, in AFHRC 145.93–101.

15. Hist Sec, AWC, General Data Re Organized Reserve Air Units (1936), in AFHRC 167.91–23; *Annual Report of the Chief of Air Service*, 1922, pp 30–32, 1925, p 8.

16. *Annual Report of the Chief of Air Service*, 1922, p 30.

17. ASNLs, Nov 18, 1921, p 17, Dec 27, 1921, pp 2, 18, Jan 6, 1922, p 19, Feb 20, 1922, pp 7, 21, Feb 28, 1922, pp 4–5, 10, May 10, 1922, pp 1–2, May 18, 1922, p 10, Dec 13, 1922, pp 6–7.

18. *Annual Report of the Chief of Air Service*, 1923, p 73.

19. ASNL, Oct 10, 1922, p 22.

20. The appropriation for active duty pay of the members of the Enlisted Reserve Corps was $100 for the entire Army in each of the Fiscal Years 1921 and 1922, $5,000 for 1923 and again for 1924, and $50,000 for 1925 and 1926. 41 *Stat* 954; 42 *Stat* 73, 723, 1381; 43 *Stat* 506, 921.

21. ASNLs, Apr 21, 1922, pp 2–3, May 23, 1923, pp 8–9; *Annual Report of the Chief of Air Service*, 1923, pp 73–74.

22. *Annual Report of the Chief of Air Service*, 1922, p 31, 1923, pp 73–74.

23. ASNLs, Jan 19, 1924, pp 1–4, Mar 3, 1925, p 8; *Annual Report of the Chief of Air Service*, 1922, p 31.

24. *Annual Report of the Chief of Air Service*, 1922, p 30, 1923, pp 74–75; ASNLs, Aug 10, 1922, p 16, Aug 23, 1922, p 9, Sep 6, 1922, pp 5, 14, Oct 10, 1922, pp 7–8.

25. *Annual Report of the Chief of Air Service*, 1924, p 134, 1925, pp 82, 82a, 1926, p 144.

26. *Annual Report of the Chief of Air Service*, 1925, p 82a.

27. *Annual Report of the Chief of Air Service*, 1923, p 75.

28. *Annual Report of the Chief of Air Service*, 1926, p 145.

29. Jones Chronology, Nov 24 and Dec 27, 1922, Aug 1 and Oct 8, 1923, Aug 6, 1925.

30. By General Menoher's calculations, the Air Service would have some 800 of these officers in a few years, when the wartime flyers passed the period of usefulness.

31. *Annual Report of the Chief of Air Service*, 1923, pp 6–8, 37–38.

32. *Annual Report of the Chief of Air Service*, 1922, p 30, 1925, pp 81–82.

33. *Annual Report of the Chief of Air Service*, 1924, p 135, 1925, p 81, 1926, pp 130–32; ASNLs, May 2, 1923, pp 9–10, Jan 7, 1924, pp 5–6, Apr 2, 1925, p 5.

34. *Annual Report of the Chief of Air Service*, 1922, p 30, 1923, p 74, 1924, p 134, 1925, pp 82–83, 1926, pp 151–52.

35. DMAWNL, Jan 29, 1919, p 5, Mar 15, 1919, p 14.

36. Rprt on ROTC by Lt Col James E. Fechet, Tng & Ops Gp, OCAS, Feb 1, 1921, in ASNL, Feb 21, 1921, pp 1–2.

37. *Annual Report of the Chief of Air Service*, 1921, p 40; ASNLs, Jul 22, 1921, p 11, Jul 28, 1921, pp 18–19, Aug 4, 1921, pp 12–13 (article by Vaughan).

38. Jones Chronicle, Nov 1, 1920 (citing ltr by Maj Horace M. Hickam to *U.S. Air Service* magazine, Dec 30, 1920); ASNLs, Apr 21, 1922, pp 4, 14, Apr 26, 1922, p 22; *Annual Report of the Chief of Air Service*, 1922, pp 34–35, 1923, p 69.

39. *Annual Report of the Chief of Air Service*, 1922, p 35; 42 *Stat* 720.

40. *Annual Report of the Chief of Air Service*, 1923, p 69, 1924, p 128.

41. 42 *Stat* 1382.

42. *Annual Report of the Chief of Air Service*, 1923, pp 69–70.

43. *Annual Report of the Chief of Air Service*, 1924, pp 130–31, 1925, pp 26, 77.

44. *Annual Report of the Chief of Air Service*, 1923, pp 68–69, 1924, pp 127–28, 1925, p 77, 1926, pp 148–49.

45. *Annual Report of the Chief of Air Service*, 1921, pp 40–41, 1922, pp 34–35, 1923, pp 68–69, 1924, pp 127–28, 1925, p 77, 1926, pp 148–49.

46. Jones Chronology, Jul 2, 1919 (citing ltr, TAG to Chiefs of Staff Bureaus), Jul 25, 1919 (citing ltr by Director of Air Service), and Sep 30, 1919 (citing ltr, Chief, Militia Bureau, to Director of Air Service).

47. Jones Chronology, Jul 2, 1918 (citing ltr, Director of Air Service to Militia Bureau, Mar 13, 1920, and related documents.

48. *Annual Report of the Chief of Air Service*, 1922, p 33, 1924, p 130, 1925, pp 79–80, 1926, p 146; ACNL, May 15, 1936, pp 1–28.

49. ASNL, Feb 17, 1922, p 5; ACNL, May 15, 1936, pp 3–10. The 135th Squadron was later redesignated 114th, and still later 106th.

50. ASNLs, Jan 31, 1923, p 7, May 2, 1923, p 12; *Annual Report of the Chief of Air Service*, 1923, p 71, 1926, p 130.

51. 41 *Stat* 783; 44 *Stat* pt 1, 1035, 1038.

52. ASNL, Oct 10, 1922, p 5; *Annual Report of the Chief of Air Service*, 1923, p 71, 1924, pp 130–31, 1925, p 79. In addition to those referred to above, there were 14 officers and 57 men in camp from the 103d Squadron of Pennsylvania, which had just been extended federal recognition. The 115th Squadron of California, also newly recognized, did not attend camp that year.

53. *Annual Report of the Chief of Air Service*, 1923, p 72, 1924, p 132, 1926, p 148.

54. *Annual Report of the Chief of Air Service*, 1926, p 147; ASNLs, Jun 8, 1925, p 2, Jun 22, 1925, p 8.

Chapter VII

Defense

1. Army Hist Div, *Order of Battle*, III, pt 1, 610–11; Clarence C. Clendenen, *Blood on the Border: The United States Army and the Mexican Irregulars* (London, 1969), pp 351–55.

2. ASNLs, Sep 3, 1919, pp 1–2, Dec 3, 1919, p 1.

3. Stacy C. Hinkle, *Wings Over the Border: The Army Air Service Armed Patrol of the United States-Mexican Border, 1919–1921* (El Paso, 1970), pp 6–9; Jones Chronology, Jun 18, 1919; Maurer, *Combat Squadrons*. Maps of the Army's patrol districts are found in Army Historical Division, *Order of Battle*, III, part 1, following pages 606, 608, and 616.

4. ASNLs, Dec 3, 1919, p 1, Dec 16, 1920, pp 3–4; Hinkle, *Wings Over the Border*, pp 42–44.

5. Stacy C. Hinkle, *Wings and Saddles: The Air and Cavalry Punitive Expedition of 1919* (El Paso, 1967), pp 8, 34; Hinkle, *Wings Over the Border*, p 8.

6. ASNL, Sep 3, 1919, pp 1–2; Hinkle, *Wings Over the Border*, pp 13–18, 22.

7. Hinkle, *Wings Over the Border*, pp 10–12; Maj Henry H. Arnold, History of Rockwell Field (1923), p 87, MS in AFHRC 168.65041.

8. ASNL, Dec 3, 1919, pp 5–6.

9. Hinkle, *Wings Over the Border*, pp 20–21.

10. *Papers Relating to the Foreign Relations of the United States, 1919* (Washington: Department of State, 1934), II, 537.

11. Hinkle, *Wings Over the Border*, pp 18–19; Burke Davis, *The Billy Mitchell Affair* (New York, 1967), pp 55–56.

12. *Foreign Relations, 1919*, II, 561–62.

13. *Ibid*, 564–65.

14. Hinkle, *Wings and Saddles, passim*; Lt Harold G. Peterson and Lt Paul H. Davis, "Held for Ransom in Mexico," *U.S. Air Service*, II (Oct 1919), 16–19.

15. Hinkle, *Wings Over the Border*, pp 47–48.

16. *Ibid*, p 46.

17. Arnold, History of Rockwell Field (1923), pp 91–93.

18. Hinkle, *Wings Over the Border*, pp 26–36; Arnold, History of Rockwell Field (1923), pp 87–91. The full story of this incident is yet to be written.

19. ASNLs, Jan 23, 1923, pp 3–6, Jan 31, 1923, p 19, Feb 20, 1923, p 4, Mar 21, 1923, p 11, Mar 29, 1923, p 14, Jun 20, 1923, pp 7–9, Aug 14, 1923, p 6; Arnold, History of Rockwell Field (1923), pp 58–62, 106–9.

20. The 12th Squadron, at Fort Bliss until 1926, sporadically flew patrols along the border during that time.

21. DMAWNL, Nov 16, 1918, p 6: Jones Chronology, May 27, 1919.

22. Jones Chronology, Apr 7, 1919, May 19, 1919.

23. Jones Chronology, May 27, 1919, Jun 3, 1919, Jun 10, 1919, Jun 18, 1919, Jul 16, 1919, Aug 16, 1919.

24. ASNLs, Apr 27, 1920, p 8, Dec 29, 1920, pp 3–4, Jan 4, 1921, pp 9, 19, Mar 15, 1921, p 26.

25. ASNLs, Nov 9, 1920, pp 17–18, Jan 28, 1921, p 8.

26. ASNL, Jan 28, 1921, pp 8–9.

27. Jones Chronology, Apr 24, 1919, citing report of meeting, Col William L. Patterson and Lt Col Leslie MacDill with Maj Gen Frank W. Coe.

28. *Annual Report of the Chief of Coast Artillery* (excerpt), in *Annual Report of the War Department*, 1921, pp 239–240,

29. The name of the board at this time was Joint Army and Navy Board on Aeronautics. It was changed soon afterwards to avoid confusion with the Joint Army and Navy Board.

30. A separate mission was stated for Marine Corps aircraft. They were to be employed with the functions of Army aviation if at an advance base and the Army was not represented in the operations, or in cooperation with the Army on shore, in which case the senior line officer present would assume command.

31. Jones Chronology, Aug 23, 1919, citing JB 349, Ser 8l; and Jan 22, 1920, citing WD GO 4, 1920; Jones Chronology, Jan 4 and Feb 12, 1920.

32. 41 *Stat* 954; Futrell, *Ideas, Concepts, Doctrine*, I, 20.

33. Ltr, CAS to TAG, Revision of Policy [Jul 1923], in AFHRC 168.65043–1.

34. Jones Chronology, Jun 5, 1920, citing Lampert hearings; *Lampert Cmte Rprt*, pp 9, 18, 36.

Chapter VIII

Planes Versus Ships

1. Alfred H. Hurley, *Billy Mitchell: Crusader for Air Power* (New York, 1964), pp 40–60; Isaac D. Levine, *Mitchell, Pioneer of Air Power*, rev ed (New York, 1958), pp 165–205; Davis, *Mitchell Affair*, pp 49–68.

2. 1st Prov Air Bde, Report on Operations in Naval Ordnance Tests (1921), Chap II, p 8, in AFHRC 248.222–69.

3. ASNL, Jul 20, 1920, pp 12–13; Hurley, *Mitchell*, p 59.

4. *Illustrated London News* 67 (Dec 11, 1920), 970; Davis, *Mitchell Affair*, pp 69–76; Levine, *Mitchell*, pp 206–221; Hurley, *Mitchell*, p 6l; ASNL, Nov 9, 1920, p 8.

5. 1st Prov Air Bde, Report on Operations in Naval Ordnance Tests, Chap I; Hurley, *Mitchell*, p 61; Levine, *Mitchell*, pp 218–220.

6. ASNLs, Mar 15, 1921, pp 26–27, Apr 1, 1921, pp 11–21, 18, Apr 15, 1921, pp 8, 11, May 16, 1921, p 4, May 23, 1921, pp 10–11, 13, 18, Jun 10, 1921, pp 11, 17, 19, Jun 16, 1921, pp 2, 8, 13, 16–17, 19, Jun 22, 1921, p 7, Jun 30, 1921, pp 3–4.

7. Milling became Chief of Staff; Capt Gerald E. Brower, G–1; Maj Joseph T. McNarney, G–2; Maj William C. Sherman, G–3; and Capt Louis R. Knight, G–4. Maj Davenport Johnson commanded the lst Provisional Bombing Wing; Maj John A. Paegelow, the 1st Provisional Airship Wing; and 1st Lt James P. Hodges, the Provisional Photographic Sections.

8. ASNLs, Jun 16, 1921, pp 7–8, Jul 28, 1921, pp 1–3; Levine, *Mitchell*, pp 228–29. Besides Lieutenant Ames, those killed were Lt Col Archie Miller; 1st Lt Cleveland W. McDermott; 1st Lt John M. Pennywill; Sgt Richard C. Blumenkranz; Mr. Amos G. Batchelder, executive secretary of the American Automobile Association; and Mr. Maurice Connolly, former U.S. representative and wartime aviator.

9. ASNL, Jul 28, 1921, pp 1–3.

10. Davis, *Mitchell Affair,* pp 88–89; Levine, *Mitchell,* pp 229–230.

11. Davis, *Mitchell Affair,* pp 88–89; Levine, *Mitchell,* pp 229–231; Hurley, *Mitchell,* pp 66–67.

12. Part of the tract acquired for Camp Eustis.

13. 1st Prov Air Bde, Report on Operations in Naval Ordnance Tests, Chap II, pp 16–17.

14. The board was made up of Navy officers. An Army board, consisting of four ordnance officers, attended the naval board, as did two newspaper correspondents, one from International News Service and the other from Associated Press.

15. Hurley, *Mitchell,* pp 64–66; 1st Prov Air Bde, Report on Operations in Naval Ordnance Tests, Chap I, p 9, and App A; CINC, Atl Fleet, to Fleet, Instructions for Exercises, May 25, 1921, in AFHRC 248.222–69, papers in AFHRC 248.222–69, Vol 7, pt 1.

16. 1st Prov Air Bde, Report on Operations in Naval Ordnance Tests, Chap V. Two other German submarines were subjected to naval gunfire and sunk.

17. 1st Prov Air Bde, Report on Operations in Naval Ordnance Tests, Chap VI; Hurley, *Mitchell,* p 67; Davis, *Mitchell Affair,* pp 96–97. This test had been postponed one day because of heavy fog at sea on June 28. Mitchell provided two airships and some seaplanes as part of the Navy's scouting force.

18. The Martins were to follow DH–4s, but there was a mixup because the Martins arrived too soon. There was some difficulty in getting it straightened out because of a breakdown in radio communications.

19. CINC, Atl Fleet, to Fleet, Instructions for Exercises, May 25, 1921; 1st Prov Air Bde, Report on Operations in Naval Ordnance Tests, Chap VII; Vice Adm Alfred W. Johnson, Bombing Operations, in AFHRC K180.058–1. Two other German destroyers were attacked and sunk by naval gunfire.

20. 1st Prov Air Bde, Report on Operations in Naval Ordnance Tests, Chap VIII; Johnson, Bombing Operations.

21. CINC, Atl Fleet, to Fleet, Instructions for Exercises, May 25, 1921; 1st Prov Air Bde, Field Order 12, Jul 18, 1921, Field Order 13, Jul 19, 1921, Field Order 14, Jul 20, 1921, in 1st Prov Air Bde, Report on Operations in Naval Ordnance Tests, App D; 1st Prov Air Bde, Report on Operations in Naval Ordnance Tests, Chap IX; Johnson, Bombing Operations; Davis, *Mitchell Affair,* pp 94–112.

22. ASNL, Aug 4, 1921, p 14.

23. Comdr, AF, to CINC, Atl Fleet, Jul 29, 1921, in AFHRC K180.058–1.

24. Johnson, Bombing Operations.

25. *Ibid*; Davis, *Mitchell Affair,* p 102. Copies of signals concerning Army aircraft in the *Ostfriesland* test, copies of the reports of the Board of Observers, and other related materials collected by Johnson are in AFHRC K180.058–1.

26. 1st Prov Air Bde, Report on Operations in Naval Ordnance Tests, Chap X (Sherman wrote this section of Mitchell's report).

27. Rprt, Bd of Obs, Jul 22, 1921, in AFHRC K180.0581.1.

28. Rprt, JB, Results of Aviation and Ordnance Tests, in AFHRC 248.222–69, Vol 7, pt 2.

29. 1st Prov Air Bde, Report on Operations in Naval Ordnance Tests, Chap XI (written by 1st Lt Clayton L. Bissell); *The New York Times,* Jul 30, 1921, p 5; 1st Lt St. Clair Streett, "14th Heavy Bombardment Squadron Attacks New York City," *U.S. Air Service* Vol V (Jul 1921), 21. Mitchell in a DH joined the formation just before it reached New York.

30. Brig Gen William Mitchell, Asst to CAS, to Maj Gen Charles T. Menoher, CAS, Report on Operations of 1st Provisional Air Brigade, Aug 29, 1921, in AFHRC 248.222–69, Vol 1.

31. *The New York Times,* Sep 14, 1921, p 1.

32. Davis, *Mitchell Affair,* pp 118–121; Hurley, *Mitchell,* pp 68–69; Levine, *Mitchell,* pp 269–273.

33. 1st Prov Air Bde, Report of Operations in Bombing of *Alabama* (1921), in AFHRC 248.222–70.

34. *Ibid.*

35. ASNLs, Oct 14, 1921, p 13, Oct 25, 1921, p 4, Nov 25, 1921, p 15.

36. ASNL, Aug 23, 1923, p 8.

37. 42 *Stat* 1397.

38. The Owl had been designed as a long-range, night, mail plane. The Air Service obtained one for evaluation.

39. Brig Gen William Mitchell, Report on Bombing Maneuvers, Sep 5, 1923, in AFHRC 248.222–71; ASNL, Aug 23, 1921, pp 6–8.

40. Mitchell, Report on Bombing Maneuvers, pp 22–26.

41. Davis, *Mitchell Affair,* pp 155–57; Levine, *Mitchell,* pp 290–94.

42. Davis, *Mitchell Affair,* pp 182–333; Levine, *Mitchell,* pp 300–370; Hurley, *Mitchell,* pp 90–109; *The New York Times,* Sep 6, 1925, p 1 (text of Mitchell's statement of Sep 5, 1925).

Chapter IX

Civil Affairs

1. R. F. Hammatt, Airplane Forest Fire Patrol in California, Aug 24, 1919, in AFHRC 168.655-7; ASNLs, Apr 19, 1919, p 3, May 24, 1919, pp 3-4, Jun 13, 1919, p 2, Aug 22, 1919, p 12, Sep 30, 1919, pp 7-8, Oct 25, 1919, p 5, Nov 7, 1919, pp 5-6, Nov 25, 1919, pp 1-5 (Arnold's report), Dec 26, 1919, pp 5-8; Eldon W. Downs, "Contributions of U.S. Army Aviation to Uses and Operation of Aircraft" (Ph.D. dissertation, University of Wisconsin, 1959), pp 96-133.

2. ASNLs, Jun 25, 1920, pp 14-15, Jul 20, 1920, p 10, Oct 22, 1920, p 20, Nov 9, 1920, p 7, Nov 28, 1920, p 17. On the gas shortage in general, see *Literary Digest,* May 25, 1920, pages 39, 115-16, July 17, 1920, pages 90-91, October 30, 1920, pages 92-93.

3. L. D. S[ullivan], Dissemination Div, Info Gp, ODAS, 1920 Forest Fire Patrol by the Air Service, in AFHRC 168.655-7; ASNLs, Jun 10, 1920, p 2, Jul 29, 1920, p 6, Aug 27, 1920, p 15, Sep 17, 1920, pp 6, 10-12. Sep 29, 1920, p 13; Downs, "Contributions," pp 108-116; Jones Chronology, Jun 1, 1919, May 16, 1920.

4. The story of the forest patrol from 1921 on is based primarily on Downs, "Contributions," pages 116-139, supplemented by Jones Chronology, June 1, 1919, and May 16, 1920; *Aircraft Year Book* (New York, 1921) pages 34-38, and other sources.

5. E. Lester Jones, "Surveying from the Air," *Journal of the Franklin Institute* 193 (Apr 1923), 461-490; Downs, "Contributions," pp 244-47.

6. ASNLs, Dec 3, 1919, p 5, Apr 20, 1920, pp 6, 18, Jun 10, 1920, pp 10-11, Jul 10, 1920, p 12, Aug 13, 1920, p 13.

7. ASNL, Oct 22, 1920, pp 23-24.

8. ASNLs, Mar 23, 1921, p 7, Dec 6, 1922, pp 7-8, Jul 10, 1923, p 13, Jan 19, 1924, pp 12-13, Apr 12, 1924, pp 9-10, Feb 14, 1925, pp 2, 9-10, Jun 22, 1925, p 12, Dec 23, 1925, pp 5-6, Mar 19, 1926, p 3, Apr 16, 1926, pp 13-14; Jones Chronology, Jun 10, 1921, Oct 1, 1922, Apr 30, 1925.

9. Downs, "Contributions," pp 268-270; 43 *Stat* 1011; *Annual Report of the Chief of Air Service,* 1925, p 69, 1926, pp 175-182; ASNL, Apr 16, 1926, pp 13-14.

10. Downs, "Contributions," pp 265-68; ASNLs, Aug 22, 1925, pp 13-14, Dec 8, 1925, pp 3-5, Dec 23, 1925, pp 8-9, Jan 19, 1926, pp 7-10, Feb 11, 1926, pp 8-13 (the last four of these citations being Stevens' report on the Rice Expedition).

11. Mary R. Self, *Reconnaissance Aircraft and Aerial Photographic Equipment, 1915-1945* (Patterson Field, Ohio, 1946), pp 27-53; Downs, "Contributions," pp 246, 264, 266, 274; 1st Lt Edward G. Plank, "Aerial Mapping," *Slipstream* VI (May 1925), 9-14, 21; ASNL, May 18, 1922, p 11; Jones Chronology, May 2, 1924; *Annual Report of the Chief of Air Service,* 1923, pp 33, 56; George W. Goddard, *Overview: A Life-Long Adventure in Aerial Photography* (Garden City, N.Y., 1969), pp 18-151; Ashley C. McKinley, *Applied Aerial Photography* (New York, 1929), pp 88-98, 102-112.

12. ASNL, Mar 9, 1922, pp 8-10; H. A. Gossard, "Aeroplane Dusting," paper read before National Nut Growers Association, Mobile, Ala., 1921, in AFHRC 168.655-7; Downs, "Contributions," pp 61-72.

13. ASNLs, Jul 19, 1923, pp 1-3, Jan 19, 1924, pp 6-7, Jun 8, 1926, p 8; Downs, "Contributions," pp 79-95.

14. ASNLs, Jun 30, 1922, p 5, Mar 28, 1924, p 3, Jun 30, 1924, p 8, Jun 5, 1925, pp 7-8, May 24, 1926, p 12; Jones Chronology, Jul 4, 1922, citing *National Aeronautical Association Review,* May 1925; *Annual Report of the Chief of Air Service,* 1923, p 58.

15. ASNLs, Nov 22, 1923, pp 8-9, Jan 7, 1924, p 10, Jan 19, 1924, p 10, Apr 30, 1924, pp 1-3 (Weddington's report), Sep 30, 1924, p 11.

16. ASNLs, Apr 26, 1922, p 6, May 3, 1922, p 5, Feb 20, 1923, pp 1-3, Jun 4, 1923, p 14, Jun 20, 1923, p 12, Sep 12, 1925, p 9; Jones Chronology, Apr 27, 1921, Mar 28, 1922, Apr 10, 1923, May 26, 1925.

17. ASNLs, Apr 1, 1921, p 10, Apr 15, 1921, pp 7, 14, May 23, 1921, pp 1-3.

18. ASNL, Oct 25, 1921, p 3.

19. ASNL, May 15, 1923, p 9.

20. ASNL, Aug 22, 1925, pp 1-4; Jones Chronology, Apr 30, 1919, Aug 7, 1919, May 1, 1925, Jul 22, 1925.

21. ASNLs, Sep 23. 1919, p 4, Jul 13, 1921, p 1, Aug 10, 1922, pp 7-8, Aug 16, 1922, pp 7-8, Mar 28, 1924, pp 1-2, Jun 12, 1924, pp 4-5, Apr 5, 1926, p 7; Jones Chronology, May 10, 1920; *Annual Report of the Chief of Air Service,* 1923, p 58.

22. ASNLs, Sep 24, 1920, p 10, Oct 8, 1920, pp 2–3, Jul 12, 1922, p 6, Feb 7, 1922, p 9.

23. ASNL, Jun 4, 1923, pp 11–13 (includes Whitehead's report); *Annual Report of the Chief of Air Service*, 1923, p 58.

24. ASNLs, Sep 21, 1920, p 17, May 24, 1926, pp 2–3; Jones Chronology, Aug 23, 1922.

25. ASNL, Aug 13, 1924, p 9.

26. ASNL, Jun 8, 1926, p 8.

27. ASNLs, May 16, 1921, p 11, Apr 5, 1926, p 7; Jones Chronology, May 1, 1923.

28. ASNLs, Feb 20, 1923, p 10, Mar 21, 1923, p 3; Jones Chronology, Feb 7, 1923, citing WD GO 8, 1934, and ACNL, Jan 15, 1935, p 13.

29. The following paragraphs are based on Maurer Maurer and Calvin F. Senning, "Billy Mitchell, the Air Service, and the Mingo War," *The Airpower Historian* XII (April 1965), 37–43.

30. Heber Blankenhorn, "Marching Through West Virginia," *The Nation* 113 (Sep 14, 1921), 288.

31. William Mitchell, *Winged Defense: The Development and Possibilities of Modern Air Power—Economic and Military* (New York, 1925), pp 75–76.

Chapter X

Aviation Facilities and Equipment

1. ASNLs, Apr 26, 1919, p 7, May 3, 1919, p 4.

2. ASNL, May 3, 1919, pp 1–5.

3. ASNLs, Sep 3, 1919, p 5, Jun 10, 1920, p 16, Jun 19, 1920, p 22, Aug 13, 1920, p 16, Sep 21, 1920, p 8.

4. ASNL, Feb 3, 1921, pp 1–3.

5. ASNLs, Feb 3, 1921, pp 1–3, Feb 21, 1921, pp 6–7, 9–10, May 6, 1921, p 4; *Annual Report of the Chief of Air Service*, 1921, pp 43–44, 1922, pp 40–43; Report of Airways Section, 1 January–30 June 1921, in AFHRC 167.401611 Jan-Dec 1921.

6. *Annual Report of the Chief of Air Service*, 1921, pp 24–25; Airways Sec, Progress Report, Dec 1921, in AFHRC 167.401611 (Jan-Dec 1921).

7. ASNL, Jun 23, 1922, pp 5–6.

8. Airways Sec, Progress Report, May 1922, in AFHRC 167.401611; ASNL, Aug 10, 1922, p 16.

9. ASNLs, Jul 12, 1922, p 11, Jul 26, 1922, pp 2–3, Aug 10, 1922, pp 15–16, Aug 16, 1922, pp 3–4, Sep 6, 1922, p 6, Sep 13, 1922, p 3, Sep 27, 1922, pp 6–8; *Annual Report of the Chief of Air Service*, 1925, p 71. The fields in mid–1925 were classified as follows: U.S. Air Service, 95; Navy, 17; Air Mail Service, 46; municipal, 255; commercial, 230; emergency, 2,817.

10. ASIC, Vol 5, Mar 1, 1923; *Annual Report of the Chief of Air Service*, 1923, pp 61–62, 1924, p 108.

11. *Annual Report of the Chief of Air Service*, 1923, pp 62–63; ASNLs, Jul 19, 1923, pp 13–14, Aug 3, 1923, pp 12–13, Mar 2, 1926, p 9; *National Aeronautical Association Review* IV (May 1926), 73, 76.

12. *Annual Report of the Chief of Air Service*, 1922, p 25, 1923, p 63, 1924, p 109; ASNL, Jan 7, 1924, p 9.

13. ACIC, Vol 5, Mar 1, 1923, pp 3, 9–13; ASNLs, Jul 23, 1923, p 23, Aug 6, 1925, pp 7–8.

14. ASNLs, May 25, 1925, p 6, Aug 6, 1925, pp 7–8; *National Aeronautical Association Review* IV (May 1926), 73.

15. ASNLs, Jun 23, 1923, pp 7–8, Aug 3, 1923, p 10, Aug 16, 1923, p 3; *Annual Report of the Chief of Air Service* , 1926, p 184.

16. *Annual Report of the Airways Section*, 1925, in AFHRC 167.401611; ASNLs, Oct 2, 1925, p 22, Dec 8, 1925, pp 1–2.

17. ASNLs, May 25, 1923, p 8, Jun 12, 1924, p 3.

18. Requests for and reports of cross-country flights, in AFHRC 167.413.3.

19. ASNL, Oct 12, 1923, p 1.

20. Jones, a captain in the Reserve, was the author of a book on *Avigation* published in New York and London in 1931.

21. Downs, "Contributions," p 337.

22. ASNL, Jun 12, 1924, p 1; Paul R. Heyl and Lyman J. Briggs, "The Earth Inductor Compass," *U.S. Air Service* VII (Nov 1922), 30–33, and VII (Dec 1922), 30–33. A rotor in the tail of the plane, driven by a fan in the slipstream and turning in the earth's magnetic field, generated an electric current which registered on the galvanometer the pilot

steered by. The rotor was connected mechanically with the device on which the navigator set the course.

23. ASNLs, Sep 29, 1923, p 4, Oct 12, 1923, p 1.

24. ASNLs, Mar 28, 1923, p 8, Apr 20, 1924, pp 7–8.

25. ASNLs, May 16, 1924, p 7, Jun 12, 1924, p 2.

26. In 1920 the Engineering Division had experimented with a British system the United States was to have used for a night bombardment program which had been planned but had not become operative during the war. A JN–4H was fitted with wire loops sewn in fabric and doped to wings and struts. There was one loop on the wings on either side of the fuselage and another at right angles around the fuselage. Leads were connected to a radio receiver in the rear cockpit where a man worked switches on the receiver and steered the plane on the maximum signal received from a ground station. (ASNLs, Sep 14, 1920, pp 1–2, Jan 12, 1921, p 2.)

27. ASNL, Sep 12, 1925, pp 7–8; *Annual Report of the Chief of Air Service*, 1926, p 91; F. H. Engle and Francis W. Dunmore, "A Directive Type of Radio Beacon and Its Application to Navigation," Scientific Papers of the National Bureau of Standards 480 (Washington, 1924); Capt William H. Murphy, "The Radio Beacon," *Slipstream* VI (Nov 1925), 21–23.

28. Herbert N. Eaton, *et al*, *Aircraft Instruments* (New York: Ronald Press Co, 1926); *cf*, Herbert N. Eaton, "Aerial Navigation and Navigating Equipment," Technical Report 131, in *Seventh Annual Report of National Advisory Committee for Aeronautics*, 1921, pp 763–802.

29. Downs, "Contributions," pp 173–74; ASNLs, Sep 17, 1920, p 14, Oct 8, 1920, p 14.

30. Downs, "Contributions," pp 171–208.

31. Jones Chronology, Aug 5, 1922; Downs, "Contributions," p 197; *Annual Report of the Chief of Air Service*, 1923, p 45.

32. Downs, "Contributions," pp 199–200; 1st Lt Donald L. Bruner and 1st Lt Harold R. Harris, "The First Night Airways," *U.S. Air Services* IX (Jun 1924), 7–13. The Air Service made information about Bruner's work available to the Air Mail Service in 1922 to use in preparing a section of the airway between Chicago and Cheyenne for a test of night operations. (ASNL, Oct 12, 1923, pp 4–5; Downs, "Contributions," pp 196–97, 200–06.)

33. *Annual Report of the Chief of Air Service*, 1925, pp 49–63; ASNLs, Jun 4, 1923, pp 2–5, Aug 3, 1923, p 20, Jan 7, 1924, pp 7–8, May 31, 1924, p 10, Apr 30, 1924, p 13, Jun 12, 1924, p 16, Oct 31, 1924, p 7, Dec 4, 1924, pp 7–8, Jan 7, 1925, pp 1–2.

34. Jones Chronology, Jun 18, 1919, Dec 5, 1919. The Air Service had authority to operate radios aboard its aircraft. All other matters pertaining to radio were the responsibility of the Signal Corps.

35. Jones Chronology, December 5, 1919, citing and quoting an exchange between Menoher and Maj Gen George O. Squier, Chief Signal Officer. The equipment used on the border was mainly the SCR–59 ground receiver and the SCR–73 airborne wireless transmitter.

36. Maurer, *The U.S. Air Service in World War I*, Vol II: *Early Concepts of Military Aviation*, 334–36; ASNL, May 5, 1925, p 8; ACM, *Communications*, revised July 1, 1928, in AFHRC 248.312–80.

37. ASNLs, May 3, 1919, p 11, Apr 20, 1920, p 15, Jul 20, 1920, p 4, Aug 13, 1920, p 6, Aug 27, 1920, p 11, Sep 17, 1920, pp 13–14, Oct 8, 1920, p 14, Oct 29, 1920, pp 5, 22, Jan 28, 1921, pp 9–10, Mar 15, 1921, p 22, Apr 1, 1921, p 16; Hinkle, *Wings Over the Border*, pp 39–40; Jones Chronology, Jul 14, 1919.

38. Downs, "Contributions," pp 294–95; Jones Chronology, Dec 5, 1919. Army Regulations 1556, amended May 10, 1920, reflected the change.

39. Downs, "Contributions," pp 296–302.

40. See, for example, 1st Provisional Air Brigade, Field Order 13, July 19, 1921, Annex 1.

41. ASNLs, Mar 29, 1923, p 16, May 15, 1923, pp 15–16, Aug 23, 1923, pp 4–5, May 31, 1924, p 11, Jun 12, 1924, p 21.

42. *Annual Report of the Chief Signal Officer* in *Annual Report of the Secretary of War*, 1922, p 258; OCSO Radio Communications Pamphlet 42, *Airplane Radio Sets Types SCR–133, –134, –135*, 1925; ACM, *Communications*, revised July 1, 1928.

43. Downs, "Contributions," pp 309–311; ACM, *Communications*, revised July 1, 1928, pp 27–33; ASNLs, Apr 2, 1925, pp 15–16, Feb 11, 1926, p 4.

44. ASNLs, May 29, 1919, pp 1–2, Aug 22, 1919, pp 2–5, Oct 29, 1920, pp 10, 23, Oct 14, 1921, p 10, Jun 30, 1922, pp 2–3.

45. *Annual Report of the Chief of Air Service*, 1921, pp 45–46, 1922, pp 37–38.

46. ASNL, Jun 30, 1922, pp 2–3.

47. *Annual Report of the Chief of Air Service*, 1922, pp 64–66, 71; ASNLs, Dec 13, 1922, p 3, Jan 19, 1926, p 16; Jones Chronology, Aug 15, 1922, citing AS Cir 75. Some places on the Model Airway still did not have weather service in 1925. Five new detachments authorized by the War Department at the end of the year were assigned to Cumberland, Uniontown, Kansas City, Muskogee, and Dallas.

48. ASNL, Feb 21, 1924, pp 1–2.

49. ASNLs, May 16, 1924, pp 1–2, Jun 30, 1924, p 2.

50. Jones Chronology, Oct 29, 1924, citing and quoting the *Baltimore Sun*, October 30, 1924. ASNLs, Nov 16, 1924, p 16, Apr 5, 1926, pp 12–13.

51. For development of parachutes during and after the World War see: Downs, "Contributions," Chapter IV; Edward O. Purtee, *Development of AAF Clothing and Other Personal Equipment Peculiar to Air Operations, 1942–1944*, 3 vols (Patterson Field, Ohio, 1945), III, in AFHRC 201–15.

52. Irvin adopted the "g" that was accidentally added to his name in papers incorporating the Irving Air Chute Company.

53. Jones Chronology, Apr 28, 1919, May 14, 1919; Downs, "Contributions," p 150; Maj Edward L. Hoffman, "Directions for the Operation of the Adopted Type of Parachutes and Comments Thereon," *U.S. Air Service* II (Dec 1919), 18–19; statement by Irving in Irving Air Chute Co, Inc, *et al*, vs Russell Parachute Co, 47 F.2d 130 (1931). There were a number of conflicting claims relating to the development of parachutes. Besides the case just cited, see, for example, Floyd Smith Aerial Equipment Company versus Irving Air Chute Company, 276 F. 834 (1921); Van Meter versus Irving Air Chute Company and Same versus United States, 27 F.2d 170 (1928); Irving Air Chute Company, Incorporated, *et al*, versus Russell Parachute Company, 41 F.2d 387 (1930).

54. Jones Chronology, May 9, 1920, citing ODAS Cir 34, Mar 9, 1920, and Cir 51, Apr 17, 1920. The requirement for two chutes, and the 1,500-foot minimum has been established in September 1919. (Jones Chronology, Aug 13, 1919, Sep 26, 1919, citing ODAS Order 44.)

55. ASNLs, Jun 25, 1920, pp 1–2, Jul 10, 1920, p 5.

56. ASNLs, Jul 10, 1921, pp 15, 18, Aug 27, 1920, p 11.

57. ASNLs, Sep 21, 1920, p 16, Sep 24, 1920, pp 2–3; Sgt Gilbert A. Shoemaker, "Lessons in Parachute Jumping," *Slipstream* VI (Feb 1925), 7–10.

58. ASNLs, Jul 10, 1920, p 12, Jul 29, 1920, p 8, Sep 24, 1920, pp 1–2, Apr 12, 1921, p 9, Nov 18, 1921, p 6, Dec 9, 1921, p 5, Aug 10, 1922, pp 2–5; Jones Chronology, Jun 7, 1920, Feb 22, 1921; Albert W. Stevens, "Parachuting from 24,000 Feet," *U.S. Air Service* VII (Jul 1922), 9–11.

59. ASNLs, Nov 9, 1920, p 8, Dec 14, 1920, p 7; Jones Chronology, Oct 23, 1920 (citing letter from Capt Norbert Carolin, the pilot of the plane from which Moon jumped), Nov 19, 1920, citing OCAS Order 21.

60. Jones Chronology, Feb 28, 1925, citing AS Cir 60–4 and 60–4A, Jones Chronology, Mar 22, 1925; ASNL, May 22, 1925, p 6.

61. ASNLs, May 6, 1921, pp 2–3, May 27, 1921, p 6; Jones Chronology, Mar 1, 1921.

62. 1st Lt Harold R. Harris, "The Value of the Parachute," *U.S. Air Services*, IX (Mar 1924), 26–29; AC, The Caterpillar Club, Nov 25, 1927, in AFHRC 248.211–123, pp 1–2 and 5–6 (the latter being a statement by Harris); Jones Chronology, Oct 20, 1922. The club, whose name probably was suggested by the fact that parachutes were made of silk, had no organization, officers, or by-laws.

63. AC, The Caterpillar Club, pp 6–7; Jones Chronology, Jan 15, 1923, citing OCAS Cir 6; *Annual Report of the Chief of Air Service*, 1923, p 33.

64. ASNLs, Mar 5, 1923, pp 12–13, Jun 4, 1923, p 10, Oct 2, 1925, p 14. Five hundred parachutes were tested during Fiscal Year 1924 and 475 were issued. (*Annual Report of the Chief of Air Service*, 1924, p 69.)

65. AC, The Caterpillar Club, pp 8–31. Lindbergh made two more jumps, in September and November 1926 while flying airmail.

66. *Annual Report of the Chief of Air Service*, 1921, pp 24–26, 1922, p 28, 1923, pp 21–22, 1924, p 9, 1925, pp 72, 72b, 72c, 1926, p 36; memo for Info Div by Medical Service, May 20, 1925, in AFHRC 200.3911–1, No. 6; AS, Statistics on Fatalities . . . CY 1924, in AFHRC 200.3911–1, No. 14; AC, Fatalities (compilation of fatal aircraft accidents for Fiscal Years 1921–1932), in AFHRC 145.91–206.

Chapter XI

Higher, Faster, Farther, and Longer

1. ASNL, Sep 23, 1919, p 4; Jones Chronology, Oct 4, 1919; *Aircraft Year Book* (New York, 1920), pp 131, 256–57; David Gregg, "The Story of the Supercharger," *Slipstream* V (Mar 1924), 17–19, 24.

2. *Aircraft Year Book* (New York, 1921), pp 120–22; ASNL, Jun 10, 1920, pp 3–6; Jones Chronology, Feb 27, 1920 (containing statements by medical officers on duty at McCook Field). Published works do not always agree on the figures they give for altitude, speed, and other records. The author has followed the best compilation he had available: Ernest Jones, "Chronology of World and American Air Records," *National Aeronautical Association Review* II (Dec 1924), 13–14, III (Jan 1925), 13–14, III (Feb 1925), 30–31, III (Mar 1925), 46–47, and updates in subsequent issues.

3. ASNL, Jun 10, 1920, pp 3–6.

4. ASNLs, Oct 14, 1921, p 6, Oct 19, 1921, pp 6–7, Dec 9, 1921, pp 1–4 (Macready's report), Feb 7, 1922, p 4, Dec 21, 1923, pp 1–3, Dec 23, 1925, pp 11–13, Feb 11, 1926, pp 1–2, Mar 2, 1926, pp 1–2; *Aircraft Year Book*, 1922, p 244, 1923, p 350, 1924, pp 316, 319–320, 1925, pp 287, 290, 1926, pp 304, 309, 1927, p 377.

5. ASNLs, Jul 16, 1920, p 26, Aug 5, 1920, pp 1–2, Aug 27, 1920, p 2, Sep 29, 1920, pp 4–5, Oct 22, 1920, p 6; Jones Chronology, Sep 28, 1920 (Jones got the story of the race from Joseph M. Pallissard, one of the mechanics in the American party); *Aircraft Year Book* (New York, 1920), pp 132–33.

6. ASNLs, Aug 13, 1920, p 3, Aug 27, 1920, pp 2, 7, Sep 29, 1920, p 6, Oct 22, 1920, pp 7–8, Nov 20, 1920, p 6, Nov 23, 1920, pp 5–6, Dec 7, 1920, pp 1–3, Dec 14, 1920, pp 1–4 (Moseley's report); *Aircraft Year Book* (New York, 1920), pp 135–38.

7. ASNL, Feb 17, 1922, pp 1–4 (includes General Patrick's address at the Aero Club banquet); *Aircraft Year Book* (New York, 1923), pp 56–61; *Annual Report of the Chief of Air Service*, 1922, p 15, 1923, p 54.

8. *Aircraft Year Book* (New York, 1923), pp 56–81, chart following p 352, and pp 353–55; ASNLs, Feb 17, 1922, pp 1–4, Sep 27, 1922, pp 4–5, Nov 9, 1922, pp 1–3; Reed Kinert, *Racing Planes and Air Races*, 4 vols (Fallbrook, Calif., 1969), I, 62–77.

9. *The New York Times*, Oct 15, 1922, p 1; *Aircraft Year Book* (New York, 1923), p 75.

10. *The New York Times*, Oct 17, 1922, p 1; ASNL, Nov 9, 1922, p 3; *Aircraft Year Book* (New York, 1923), p 76.

11. ASNL, Nov 9, 1922, pp 3–5 (included the report of the official observer, Jesse G. Vincent, and the timer's sheet); *The New York Times*, Oct 19, 1922, p 2; *Aircraft Year Book* (New York, 1923), p 76; Jones Chronology, Oct 18, 1922.

12. *Annual Report of the Chief of Air Service*, 1923, p 54; *Aircraft Year Book* (New York, 1924), pp 221–22. The distance for such trials for official records changed from one to three kilometers on April 1, 1924.

13. ASNLs, Sep 29, 1923, p 7, Oct 27, 1923, p 4.

14. ASNLs, Sep 29, 1923, p 7, Oct 12, 1923, pp 21–22, Oct 27, 1923, pp 1–5; *Aircraft Year Book* (New York, 1924), pp 147–175; Kinert, *Racing Planes*, I, 82–91.

15. Kinert, *Racing Planes*, I, 91; Jones, "Chronology of Air Records," III (Jan 1925), 14.

16. Canceled because there were no entries from outside the United States.

17. ASNL, Oct 20, 1924, pp 1–11; Accident Report, Oct 4, 1924 (Skeel).

18. Report of Committee Selected by the Chief of Air Service to Make a Survey of the Engineering Division, Mar 9, 1925, in AFHRC 168.63–4. The other members of the committee were Charles M. Manly, who had been Dr. Samuel P. Langley's chief assistant in aviation at the Smithsonian Institution, and Prof. Edwin B. Wilson, mathematician, Harvard University.

19. ASNLs, Apr 20, 1925, pp 12–13, Sep 12, 1925, pp 4–6.

20. ASNLs, Aug 6, 1925, p 8, Aug 22, 1925, pp 18–19, Sep 12, 1925, pp 6–7, Oct 20, 1925, pp 5–6.

21. ASNLs, Oct 2, 1925, pp 12–13, Nov 16, 1925, pp 1–4.

22. ASNLs, Feb 17, 1922, p 5, May 18, 1922, p 3, Sep 27, 1922, pp 1–4 (Westover's report); *Annual Report of the Chief of Air Service*, 1923, p 54.

23. ASNLs, Jul 10, 1923, p 3, Aug 3, 1923, pp 1–3, Sep 29, 1923, pp 1–3, Oct 27, 1923, p 5.

24. ASNLs, Dec 3, 1919, pp 2–3, Jul 10, 1920, p 20, Jul 16, 1920, pp 1–5, Jul 20, 1920, p 7, Jul 29, 1920, pp 10, 15–16, Aug 13, 1920, pp 5–6, Aug 27, 1920, p 3; Oct 22, 1920, pp 10–12, Oct 29, 1920, pp 1–3, Nov 9, 1920, pp 1–4 (Nutt's narrative), Nov 20, 1920, p 12 (log and statistical summary); Capt St. Clair Streett, "The First Alaskan Air Expedition," *National Geographic Magazine* XLI (May 1922), 445–552.

25. ASNLs, Jan 12, 1921, p 7, Jan 28, 1921, pp 13–14.

26. ASNLs, May 27, 1921, pp 14–15, Jul 22, 1921, pp 1–3; Hinkle, *Wings Over the Border*, pp 55–61.

27. ASNLs, Jan 28, 1921, pp 13–14, Mar 15, 1921, p 21; Mar 23, 1921, pp 1–3, Apr 1, 1921, pp 1–4, Apr 12, 1921, pp 6–7, May 6, 1921, p 16; 1st Lt William D. Coney, "Flying Across the Continent in Twenty-two Hours," *U.S. Air Service* V (Apr 1921), 12–15.

28. ASNLs, Jun 23, 1922, pp 1–2, Sep 6, 1922, p 14, Sep 13, 1922, p 2, Oct 10, 1922, pp 1–4, Oct 19, 1922, p 9.

29. ASNLs, Nov 1, 1922, pp 1–4, Nov 16, 1922, pp 1–6 (report by Macready and Kelly); Louis S. Casey, *The First Nonstop Coast-to-Coast Flight and the Historic T-2 Airplane* [Smithsonian Annals of Flight] (Washington: Smithsonian Institution Press, 1964), I, 1–11.

30. ASNL, Dec 6, 1922, pp 1–7 (report by Macready and Kelly); Casey, *The First Nonstop Coast-to-Coast Flight*, 11–17.

31. ASNL, Feb 20, 1923, pp 7–8.

32. ASNLs, Mar 21, 1923, pp 8–9; May 15, 1923, pp 1–4.

33. ASNL, May 15, 1923, p 4.

34. ASNLs, Jun 4, 1923, pp 1–5, Jun 20, 1923, pp 13–14; Casey, *The First Nonstop Coast-to-Coast Flight*, 19–32.

35. ASNLs, Jul 19, 1923, pp 7–8, Aug 3, 1923, p 14, Oct 27, 1923, pp 5–8; Jones Chronology, Jun 28, 1923.

36. He figured from the beginning of twilight, which in mid-June comes about two hours before sunrise at New York; to the end of twilight, which comes about two hours after sunset at San Francisco.

37. ASNL, Aug 3, 1923, pp 6–9.

38. Checking the almanac, Maughan found that dawn (that is, the beginning of observational twilight) would come at 0222 Eastern Standard Time, and dusk at 2141 Pacific Time. Because of clouds, however, he waited until 0258 to get a little more light for takeoff.

39. ASNLs, Jul 15, 1924, pp 1–4, Jun 12, 1924, pp 10–11, Sep 30, 1924, pp 1–4 (Maughan's report).

40. 1st Lt Lowell H. Smith, Report of the Commander of the World Flight, in AFHRC 168.3952–155; Press Relations, G-2, WDGS, Memoranda for the Press, Apr 7–Sep 15, 1924 (includes radiograms between flight commander and CAS), in same file; ASNLs, Apr 12, 1924, pp 1–3, Apr 30, 1924, pp 5–6, May 16, 1924, pp 2–3, May 31, 1924, pp 1–2, Jun 12, 1924, pp 12–14, Jun 30, 1924, pp 5–8, 23, Jul 15, 1924, pp 11–13, Jul 31, 1924, pp 1–3, Aug 13, 1924, pp 1–8, Sep 15, 1924, pp 1–4, Sep 30, 1924, pp 4–8, Oct 20, 1924, pp 13–16; *Aircraft Year Book* (New York, 1925), pp 66–95. The best account of the flight is Lowell Thomas, *The First World Flight* (Boston, 1925).

41. ASNLs, Oct 20, 1924, pp 13–16, Jan 30, 1925, p 6, Feb 14, 1925, p 8, Mar 5, 1925, p 21, Mar 17, 1925, pp 9–10, Apr 2, 1925, p 3, Jun 5, 1925, p 14; 43 *Stat* 979.

42. *Aircraft Year Book* (New York, 1925), pp 94–95.

Chapter XII

The New Air Corps and the Five-Year Program

1. *The New York Times*, Jul 3, 1926, p 1, Nov 16, 1974, p 34; *U.S. Air Services*, XI (Aug 1926), 13–14; 44 *Stat* 784.

2. Statements by F. Trubee Davison, Asst SW for Air, and others before Baker Board, May 15, 1934, in "Transcript of Shorthand Report, the Special Committee on Army Air Corps and Air Mail" [hereafter cited as Baker Bd Transcript], in AFHRC 167.66–1.

3. 44 *Stat* 780.

4. 44 *Stat* 782.

5. Ltr, The Adjutant General to Chief of Air Service, Dec 23, 1925, in AFHRC 145.91–110.

6. The Army plan, designated "Major Army Project No. 4," had been revised June 30, 1925 (AFHRC 145.91–119) and would be

revised again in May 1926 to become the 5-year program (AFHRC 145.91–108).

7. 1st Ind, Maj Gen Mason M. Patrick, Chief of Air Service, to TAG, Jan 8, 1926, in AFHRC 145.91–110.

8. Memo, Brig Gen Hugh A. Drum, ACS, G–3, for Army Chief of Staff, Jan 30, 1926, in AFHRC 145.91–116; Secretary of War approval, Mar 5, 1926, in AFHRC 145.91–108.

9. H.R. 7916, 69th Cong, lst sess, offered by Representative John M. Morin of Pennsylvania.

10. Ltr, TAG to CAS, Mar 11, 1926, in AFHRC 145.93–18.

11. Memo, Maj Herbert A. Dargue, Chief, War Plans Section, Air Service, to Maj Millard F. Harmon, Jr., head of G–3 Air Section, with Tables I and II, Mar 30, 1926, in AFHRC 145.93–18.

12. Memo, Maj Gen Malin Craig, ACS, G–3, for Maj Gen John L. Hines, Army Chief of Staff, May 21, 1926, in AFHRC 145.91–108.

13. Memo, Craig for CSA, May 21, 1926, with note added by Hines, Jul 23, 1926, in AFHRC 145.91–108.

14. *Annual Report of the Chief of Air Corps*, 1927, pp 11, 16; Hearings before the Committee on Military Affairs, House of Representatives, *Air Corps, Progress under Five-Year Program*, Jan 19, 1927, 69th Cong, 2d sess (Washington, 1927), p 33, in AFHRC 145.91–116. The figures are for June 30, 1926. The airplanes do not include 140 for the National Guard and 36 for service tests.

15. Army Project 4, Table 1, May 21, 1926, in AFHRC 145.91–108.

16. Army Project 4, May 21, 1926, in AFHRC 145.91–108.

17. 1st Ind, Patrick to TAG, Dec 14, 1926, in AFHRC 145.91–120.

18. Ltr, Secretary of War to Chairman, Cmte on Appropriations, HR, Apr 20, 1926, in AFHRC 145.91–116.

19. 44 *Stat* 268.

20. *Annual Report of the Chief of Air Corps*, 1927, p 63; ltr, Patrick to TAG, Oct 20, 1926, in AFHRC 145.93–18; memo, subj: Tentative Estimate . . . FY 1927, in AFHRC 145.91–120.

21. Ltr, SW Dwight F. Davis to President, Oct 6, 1926, in AFHRC 145.93–18; HR, Rprt 700, 69th Cong, 1st sess.

22. Ltr, Patrick to TAG, Oct 28, 1926; memo, TAG to CAC, Oct 29, 1926, in AFHRC 145.93–18.

23. Memo, Maj Gen Mason M. Patrick, CAC, to Deputy CSA Maj Gen Dennis E. Nolan, in AFHRC 145.93–18.

24. 44 *Stat* 784; memo, Patrick to F. Trubee Davison, Asst SW for Air, Nov 19, 1926, in AFHRC 145.93–18.

25. Armed Forces Information School, *The Army Almanac* (Washington, 1950), pp 692–93; John W. Killigrew, "The Impact of the Great Depression on the Army, 1929–1936" (Ph.D. dissertation, Indiana University, 1960), pp I–17 to I–18; War Department, Special Committee on Army Air Corps, *Final Report*, Exh 4 (Washington: War Department, 1934) [hereafter cited as Baker Bd *Final Report*].

26. 44 *Stat* 975; Killigrew, "Impact of the Great Depression," pp 8–13.

27. 44 *Stat* 1119; *Annual Report of the Chief of the Air Corps*, 1928, pp 1, 67.

28. *Annual Report of the War Department*, 1928, p 68.

29. 45 *Stat* 338; *Annual Report of the Chief of the Air Corps*, 1929, pp 72, 95; *Annual Report of the War Department*, 1929, pp 88, 90.

30. *Annual Report of the Chief of the Air Corps*, 1930, p 66; 45 *Stat* 1360.

31. *Annual Report of the Chief of the Air Corps*, 1930, p 83; *Annual Report of the War Department*, 1930, pp 2, 57.

32. Killigrew, "Impact of the Great Depression," Chap II.

33. Ltr, CAC to TAG, Nov 5, 1929, in AFHRC 145.93–19.

34. 46 *Stat* 444; *Annual Report of the Chief of the Air Corps*, 1931, p 69; *Annual Report of the War Department*, 1931, pp 79–80.

35. Killigrew, "Impact of the Great Depression," pp III–7 to III–12.

36. *Annual Report of the War Department*, 1932, p 39.

37. 46 *Stat* 1289; *Annual Report of the War Department*, 1932, pp 1, 5; *Annual Report of the Chief of the Air Corps*, 1932, p 84; Killigrew, "Impact of the Great Depression," pp III–18 to III–23.

38. Alfred Goldberg, ed, *A History of the United States Air Force, 1907–1957* (Princeton, N.J., 1957), p 37.

39. *Annual Report of the War Department*, 1932, p 39.

40. Goldberg, *History*, p 37.

41. *Annual Report of the War Department*, 1932, p 40.

42. 42 *Stat* 840.

43. 44 *Stat* 783.

44. Report of Board of Officers to Revise 5-Year Program, Jul 13, 1929, in AFHRC 145.93-21.

45. Memo, Brig Gen Campbell King, ACS, G-1, to CSA, Aug 20, 1928, in AFHRC 145.93-19.

46. *Annual Report of the Chief of the Air Corps*, 1929, pp 2-3, 5; *Annual Report of the War Department*, 1929, pp 88-89.

47. *Annual Report of the War Department*, 1930, p 59, 1931, p 32, 1932, pp 42-43. An exception to the last statement is 2d Lt Elvin F. Maughan, appointed in the Regular Army on June 18, 1930, and accepted on July 2.

48. 44 *Stat* 780. Ninety percent could be given duty for not more than one year; ten percent for not more than two years.

49. *Annual Report of the Chief of the Air Corps*, 1928, p 15; *Annual Report of the War Department*, 1928, p 68.

50. *Annual Report of the Chief of the Air Corps*, 1930, p 14, 1931, p 15, 1932, p 20; *Annual Report of the War Department*, 1930, p 69, 1931, p 35, 1932, p 43.

51. *Annual Report of the War Department*, 1927, p 44, 1932, p 42; memo, Brig Gen James E. Fechet, Asst CAC, for F. Trubee Davison, Asst SW for Air, Dec 1, 1927, in AFHRC 145.93-14; *Annual Report of the Chief of the Air Corps*, 1931, p 7; AC, Resume of 5-Year Program, in AFHRC 145.93-19; Hearings before the Committee on Military Affairs, HR, *Air Corps, Progress under Five-Year Program*, Jan 19, 1927, in AFHRC 145.91-116.

52. *Annual Report of the Chief of the Air Corps*, 1931, p 7; ACNL, Jul 28, 1930, pp 218-19, Aug 19, 1931, p 287; George W. Cullum, *Biographical Register of the U.S. Military Academy*, Supplement, Volume VIII (Chicago: R. R. Donnelley & Sons, Co, 1940), 876-935.

53. Memo, Fechet for Davison, Dec 14, 1927, in AFHRC 145.93-19.

54. *Annual Report of the Chief of the Air Corps*, 1928, p 41, 1929, pp 42-44, 1930, pp 39-40, 1931, p 48, 1932, p 58, 1933, p 33, 1934, p 28, 1935, p 26, 1936, p 30.

55. The Air Corps Act required ninety percent of the commissioned officers to be flyers, who in peacetime would have to be qualified airplane pilots (44 *Stat* 781).

56. Memo, Drum for CSA, Jan 30, 1926; Army Project 4, May 21, 1936, in AFHRC 145.91-108. The Air Corps was estimating that one-third of the cadets and enlisted men, and half of the officers entering primary training would graduate from the Advanced Flying School. The number of replacement pilots needed would be about 350 during the 5-year period. [Notes on memo form of 1st Lt Horace S. Kenyon, and related papers, in AFHRC 145.91-102.]

57. 44 *Stat* 781; HR, Rprt 700, 69th Cong, 1st sess, pp 4-5; *Annual Report of the Chief of Air Corps*, 1926, p 46. Davison reported on January 19, 1927, that 14.5 percent of the pilots of tactical units were enlisted men. [Hearings before the Committee on Military Affairs, HR, *Air Corps, Progress under Five-Year Program*, Jan 19, 1927, in AFHRC 145.91-116.]

58. Memo, Chief, Training and War Plans Division, to Executive, Dec 21, 1926, in AFHRC 145.93-18.

59. *Annual Report of the War Department*, 1928, p 2.

60. Duncan Field, the home of the San Antonio Air Intermediate Depot, had formerly been part of Kelly Field.

61. Hist, AAF CFTC, Jan 1, 1939-Dec 7, 1941, Vol 4, Doc I, 2, in AFHRC 223.01.

62. Ltr, Brig Gen Frank P. Lahm to Maj James E. Chaney (identical ltr to Maj Frank M. Andrews), Aug 13, 1926, in hist, AAF CFTC, Jan 1, 1939-Dec 7, 1941, Vol 4, Doc I, 8.

63. The term "transition" was beginning to replace "transformation" for this kind of training.

64. Ltr, Brig Gen James E. Fechet, Asst CAC, Mar 9, 1927, in AFHRC 145.91-120.

65. Ltr, Patrick to TAG, Mar 26, 1927, 1st Ind, TAG to CAC, Apr 5, 1927, in AFHRC 145.91-120. The General Staff made only one change. Patrick, following Fechet's proposal, had divided flying instruction at the basic schools into two phases: "Primary Flying (Primary training type aircraft)," and "Basic Flying (Observation type aircraft)." The General Staff combined the two as "Flying (Training and Observation type airplanes)."

66. Air Corps documents of this period are sometimes confusing because the terms "primary" and "basic" had not yet attained precise definition and consistent usage in connection with flying training except as related to designation of primary training (PT) and basic training (BT) types of airplanes.

67. *Annual Report of the Chief of the Air Corps*, 1928, p 44. Reserve officers, members of the National Guard, and students from foreign countries brought the total number entering to 647. The figures do not include

some students in training at the beginning of the year or those still in training at the end of the year.

68. Memo, G–1 for CSA (prepared by Lt Col Roy C. Kirtland), Aug 20, 1928, in AFHRC 145.93–19.

69. Memo, Maj Gen James E. Fechet, CAC, for G–1, Nov 22, 1928, in AFHRC 145.93–19; *Annual Report of the Chief of the Air Corps*, 1928, pp 46–47.

70. Memo, Fechet for G–1, Nov 22, 1928, in AFHRC 145.93–19.

71. Ltr, Maj Robert E. M. Goolrick, Comdr, 2d Cav Div Aviation, to CAC, Apr 29, 1929, in AFHRC 145.93–19.

72. 1st Ind, Commandant, Fort Riley, Kans., to VII Corps Area, May 1, 1929, 2d Ind, VII CA to TAG, May 7, 1929, 3d Ind, TAG to Fechet, Jun 15, 1929, 4th Ind, Fechet to TAG, Jul 11, 1929, in AFHRC 145.93–19.

73. Jones Chronology, Aug 14, 1929 (quoting WD Cir 47); ACNL, Sep 23, 1929, p 322; *Annual Report of the Chief of the Air Corps*, 1930, p 10.

74. Rossi L. Selvaggi, "A History of Randolph Air Force Base" (MA thesis, University of Texas, 1958); Program for Dedication of Randolph Field, June 20–21, 1930, in AFHRC K287.86 (1930–1955); *Annual Report of the Chief of the Air Corps*, 1932, p 59.

75. Memo, Fechet for G–1, Nov 22, 1928, in AFHRC 145.93–19.

76. Report of Board of Officers to Revise 5–Year Program, Jul 13, 1929, in AFHRC 145.93–21.

77. *Annual Report of the Chief of Air Corps*, 1926, p 46, 1927, p 19, 1928, pp 9–10, 1930, p 9, 1931, p 12, 1932, p 16, 1933, p 14.

78. *Annual Report of the Chief of the Air Corps*, 1928, p 44, 1929, following p 44, 1930, following p 38, 1931, following p 48, 1932, p 16 and following p 58; AC, Strength Report, Jun 30, 1932.

79. 1st Ind, Patrick to TAG, Sep 14, 1926; memo, Actg CAC for Davison, Dec 22, 1926, in AFHRC 145.93–18; 44 *Stat* 257, 283.

80. Memo, Exec, Tng & Ops Div, to Ch, Pers Div, Apr 11, 1927; ltr, Exec, OCAC, to TAG, Jun 7, 1927; ltrs, Fechet to TAG, Jul 5, 1927; ltr, TAG to CAC, Aug 17, 1927, in AFHRC 145.93–18; Report of Board of Officers to Revise 5-Year Program, Jul 13, 1929, in AFHRC 145.93–21.

81. Memo, Capt Charles B. B. Bubb, AC War Plans Sec, to Exec, OCAC, Sep 22, 1928, in AFHRC 145.93–19.

82. *Annual Report of the Chief of the Air Corps*, 1930, p 6.

83. *Annual Report of the War Department*, 1928, p 184, 1929, p 194, 1930, pp 3, 308, 1931, p 172; ACNL, Sep 23, 1929, p 318.

84. *Annual Report of the War Department*, 1932, p 44; *Annual Report of the Chief of the Air Corps*, 1928, p 11, 1932, p 12. The figures do not include flying cadets.

85. Ltr, TAG to CG, II CA, Sep 27, 1927, in AFHRC 145.93–18.

86. Base pay for all grades increased five percent with each four years of service.

87. 44 *Stat* 781; WD Cir 55, Oct 5, 1926.

88. *Annual Report of the Chief of Air Corps*, 1927, p 16.

89. Ltr, TAG to CAC, Sep 1, 1926, in AFHRC 145.93–18.

90. 1st Ind, Patrick to TAG, Sep 14, 1926, in AFHRC 145.93–18.

91. *Annual Report of the Chief of the Air Corps*, 1928, p 16, 1929, p 7, 1930, p 8, 1931, p 11, 1932, p 14.

92. Ltr, TAG to CAC, Aug 17, 1927, in AFHRC 145.93–18; memo, Fechet for Davison, Dec 14, 1927, in AFHRC 145.93–19.

93. *Annual Report of the Chief of the Air Corps*, 1928, pp 14–15.

94. *Annual Report of the War Department*, 1928, p 67, 1929, p 88; memo, G–3 to G–4, Dec 22, 1928, in AFHRC 145.93–19.

95. *Annual Report of the Chief of the Air Corps*, 1929, pp 7–8, 1930, p 8; *Annual Report of the War Department*, 1930, pp 59–60.

96. *Annual Report of the Chief of the Air Corps*, 1931, p 14, 1932, p 19; *Annual Report of the War Department*, 1931, p 32, 1932, p 44.

97. 1st Ind, Patrick to TAG, Jan 8, 1926, in AFHRC 145.91–110.

98. Memo, Drum for CSA, Jan 30, 1926, in AFHRC 145.91–108.

99. Memo, Patrick to Davison, Jul 31, 1926, in AFHRC 145.91–116; ltr, Patrick to TAG, Nov 30, 1927, in AFHRC 145.93–19.

100. Memo, Patrick to G–3, May 8, 1926, in AFHRC 145.91–111; Army Project 4, May 21, 1926, in AFHRC 145.91–108.

101. 44 *Stat* 784.

102. Memo, Patrick to Davison, Jul 31, 1926, in AFHRC 145.91–116.

103. Ltr, Patrick to TAG, Oct 26, 1926, in AFHRC 145.93–19.

104. *Ibid.*; ltr, Patrick to TAG, Nov 30, 1927, in AFHRC 145.93–19; Table of Comparison, Secs A and C, in AFHRC 145.91–346.

105. Table of Comparison, Secs D and E, in AFHRC 145.91–346.

106. Ltr, Partick to TAG, Nov 30, 1927, in AFHRC 145.93–19.

107. Memo, Davison to Davis, Feb 17, 1928; 1st Ind, TAG to CAC, Mar 9, 1928, in AFHRC 145.93-19.

108. Ltr, Fechet to TAG, Nov 2, 1928, in AFHRC 145.93-19.

109. Report of Board of Officers to Revise 5-Year Program, Jul 13, 1929, in AFHRC 145.93-21; *Annual Report of the War Department*, 1929, p 90.

110. *Annual Report of the Chief of the Air Corps*, 1930, p 66, 1931, p 69; *Annual Report of the War Department*, 1930, p 61.

111. *Annual Report of the War Department*, p 31.

112. Ltr, OCAC to TAG, May 28, 1929, in AFHRC 145.93-19.

113. Memo, Maj Delos C. Emmons, Exec to Asst SW for Air, for CAC, Jul 31, 1929; memo, Emmons for CAC, Aug 2, 1929; memo, Exec, OCAC, for Asst SW, Aug 5, 1929; memo, Exec, OCAC, for Ch, Tng & Ops Div, Aug 5, 1929; ltr, OCAC to TAG, Aug 10, 1929, 1st Ind, TAG to CAC, Aug 24, 1929, in AFHRC 145.93-19.

114. Ltr, CAC to TAG, Oct 4, 1929, 5th Ind, TAG to CAC, Jan 11, 1930; ltr, Judge Advocate General to Asst SW, Dec 2, 1929, in AFHRC 145.93-19; *Annual Report of the War Department*, 1930, p 61. The Attorney General of the United States supported the Judge Advocate General's opinion. [36 *Opinions of the Attorney General*, 418.]

115. *Annual Report of the Chief of the Air Corps*, 1932, pp 83-84; 46 *Stat* 1289.

116. *Annual Report of the Chief of the Air Corps*, 1933, p 57, 1934, p 47; *Annual Report of the War Department*, 1932, p 42.

117. *Annual Report of the War Department*, 1932, p 42.

118. Robert W. Ackerman, *The Maintenance of Army Aircraft in the United States, 1921-1929* (Patterson Field, Ohio, 1945), pp 30-43.

119. The first nine P-6s had water-cooled engines; the remainder used Prestone.

120. Peter M. Bowers, *The Curtiss Army Hawks* [*Aircraft in Profile*, No. 45], pp 8-9; Swanborough, *U.S. Military Aircraft*, pp 169-172.

121. Adopted in 1926, the previous markings being three vertical stripes—blue, white, and red.

122. Bowers, *The Curtiss Army Hawks*, pp 10-12, and accompanying pictures in color.

123. Peter M. Bowers, *The Boeing P-12E* [*Aircraft in Profile*, No. 2]; Swanborough, *U.S. Military Aircraft*, pp 65-67; Aircraft Characteristics, Jan 1, 1931, Sep 1, 1932.

124. Proceedings of a Board of Officers . . . for the Purpose of Testing the XP-16 . . . , Nov 3, 1930, in AFHRC 248.122-3; ACNL, Oct 18, 1932, p 405; *Annual Report of the Chief of the Air Corps*, 1932, p 50; Lloyd S. Jones, *U.S. Fighters* (Fallbrook, Calif.: Aero Publishers, 1975), pp 50-52, 63-67; Swanborough, *U.S. Military Aircraft*, pp 130-31, 483. The P-16 later became the PB-1, and the P-30 the PB-2, the PB standing for pursuit biplace.

125. Swanborough, *U.S. Military Aircraft*, pp 178-79.

126. Aircraft Characteristics, Jan 1, 1931, Mar 15, 1935; Swanborough, *U.S. Military Aircraft*, pp 277-281, 494; Fahey, *U.S. Army Aircraft*, pp 22-29.

127. *Annual Report of the Chief of Air Corps*, 1927, p 44.

128. AC, Model Designations of Airplanes, Mar 1936, in AFHRC 218.9-1; Fahey, *U.S. Army Aircraft*, pp 24, 28; Swanborough, *U.S. Military Aircraft*, pp 268, 497, 507, 510, 528.

129. AC, Procurement Data, Training Airplanes (chart) [*ca.* 1939], in AFHRC 145.91-508; Swanborough, *U.S. Military Aircraft*, pp 126-29; Fahey, *U.S. Army Aircraft*, pp 24, 35.

130. Ltr, Patrick to TAG, Mar 26, 1927, 1st Ind, TAG to CAC, Apr 5, 1927, 8th Ind, Chaney to Andrews, May 19, 1927, 9th Ind, Andrews to Lahm, Jun 7, 1927, in AFHRC 145.91-120.

131. AC, Procurement Data, Training Airplanes; Fahey, *U.S. Army Aircraft*, p 24.

132. AC, Procurement Data, Training Airplanes; Swanborough, *U.S. Military Aircraft*, pp 166, 168; Fahey, *U.S. Army Aircraft*, p 21.

133. *Annual Report of the Chief, Materiel Division*, 1932, pt II, pp 19-25. The A-3 and A-3A were classified "limited standard" and the A-3B "substitute standard."

134. Memo, Ch, Tng & War Plans Div, for Exec, Dec 21, 1926; ltr, Exec, Tng & Ops Div, to Ch, Finance Sec, Mar 12, 1927; ltr, Exec to TAG, Jun 7, 1927, in AFHRC 145.93-18; *Annual Report of the Chief of Air Corps*, 1927, pp 7-8.

135. *Annual Report of the War Department*, 1932, pp 40-41; *The New York Times*, Mar 2, 1933, p 13; Maurer, *Combat Squadrons*, pp 146, 174-76, 270-71, 281-82, 523-25.

136. Report of Board of Officers to Revise 5-Year Program, Jul 13, 1929, in AFHRC 145.93-21.

137. Air Corps 5-Year Program, revised Jul 1, 1927, in AFHRC 145.91-346; Air Corps Strength Reports, Jun 30, 1927, thru Jul 31, 1932.

138. Air Corps 5-Year Program, revised Jul 1, 1927, in AFHRC 145.91-436; Report of Board of Officers to Revise 5-Year Program, Jul 13, 1929, in AFHRC 145.93-21; *Annual Report of the Chief of the Air Corps*, 1929, p 1, 1930, p 1.

139. *Annual Report of the War Department*, 1932, p 41; AC, Strength Report, Aug 31, 1933.

140. 1st Ind, AC to TAG, Dec 14, 1926, in AFHRC 145.91-120; 3d Wg, History of Barksdale Field (1937), in AFHRC 280.58-1, Vol 2.

141. Hist, 20th Ftr Gp, Nov 15, 1929-Jul 17, 1944; in AFHRC GP-20-Hi (Ftr); 3d Wg, History of Barksdale Field (1937), in AFHRC 280.58-1, Vol 2.

142. See note above.

143. AC, Strength Reports, Jun 30, 1926, Jun 30, 1923. Air Corps units were still organized under Tables of Organization approved in the early 1920s (AFHRC 167.412-1). These were revised in 1930 and approved for war planning and mobilization but not for reorganization of the Air Corps.

144. Memo, Exec for Ch, Materiel Div, Apr 9, 1926, in AFHRC 145.93-18.

145. Ltr, Patrick to TAG, Jun 3, 1927, in AFHRC 145.91-120.

146. AC, Strength Report, Jun 30, 1926.

147. *Annual Report of the War Department*, 1930, p 58; AC, Strength Report, Jun 30, 1930.

148. *Annual Report of the Assistant Secretary of War*, 1927, pp 43-44, 1928, p 67, 1929, p 87, 1931, p 32; *Annual Report of the War Department*, 1927, p 53, 1932, pp 6, 64-66; *Annual Report of the Chief of the Air Corps*, 1929, p 7.

149. 44 *Stat* 782.

150. Hearings before the Committee on Military Affairs, House of Representatives, *Air Corps, Progress under Five-Year Program*, Jan 19, 1927, in AFHRC 145.91-116.

151. *Annual Report of the War Department*, 1930, pp 58-59; AC, Resume of 5-Year Program, n.d., in AFHRC 145.93-19.

Chapter XIII

Tactical Training

1. *Annual Report of the Chief of Air Corps*, 1927, p 43; ASNL, Nov 10, 1927, p 321.

2. *Annual Report of the Chief of the Air Corps*, 1929, pp 31-32.

3. *Annual Report of the Chief of the Air Corps*, 1931, p 43.

4. ACNLs, Mar 31, 1937, pp 84-85, Jun 27, 1927, pp 188-89. Oct 18, 1932, p 408; *Annual Report of the Chief of Air Corps*, 1927, p 29.

5. ACNLs, May 22, 1928, pp 173-74, 177, 180, Jun 20, 1928, pp 236-37, Oct 18, 1932, p 408; *Annual Report of the Chief of the Air Corps*, 1928, pp 31-32.

6. ACNLs, Mar 26, 1929, p 179, Jun 7, 1929, p 207, Oct 18, 1932, p 408; *Annual Report of the Chief of the Air Corps*, 1929, pp 28-29.

7. ACNLs, May 13, 1930, p 143, Sep 30, 1930, p 260, Oct 20, 1930, pp 293, 296, Oct 18, 1932, p 208; *Annual Report of the Chief of the Air Corps*, 1931, p 41.

8. ACNLs, Sep 18, 1931, p 325, Oct 17, 1931, pp 355-56, Oct 18, 1932, p 408; *Annual Report of the Chief of the Air Corps*, 1932, p 44.

9. ACNLs, Aug 12, 1932, p 313, Oct 18, 1932, pp 405-408; Jones Chronology, Jun 9, 1928.

10. *Annual Report of the Chief of the Air Corps*, 1928, pp 26-27, 1931, p 41, 1932, p 44.

11. *Annual Report of the Chief of the Air Corps*, 1928, p 27.

12. ACNLs, Jan 27, 1928, pp 47-49, Feb 21, 1928, pp 64-65; *Annual Report of the Chief of the Air Corps*, 1928, pp 26-27.

13. *Washington Herald*, Aug 19, 1931.

14. ACNLs, Jul 21, 1931, p 254; *The New York Times*, Aug 10, 1931, p 8, Aug 12, 1931, p 39, Aug 13, 1931, p 24, Aug 14, 1931, p 1, Aug 15, 1931, p 3; Bombing of Mount Shasta (scrapbook of clippings, photographs, and charts, with marginal comments by 1st Lt James G. Pratt, who participated in the operation), in AFHRC 248.222-29A, pt l; Murray Green, "'The *Shasta* Disaster': Forgotten Les-

sons in Interservice Relations," *Air University Review* XXX (Mar-Apr 1979), 68–74. Dargue's statements quoted in the text are from an unidentified clipping in Pratt's scrapbook.

15. ACNL, Apr 8, 1932, p 123.

16. ACNLs, Apr 26, 1929, p 138, May 16, 1929, pp 175–76; *Annual Report of the Chief of the Air Corps*, 1929, pp 30–31.

17. ACNL, Jan 16, 1930, pp 5, 12–13.

18. ACNL, Dec 8, 1930, p 351.

19. ACNL, Jun 9, 1930, p 153.

20. ACNL, Oct 20, 1930, p 299.

21. ACNL, Nov 9, 1931, p 397.

22. ACNL, Dec 4, 1931, pp 419–421. The flying suits were B–7s, the goggles B–6s.

23. ACNLs, Aug 27, 1932, p 339, Dec 30, 1932, p 486.

24. *Signal Corps Bulletin* 41, Dec 1927, pp 23–24.

25. *Airplane Radio Sets Types SCR–133, –134, –135* (Washington: OCSO, 1925), pp 1–9, 47–48, 84–86; ACM, Communications, revised Jul 1, 1928, pp 33–40, in AFHRC 248.312–80; ACNL, Apr 10, 1931, p 45. 1st Lt Herbert G. Messer, "Signal Corps Aircraft Radio Laboratory," in *Signal Corps Bulletin*, November-December 1930, pages 5–11, reviews the work of the laboratory at Wright Field, including the Air Corps planes fitted as flying laboratories.

26. Capt Fred G. Borden, "Communications to and from Airplanes in Flight," in *Signal Corps Bulletin* 52, Jan-Feb 1930, pp 1–5.

27. Proceedings of Board of Officers, Nov 1928, in AFHRC 248.312–68; *Signal Corps Bulletin* 48, May-Jun 1929, pp 66–67; *Annual Report of the Chief of the Air Corps*, 1931, p 45.

28. ACNL, Apr 18, 1930, p 107.

29. Report on Air Corps Field Exercises and Demonstrations, 1930, in AFHRC 248.2122–1 (Mar-Apr 1930).

30. *Ibid.*

31. *Signal Corps Bulletin* 60, May-Jun 1931, pp 57–60.

32. ACNL, May 7, 1931, pp 173–74.

33. ACNL, Apr 10, 1931, p 145; *Annual Report of the Chief of the Air Corps*, 1931, p 45.

34. ACNL, Jul 21, 1931, p 269.

35. ACNL, Jun 5, 1931, p 206.

36. ACNLs, Jun 5, 1931, p 209, Jun 30, 1931, p 238, Nov 9, 1931, p 395; *Annual Report of the Chief of the Air Corps*, 1932, pp 54–55; *Signal Corps Bulletin* 59, Mar–Apr 1931, p 50; *Signal Corps Bulletin* 66, May–Jun 1932, pp 43–44.

37. ACNL, Oct 6, 1928, pp 365–66.

38. ACNLs, Aug 12, 1929, p 282, Sep 23, 1929, pp 330, 338. On the trip west, one LB–7 was forced down at Winslow, Arizona, by trouble with the gasline and did not arrive at Rockwell Field until August 7.

39. *Signal Corps Bulletin* 52, Jan–Feb 1930, pp 36–37.

40. ACNL, Mar 31, 1930, pp 75–77.

41. ACNLs, Feb 17, 1927, pp 43–44, Mar 10, 1927, pp 58–62.

42. ACNLs, Dec 21, 1929, pp 435–36, Jan 16, 1930, p 13, Feb 6, 1930, pp 31–37, Mar 5, 1930, pp 54–55; *Annual Report of the Chief of the Air Corps*, 1930, p 27.

43. ACNL, Feb 6, 1932, p 32.

44. ACNL, Feb 6, 1930, p 31.

45. ACNL, Jun 5, 1930, pp 154–56.

Chapter XIV

Exercises and Maneuvers

1. HQ VIII CA, Report on Air-Ground Maneuvers, May 15–21, 1927, in AFHRC 248.2122; *Annual Report of the Chief of Air Corps*, 1927, pp 27–28; ACNL, Mar 31, 1927, pp 79–80, 147, 156–59.

2. Directives, operations orders, and related papers, in AFHRC 248.2122 (1927–28); *Annual Report of the Chief of the Air Corps*, 1928, pp 32–34; ACNLs, Jun 5, 1928, pp 208, 210–11, Jun 20, 1928, pp 223–24, 237.

3. AC Tac Sch Memo 22, Umpire Rules for Air Corps Maneuvers, May 6, 1929, in AFHRC 248.212/1927–34.

4. *The New York Times*, May 19, 1929, p 3.

5. The Air Corps Tactical School's file on the maneuvers (AFHRC 248.2122/1929) contains plans, orders, and other valuable materials, including the critique by Maj Walter H. Frank, Assistant Commandant, and an anonymous statement critical of the way the

maneuvers were conducted. Another Tactical School folder (AFHRC 248.2122/May 12–26, 1929) contains the translation of an article about the maneuvers published in a Russian magazine. *The New York Times* gave full coverage to the maneuvers: May 16, 1929, p 5; May 17, 1929, p 2, May 18, 1929, p 9, May 19, 1929, p 1, May 20, 1929, p 8, May 21, 1929, p 3, May 22, 1929, p 1, May 23, 1929, pp 1, 28 (editorial), May 24, 1929, p 19, May 25, 1929, p 9, May 26, 1929, p 17, May 27, 1929, p 17; ACNL, Jun 7, 1929, pp 209–210; *Annual Report of the Chief of the Air Corps*, 1929, p 26; *Annual Report of the War Department*, 1929, pp 1–2, 92; Accident Report, May 18, 1929 (2d Lt Edward L. Meadow), in AFHRC 3912–1.

6. Brig Gen William E. Gillmore, Asst to CAC, Report on Air Corps Field Exercises and Demonstrations, 1930, in AFHRC 248.2122–1/Mar-Apr 1930; *Annual Report of the Chief of the Air Corps*, 1930, pp 29–30; ACNLs, Mar 31, 1930, pp 75–77, Apr 18, 1930, p 107, May 13, 1930, pp 126–29, Jun 5, 1930, pp 154–56.

7. Gillmore, Report.

8. The 16th Observation Squadron, committed to work with Army service schools, did not participate.

9. Brig Gen Benjamin D. Foulois, Asst to CAC, Report of the Air Corps Field Exercises of 1931; G–1, G–2, G–3, and G–4 rprts, and related documents in AFHRC 248.2122–2; *Annual Report of the War Department*, 1931, p 32; *Annual Report of the Chief of the Air Corps*, 1931, pp 40–41, 47; ACNLs, Jan 15, 1931, p 51, Apr 10, 1931, p 138, Jun 5, 1931, pp 190–95, Jun 30, 1931, pp 224–28, 231, 242, Jul 21, 1931, pp 252–54.

10. G–2 rprt, lst Air Division (Provisional), in AFHRC 248.2122–2.

11. *The New York Times*, Apr 26, 1931, Sec III, p 6, May 16, 1931, p 21, May 24, 1931, p 28, May 23, 1931, p 3, May 25, 1931, pp 1, 12.

12. *The New York Times*, May 28, 1931, p 1, May 29, 1931, p 23.

13. *The New York Times*, Jun 2, 1931, p 28.

14. *The New York Times*, Jun 3, 1931, p 27.

15. Asst SW for Air, Press Release, May 31, 1931, in AFHRC 248.2122–2 (G–3 exhibits).

16. Foulois, Report.

17. *Annual Report of the War Department*, 1932, p 5.

18. ACNL, Dec 21, 1929, pp 439–440.

19. Ltr, Maj Carl Spatz, Comdr, 1st Bomb Wg, to CG, IX CA, subj: Report on Night Attack on Fort MacArthur, Apr 29, 1932, 4th Ind, Lt Col Howard S. Miller, HQ Harbor Defense, Los Angeles, Jun 2, 1932, 7th Ind, Lt Col Henry H. Arnold, Comdr, March Field, Calif., Jun 16, 1932, in AFHRC 248.2122–2A.

20. ACNL, Apr 8, 1932, p 124.

21. ACNL, May 28, 1932, pp 193–94.

22. ACNL, Apr 8, 1932, pp 120–22.

Chapter XV

Flying

1. Herbert A. Dargue, Personal Papers, in AFHRC; Herbert A. Dargue, "How Latin America Looks from the Air," *National Geographic Magazine* LII (Oct 1927), 451–502; *Aero Digest* 10 (Jan 1927), 30, (Feb 1927), 118, (Mar 1927), 220, (Apr 1927), 314, (May 1927), 372–73, 485; *U.S. Air Services* XII (Jan 1927), 18, (Feb 1927), 32–34, (Mar 1927), 19–20, (Apr 1927), 30, (May 1927), 15–16, (Jun 1927), 25–28; Wesley P. Newton, *The Perilous Sky: U.S. Aviation Diplomacy and Latin America, 1919–1931* (Coral Gables, Fla.: University of Miami Press, 1978).

2. *The New York Times*, May 26, 1927, p 1. The following paragraphs on the first flight to Hawaii are based principally on accounts written by two flyers: Albert F. Hegenberger, California-Hawaiian Flight [1927], in AFHRC K239.0512–854; Lester J. Maitland, *Knights of the Air* (Garden City, N.Y., 1929), Chapter XXVII; and on stories in *Air Corps News Letters* as follows—June 27, 1927, pages 181, 196, July 19, 1927, pages 205–7, August 9, 1929, pages 238–241, October 15, 1927, pages 293–94.

3. ACNL, Aug 9, 1927, p 240.

4. ACNL, Jun 27, 1926, pp 181, 186.

5. *Aero Digest* 11 (Jul 1927), 16.

6. The time difference between California and Hawaii was two and a half hours.

7. Historical Office of the Army Air Forces, *The Official Pictorial History of the AAF,* p 72.

8. ACNL, Aug 9, 1927, p 240.

9. ACNLs, Jan 10, 1929, pp 16–17, Jan 31, 1929, pp 22–27, 38–41 (quotations from newspaper editorials), 43, Feb 23, 1929, p 58 (quoting Spatz's report); *Aviation* XXVI (Jan 19, 1929), pp 158–162; *The New York Times,* Jan 2, 1929, p 29, Jan 3, 1929, p 7, Jan 5, 1929, p 1, Jan 6, 1929, p 22, Jan 7, 1929, p 1 (*The New York Times* printed many parts of the log); *Aircraft Year Book* (New York, 1930), pp 145–48; WD Press Release, Dec 17, 1928, clippings, and miscellaneous papers in AFHRC K141.2421 (Spaatz); intvw, Arthur K. Marmor, hist, USAF Hist Div, with Lt Gen Ira C. Eaker, Ret, Jan 1966, pp 24–26, in AFHRC K239.0512–626.

10. ACNL, Feb 23, 1929, p 58.

11. ACNL, Jan 31, 1929, pp 38–41.

12. ACNL, Aug 12, 1929, pp 277–280; *Aircraft Year Book* (New York, 1930), pp 145–159, 605; *Aircraft Year Book* (New York, 1931), pp 113–14.

13. Program, National Air Races, 1926; Report of Contest Committee, National Air Races, 1926, in AFHRC 168.657–7; ACNL, Sep 22, 1926, pp 1–5; *Aircraft Year Book* (New York, 1927), pp 184–89; Kinert, *Racing Planes,* II, 26–27; *The New York Times,* Sep 10, 1926, p 23.

14. ACNL, Oct 15, 1927, pp 290–92; Kinert, *Racing Planes,* II, 34–36; Bowers. *The Curtiss Army Hawks,* p 8.

15. ACNL, Sep 23, 1929, pp 324–25; Kinert, *Racing Planes,* II, 50–57; Bowers, *The Curtiss Army Hawks,* p 7.

16. ACNL, Oct 17, 1929, pp 348–49; Accident Report, Sep 10, 1928 (2d Lt John J. Williams), in AFHRC 200.3912–1.

17. ACNL, Jul 10, 1928, pp 248–250 (2d Lt William O. Eareckson's story), 255–57 (report by 1st Lt Uzal G. Ent), 264–65, Sep 13, 1928, pp 315–18.

18. ACNL, Jun 22, 1932, pp 227–231 (includes accounts by 2d Lt Wilfred J. Paul, MSgt Joseph H. Bishop, and Capt William J. Flood); *Aircraft Year Book* (New York, 1933), pp 87, 100, 398–99.

19. ACNLs, Jul 1, 1936, pp 3, 20, Jul 15, 1936, p 14; *National Aeronautic Magazine,* Jul 1936, p 30, Aug 1936, p 36; *Aircraft Year Book* (New York, 1937), p 290.

20. ACNL, Apr 26, 1927, p 107.

21. ACNLs, Mar 31, 1927, pp 80–81, Jun 8, 1927, p 164, Apr 26, 1927, pp 107–9 (includes Capt Hawthorne C. Gray's report). This report became official. [*Aircraft Year Book* (New York, 1928), p 517.]

22. ACNL, Jun 8, 1927, pp 164–66; *Annual Report of the Chief of Air Corps,* 1927, p 30.

23. ACNL, Jan 27, 1928, pp 38–39.

24. ACNLs, Nov 10, 1927, pp 318–19, Jan 7, 1928, p 5, Jan 27, 1928, pp 34–41; Jones Chronology, Nov 4, 1927 (citing ACNLs, files, and WD GO 5, 1928, and quoting Capt Harrison H. Fisher on the findings).

25. ACNL, Jul 10, 1928, p 260.

26. *Ibid.*

27. ACNL, Jul 10, 1928, pp 258–260.

28. ACNL, Oct 29, 1928, pp 382–83.

29. *Ibid,* p 381.

30. ACNL, Jan 10, 1929, p 2.

31. ACNL, Oct 29, 1928, p 382.

32. ACNL, Oct 29, 1928, pp 380–82 (including Capt Albert W. Steven's account of the flight), Jan 10, 1929, pp 1–2 (including Capt St. Clair Streett's account).

33. ACNLs, Oct 29, 1928, p 382, Jan 10, 1929, p 1.

34. ACNL, Mar 15, 1929, pp 87, 97–98.

35. ACNL, Apr 18, 1980, p 101; Peyton, *50 Years,* pp 39–40, 42, 47, 58, 87; Harry G. Armstrong, *Principles & Practice of Aviation Medicine* (Baltimore, 1952), pp 29–34.

36. Capt David A. Myers, "The Medical Contributions to the Development of Blind Flying," in ACNL, Jul 15, 1936, pp 1–4; ACNL, Aug 1, 1926, pp 7–10; William C. Ocker and Carl J. Crane, *Blind Flying in Theory and Practice* (San Antonio, 1932), pp 9–14; James H. Doolittle, "Early Blind Flying," Third Lester Gardner Lecture at MIT, Cambridge, Mass., Apr 28, 1961, in AFHRC 168.3–67; Downs, "Contributions," pp 340–41; Peyton, *50 Years,* pp 50–71; ACNLs, Mar 10, 1927, pp 63–64, Jan 15, 1931, p 7, Aug 15, 1937, pp 16–18; 46 *Stat* 94.

37. Intvw, Lt Col Robert M. Burch, Maj Ronald R. Fogleman, and Capt James P. Tate, USMA, with Gen. James H. Doolittle, Ret, Sep 26, 1971, in AFHRC K239.0512–793; Doolittle, "Early Blind Flying'; Daniel Guggenheim Fund for the Promotion of Aeronautics, *Solving the Problem of Fog Flying* (New York, 1929); Daniel Guggenheim Fund for the Promotion of Aeronautics, *Equipment Used in Experiments to Solve the Problem of Fog Flying* (New York, 1930); Downs, "Contributions," pp 345–350; ACNL, Jun 22, 1932, p 238.

38. ACNLs, May 28, 1932, pp 207–8, Jun 22, 1932, p 237; Downs, "Contributions," 350–57; Jones Chronology, May 9, 1932.

Chapter XVI

GHQ Air Force Headquarters

1. Maurer, *The U.S. Air Service in World War I*, Vol II: *Early Concepts of Military Aviation*, Docs 17–18, 22–23, 27, 29–30, 38–39, 43, and *passim*.

2. WD Training Regulation 440–15, Jan 26, 1926, in AFHRC 170.108; AR 95–10, Mar 10, 1928, in AFHRC 170.109095–10.

3. Baker Bd Transcript, pp 3965–66; John F. Shiner, "The Army Air Arm in Transition: General Benjamin D. Foulois and the Air Corps, 1931–1935" (Ph.D. dissertation, The Ohio State University, 1975), p 145.

4. WPD, Duplication by the Army and Navy in the Provision of Land-Based Aircraft, Feb 1930, in AFHRC 145.91–565.

5. Ltr, SW to President, Feb 25, 1930; ltr, SN to President, Apr 14, 1930; USN, WPD, Procurement of Land-Based Aircraft [Apr 1930]; ltr, SW to President, May 29, 1930; ltr, SN to President, Aug 19, 1930; ltr, Actg SW to President, Sep 10, 1930, in AFHRC 145.91–565.

6. WD, Press Release, Jan 9, 1931, in AFHRC 168.3952–191.

7. *Annual Report of the War Department*, 1931, p 31; *Annual Report of the Chief of the Air Corps*, 1931, p 35; ACNL, Jan 15, 1931, p 15; WD, Press Release, Jan 9, 1931.

8. Hughes in Baker Bd Transcript, pp 3670–74.

9. *Ibid.*

10. Norris B. Harbold, *The Log of Air Navigation* (San Antonio: The Naylor Co, 1970), pp 20–21, 26–30, 31–38, 44, 60; Frank in Baker Bd Transcript, pp 455–56.

11. Frank in Baker Bd Transcript, p 2730.

12. Maj Gen Benjamin D. Foulois, CAC, Remarks before Harbor Defense Board, Apr 11, 1932, in AFHRC 145.81–125, Vol 8.

13. OCAC, Plan for Defense of Our Seacoast Frontiers, Jan 21, 1932, in AFHRC 145.81–125, Vol 8.

14. OCAC, Plan for Defense of Our Seacoast Frontiers, Jun 21, 1932; ltr, President, Harbor Defense Board, to TAG, Jul 7, 1932; draft memo, WPD for CSA, Sep 1932; memo, Foulois for WPD, Oct 11, 1932; memo, Maj Walter H. Frank for Foulois, Oct 18, 1932; memo, Lt Col Leroy P. Collins, WPD, for Brig Gen Charles E. Kilbourne, Ch, WPD, Oct 19, 1932; memo, Foulois for WPD, Nov 14, 1932; memo, WPD for CSA, Nov 28, 1932; ltr, Gen Douglas MacArthur, CSA, to CGs all Armies, CAs, and Depts, Jan 3, 1933, in AFHRC 145.81–125, Vol 8; Collins in Baker Bd Transcript, pp 4135–38.

15. Ltr, MacArthur to CGs, Jan 3, 1933, in AFHRC 145.81–125, Vol 8; Collins in Baker Bd Transcript, p 4144.

16. Swanborough, *U.S. Military Aircraft*, pp 210–11; AC Materiel Division, Airplane Reports, Aug 15 thru Oct 30, 1933.

17. Killigrew, "Impact of the Great Depression," p XI–23.

18. James N. Eastman, Jr., "The Development of Big Bombers," *Aerospace Historian* 25 (Dec 1978), 213–14.

19. *Ibid*, 214–15.

20. Ltr, MacArthur to CGs of CAs, Aug 9, 1932; ltr, MacArthur to CGs of Field Armies, Oct 22, 1932, both in AFHRC 145.93–238; ltr, MacArthur to CGs of Armies and CAs, Feb 8, 1933, in AFHRC 145.93–234; ltr, MacArthur to CGs of Armies and CAs, Feb 28, 1933, in AFHRC 168.65043–17.

21. Shiner, "Foulois," pp 128–170. See also, for example, memorandum, Foulois for Kilbourne, March 13, 1933, in AFHRC 167.9–5, in which Foulois says the delay in furnishing data requested on December 27, 1932, was the result of work on the Army appropriation bill and "several absences from Washington on official inspection trips."

22. Report of GHQAF (Prov), 1933, in AFHRC 248.2122–3.

23. *Ibid*, ACNL, Jun 30, 1933, p 153.

24. Mooney and Layman, *Organization, 1907–1935*, pp 89–90.

25. Walter Lippman, ed, *The United States in World Affairs, 1933* (New York: Harper & Brothers, 1934), pp 172–79.

26. Ltr, TAG to CAC, Aug 12, 1933, in AFHRC 145.93–81.

27. Ltr, Lt Col James E. Chaney, Ch, AC Plans Div, to TAG, Aug 15, 1933, 1st Ind, TAG to CAC, Aug 22, 1933, in AFHRC 145.81–124.

28. 2d Ind (basic ltr, TAG to CAC, Aug 22, 1933), Foulois to TAG, Aug 23, 1933, in AFHRC 145.93–81.

29. 1st Ind (basic ltr, TAG to CAC, Aug 12, 1933), Foulois to TAG, Aug 23, 1933, in AFHRC 145.93–81.

30. 3d Ind, TAG to CAC, Aug 28, 1933; 4th Ind, Westover to TAG, Aug 30, 1933; 3d Ind, TAG to CAC, Sep 1, 1933, all in AFHRC 145.93-81.

31. Ltr, TAG to CAC, Sep 2, 1933, in AFHRC 145.93-81.

32. *Army and Navy Journal* LXXI (Sep 9, 1933), 21, (Sep 16, 1933), 41.

33. Ltr, TAG to QMC, *et al*, Sep 20, 1933; memo, Westover to ACS, WPD, Sep 7, 1933; ltr, Exec, Fort Crockett, Tex., to CAC, Sep 18, 1933; ltr, CO, Langley Field, Va., to CAC, Sep 20, 1933; ltr, Capt Arthur L. McCullough to CAC, Sep 22, 1933, all in AFHRC 145.91-124.

34. Ltr, CO, 8th Pursuit Gp, to CO, Langley Field, Sep 16, 1933, in AFHRC 145.81-124.

35. Draft ltr, OCAC to TAG, Sep 22, 1933, in AFHRC 145.93-47.

36. New England, Chesapeake Bay, Southeast, Great Lakes, Northwest, San Francisco Bay, and Los Angeles-San Diego.

37. Report of Special Committee on Employment of Army Air Corps, in AFHRC 145.93-47; AC Materiel Division, Airplane Report, Sep 15, 1933; Futrell, *Ideas, Concepts, Doctrine*, I, 34-35; Mooney and Layman, *Organization, 1907-1935*, pp 90-92; Greer, *Development of Air Doctrine*, p 71.

38. 1st Ind, TAG thru CAC to CG, III CA, Jan 31, 1934; memo, Capt Edwin J. House for Westover, May 17, 1934; ltr, Orlo H. Quinn to CAC, Apr 9, 1934, all in AFHRC 145.93-8; Baker Bd Transcript, p 4155; AC Strength Reports, monthly, 1933-34; Shiner, "Foulois," pp 159-160.

39. Memo by Maj Gen Hugh A. Drum, Dep CSA, Oct 10, 1933, quoted in Baker Bd Transcript, p 4155.

Chapter XVII

Airmail

1. Benjamin D. Foulois and Carroll V. Glines, *From the Wright Brothers to the Astronauts: The Memoirs of Major General Benjamin D. Foulois* (New York, 1968), p 235.

2. Foulois, *Memoirs*, pp 235-261; Foulois in Baker Bd Transcript, pp 1144-47; Paul Tillett, *The Army Flies the Mails* (Tuscaloosa, Ala., 1955); Norman E. Borden, Jr., *Air Mail Emergency, 1934* (Freeport, Maine, 1938); Downs, "Contributions," pp 366-417; Shiner, "Foulois," pp 234-276; John F. Shiner, "General Benjamin Foulois and the 1934 Air Mail Disaster," *Aerospace Historian* 25 (Dec 1978), 221-230; Baker Bd *Final Report*.

3. Foulois, *Memoirs*, p 257; Tillett, *The Army Flies the Mails*, p 55; Maj Byron Q. Jones, Report of Eastern Zone [*ca* Apr 1934], Atch A, Exh A; Westover in Baker Bd Transcript, p 400.

4. Msg, Foulois to Jones, Feb 11, 1934, in Jones, Report of Eastern Zone, Exh 6; AR 95-15, Apr 21, 1930, cited and quoted by Westover in Baker Bd Transcript, pp 313-18.

5. Jones, Report of Eastern Zone, Exh 3, Sheet 1.

6. HQ Eastern Zone, msgs 271 and 280, Feb 22, 1934, in Jones, Report of Eastern Zone, Annex IV, Encls 2 and 3.

7. Msg, Foulois to Jones, Feb 26, 1934, in Jones, Report of Eastern Zone, Exh 3; *The New York Times*, Feb 28, 1934, p 8.

8. Foulois, *Memoirs*, pp 256-57; Shiner, "Foulois," pp 266-68; Tillett, *The Army Flies the Mails*, pp 51-52.

9. Msg, Foulois to Jones, Mar 12, 1934, in Jones, Report of Eastern Zone, Exh 3, Sheet 5.

10. Jones, Report of Eastern Zone, Annex VII, Encls 3 and 4.

11. Jones, Report of Eastern Zone, Annex VII, Encl 3, pp 1-2, Exh 11, pp l, 5.

12. Shiner, "Foulois," pp 271-72; Foulois, *Memoirs*, p 275; Arnold in Baker Bd Transcript, p 800; msgs, Westover to Jones, Mar 17 and 18, 1934, in Jones, Report of Eastern Zone, Exh 3, Sheets 5 and 6.

13. Msg, Jones to sections and stations, Mar 19, 1934, in Jones, Report of Eastern Zone, Exh 4.

14. Arnold in Baker Bd Transcript, pp 787-88.

15. Foulois in Baker Bd Transcript, p 440.

16. Jones in Baker Bd Transcript, p 497.

17. Jones, Report of Eastern Zone, Annex III, p 6.

18. Lt Col Horace M. Hickam in Baker Bd Transcript, p 600.

19. Arnold in Baker Bd Transcript, p 790.

20. Accident Report, Feb 22, 1934 (2d Lt Charles P. Hollstein); *The New York Times* , Feb 23, 1934, p 11.

21. Hickam in Baker Bd Transcript, p 591.

22. Jones, Report of Eastern Zone, Annex III, App A, p 50.

23. Arnold in Baker Bd Transcript, p 784.

24. Capt Caleb V. Haynes, in Jones, Report of Eastern Zone, Exh 11, p 24.

25. Jones in Baker Bd Transcript, p 512.

26. Hickam in Baker Bd Transcript, p 606.

27. Arnold and Little in Baker Bd Transcript, pp 211, 825, 828.

28. Jones and Arnold in Baker Bd Transcript, pp 513, 778.

29. Arnold in Baker Bd Transcript, p 824.

30. Jones, Report of Eastern Zone, pp 11, 16.

31. Baker Bd Transcript, pp 245–46, 339–344; Jones, Report of Eastern Zone, p 17 and Exhs 19 and 20.

32. Beirne Lay, Jr., *I Wanted Wings* (New York, 1937), pp 201–229.

33. Msg, Foulois to Jones, Feb 11, 1934, in Jones, Report of Eastern Zone, Exh 6; Capt Caleb V. Haynes, Report (Extracts), in Jones, Report of Eastern Zone, Atch C and Exh, pp 1, 5, 11, 13; Hickam in Baker Bd Transcript, pp 569–571.

34. Accident Report, Apr 2, 1934 (2d Lt William J. McCray); Hickam in Baker Bd Transcript, p 616.

35. Accident Report, Apr 23, 1934 (Maj Charles B. Oldfield).

36. Accident Report, Apr 20, 1934.

37. *The New York Times*, Mar 19, 1934, p 24, Mar 19, 1934, p 3; Arnold in Baker Bd Transcript, p 800.

38. Accident Report, Mar 9, 1934 (1st Lt F. L. Howard).

39. Accident Report, Mar 17, 1934 (1st Lt H. C. Richardson).

40. Arnold in Baker Bd Transcript, pp 800–801, 835, 866; see note 38.

41. Arnold in Baker Bd *Final Report*, p 893.

42. Jones, Report of Eastern Zone, Annex VII, pp 3–5, Exh 11, p 4.

43. Arnold in Baker Bd Transcript, p 803.

44. Accident Report, Mar 9, 1934, (1st Lt Otto Wienecke).

45. Jones, Report of Eastern Zone, Annex VII, pp 3–5, Exh 11, p 4; Westover and Maj Hugh J. Knerr in Baker Bd Transcript, pp 255, 284.

46. *Air Commerce Bulletin* 5 (Jul 15, 1933), 12–13, (Sep 15, 1933), 89–90, (Apr 15, 1934), 251.

47. Westover and Hickam in Baker Bd Transcript, pp 261, 656.

48. Maj Hugh Mitchell, Signal Corps, in Baker Bd Transcript, p 1946; *Signal Corps Bulletin* 66, May–Jun 1934, pp 43–44; *Air Commerce Bulletin* 5 (Aug 15, 1933), 57, (Sep 15, 1933), 89–90; Jones, Report of Eastern Zone, Annex VI, App A.

49. Westover and Arnold in Baker Bd Transcript, pp 261, 783; Jones, Report of Eastern Zone, Annex VI, pp 3–4.

50. *Signal Corps Bulletin* 59, Mar–Apr 1931, p 50; *Air Commerce Bulletin* 5 (Sep 15, 1933), 89–90; Maj Gen Irving J. Carr, Chief Signal Officer, in Baker Bd Transcript, p 1905.

51. Arnold in Baker Bd Transcript, p 818.

52. 2d Lt James M. Goodbar in Baker Bd Transcript, pp 712–17.

53. Tillett, *The Army Flies the Mails*, pp 59–62; Jones Chronology, Feb 9, 1934.

54. Baker Bd *Final Report*, pp 61–75.

55. Memo, Westover for Plans Division, Jul 23, 1934, in AFHRC 248.93–95.

56. Ltr, Postmaster General James A. Farley to Foulois, in *Army and Navy Journal*, May 12, 1934, p 743.

57. Foulois, Comments on Tillett's *The Army Flies the Mails*, in AFHRC 168.68–11B.

58. Foulois, *Memoirs*, p 260.

59. Colonel Shiner used both characterizations, one in his dissertation, the other in his article in the *Aerospace Historian*.

Chapter XVIII

GHQ Air Force

1. Baker Bd Transcript, pp 2032–33, 3544, 3593, and *passim*. The stenographic report has 1st Lt Lawrence J. Carr saying "cannot" (p 3593).

2. Baker Bd Transcript, pp 2675–76, 3358–3362.

3. *Ibid*, pp 947–48.

4. *Ibid*, p 948.

5. *Ibid*, pp 2742, 2760.

6. *Ibid*, p 3451.

7. *Ibid*, pp 3396, 3398.

8. *Ibid*, p 3485.

9. *Ibid*, p 2167.

10. *Ibid*, pp 3731–33.

11. *Ibid*, pp 1268, 1277.

12. *Ibid*, pp 2494–96.

13. *Ibid*, pp 1788–89.

14. *Ibid*, pp 1330, 1336–39, 1345, 1351.

15. *Ibid*, p 3937.

16. *Ibid*, pp 3654–55.

17. *Ibid*, p 3718.

18. Doolittle signed the report along with all the other members but filed a dissent on this point.

19. Baker Board *Final Report*, pp 29, 66–67.

20. *Report of the Federal Aviation Commission*, S. Doc. 15, 74th Cong, 1st sess, Jan 30, 1935, pp 119–120. Other members of the commission were Edward P. Warner, former Assistant Secretary of Navy for Aeronautics; Albert J. Berres, a labor official; Jerome C. Hunsaker, director of the department of mechanical engineering at Massachusetts Institute of Technology; and Franklin K. Love, Jr., a lawyer. J. Carroll Cone of the Department of Commerce was executive secretary.

21. *The New York Times*, Sep 1, 1934, p 2, Sep 2, 1934, Sec II, p 2.

22. This and succeeding paragraphs concerning the command post exercise of 1934 are based on orders, instructions, reports, and other documents, including maps, in AFHRC 145.93–113 (7 volumes), 145.93–239, and 168.7032–1, plus coverage of the exercise by *The New York Times*, September 1–9, 1934.

23. Maj Gen Hugh A. Drum, Dep CSA, Statement at Critique, in AFHRC 145.93–239.

24. WD, Press Release, Dec 27, 1934, in AFHRC 145.93–81.

25. Ltr, TAG to CGs all CAs, *et al*, Dec 31, 1934, in AFHRC 145.93–81.

26. ACNLs, Feb 15, 1935, p 48, Mar 1, 1935, pp 86, 91, Apr 1, 1935, p 153.

27. Brig Gen Frank M. Andrews, CG, GHQAF, Report of 1935 Service Test of GHQ Air Force, p 1, in AFHRC 168.6008–23.

28. HQ GHQAF Bulletin 1, Apr 24, 1934, copy in Jones Chronology.

29. *Ibid*.

30. Ltr, TAG to CAC, Jul 21, 1934; memo, Westover for Plans Division, Jul 23, 1934; 1st Ind, Foulois to TAG, Aug 2, 1934, in AFHRC 145.93–95.

31. Ltrs, TAG to CGs all CAs, *et al*, Dec 31, 1934, and Feb 19, 1935, in AFHRC 145.91–300.

32. See note 28. The tables of organization being tested were prepared in June 1934 by an Air Corps board headed by Andrews, reworked in October by the board, and revised in the Office of the Assistant Chief of Staff, G–3, War Department General Staff.

33. ACNLs, Mar 15, 1935, p 124, Apr 1, 1935, p 139.

34. ACNLs, Feb 1, 1935, p 24, Mar 15, 1935, pp 101–102.

35. ACNLs, Mar 15, 1935, pp 101–102, 117, 123, May 1, 1935, pp 11–13.

36. ACNL, Apr 1, 1935, p 152.

37. See note 28.

38. 49 *Stat* 506; ACNL, Oct 1, 1935, pp 10, 15–16.

39. Tables of Organization for Service Test, 1935, in AFHRC 145.91–286; Materiel Division, Airplane Report. Feb 28, 1935, in AFHRC 218–5.

40. ACNL, Apr 15, 1935, p 163.

41. Materiel Division, Airplane Reports, semimonthly, Feb 28 thru Jun 30, 1935, in AFHRC 218–5. The transfers were more complicated than indicated because the swaps involved sending the P–26As to Boeing for installation of flaps.

42. Tables of Organization for Service Test, 1935, in AFHRC 145.91–286; AC, Strength Report, Mar 31, 1935.

43. ACNLs, Oct 15, 1935, p 18, Nov 1, 1935, pp 2–3, Dec 1, 1935, p 4.

44. The project began with a plan for control of GHQ Air Force in the command post exercise in September 1934. Memo, WPD for CG, GHQAF, *et al*, Jun 13, 1934; memo,

Westover thru CAC for WPD, Jun 26, 1934, in AFHRC 145.93-81.

45. WD Training Regulation 440-15, Employment of the Air Forces of the Army, Oct 15, 1935.

46. AC Tac Sch, A Study of Prepared Air Corps Doctrine, Jan 31, 1935, in AFHRC 248.211-65.

47. WPD, Doctrines of Army Air Corps (draft), Dec 21, 1934, in AFHRC 248.211-65.

48. See note 46.

49. See note 46.

50. See note 45.

51. See note 45.

52. ACNLs, Mar 1, 1935, pp 73-75, Apr 1, 1935, pp 131-33 (Lt Col Byron Q. Jones' report). The airship squadron returned to Langley at the end of the wing's concentration at Miami.

53. ACNL, Nov 1, 1935, p 10.

54. ACNL, Nov 15, 1935, p 5.

55. ACNL, Nov 1, 1935, pp 8-9.

56. ACNL, Oct 15, 1935, p 6.

57. ACNLs, Sep 15, 1935, pp 5-6, Nov 1, 1935, pp 19-20, Dec 1, 1935, p 4.

58. ACNL, Apr 15, 1935, p 170.

59. ACNLs, Mar 1, 1935, p 95, May 1, 1935, pp 14, 21, May 15, 1935, p 11.

60. ACNL, Aug 15, 1935, p 20.

61. ACNL, Sep 15, 1935, p 4.

62. ACNLs, Oct 15, 1935, p 2, Feb 1, 1936, p 23.

63. HQ GHQAF, Air Force Plan B, Oct 21, 1935; HQ GHQAF, Letter of Instruction 1, Nov 12, 1935, in AFHRC 248.2122-5; Andrews, Report of 1935 Service Test of GHQ Air Force, pp 45-46, in AFHRC 168.6008-23; ACNLs, Dec 15, 1935, pp 11, 13, Jan 1, 1936, pp 20, 22, Jan 15, 1936, pp 11-15, Feb 1, 1936, pp 8, 19-22, 26.

64. ACNLs, Jan 1, 1936, p 21, Jan 15, 1936, p 15; 49 *Stat* 1525.

65. Andrews, Report of 1935 Service Test of GHQAF, in AFHRC 168.6008-23.

66. ACNL, Jan 1, 1936, p 20.

67. Fort Hunt, situated just above Mount Vernon, was not then garrisoned. The nearest airfields were Washington Airport and Bolling Field.

68. Report of Browning Board, Jan 7, 1936, extract in AFHRC 145.91-302.

69. Memos, Maj William E. Lynd to CAC, Mar 19 and 20, 1935, in AFHRC 145.91-302.

70. Chase C. Mooney and Edward C. Williamson, *Organization of the Army Air Arm, 1935-1945* (USAF Hist Study 10, Maxwell AFB, Ala., 1956), p 3; ltr, TAG to CGs all CAs, *et al*, May 8, 1936, in AFHRC 145.91-302.

71. Previously under corps area control.

72. Ltrs, TAG to CGs all CAs, *et al*, Aug 12 and 17, 1936; ltr, Lt Col Joseph T. McNarney, Actg Chief of Staff, GHQAF, to CGs all Wings, Aug 19, 1936, in AFHRC 145.91-302.

73. Tables of Organization for Service Test, 1935, in 145.91-286. Tables of organization dated June 1, 1936, published by Headquarters GHQAF on August 19, 1936, are in AFHRC 145.91-289.

74. Base headquarters and air base squadrons were organized in administrative, operations, and materiel sections. The "post commander" (the term used during the 1920s) became "Commanding Officer, Air Base, Selfridge Field, Michigan" (or whatever the field was). The office of the post commander became "Air Base Headquarters, Selfridge Field." This resulted in the use of terms like "Selfridge Field Base" or "Selfridge Field Air Base," which inevitably caused confusion. Maj Gen Henry H. Arnold and Maj Gen Delos C. Emmons tried to straighten this out in 1939 but did not succeed. The confusion, which became even worse during World War II, continued until after creation of the U.S. Air Force in 1947.

75. Mooney and Williamson, *Organization, 1935-1945*, pp 3-4.

76. *Ibid.* On October 30, 1940, Arnold became Deputy Chief of Staff for Air as well as Chief of the Air Corps. On November 19, 1940, GHQ Air Force was removed from the control of the Chief of the Air Corps, but on June 20, 1941, both the Air Corps and GHQ Air Force were placed under Arnold, who became Chief of Army Air Forces.

Chapter XIX

Building an Air Force

1. Materiel Division, Airplane Reports, semimonthly, June 30, 1930 to June 30, 1939; Air Corps Strength Reports, monthly, June 30, 1932 to June 30, 1939.

2. *Annual Report of the Chief of the Air Corps*, 1933, pp 57–58; *Annual Report of the War Department*, pp 1–2, 60; Phillipson in Baker Bd Transcript, pp 2543–44.

3. *Annual Report of the Chief of the Air Corps*, 1934, pp 47–48.

4. *Ibid*, pp 47–48, 50; *Annual Report of the War Department*, 1933, pp 3, 15–17; Killigrew, "Impact of the Great Depression," pp X–8, X–14, X–22, X–24; Phillipson in Baker Bd Transcript, pp 2544–47.

5. *Annual Report of the Chief of the Air Corps*, 1935, p 49; *Annual Report of the War Department*, 1935, p 53; Killigrew, "Impact of the Great Depression," pp XI–2, XI–10, XI–20, XI–23–24.

6. *Annual Report of the Chief of the Air Corps*, 1935, pp 47–48; Phillipson in Baker Bd Transcript, pp 2550–51; Killigrew, "Impact of the Great Depression," pp XIV–5, XIV–6.

7. 47 *Stat* 407, 1514; 48 *Stat* 14.

8. 48 *Stat* 13, 521; Executive Orders 6085, Mar 28, 1933, 6188, Jul 3, 1933, 6553, Jan 9, 1934, and 6791, Jul 6, 1934; *Annual Report of the War Department*, 1933, pp 40–42.

9. *Military Laws of the United States*, 1929, Sec 350.

10. 47 *Stat* 1516.

11. ACNLs, Apr 28, 1933, p 102, Jun 30, 1933, p 160.

12. 48 *Stat* 816.

13. 47 *Stat* 1519.

14. *Annual Report of the Chief of the Air Corps*, 1934, pp 46–48; Westover in Baker Bd Transcript, p 311.

15. ACNL, Apr 15, 1935, p 180.

16. *Annual Report of the Chief of the Air Corps*, 1934, p 22, 1935, p 21.

17. *Annual Report of the Chief of the Air Corps*, 1934, p 7.

18. *Ibid*, pp 10–11; ACNLs, Aug 29, 1933, p 197, Jan 15, 1935, p 10, Feb 1, 1935, p 36, Mar 1, 1935, p 90; Foulois in Baker Bd Transcript, p 890.

19. ACNL, Jul 31, 1933, p 171.

20. ACNL, May 29, 1933, p 119.

21. See note 19; ACNL, Aug 29, 1933, p 198.

22. ACNL, Aug 29, 1933, p 198.

23. ACNL, Oct 31, 1933, p 235.

24. ACNLs, Jun 30, 1933, p 151, Jul 31, 1933, pp 169, 171, Sep 30, 1933, p 222; Arnold in Baker Bd Transcript, pp 756, 833.

25. *Annual Report of the War Department*, 1933, pp 9–10.

26. Accident Report, Sep 23, 1933 (2d Lt George F. Schlatter), in AFHRC 200.3912–1.

27. ACNL, Oct 31, 1933, p 235; *Annual Report of the War Department*, 1933, pp 3–4, 13–14. One of the Reservists called to active duty, 1st Lt (later Capt) Albert W. Jernberg, wrote a book about his experiences: *My Brush Monkeys: A Narrative of the CCC* (New York, 1941).

28. ACNLs, Oct 31, 1933, p 236, Feb 1, 1935, p 29, Mar 1, 1935, p 75, Apr 1, 1935, p 141, Apr 15, 1935, p 161, Aug 15, 1935, pp 3–4, Nov 1, 1935, pp 1–2, Jan 1, 1936, p 6, Feb 1, 1936, p 26, Mar 15, 1936, p 20; *Annual Report of the Chief of the Air Corps*, 1934, pp 3–4, 1935, p 2.

29. AAF, Statistical Digest, World War II, p 297; *Congressional Record*, 76th Cong, 3d sess, Jan 3, 1940–Jan 3, 1941, pp 556–57.

30. *Annual Report of the War Department*, 1935 to 1939, personnel tables; 52 *Stat* 217; 53 *Stat* 558; Elias Huzar, *The Purse and the Sword: Control of the Army by Congress Through Military Appropriations, 1933–1950* (Ithaca, N.Y., 1950), p 143.

31. Air Corps Strength Reports, monthly, Jun 30, 1932 to Jun 30, 1936.

32. Andrews, Report of 1935 Service Test of GHQ Air Force, pp 7–9, in AFHRC 168.6008–23.

33. ACNLs, May 1, 1937, p 3, Jun 15, 1937, p 14, Jul 1, 1937, p 10, Jul 15, 1937, p 10.

34. 49 *Stat* 1524; ACNL, May 15, 1936, p 35.

35. ACNLs, Oct 15, 1938, p 6, Jan 1, 1939, pp 7–8, Jan 15, 1939, p 9, Feb 1, 1939, p 18, May 1, 1939, p 2.

36. ACNL, Oct 15, 1936, p 5; *Annual Report of the War Department*, 1936, p 3.

37. *Annual Report of the War Department*, 1936 to 1939, personnel charts; Air Corps Strength Reports, 1936 to 1939; ltr, Maj Gen Delos C. Emmons to TAG, Jul 28, 1939, in AFHRC 415.01.

38. *Annual Report of the War Department*, 1934, pp 15–16, 1935, pp 8, 66, 1936, pp 4, 38, 1937, p 5, 1938, p 34; ACNLs, Jan 15, 1937, p 1, Jul 1, 1938, p 20; 49 *Stat* 1907; Killigrew, *Impact of the Great Depression*, pp XV–8, XV–13. While the number of planes increased 40 percent between June 1938 and June 1939, the percentage of serviceable planes classified as obsolete dropped from 27 to 10 percent of the number on hand (Materiel Division, Airplane Reports, June 30, 1938, June 30, 1939).

39. Swanborough, *U.S. Military Aircraft*, pp 300–332.

40. *U.S. Air Services* XIV (Sep 1934), 13–15 (the Arnold quotation is on page 15); ACNLs, Jan 15, 1935, p 14, Apr 15, 1935, p 170; *Annual Report of the Chief of the Air Corps*, 1935, p 21; Henry H. Arnold, *Global Mission* (New York, 1949), pp 145–46.

41. ACNL, Jan 15, 1935, p 16; Arnold, *Global Mission*, pp 146–47; *The New York Times*, Sep 4, 1934, p 8, Sep 5, 1934, p 24.

42. Edward Jablonski, *Flying Fortress* (Garden City, N.Y., 1965), pp 3–9; ACNL, Sep 15, 1935, pp 3–4; *The New York Times*, Jul 5, 1935, p 1, Jul 6, 1935, p 3, Jul 19, 1935, p 5, Jul 29, 1935, p 16; Bowers, *Boeing Aircraft Since 1916* (London, 1966), pp 244–45, 251–52.

43. Jablonski, *Flying Fortress*, pp 6, 9–10; ACNL, Nov 15, 1935, p l; WD, Press Releases, Nov 12, 1935, Jan 16, 1936, in AFHRC 200.3912–1, Oct 30, 1935 (Hill).

44. Airplane Characteristics, Jan 1, 1939; ACNLs, Jan 15, 1937, p 10, Apr 1, 1937, pp 8–9, Jun 1, 1937, p 13; Swanborough, *U.S. Military Aircraft*, p 75.

45. ACNL, Apr 1, 1937, p 9.

46. ACNL, Jun 15, 1937, p 16.

47. ACNL, Oct 15, 1937, pp 1–3.

48. ACNL, Mar 1, 1939, p 9.

49. ACNL, Feb 15, 1938, pp 6–8; *The New York Times*, Jan 7, 1938, p 3, Jan 10, 1948, p 3.

50. ACNLs, Feb 15, 1938, p 19, Mar 1, 1938, pp 1–2, 4, 18–19, Apr 1, 1938, pp 3, 6, May 15, 1938, p 8; *Aircraft Year Book* (New York, 1940), p 45.

51. Ltr, Maj Vincent J. Meloy, to CO, 2d Bombardment Gp (rprt of flight), Aug 18, 1938, in AFHRC 285.49–2, Vol 2, App 48; ACNLs, Aug 1, 1938, pp 8, 20, Aug 15, 1938, pp 1–3.

52. ACNLs, Apr 1, 1938, p 12, Sep 1, 1938, p 3; Bowers, *Boeing Aircraft*, pp 199–201.

53. This account of the flight to Santiago is based on the 2d Wing's project file (AFHRC WG–2–SU–OP, Feb 1939) and on Air Corps News Letters, February 15, 1939, pages 3–4, and March 1, 1939, pages 1–5 (article by Capt William D. Old).

54. History of Langley Field, Mar 1, 1935 to Dec 7, 1941, Vol 1, 88.

55. Air Force Pamphlet 190–2–2, Sep 1, 1965, p 33; *Aircraft Year Book* (New York, 1940), p 434.

56. 49 *Stat* 1907.

57. Jean H. DuBuque and Robert F. Gleckner, *The Development of the Heavy Bomber, 1918–1944* (USAF Hist Study 6, Maxwell AFB, Ala., 1951), pp 18–23; Eastman, "The Development of Big Bombers," pp 215–16.

58. Memo rprt, Engineering Sec, Materiel Div, Comparative Performance of B–10B, YB–17, and B–18 Airplanes, Aug 6, 1936, in AFHRC 168.6008–30.

59. Documents, working papers, and notes (Whitehead's file) in AFHRC 168.6008–30.

60. DuBuque and Gleckner, *Development of the Heavy Bomber*, p 19.

61. Swanborough, *U.S. Military Aircraft*, pp 218–220; Materiel Division, Airplane Reports, Oct 15, 1937 thru Aug 31, 1939, in AFHRC 218.5; Aircraft Characteristics, Jan 1, 1939, in AFHRC 203–17.

62. Ltr, Lt Col Robert Olds, CO, 2d Bombardment Gp, to CG, 2d Wing (on service test of B–17s), Jul 20, 1937, in AFHRC GP–2–SU; Bowers, *Boeing Aircraft*, pp 255–56.

63. Materiel Division, Airplane Report, Aug 31, 1939, in AFHRC 218–5.

64. Report of GHQAF (Prov), 1933, p 12, in AFHRC 248.2122–3.

65. Baker Bd Transcript, pp 929–932.

66. Baker Bd Transcript, pp 852–55.

67. Ltr, Arnold to CAC, Nov 26, 1934, in AFHRC 248.211–65C.

68. ACNLs, May 15, 1935, p 34, Jan 15, 1937, p 1; Materiel Division, Airplane Reports, semimonthly, Jun 30, 1932 to Aug 31, 1939.

69. Intvw, Hugh N. Ahmann, hist, Albert F. Simpson Historical Research Center, and Thomas A. Sturm, Office of Air Force History, with Gen Earle E. Partridge, Ret, Apr 23–25, 1974, in AFHRC K239–0512–729.

70. Accident Reports, Jun 20, 1933 (31–597) and Nov 5, 1933 (Gaughen), in AFHRC 200.3912–1; Materiel Division, Air-

plane Reports, Jan 1, 1932 thru Dec 31, 1934; Aircraft Record Cards, 31–520/515, 31–597, 32–221/230, in AFHRC.

71. Jones, *U.S. Fighters,* p 74.

72. Ltr, Arnold to CAC, Nov 26, 1934, in AFHRC 248.211–65C.

73. Materiel Division, Airplane Reports, Jan 15, 1936 thru Aug 31, 1939.

74. See note 72.

75. Claire C. Chennault, The Role of Defensive Pursuit (mimeographed copy in AFHRC 168.7001–114), consisting of three articles published in *The Coast Artillery Journal,* LXXVI (November–December 1933), 411–17, LXXVII (January–February 1934), 7–11, (March–April 1934), 87–93. Also see Chennault, "Special Support for Bombardment," *U.S. Air Services* XIX (January 1934), 18–21, and Chennault, *Way of a Fighter* (New York, 1949), pages 18–31.

76. ACNL, Aug 29, 1933, p 190.

77. ACNLs, Sep 15, 1937, p 6, Jun 1, 1938, p 24, Jun 15, 1939, p 5; Airplane Characteristics, Jul 1, 1940; Materiel Division, Airplane Report, Aug 31, 1939; Swanborough, *U.S. Military Aircraft,* p 482; Greer, *Development of Air Doctrine,* pp 65–66.

78. Airplane Characteristics, Jan 1, 1940.

79. ACNL, Oct 1, 1938, p 17.

80. ACNLs, Feb 1, 1938, p 12, Apr 1, 1938, p 6, Dec 1, 1938, p 8; Swanborough, *U.S. Military Aircraft,* p 24.

81. Materiel Division, Airplane Reports, Jan 15, 1938 through Aug 31, 1939.

82. ACNL, May 15, 1936, p 26; Airplane Characteristics, Jan 1, 1949; Materiel Division, Airplane Reports, for calendar years 1936 and 1937; Swanborough, *U.S. Military Aircraft,* pp 388–89.

83. ACNL, Aug 1, 1936, p 23; Materiel Division, Airplane Reports, calendar years 1937, 1938, 1939.

84. In 1940 the War Department disposed of 93 A–17As for resale to Great Britain and France (Swanborough, *U.S. Military Aircraft,* p 388).

85. ACNLs, Jun 1, 1938, pp 1–2, Oct 1, 1938, pp 3–4; Aircraft Record Card, 36–349, in AFHRC.

86. Greer, *Development of Air Doctrine,* p 88; Swanborough, *U.S. Military Aircraft,* pp 230–36.

87. Maj Hugh J. Knerr, Remarks at Engineer-Supply Conference at Wright Field, Oct 1932, quoted by Capt Ralph B. Walker, Transport Squadrons, Jun 8, 1934, p 4, in AFHRC 205–11–101.

88. Walker, Transport Squadrons, pp 4–5; Genevive Brown, *Development of Transport Airplanes and Air Transport Equipment, 1917–1945* (Patterson Field, Ohio, 1946), pp 50–51, in AFHRC 201.7.

89. 1st Ind, Col Frederick L. Martin, Exec, Materiel Div, to CAC, Sep 23, 1936, in AFHRC 145.91–316.

90. Ltr, Brig Gen Augustine W. Robins, Ch, Materiel Div, to CAC, Apr 13, 1937, in AFHRC 145.91–136; Maurer, *Combat Units,* pp 52–53; Maurer, *Combat Squadrons,* pp 9–10, 17–18, 25, 31–32.

91. The squadron at Dayton with two flights, one at Wright Field and the other at Patterson Field, was larger.

92. Ltr, Brig Gen Augustine W. Robins, Ch, Materiel Div, to CAC, Aug 19, 1937, in AFHRC 145.91–316.

93. Memos, Col Rush B. Lincoln for CAC, Apr 14, 1937, and May 27, 1937; ltr, Westover to TAG, Mar 25, 1937, in AFHRC 145.91–316.

94. Ltrs, Brig Gen Augustine W. Robins, Ch, Materiel Div, to CAC, Apr 22, 1936, and Aug 19, 1937, in AFHRC 145.91–316.

95. ACNLs, Aug 1, 1936, pp 23–24, Oct 15, 1937, p 4, Nov 15, 1937, p 4, Dec 1, 1937, p 13, Jan 15, 1938, p 5, Jun 15, 1938, p 12, Oct 1, 1938, p 7, Jun 1, 1939, p 9; Materiel Division, Airplane Reports, calendar years 1936–1939.

96. Donald Duke, *Airports and Airways* (New York, 1927), pp 17–20; Sterling R. Wagner, "The Modern Airport," *Bulletin of New York State College of Forestry* IV (Jun 1931); *Report of Commission on Airport Drainage and Surfacing* (Washington: Department of Commerce, 1932); History of Langley Field, Mar 1, 1935 to Dec 7, 1941, Vol 1, 1, in AFHRC 285.42–2.

97. *Annual Report of the Chief of the Air Corps,* 1934, p 5.

98. *Annual Report of the Chief of the Air Corps,* 1936, p 1; History of Buildings and Grounds, Office of Army Air Corps, 1918–1945, pp 19–20, in AFHRC 144.01; Andrews, Report of 1935 Service Test of GHQ Air Force, p 2, in AFHRC 168.6008–23; 5th Ind, Arnold to TAC, Oct 10, 1936, in AFHRC 145.91–547.

99. *Annual Report of the Chief of the Air Corps,* 1932, pp 5, 7, 1934, p 3; ACNLs, Sep 30, 1933, p 214, May 1, 1936, p 3.

100. Constructing Quartermaster, Report for Paved Runways and Drains at Langley Field, Sep 21, 1938; Table, Runway Con-

struction at Langley Field, 1936–1940, in AFHRC 245.49–2, Vol 2, Apps 2 and 5.

101. ACNLs, Jun 1, 1936, p 8, Feb 1, 1937, p 7.

102. ACNL, Jul 1, 1936, p 13.

103. ACNL, Aug 1, 1938, p 13.

104. Ltr, CO, Bolling Field, to OCAC, Nov 17, 1941, in AFHRC 280.80–1, 1939–1945, App 101; ACNL, May 1, 1936, p 7.

105. ACNL, Nov 15, 1937, p 2.

106. Table, Runway Construction at Langley Field, 1936–1940, in AFHRC 285–49–2, Vol 2, App 5; ltrs, CO, Bolling Field, to OCAC, Jun 23, 1939, Sep 18, 1939, and Nov 17, 1939, in AFHRC 280.90–1, 1939–1945, Apps 101–103.

107. Executive Order 7215, Oct 26, 1935; ltr, SW Harry H. Woodring to SN, Oct 16, 1935, in AFHRC 145.91–378; ACNLs, Oct 1, 1935, p 7, Oct 15, 1935, pp 9–10, Apr 1, 1936, p 2, Jun 15, 1937, p 20, Oct 1, 1937, p 9, Jun 1, 1938, p 12; *Annual Report of the Chief of the Air Corps*, 1936, p 52.

108. ACNL, Aug 1, 1936, p 4.

109. ACNLs, Apr 15, 1937, p 13, Feb 15, 1938, p 13, Mar 1, 1938, p 7; Wesley F. Craven and James L. Cate, *The Army Air Forces in World War II*, 7 vols (Chicago, 1948–1958), VI, 123.

110. 49 *Stat* 610; ACNLs, Feb 15, 1938, p 13, Jul 15, 1939, p 7; Craven and Cate, *The Army Air Forces in World War II*, VI, 122, 126, 128.

111. Foulois, Testimony at *Hearings before a Joint Committee to Investigate Dirigible Disasters*, 63d Cong, 1st sess, May 22–Jun 6, 1933, p 627; Westover in Baker Bd Transcript, pp 1189–1190; Air Corps Strength Report, Feb 28, 1933; 47 *Stat* 1582; 48 *Stat* 626; 49 *Stat* 133, 1290. The envelope of a dirigible lasted but a short time, the car much longer. Congress prohibited repeated buying of $60,000 envelopes to go with a $40,000 car, as the Air Corps wanted to do (Jones Chronology, June 18, 1933, citing the law and opinion of the Judge Advocate General).

112. Westover in Baker Bd Transcript, pp 1189–1190; ACNL, Jul 1, 1936, p 14; memo, Lt Col Arnold N. Krogstad, Ch, Personnel Div, for Westover, Jul 12, 1935, in AFHRC 145.93–81.

113. *Akron* and *Macon*, 6,850,000 cubic feet; *Shenandoah*, 2,235,000; *Graf Zeppelin*, 3,955,000; *Hindenburg*, 7,062,100.

114. Mitchell, Testimony at *Hearings before a Joint Committee to Investigate Dirigible Disasters*, pp 686–89.

115. *Hearings before a Joint Committee to Investigate Dirigible Disasters*, pp 629–630.

116. 1st Ind, Westover to TAG, Jul 23, 1935, in AFHRC 145.93–81; Jones Chronology, Apr 9, 1935.

117. Westover in Baker Bd Transcript, pp 1194, 1201.

118. Ltr, Maj Clarence B. Lober, to Arnold, Jan 27, 1936, in AFHRC 248.211–31; AC Tac Sch, Committee Study, Lighter-than-Air, Recommendation for Program and Details of Employment, May 1936, in AFHRC 248.211–31; Capt Edmund C. Lynch, Individual Research (AC Tac Sch), Lighter-than-Air, Recommendation for Program and Details of Employment, May 1, 1937, in AFHRC 248.211–31A.

119. Ltr, Andrews thru CAC to TAG, Jun 21, 1935, in AFHRC 145.93–81; Andrews, Report of 1935 Service Test of GHQ Air Force, p 72. Lighter-than-air companies became squadrons on October 1, 1933.

120. Ltr, TAG to CG, VI CA, May 25, 1937, in AFHRC; 21st Airship Gp, unit record card, in AFHRC; Jones Chronology, Jun 8, 1937. The 9th Airship Squadron became the 1st Observation Squadron at Marshall Field, Fort Riley, Kansas.

121. After fire destroyed 3 balloons, killed 2 men, and seriously injured 2 others in less than a year, Westover in July 1937 ordered helium used in balloons if possible (Jones Chronology, July 16, 1939), but the C–3s on hand were not constructed to use helium.

122. ACNLs, Mar 15, 1937, p 1, Sep 15, 1937, p 19, Feb 15, 1938, p 13, May 1, 1938, p 16, Jul 1, 1938, p 7; AC Tac Sch, Committee Study, Lighter-than-Air, Recommendation for Program and Details of Employment, May 1936, in AFHRC 248.211–31.

123. Ltr, OCAC to TAG, Jul 13, 1938, 1st Ind, TAG to CAC, Jul 25, 1938, in AFHRC 145.91–430; Jones Chronology, Dec 15, 1938.

124. Ltr, OCAC to TAG, Jul 13, 1938, 1st Ind, TAG to CAC, Jul 25, 1938; draft memo, Lt Col William E. Farthing for G–3, Aug 9, 1938, in AFHRC 145.91–430; 21st Airship Gp, unit record card, in AFHRC.

125. The War Department disbanded the squadrons in February 1942.

126. Robert L. Thompson, *Barrage Balloon Development in the United States Army Air Corps, 1923–1942* (AAF Hist Study 3, Maxwell Field, Ala., 1943), pp 3–10; ltr, TAG to CG, Panama Canal Department, Jun 15, 1940, in AFHRC 145.93–72.

127. Thompson, *Barrage Balloon Development*, pp 14–17, 19–20.

128. Craven and Cate, *The Army Air Forces in World War II*, VI, 111–12; Stetson Conn, Rose C. Engelman, and Byron Fairchild, *Guarding the United States and Its Outposts* [U.S. Army in World War II: The Western Hemisphere] (Washington, 1964), pp 62–63. The Air Corps began a barrage balloon program in 1940. Unsuccessful in wresting operational control from the Coast Artillery, it got out of the business in 1942, with the Corps of Engineers assuming responsibility for development and procurement.

Chapter XX

Crew Training

1. Baker Bd *Final Report*, pp 46–47, 69; intvw, William R. Porretto, with Lt Gen Barney M. Giles, Ret, Oct 1966, pp 13–14, in AFHRC K239.0502–779.

2. ACNLs, Feb 15, 1935, pp 24–25, Jun 1, 1935, pp 13–14, Jul 15, 1935, p 15; Jones Chronology, Jan 24, 1935, and *passim*, citing and quoting WD Cirs 6, Jan 24, 1935, 8, Feb 12, 1935, 20, Apr 19, 1935, 27, May 25, 1935, 40, Jul 3, 1935, 54, Sep 30, 1935, 69, Nov 25, 1935; ltr, Lt Col Charles B. Oldfield to CG, GHQAF, Sep 5, 1936, in 285.49–2, Vol 2, Doc 20; AC Cir 50–12, Apr 1, 1939.

3. ACNL, Mar 1, 1935, p 86 (story contributed by Albert D. Cannon).

4. Materiel Command, Case History of Link Trainers, 1944, in AFHRC 32–996; Tng Aids Div, HQ AAF, History of Link Training, 1945, in AFHRC 143.512–4D.

5. ACNL, Oct 1, 1936, p 2; *Aircraft Year Book* (New York, 1938), pp 209–216.

6. ACNL, Apr 15, 1938, p 9.

7. AC Cir 50–12, Apr 1, 1939; ACNL, Jun 15, 1936, p 15.

8. HQ GHQAF Memo 5, Personnel and Training, Mar 6, 1937, in AFHRC 285.49–2, Vol II, Doc 27; Maj Gen Frank M. Andrews, Army War College Lecture, Oct 9, 1937; HQ 19th Bombardment Gp, Ops Memo 3, Aug 30, 1938, in AFHRC B0086.

9. Jones Chronology, Oct 15, 1934, citing and quoting AC Cir 50–10A, Jun 13, 1937.

10. Jones Chronology, Apr 1, 1939, citing and quoting AC Cir 50–10, Apr 1, 1939. The Air Corps also revised its ratings for airplane observers and for lighter-than-air aviation. As in the past, it provided ratings for navigators, bombers, gunners, and other specialties but did not issue requirements and thus did not authorize use of these ratings. The Air Corps revised its ratings again the following November, the changes including the counting of pilot and copilot time in commercial planes toward the senior and command ratings (Jones Chronology, Nov 21, 1939, citing and quoting AC Cir 50–10, Nov 21, 1939).

11. An interdepartmental committee organized by the National Advisory Committee on Aeronautics to standardize nomenclature did not think such a distinction necessary. The Secretary of War on March 27, 1935 approved "air navigation," recommended by the committee (ACNL, Apr 1,1935, p 156).

12. Ltr, Col Hugh J. Knerr, Chief of Staff, GHQAF, to CG, 2d Wg, Jul 29, 1935, in AFHRC 285.49–2, May 1933–Dec 1941, Vol II, Doc 23; Capt Norris B. Harbold, log, pp 60–66; ACNLs, Jan 15, 1935, p 17, Feb 1, 1935, p 26, Feb 15, 1936, p 5.

13. Ltr, Knerr to CG, 2d Wg, Jul 9, 1935; ltr, Lt Col Charles B. Oldfield to CG, 2d Wg, Oct 9, 1935, in AFHRC 285.49–2, May 1935–Dec 1941, Vol II, Docs 23–24.

14. ACNL, Feb 1, 1936, p 24.

15. History of Langley Field, in AFHRC 285–49–2, May 1935–Dec 1941, Vol I, 55–57.

16. The Air Corps Tactical School's manual on bombardment used "bomber" in 1935, "bombardier" in 1938.

17. AC Tac Sch, Bombardment Aviation, Jan 1, 1938, p 9, in AFHRC 248.101–9.

18. Andrews, Army War College Lecture, Oct 9, 1937; HQ 19th Bombardment Gp, Ops Memo 3, Aug 30, 1938, in AFHRC B0086.

19. Ltr, Lt Col Adlai H. Gilkeson to CG, GHQAF, Sep 5, 1936, in AFHRC 285.49–2, Vol 2, Doc 19; ACNL, Jun 1, 1936, p 4.

20. Ltr, Lt Col Harvey S. Burwell, CO, 19th Bombardment Gp, to CG, 1st Wg, Feb 8, 1939 (a detailed analysis of the 19th Group's training from July 1, 1937 to November 30, 1938), in AFHRC B0086.

21. 2d Bombardment Gp, Manual for Bombardiers, Jan 1, 1939, in AFHRC GP–2–SU–MA.

22. Ltr, Arnold to CAC, Jul 26, 1932; memo, Maj John H. Pirie for CAC, Aug 24, 1932; ltr, Maj Frank D. Lackland, AC Plans Div, to CAC, Aug 2, 1934, in AFHRC 145.93–282; ACNLs, Feb 15, 1935, p 63, Apr 1, 1935, p 152, Apr 15, 1935, pp 179–180, Nov 1, 1935, p 5, Nov 15, 1935, p 17.

23. Ltr, Smith to CO, March Field, Jun 5, 1934; 2d Ind, Arnold to CG, 9th CA, Jun 13, 1934; 6th Ind, Lt Col Clarence L. Tinker to CG, 9th CA, Dec 31, 1934; 11th Ind, Andrews to TAG, Feb 9, 1935; ltr, Carl B. Sturzenacker to John F. Dockweiler, Representative from California, Mar 15, 1935; ltr, Dockweiler to SW George H. Dern, Mar 23, 1935, and related correspondence in AFHRC 145.93–282.

24. ACNLs, May 15, 1935, p 21, Jul 1, 1935, p 21, Aug 1, 1935, p 3, Aug 15, 1935, p 20, Jun 1, 1936, p 15, Apr 1, 1937, p 16, Oct 15, 1937, p 10.

25. Ltr, Base Ops Officer, Langley Field, to Hydrographic Office, Norfolk, Sep 6, 1937, in AFHRC WG–2–SU, 1933–1941; ACNLs, Jun 1, 1935, p 21, Jul 15, 1935, p 7, Aug 1, 1935, p 19, Aug 15, 1935, p 4, Jul 1, 1936, p 12, Sep 1, 1936, p 7, Jun 1, 1937, p 24, Oct 1, 1937, p 19, Nov 15, 1937, p 7, Sep 1, 1939, p 16.

26. Ltr, Lt Col Carl W. Connell, CO, 9th Bombardment Gp, to CO, Mitchel Field, Sep 9, 1937; ltr, HQ 2d CA to CO, Mitchel Field, Nov 22, 1937, in AFHRC WG–2–SU, 1937–1940. Map showing Mitchel Field ranges, Apr 11, 1939, in AFHRC WG–2–SU, 1933–1941.

27. ACNLs, Jun 1, 1936, p 11, Jun 15, 1936, p 14, Dec 15, 1936, p 4, Jan 15, 1937, p 9, Apr 1, 1937, p 18, Aug 15, 1937, p 4.

28. ACNL, May 15, 1935, p 12.

29. ACNLs, Feb 15, 1935, p 65, Mar 1, 1935, pp 94–95, Jul 1, 1935, p 142.

30. ACNL, May 1, 1937, p 13.

31. Ltr, Col Gerald C. Brant, CG, 3d Wg, to GHQAF, Jun 7, 1935, and related papers in AFHRC 145.93–282; ACNL, Jul 15, 1939, p 2.

32. Ltr, Gen Douglas MacArthur, CSA, to Douglas, Jul 20, 1934; ltr, SW Harry W. Woodring to Attorney General, Jul 23, 1934, and related papers in AFHRC 145.93–282; ACNLs, Apr 28, 1933, p 81, Feb 15, 1935, p 65, Mar 1, 1935, pp 85, 94–96, Apr 1, 1935, p 142, Nov 15, 1935, p 9.

33. Memo, Col Arthur G. Fisher, Commandant, Air Tac Sch, for Brig Gen Oscar Westover, Asst to CAC, Mar 29, 1935, and related papers in AFHRC 145.93–282; Notice of Renewal of Lease, Jun 18, 1936; ltr, Woodring to Valparaiso Realty Co, Mar 31, 1937, in AFHRC 240.01, 1933–1944, pt I, Exhs 2 and 3; ACNLs, Apr 15, 1936, p 15, May 1, 1936, p 6, Aug 15, 1936, p 6, Sep 15, 1936, p 8, Apr 1, 1937, p 20, May 15, 1937, p 19, Oct 1, 1937, p 10, Jan 1, 1939, p 11, Apr 15, 1939, p 11, Jun 1, 1939, p 4.

34. ACNLs, May 15, 1935, p 8, Sep 15, 1935, p 11, Dec 1, 1935, p 11, Oct 1, 1938, p 17, Oct 15, 1938, p 9, Dec 15, 1938, p 17, Jun 15, 1939, p 3.

35. ACNLs, May 15, 1935, p 23, Jun 1, 1935, pp 4–5, Jul 1, 1935, p 26, Dec 15, 1935, p 11.

36. ACNL, Jan 1, 1939, p 10.

37. ACNLs, Nov 1, 1936, p 5, Dec 15, 1938, p 17, Apr 15, 1939, p 12, Oct 15, 1939, p 13; see note 21.

38. ACNLs, Aug 15, 1937, p 11, Feb 15, 1939, p 15.

39. Report of Armament Officer, 96th Bombardment Sq, and Report of Armament Section, 59th Service Sq, in Report of the Field Exercises Held at Byrd Airport, in AFHRC WG–2–SU–TB (Bomb), Vol I.

40. Report of Ordnance Service for 2d Wing Bombing Exercise, Dec 23, 1938, in AFHRC WG–2–SU.

41. ACNLs, Aug 15, 1937, p 11, Jan 15, 1938, p 11, May 15, 1938, p 12.

42. See note 40.

43. Virginia G. Toole, *Development of Bombing Equipment in the Army Air Forces* (Patterson Field, Ohio, 1945), *passim*, especially Chaps II, V, and VI.

44. Memo, Maj Gen Benjamin D. Foulois, CAC, to Dep CSA, Nov 23, 1934; ltr, Foulois to Ch, Materiel Division, Mar 30, 1935, in AFHRC 202.2–35, Docs 3 and 5.

45. Memo, Foulois to Dep CSA, Nov 23, 1934, in AFHRC 220.2–35, Doc 5.

46. Ltr, Brig Gen Augustine W. Robins, Ch, Materiel Division, to CAC, Jan 18, 1936, in AFHRC 220.2–35, Doc 11.

47. See note 45.

48. Memo, Brown to CAC, Nov 13, 1934; memo, Foulois to Dep CSA, Nov 23, 1934, in AFHRC 202.2–35, Docs 4 and 5.

49. Ltr, Maj Gen Hugh A. Drum, CG, Hawaiian Department, to TAG, Apr 29, 1935, 1st Ind, TAG to CAC, May 16, 1935, in AFHRC 202.2–35, Doc 8.

50. 1st Ind, Robins to CAC, Jun 12, 1935, in AFHRC 202.2–35, Doc 9.

51. Memo, Rear Adm Harold R. Stark, Ch, Navy Bureau of Ordnance, to CAC, May 16, 1935, in AFHRC 202.2–35, Doc 10.

52. Memo, Robins to section chiefs, Jun 4, 1935, in AFHRC 202.2–35, Doc 10.

53. lst Ind, Arnold to Ch, Materiel Division, Jan 20, 1936, in AFHRC 202.2–35, Doc 12.

54. See notes 51 and 52; ltr, Robins to CAC, Jan 18, 1936; ltr, Andrews to CAC, Jan 21, 1936; ltr, Westover to CG, GHQAF, Jan 24, 1936; Engrg Sec Memo Rprt R–53–512, Mar 1, 1937; Echols, Report on Conference, Aug 10, 1937; Stace, Conference on Mark XV (M–1) Bombsights, Oct 20, 1937, in AFHRC 202.2–35, Docs 10–11, 13–14, 19, 28–29.

55. ACNL, Aug 1, 1936, p 1.

56. 2d Bombardment Gp, Manual for Bombardiers, Jan 1, 1939, Sec I, Chap II, in AFHRC GP–2–SU–MA.

57. The Navy called it "stabilized bombing approach equipment," a term it classified Secret. The Air Corps adopted a different term to avoid Navy restrictions. The Air Corps put "automatic flight control equipment" into general use as an automatic pilot on planes with Norden sights, later designating it "C–1 Automatic Pilot."

58. Engrg Sec Memo Rprt R–53–512, Mar 1, 1937; ltr, Col Hugh J. Knerr, Chief of Staff, GHQAF, to CAC, Jun 17, 1937, 1st Ind, OCAC to CG, GHQAF, Jul 17, 1937, 2d Ind, Knerr to CAC, Jul 26, 1937, 1st Wrapper Ind, Exec, OCAC, to Ch, Materiel Div, Aug 3, 1937, 2d Wrapper Ind, Robins to CAC, Aug 21, 1937, in AFHRC 202.2–35, Docs 19, 21–25.

59. Ltr, Lt Col Robert Olds, CO, 2d Bombardment Gp, to CG, 2d Wg, Jul 20, 1938, in AFHRC 202.2–35, Doc 30.

60. 1st Wrapper Ind, Exec, OCAC, to Ch, Materiel Div, Aug 3, 1937, 2d Wrapper Ind, Robins to CAC, Aug 21, 1937; Engrg Sec Memo Rprt R–53–766, Dec 14, 1938, AFHRC 202.2–35, Docs 24–25, 34.

61. The manuals, under continual revision, changed considerably over the years. Compare, for example, the pursuit manual of September 1933 with that for October 1937 and September 1939, in AFHRC 248.101–8.

62. AC Tac Sch, Bombardment Aviation, Jan 1, 1938, para 31 and Table I, in AFHRC 248.101–9.

63. See note 20.

64. See note 20.

65. Report of 2d Wing Bombing Exercise, Dec 13, 1938, in AFHRC WG–2–SU.

66. ACNL, Aug 1, 1936, p 12.

67. See note 20.

68. AC Tac Sch, Bombardment Aviation, Nov 1, 1935, and Jan 1, 1938, paras 32–34 and Sec VII, in AFHRC 248.101–9.

69. ACNL, Nov 1, 1936, p 9.

70. ACNL, May 1, 1937, pp 7–8.

71. ACNL, Nov 15, 1935, p 2.

72. ACNL, Oct 1, 1935, pp 11–12.

73. ACNL, May 1, 1937, p 13.

74. ACNL, Jun 15, 1936, p 9.

75. See note 20.

76. ACNL, Feb 1, 1937, p 14.

77. *Ibid.*

78. ACNL, Oct 1, 1935, pp 5–6.

79. 2d Bombardment Gp, Manual for Bombardiers, Jan 1, 1939, in AFHRC GP–2–SU–MA; 2d Wrapper Ind, Lt Col Robert Olds, CO, 2d Bombardment Gp, to CG, 2d Wg, Jun 6, 1939, in AFHRC WG–2–SU.

80. See note 20.

81. 2d Wrapper Ind, Lt Col Carl W. Connell, CO, 9th Bombardment Gp, to CG, 2d Wg, Aug 19, 1939, in AFHRC WG–2–SU.

82. Ltr, GHQAF, to CG, 2d Wg, May 16, 1939, in AFHRC WG–2–SU.

83. Hist, Army Airways Communications System, Nov 15, 1938–Sep 1945, Vol 1, 29–33, 39–49, 64–91, in AFHRC 380.01.

84. ACNLs, Jul 1, 1937, p 1, Jul 15, 1937, p 12, Oct 15, 1937, p 17.

85. ACNLs, Jul 1, 1937, p 1, Oct 1, 1937, p 11, Mar 1, 1938, pp 3, 11, Jun 1, 1938, p 18, Sep 1, 1938, p 2, May 15, 1939, p 9; Craven and Cate, *The Army Air Forces in World War II*, VII, 313–14.

Chapter XXI

Operations

1. Critique, Third Army CPX, 1936, in AFHRC 284.2123–6; Report of Fourth Army Maneuvers, 1937, Vol I, Sec XV (Report of Air Officer), in AFHRC 248.2123–7A; Folder, Fourth Army CPX, Jul 22–Aug 18, 1939, in AFHRC 168.7032.43.

2. ACNLs, Sep 1, 1936, pp 7, 12–13, 21, Sep 15, 1936, p 13, Oct 1, 1936, p 5, May 15, 1937, pp 1–2.

3. Reports on GHQAF Exercises, May 1937, in AFHRC 248.2122–7; ACNLs, Apr 15, 1937, p 3, May 15, 1937, p 6, Jun 1, 1937, pp 1–4, Jun 15, 1937, pp 5–6, 16–18.

4. ACNLs, Nov 13, 1936, p 3, Apr 1, 1937, p 5, Dec 15, 1937, p 8. The standard in the 5th Bombardment Group in 1938 was 1 mile in 60 and 1.5 minutes in an hour (5th Bombardment Gp, Training Directive, Jul 1, 1938–Jul 1, 1939, in AFHRC GP–5–SU–DI).

5. ACNL, Sep 1, 1936, p 22.

6. ACNL, Oct 1, 1936, p 17.

7. ACNL, Mar 1, 1938, p 6.

8. ACNL, Sep 1, 1936, p 22.

9. ACNL, Mar 15, 1937, pp 7–10 (article by Maj Jasper K. McDuffie).

10. *Ibid.*

11. *Ibid.*

12. *Ibid*; ltr, Maj Jasper K. McDuffie to CO, 2d Bombardment Gp (Report of Panama Flight), Feb 23, 1937, in AFHRC 285.49–2, Vol 2, App 47.

13. ACNLs, Feb 1, 1938, pp 14–15, Aug 15, 1938, p 3; HQ 2d Bombardment Gp, Support Plan, Flight to Santiago, Jan 27, 1939, in AFHRC WG–2–SU–OP Feb 1939.

14. Aircraft Characteristics, Jan 1, 1939; ACNL, Aug 15, 1938, p 3.

15. Ltr, TAG to CG, GHQAF, Sep 9, 1936, in AFHRC 168.7032–34.

16. Ltr, CNO to Comdt, First Naval District, *et al*, Feb 13, 1939, in AFHRC 180.3717, Apr 17–21, 1939.

17. Paragraphs concerning operations on August 12, 13, and 14, 1937, are based on: Ltr, Rear Adm Arthur St. Clair Smith, Comdt, Twelfth Naval District, to CNO, Aug 31, 1937; ltr, Rear Adm Ernest J. King to Senior Comdt Blue (Smith); USS *Utah*, Diary, Coastal Frontier Defense Joint Exercise #4, Aug 12, 13, and 14, 1937; USS *Utah*, Report of Bomb Drops and Hits, U.S. Army Airplanes (Aug 13, 1937), Aug 18, 1937; USS *Utah*, Report of Bomb Drops and Hits, U.S. Army Airplanes (Aug 14, 1937), Aug 19, 1937, in AFHRC 285.49–2, Vol 2, Exhs 52–55; Andrews, Army War College Lecture, Oct 9, 1937, in AFHRC 248.211–62G; ACNLs, Sep 1, 1937, p 9, Sep 15, 1937, p 17; Jones Chronology, Aug 12, 1937.

18. It changed 126° to 127°.

19. Ltr, Smith to CNO, Aug 31, 1937.

20. 1st Lt Earl D. Lyon, Report of Conversation with Haynes, Nov 30, 1944, in AFHRC 420.457, Aug 1937.

21. USS *Utah*, Diary, Coastal Frontier Defense Joint Exercise #4, Aug 13, 1937, in AFHRC 285.49–2, Exh 52.

22. 1st Ind, Maj Gen Frank M. Andrews, CG, GHQAF, to TAG, Aug 26, 1937, quoted in History of Langley Field, Mar 1, 1935 to Dec 7, 1941, Vol 1, 107.

23. ACNL, May 15, 1938, p 7.

24. *New York Herald Tribune*, May 13, 1939, p 1. The clock method of indicating relative bearing, borrowed from artillery, was fairly new in the Air Corps. C. B. Allen explained to readers of the *Herald Tribune* what "twelve o'clock" meant.

25. ACNLs, Jun 1, 1938, pp 9–10, Jun 15, 1938, p 3; Goddard, *Overview*, pp 254–260; George W. Goddard, *Pioneering Years in Aerial Photography* (n.p., 1969), pp 31, 47; *The New York Times*, May 13, 1938, p 3.

26. Arnold, *Global Mission*, p 176.

27. *Ibid*, p 177.

28. See note 26.

29. Memo, A. M. Miller for Maj Gen Earle E. Partridge, CG, 8th AF, Nov 20, 1945, in AFHRC 170.689–1; Jones Chronology, Jun 12, 1938 (citing letter from Lt Gen Ira C. Eaker, Ret, to Col Ethelred L. Sykes, Jan 7, 1949); Greer, *Development of Air Doctrine*, p 91 (citing Greer's interview with Eaker on July 31, 1952). Miss Miller, an employee on the War Department General Staff, made an extensive search at General Partridge's request in 1945 to find documents relating to the 100–mile limit. Gen Otto P. Weyland had another search made in 1946 for Gen Lauris Norstad to reply to Hanson Baldwin of *The New York Times* about the 100–mile limit. The conclusion in both instances was that the Army rather than the Navy imposed this constraint. Documents (AFHRC 170.689–1) collected more recently by Thomas A. Sturm, formerly of the Office of Air Force History, support that conclusion. Also see Army Air Forces Historical Office, Alleged 100–Mile Limitation on Army Aviation in Coast Defense (1931–1939), November 1945, in AFHRC 105–17.

30. Memo, Weyland for Norstad, Jul 22, 1946, in AFHRC 170.689–1.

31. Memo, Norstad for Weyland, Jul 11, 1946, in AFHRC 170–689–1.

32. Memo, Miller for Partridge, Nov 20, 1945, in AFHRC 170.689–1.

33. Unsigned memo (1945) summarizing results of research in documents of the Joint Board, in AFHRC 170.689–1.

34. Ltr, TAG to CG, GHQAF, Sep 1, 1936, extract in AFHRC 170.689–1; ltr, TAG to CG, GHQAF, Sep 9, 1939, in AFHRC 168.7032–34.

35. HQ 4th CA District, Plan for Joint Exercises, Oct 27, 1936; Capt John P. Doyle, Jr., Final Report, Nov 10, 1936, in AFHRC 170.689–1.

36. See note 33.

37. 2d Wg, Annex 1 to FO 2, Oct 28, 1937; ltr, Brig Gen Gerald C. Bryant, CG, 2d Wg, to CG, GHQAF, Dec 1, 1937, in AFHRC 170.689–1.

38. 2d Wg, Annex 3 to FO 1, May 18, 1938, in AFHRC 170.689–1.

39. ACNL, Jul 1, 1938, p 6.

40. ACNLs, Oct 15, 1938, p 15, Apr 1, 1939, pp 6–7.

41. ACNL, Sep 15, 1938, p 8.

42. Resume, Aug 4, 1938, in AFHRC 170.689–1.

43. *The New York Times*, Oct 12, 1938, p 17.

44. 2d Wg Ops Instruc 1, Annex 1, FO 1, Oct 21, 1938; Change 1, Nov 7, 1938, to 2d Wg Ops Instruc 1, Annex 1, FO 1, Oct 21, 1938; ltr, HCD to Exec, OCAC, Oct 28, 1938; msg, Maj Gen Henry H. Arnold, CAC, to CG, GHQAF, Oct 28, 1938; ltr, Exec, OCAC, to TAG, Oct 31, 1938; ltr, Brig Gen Arnold N. Krogstad, CG, 2d Wg, to CG, GHQAF, Nov 18, 1938, in AFHRC 170.689–1.

45. Memo, Col George Brett, Chief of Staff, GHQAF, to Brig Gen George C. Marshall, Dep Ch, WDGS, Jan 12, 1939, in AFHRC 170.689–1.

46. Ltr, TAG to CG, GHQAF, Dec 23, 1938, with encl, ltr, CNO to Comdt, First Naval District; memo, Brett for Marshall, Jan 12, 1939; ltr, Marshall to Brett, Jan 17, 1939, 1st Ind, Brett to TAG, Jan 17, 1939; memo, Brig Gen George V. Strong, Ch, WPD, for CG, GHQAF, Jan 18, 1939, in AFHRC 170.689–1.

47. Bowers, *Boeing Aircraft*, p 248, said the State Department received protests from foreign governments about interception of their ships.

48. Ltr, Brett to TAG, Jan 23, 1939; memo, Maj Gen Robert McC. Beck, Jr., ACS, G–3, for CAC, Feb 16, 1939; memo, Arnold for ACS, G–3, Feb 23, 1939.

49. 1st Ind, TAG to CAC, Mar 16, 1939, in AFHRC 170.689–1.

50. Ltr, Exec, 2d Wg, to CG, GHQAF, Mar 21, 1939; ltr, Exec, OCAC, to TAG, Apr 7, 1939; ltr, Exec, GHQAF, to CG, First Army, May 16, 1939, in AFHRC 170.689–1.

51. Rear Adm William T. Tarrant, Report of Minor Joint Air Exercise No. 1, Apr 17–21, 1939, in AFHRC 180.3717.

52. General Marshall served as Acting Chief of Staff for two months while General Craig was on terminal leave before retiring on August 31.

53. Forrest C. Pogue, *George C. Marshall*, Vol I: *Education of a General, 1880–1939* (New York, 1963), 420, Vol II: *Ordeal and Hope, 1939–1942* (New York, 1966), 85.

54. OCAC Cir 60–1, Aug 24, 1939. Lt Col Ira C. Eaker, as executive, signed the circular which the National Guard Bureau had coordinated and approved. Issued in this fashion, it applied to the Air Corps, GHQ Air Force, Air Corps Reserve, and National Guard, but not to aviation units of the overseas departments. Writing after World War II, Arnold said that as far as he knew the 100–mile limit had not been rescinded. Also note his comment as to a literal interpretation of the 100–mile limit (Arnold, *Global Mission*, p 177).

55. *Congressional Record*, 74th Cong, 1st sess, Jan 3, 1935–Aug 26, 1935, p 4573.

56. Maj Gen Oscar Westover, "Defense Against Air Attack," in ACNL, Feb 1, 1936, pp 1–2; Westover, Talk on Air Defense of the United States (1934), in AFHRC 168.7089–10.

57. AC Tac Sch, Text on Antiaircraft Defense, 1938–1939, in AFHRC 248.2019E.

58. See note 55.

59. ACNLs, Apr 28, 1933, p 99, Jun 30, 1933, pp 153–54, Aug 29, 1933, p 190; *Annual Report of the Chief of the Air Corps*, 1933, p 30; Capt Claire L. Chennault, The Role of Defensive Pursuit, 1933, Article III, in AFHRC 168.7001–114; hist, 1st Pursuit Gp, 1918–1941, in AFHRC GP–1–HI, pp 281–83; *Signal Corps Bulletin* 75, Nov–Dec 1933, pp 1–13; *National Aeronautic Magazine*, Jun 1933, pp 12–13; ltr, Brig Gen Ross G. Hoyt, Ret, to Maurer Maurer, Feb 9, 1981.

60. Air Warning Service at GHQAF Maneuvers of 1937 at Muroc Lake (Incl 7 to ltr, TAG to CG, First Army, May 23, 1940), in AFHRC 410.01, Feb 26, 1940–Jun 2, 1941, App 4; ACNLs, Jun 1, 1937, p 3, Jun 15, 1937, p 6.

61. Summary of Air Warning Service in Southern California Sector Exercise of April

1938 (Incl 8 to ltr, TAG to CG, First Army, May 23, 1940), in AFHRC 410.01, Feb 26, 1940–Jun 2, 1941, App 5.

62. Air Warning Service at GHQAF Maneuvers of 1937 at Muroc Lake, and Summary of Air Warning Service in Southern California Sector Exercise of April 1938, in AFHRC 410.0l, Feb 26, 1940–Jun 2, 1941, Apps 4 and 5.

63. Final Report, Joint Antiaircraft-Air Corps Exercise, Fort Bragg, N.C., 1938 (including reports by Gardner, Bryden, Krogstad, Jarman, Kepner, Craig (CO, 18th Reconnaissance Sq), *et al*, in AFHRC 248.2124–18; ACNLs, Aug 1, 1938, p 7, Oct 15, 1938, pp 4, 14; *The New York Times*, Oct 2, 1938, p 27, Oct 3, 1938, p 10, Oct 9, 1938, p 36, Oct 10, 1938, p 10, Oct 11, 1938, p 24, Oct 12, 1938, p 17, Oct 14, 1938, p 11, Oct 15, 1938, p 9, Oct 16, 1938, p 29, Oct 17, 1938, p 34, Oct 18, 1938, p 26, Oct 19, 1938, p 10.

64. ACNL, Jun 1, 1938, pp 3, 9; 19th Bombardment Gp, Analysis of Bombardment Training, Feb 8, 1939, in AFHRC GP–19–SU, Vol I, 21.

65. Final Report, Joint Antiaircraft-Air Corps Exercise, Fort Bragg, N.C., 1938, Annex 3, App 4, in AFHRC 248.2124–18; ACNL, Oct 15, 1938, p 4; *The New York Times*, Oct 14, 1939, p 11.

66. Harold R. Shaw, "Blackout in Hawaii," *The Reserve Officer* XVI (Oct 1939), 3–5; ACNL, Jun 15, 1939, p 5.

67. ACNL, Nov 1, 1939, p 3.

68. ACNLs, Jan 15, 1935, p 15, Sep 15, 1935, pp 1–2, Sep 15, 1936, p 12, Oct 1, 1936, p 16, Sep 15, 1937, p 1, Sep 1, 1938, p 6.

69. *The New York Times*, Jan 12, 1935, p 7 (which reported the disciplinary action); ACNLs, Jan 15, 1935, p 17, Feb 1, 1935, p 30, Feb 15, 1935, pp 47, 59, 69.

70. *U.S. Air Services* XIX (Oct 1934), 32–33.

71. ACNLs, Feb 1, 1935, p 35, Jan 15, 1936, p 19, Aug 15, 1936, p 6.

72. ACNL, Aug 15, 1936, pp 6–7.

73. ACNL, Sep 15, 1937, p 1; *U.S. Air Services* XXII (Oct 1937), 12–13.

74. ACNLs, Feb 1, 1936, p 2, Feb 15, 1936, pp 20–21; Gordon A. Macdonald, *Volcanoes* (Englewood Cliffs, N.J.: Prentice-Hall, 1972), pp 419–421.

75. ACNLs, Mar 27, 1933, p 62, Oct 31, 1933, p 240; *Annual Report of the Chief of the Air Corps*, 1935, p 20.

76. Ltr, Capt Albert W. Stevens to CAC, Dec 18, 1933 (citing ltr, Feb 25, 1933); lst Ind, Brig Gen Henry C. Pratt to CAC, Jan 2, 1934; 2d Ind, Westover to Chief, Materiel Div, Jan 9, 1934, in AFHRC 168.7006–10; Mark P. Friedlander, Jr., and Gene Gurney, *Higher, Faster, and Farther* (New York: William Morrow and Co, 1973), p 281.

77. *The National Geographic Society—U.S. Army Air Corps Stratosphere Flight of 1934 in the Balloon "Explorer"* [Contributed Technical Papers, Stratosphere Series No. 1] (Washington: National Geographic Society, 1935) which, besides scientific papers, contains Lt Col William E. Kepner's report and an article Capt Albert W. Stevens wrote for the *National Geographic Magazine* of October 1934.

78. *The National Geographic Society—U.S. Army Air Corps Stratosphere Flight of 1935 in the Balloon "Explorer II"* [Contributed Technical Papers, Stratosphere Series No. 2] (Washington: National Geographic Society, 1936), which contains reports by Stevens and Capt Orvil A. Anderson, an article Stevens wrote for the *National Geographic Magazine* of January 1936, and numerous scientific papers. Also see Anderson's personal papers in AFHRC 168.7006–10.

Chapter XXII

Mobilization

1. *Annual Report of the War Department*, 1938, pp 1–2.

2. Mark S. Watson, *Chief of Staff: Prewar Plans and Preparations* [U.S. Army in World War II: The War Department] (Washington, 1950), pp 23–29; Protective Mobilization Plan, Apr 23, 1937, Annex 1, in AFHRC 145.92–249, pt II.

3. The plan withheld the l9th Bombardment Group, with two squadrons, from the initial protective force to reinforce overseas garrisons. The 20th Pursuit Group and 38th Reconnaissance Squadron gave up their people to build other units but formed again by M-day plus 120.

4. Smith in Baker Bd Transcript, pp 4195, 4203; *Annual Report of the Chief of Air Corps*, 1927, p 45; *Annual Report of the Chief of the Air Corps*, 1928, p 48, 1933, p 34; Materiel Division, Airplane Reports, semimonthly, 1928–1939.

5. *Annual Report of the Chief of the Air Corps*, 1932, p 62; ACNLs, Mar 1, 1935, p 75, Mar 15, 1935, p 118, Apr 1, 1935, p 141, May 15, 1936, pp 29–30, Jan 15, 1938, p 15, Oct 15, 1938, p 5.

6. ACNLs, Mar 1, 1935, p 75, Apr 1, 1935, p 146, Jul 15, 1935, pp 5–6, 19–20.

7. ACNL, Oct 1, 1936, pp 7, 9.

8. Report of Fourth Army Maneuvers, 1937, Vol I, Sec XV (Report of Air Officer), in AFHRC 248.2123–7A.

9. Folder, Fourth Army CPX, Jul 22–Aug 18, 1939, in AFHRC 168.7032–43.

10. Baker Bd Transcript, p 4183.

11. *Ibid*, p 4184.

12. *Annual Report of the Chief of the Air Corps*, 1931, p 91.

13. Maj Gen George E. Leach, Chief of the National Guard Bureau, in Baker Bd Transcript, pp 3754–55; memo [draft?], Plans Sec, OCAC, to CSA, Aug 10, 1937, in AFHRC 145.91–340.

14. See, for example, Routing and Record Sheet, Olds to Wilson, OCAC, Feb 27, 1940, in AFHRC 145.91–340.

15. *Annual Report of the Chief of Air Corps*, 1926, p 49; *Annual Report of the Chief of the Air Corps*, 1928, p 49, 1929, pp 49–50, 1930, p 45, 1931, p 52, 1932, p 62; *Annual Report of the War Department*, 1936, pp 91–92; Maj Gen James E. Fechet, Ret, in Baker Bd Transcript, pp 2677–2682.

16. Memo, Fechet, CAC, for ACS, G–1, Nov 22, 1928, in AFHRC 145.93–19; *Annual Report of the Chief of the Air Corps*, 1930, p 41, 1931, pp 49–50.

17. Jones Chronology, Feb 19, 1930, and Oct 15, 1934, citing AC Cir 50–10; ACNL, Apr 1, 1937, p 11; Baker Bd Transcript, p 889; memo, Young for Yount, Aug 21, 1939, in AFHRC 145.93–248.

18. *Annual Report of the Chief of the Air Corps*, 1937, p 21, 1938, p 25, 1939, p 26.

19. *Annual Report of the Chief of the Air Corps*, 1929, pp 46–47, 1930, p 42; Air Corps Strength Report, Jun 30, 1930; OCAC, Study of Reorganization of Air Corps Reserve [1937], in AFHRC 145.93–248.

20. ACNL, Nov 1, 1936, p 9.

21. *Annual Report of the Chief of the Air Corps*, 1934, pp 10, 13–14, 1936, pp 20–21; Air Corps Strength Reports, monthly, Jun 30, 1930–Jun 30, 1937.

22. *Annual Report of the Chief of the Air Corps*, 1928, p 48, 1930, p 42, 1933, pp 33–34; Materiel Division, Airplane Reports, semimonthly, 1929–1936.

23. Baker Bd Transcript, p 2405; *Annual Report of the Chief of the Air Corps*, 1934, p 20.

24. *The New York Times*, May 15, 1932, Sec 9, p 6; Baker Bd Transcript, pp 2405–6.

25. Materiel Division, Airplane Reports, semimonthly, 1936–1939; ACNLs, Oct 1, 1936, p 3, Feb 1, 1937, p 27, Oct 1, 1937, p 6. In August 1939, North American held contracts for forty BC–1As for the Reserve and nineteen more for the National Guard (Materiel Division, Airplane Report, Aug 31, 1939).

26. Baker Bd Transcript, p 2402.

27. *Ibid*, p 882.

28. *Ibid*, p 2442.

29. Charts on Reserve flying, in *Annual Report of the Chief of the Air Corps*, 1931–1939.

30. *Annual Report of the Chief of the Air Corps*, 1930, p 42, 1935, p 22; Baker Bd Transcript, p 2410; *The New York Times*, Oct 23, 1932, p 13.

31. ACNL, Oct 15, 1935, p 14.

32. ACNL, Oct 15, 1936, pp 9, 12.

33. Chennault, The Role of Defense Pursuit, 1933, in AFHRC 168.7001–114; hist, 1st Pursuit Gp, in AFHRC GP–1–HI, Jan 16, 1918–Dec 17, 1937, p 281.

34. ACNL, Nov 15, 1935, p 7.

35. Report of Fourth Army Maneuvers, 1937, Vol I, Sec XV, in AFHRC 248.2123–7A.

36. See, for example, Baker Bd Transcript, p 2429.

37. OCAC, Study on Reorganization of Air Corps Reserve [1937], in AFHRC 145.93–248.

38. ACNL, Oct 15, 1937, p 8.

39. See note 37.

40. Draft memo for CAC [Oct 1939], in AFHRC 145.93–248.

41. Memo (draft by Reserve Division) for CAC, Oct 1939, in AFHRC 145.93–248.

42. Memo, Young for Yount, Aug 21, 1939, in AFHRC 145.93–248.

43. Fourth Army CPX, Jul 22–Aug 18, 1939, in AFHRC 168.7032–43.

44. *The Times*, Oct 1, 1938, p 14.

45. Watson, *Chief of Staff: Prewar Plans and Preparations*, pp 126–143; Franklin D. Roosevelt, *Public Papers and Addresses*, 1939

Volume (London: Macmillan and Co, 1941), pp 70–74; 53 *Stat*, 555, 592.

46. OCAC, 24–Group Program [1939], in AFHRC 168.603–64.

47. OCAC, Airplane Objective, Jan 13, 1939, in AFHRC 145.93–248; ACNLs, May 1, 1939, pp 11–12, Jun 1, 1939, p 15; Materiel Division, Airplane Reports, Apr 15, 1939, May 15, 1939.

48. ACNL, Jun 1, 1939, pp 3, 13.

49. *Ibid*, pp 13–14.

50. *Annual Report of the Chief of the Air Force*, 1937, 1938, 1939, charts on students trained; OCAC, Heavier-than-Air Pilot Training (schedule for classes 40A thru 41D), in AFHRC 145.93–248.

51. ACNLs, May 15, 1939, pp 15–16, Aug 15, 1939, pp 17–20.

52. Robert F. Futrell, *Development of AAF Base Facilities in the United States, 1939–1945* (USAF Hist Study 69, Maxwell AFB, Ala., 1951), pp 13–14; OCAC, weekly memos to SW or CSA on status of construction program, Jul 27–Dec 28, 1939, in AFHRC 145.93–53.

53. 49 *Stat* 610; 53 *Stat* 994, 1327; Futrell, *Development of AAF Base Facilities*, pp 11–12.

54. Craven and Cate, *The Army Air Forces in World War II*, I, 101–193, 313–328, II, 8–27, 80–95, 119–145, 171–192, 263–298, 299–314, 427–453, 454–487, VII, 5–9, 340–45; Watson, *Chief of Staff: Prewar Plans and Preparations.*

Glossary

Glossary

AAF	Army Air Forces
AC	Air Corps
ACM	Air Corps Manual
ACNL	*Air Corps News Letter*
ACS	Assistant Chief of Staff
ACS, G–1	Assistant Chief of Staff, Personnel, War Department General Staff
ACS, G–2	Assistant Chief of Staff, Intelligence, War Department General Staff
ACS, G–3	Assistant Chief of Staff, Operations and Training, War Department General Staff
ACS, G–4	Assistant Chief of Staff, Supply, War Department General Staff
actg	acting
AEF	American Expeditionary Force
AFB	Air Force Base
AFHRC	United States Air Force Historical Research Center
AFROTC	Air Force Reserve Officers' Training Corps
amph	amphibious
app	appendix
AR	Army Regulations
AS	Air Service
AS Cir	Air Service Circular
ASIC	*Air Service Information Circular*
ASNL	*Air Service News Letter*
atch	attachment
atck	attack
Atl	Atlantic
AWC	Army War College
BAP	Bureau of Aircraft Production
bd	board
ca.	about, approximately
CA	Corps Area
CAC	Chief of Air Corps, Chief of the Air Corps
C&GSS	Command and General Staff School
CAS	Chief of Air Service
Cav	Cavalry
CCC	Civilian Conservation Corps
CFTC	Central Flying Training Command
CG	commanding general
ch	chief
chap	chapter
CINC	commander in chief
cmte	committee
CNO	Chief of Naval Operations

co	company
CO	commanding officer
Cong	Congress of the United States
CPX	command post exercise
CSA	Army Chief of Staff (War Department Chief of Staff)
CW	continuous wave
dep	deputy
dept	department
det	detachment
div	division
DMA	Division of Military Aeronautics
DMAWNL	*Division of Military Aeronautics Weekly News Letter*
doc	document
ed	edition, editor
encl	enclosure
engrg	engineering
et al.	and others
exec	executive
exh	exhibit
FA	Field Artillery
FAI	Federation Aeronautique Internationale
FO	field order
ftr	fighter
GHQ	General Headquarters
GHQAF	General Headquarters Air Force
GO	general orders
gp	group
HD	Hawaiian Department
H. Doc	House Document (with number)
hist	historical, history, historian
HQ	headquarters
HR	House of Representatives
H.R.	House Bill (with number)
(I)	inactive
ibid.	*ibidem*, in the same place
ind	indorsement
Inf	Infantry
info	information
instruc	instruction
intvw	interview
JAP	junior airplane pilot
JB	Joint Army and Navy Board
JMA	junior military aviator
Joint Board	A joint Army-Navy board created in 1903 to provide for close cooperation between the services.
KP	kitchen police

LB	light-bombardment
L.I.	Long Island
LR	long-range
ltr	letter
MA	military aviator
memo	memorandum
MIT	Massachusetts Institute of Technology
MS	manuscript
msg	message
n.d.	no date
NG	National Guard
no.	number
n.p.	no place, no publisher
obs	observer(s)
obsn	observation
OCAC	Office of the Chief of Air Corps, Office of the Chief of the Air Corps
OCAS	Office of the Chief of Air Service
OCSO	Office of the Chief Signal Officer
ODAS	Office of the Director of Air Service
opr	operator
ops	operations
OR	Organized Reserve
ORC	Officers' Reserve Corps
Panagra	Pan American Grace Airways
para	paragraph
passim	throughout
PCD	Panama Canal Department
PD	Philippine Department
PDI	pilot direction indicator
pers	personnel
P.I.	Philippine Islands
prov	provision, provisional
PT	primary training
pt	part
pur	pursuit
QMC	Quartermaster Corps
RA	Regular Army
rev ed	revised edition
RMA	Reserve military aviator
ROTC	Reserve Officers' Training Corps
rprt	report
S.	Senate Bill (with number)
sch	school
S. Doc.	Senate Document (with number)
sec	second
ser	serial, series
sess	session
SN	Secretary of the Navy

sq	squadron
stat	U.S. *Statutes at Large*
subj	subject
sup	supply
SW	Secretary of War
tac	tactical
TAG	The Adjutant General
T.H.	Territory of Hawaii
tng	training
USA	United States Army
USAF	United States Air Force
USMA	United States Military Academy
USN	United States Navy
USS	United States Ship
vol	volume
WD	War Department
WDGS	War Department General Staff
WPD	War Plans Division, War Department General Staff

Bibliographic Note

Bibliographic Note

This book is based principally on historical records in the United States Air Force Historical Research Center (AFHRC), formerly the Albert F. Simpson Historical Research Center, at Maxwell Air Force Base, Alabama, and on books and periodicals in the Air University Library at the same place. The Library's 370,000 books and bound periodical volumes include many of use in the study of U.S. Army aviation in the 1920s and 1930s. Especially valuable are the books and magazines which librarian Geraldine Carlisle obtained for the Tactical School during that period and which are now dispersed through the stacks of the Air University Library.

The AFHRC is the U.S. Air Force's chief repository for historical documents. Its archival collection, begun in Washington during World War II as part of the U.S. Army Air Forces' historical program, moved to Maxwell Air Force Base in 1949. There it received the many documents Mrs. Carlisle had assembled and maintained as part of the Tactical School Library, first at Langley Field, Virginia, and afterwards at Maxwell Field, Alabama. Since then the Center's collection has burgeoned, and includes much excellent material pertaining to the years between the First and Second World Wars.

These repositories afford sufficient materials for a work of the kind originally conceived. Others may wish to delve deeper into national policy; political, social, and economic factors; public opinion; technology; relationships between military and civil aviation; or conflicts between the Army and the Navy and between Army airmen and the General Staff—all mentioned but none treated in depth in this book. They should extend their research to the National Archives, the Library of Congress, and other repositories of published and documentary materials.

Bibliographies

The best bibliography of U.S. Air Force History is Robert F. Futrell's essay in Robin Higham, ed, *A Guide to the Sources of United States Military*

History (Hamden, Conn.: Archon Books, 1975), with Robert T. Finney's continuation in *Supplement I* (1981). The Office of Air Force History, United States Air Force, has published a number of useful guides. The following were compiled by staff members of that office in Washington: Samuel Duncan Miller, *An Aerospace Bibliography* (1978); Lawrence J. Paszek, *United States Air Force History: A Guide to Documentary Sources* (1973); and Jacob Neufeld, *United States Air Force History: A Guide to Monographic Literature, 1943–1974* (1977), which lists studies written in the historical offices throughout the Air Force. Three other guides, issued jointly by the Office of Air Force History and the Albert F. Simpson Historical Research Center, were the work of the Center's staff: *U.S. Air Force Oral History Catalog* (1982), listing interviews held by the Center; Richard E. Morse and Thomas C. Lobenstein, *Personal Papers in the Albert F. Simpson Historical Research Center* (1980); and Richard E. Morse, *et al*, *Numbered USAF Historical Studies Prepared by the Albert F. Simpson Historical Research Center*, 3d ed (1981), which in addition to studies written in the Center and its predecessors, includes some from the historical office of the Air Materiel Command (now Air Force Logistics Command).

Two works published by the Air University Library are helpful for periodical literature: Helen J. Hopewell, ed, *Union List of Military Periodicals* (1960); and Raymond Estep's five increments of *An Air Power (Aerospace) Bibliography* (1956–1967), containing references to periodical literature from 1950 through 1966. Less useful for this work was the Air University Library's quarterly *Air University Periodical Index*. August Hanniball, *Aircraft, Engines and Airmen: A Selective Review of Periodical Literature, 1930–1969* (Metuchen, N.J.: Scarecrow Press, 1972), is worth consulting.

Anyone interested in the history of aviation in the 1920s and 1930s will find many useful references in *Bibliography of Aeronautics, 1909–1932*, 14 vols (Washington: National Advisory Committee for Aeronautics, 1921–1936); *Bibliography of Aeronautics*, 50 pts (New York: U.S. Public Works Administration, 1936–1940); and William B. Gamble, *History of Aeronautics: A Selected List of References to Materials in the New York Public Library* (New York: New York Public Library, 1938), containing more than 5,500 entries citing U.S. and foreign books and periodicals.

Reference Works

The best available chronology of American aviation for the years between the First and Second World Wars is that published annually in *The Aircraft Year Book* (New York: Manufacturers Aircraft Association, 1919–1921; Aeronautical Chamber of Commerce of America, 1922–), which also gives more detailed coverage of major events and developments, and contains data on aviation records. The U.S. Air Force's *A Chronology of American Aerospace Events*, Air Force Pamphlet 190–2–2, Vol II, which went through several editions before being terminated, was very thin and suffered from poor editing. The vast, detailed, and extremely valuable chronology to which Ernest L. Jones (the U.S. Army air arm's first historian) devoted many years, exists only in manuscript in the AFHRC. It is useful both for the historical data and the references it contains. Moreover, it is supported by the large body of historical material assembled by Colonel Jones for his use.

A number of published works furnish biographical data of various sorts on officers of the Army's air arm: The *Official Army Register*, published annually by the War Department, contains details on the military status of Regular Army officers; the *Army List and Directory*, issued periodically by the War Department, shows where each officer on active duty was stationed at a particular time; the *Official National Guard Register*, issued by the Militia (National Guard) Bureau, supplies information on National Guard officers; The Adjutant General's *Army Directory, Reserve and National Guard Officers on Active Duty, July 31, 1941* (1941), helps with identification of both Reservists and Guardsmen of the twenties and thirties. George W. Cullum's monumental *Biographical Register of the Officers and Graduates of the U.S. Military Academy*, 3d ed, 3 vols (Boston: Houghton Mifflin Co, 1891) and *Supplement*, Vols IV–VIII (various publishers, 1901–1940) provides data on promotions and assignments of West Point graduates. Flint O. DuPre, *U.S. Air Force Biographical Dictionary* (New York: Franklin Watts, 1965), is based chiefly on biographical sketches prepared by the Air Force for release to the press. Especially useful for information about American aviators of the early 1920s is Lester D. Gardner, *Who's Who in American Aeronautics*, 2d ed (New York: The Gardner Publishing Co, 1925). Lists of key personnel of the War Department and General Staff are found in *The Army Almanac: A Book of Facts Concerning the Army of the United States* (Washington: Department of the Army, 1950).

From the many works dealing with aircraft, these are especially useful for reference: Frederick G. Swanborough and Peter M. Bowers, *United States Military Aircraft Since 1909* (London and New York: Putnam, 1963); *Janes' All the World's Aircraft* (1909–), annually, with various editors and publishers; and James C. Fahey, *U.S. Army Aircraft (Heavier-than-Air), 1908–1946* (New York: Ships and Aircraft, 1946).

Dictionaries of military terms are indispensable to a study of this kind, the following being particularly helpful: Edward S. Farrow, *A Dictionary of Military Terms*, rev ed (New York: Thomas Y. Crowell Co, 1918); Max B. Garber, *A Modern Military Dictionary* (Washington: Garber, 1936); Woodford Agee Heflin, ed, *The United States Air Force Dictionary* (Maxwell Air Force Base, Ala.: Research Studies Institute, Air University, 1956); and

Elbridge Colby, *Army Talk: A Familiar Dictionary of Soldier Speech* (Princeton: Princeton University Press, 1942).

Brief historical sketches of U.S. Army squadrons, groups, and wings active during the period between world wars may be found in Maurer Maurer, ed, *Air Force Combat Units of World War II* (Maxwell Air Force Base, Ala.: USAF Historical Division, Air University, 1961, reprinted by Franklin Watts, 1963, Arno Press, 1979, and Zenger Publishing Co, 1980); and in Maurer Maurer, ed, *Combat Squadrons of the Air Force, World War II* (Maxwell Air Force Base, Ala.: USAF Historical Division, Air University, 1969). Information on many of the air bases of the 1920s and 1930s is available in Robert Mueller, *Air Force Bases, Vol I, Active Air Force Bases within the United States of America on 1 January 1974* (Maxwell Air Force Base, Ala: Research Division, Albert F. Simpson Historical Research Center, 1982).

General Works

Maurice Matloff, ed, *American Military History* [Army Historical Series] (Washington: Office of the Chief of Military History, 1969), and Russell F. Weigley, *History of the United States Army* (New York: Macmillan Publishing Co, 1967), have excellent chapters on the Army in the 1920s and 1930s. The best summaries of Army aviation between wars are by James L. Cate and Wesley F. Craven in the first volume of Craven and Cate, eds, *The Army Air Forces in World War II*, 7 vols (Chicago: University of Chicago Press, 1948–1958); by Alfred Goldberg in Goldberg, ed, *A History of the United States Air Force, 1907–1957* (Princeton: D. Van Nostrand Co, 1957); and by Thomas E. Mackin in an AFROTC textbook, *U.S. Air Power: Ascension to Prominence* (Maxwell Air Force Base, Ala., 1974). Others include those by Herbert M. Mason, Jr., *The United States Air Force: A Turbulent History* (New York: Mason-Charter, 1976), and Carroll V. Glines, *The Compact History of the United States Air Force* (New York: Hawthorn Books, 1963). Monro MacCloskey gave six pages to the interwar period in *The United States Air Force* (New York: Frederick A. Praeger, 1967), and two in a book entitled *From Gasbags to Spaceships: The Story of the U.S. Air Force* (New York: Richards Rosen Associates, 1968). An illustrated capsule history of the period is available in *The Official Pictorial History of the AAF*, prepared by the Historical Office of the Army Air Forces (New York: Duell, Sloan and Pearce, 1947). *The American Heritage History of Flight* edited by Alvin M. Josephy, Jr. (New York: American Heritage, 1962) contains excellent illustrations of aviation in the twenties and thirties.

Biographical and Autobiographical Works

A few Army aviators have told of their service in the 1920s and 1930s in published memoirs. Henry H. Arnold took with him into retirement a large collection of documents which he had assembled to write his autobiography but which he did not always consult to verify his memory before publishing *Global Mission* (New York: Harper and Brothers, 1949), which nonetheless is of great interest and value. Arnold deserves a much fuller biography than that written by Flint O. DuPre, *Hap Arnold, Architect of American Air Power* (New York: The Macmillan Co, 1972). Thomas M. Coffey, *Hap: The Story of the U.S. Air Force and the Man Who Built It, General Henry H. "Hap" Arnold* (New York: The Viking Press, 1982) is a popular biography but not a definitive one.

Persons interested in Arnold will not be wasting their time if they take a look at the stories Arnold wrote for his young son, William Bruce Arnold. The six volumes in this series of boys' books are titled *Bill Bruce and the Pioneer Aviators, Bill Bruce the Flying Cadet, Bill Bruce Becomes an Ace, Bill Bruce on Border Patrol, Bill Bruce in the Transcontinental Race, and Bill Bruce on Forest Patrol* (New York: A. L. Burt, 1928).

Benjamin D. Foulois, who spent much effort during a long lifetime defending himself, chiefly his actions during World War I and as Chief of the Air Corps, continued in the same vein in *From the Wright Brothers to the Astronauts: The Memoirs of Major General Benjamin D. Foulois* (New York: McGraw-Hill Book Co, 1968), written with the assistance of Carroll V. Glines three decades after retirement. An objective view of Foulois' service as Chief of the Air Corps, 1931–1935, has been provided by John F. Shiner in *Foulois and the U.S. Army Air Corps, 1931–1935* (Washington: Office of Air Force History, 1983).

Claire L. Chennault's *Way of a Fighter: The Memoirs of Claire Lee Chennault* (New York: G. P. Putnam's Sons, 1949) reflects the bitterness that long survived his unhappy experience in Army aviation in the 1930s. Donald Wilson's *Wooing Peponi: My Odyssey Through Many Years* (Monterey, Calif.: Angel Press, 1973) proved of little use in this particular work. Other autobiographies of value for the period between wars include Curtis E. LeMay and MacKinlay Kantor, *Mission with LeMay: My Story* (Garden City, N.Y.: Doubleday and Co, 1965); George W. Goddard (with Dewitt C. Copp), *Overview: A Lifelong Adventure in Aerial Photography* (Garden City, N.Y.: Doubleday and Co, 1969); Goddard, *Pioneering Years in Aerial Photography* (a scrapbook of various photocopied materials, especially newspaper clippings, issued in a limited edition, no publisher indicated, 1969); Norris B. Harbold, *The Log of Air Navigation* (San Antonio: The Naylor Co, 1970).

Charles A. Lindbergh told of his days as a flying cadet and as a young Reserve officer in *We* (New York: G. P. Putnam's Sons, 1927), as did biographers Kenneth S. Davis, in *The Hero: Charles A. Lindbergh and the American Dream* (Garden City, N.Y.: Doubleday and Co, 1959), and Walter S. Ross in *The Last Hero: Charles A. Lindbergh* (New York: Harper and Row, 1976). Beirne Lay, Jr., recounted his experiences as a cadet and a lieutenant on active duty in the early 1930s in *I Wanted Wings* (New York: Harper and Brothers, 1937). Lester J. Maitland's *Knights of the Air* (Garden City, N.Y.: Doubleday, Doran, and Co, 1929) contains an autobiographical chapter about his historic flight from California to Hawaii with 1st Lt. Albert F. Hegenberger. Albert W. Jernberg, an Air Corps Reserve officer called to duty during the Great Depression, told of his assignment with the Civilian Conservation Corps in *My Brush Monkeys: A Narrative of the CCC* (New York: Richard R. Smith, 1941). Edward V. Rickenbacker gave his view of the airmail controversy of 1931 in *Rickenbacker: An Autobiography* (Englewood Cliffs, N.J.: Prentice-Hall, 1967).

Supplements thus far published to the *Dictionary of American Biography* (New York: Charles Scribner's Sons, 1928–) provide biographical sketches of Henry H. Arnold, William Mitchell, and Mason M. Patrick, as well as Newton D. Baker, Hugh A. Drum, and John J. Pershing. Few Army airmen of the twenties and thirties have been the subjects of full-length biographies. The best studies of William Mitchell are those by Alfred F. Hurley and Burke Davis. Hurley's work, *Billy Mitchell: Crusader for Air Power* (New York: Franklin Watts, 1964), new edition (Bloomington, Ind.: Indiana University Press, 1975) deals mainly with Mitchell's ideas. His book grew out of a doctoral dissertation, "The Aeronautical Ideas of General William Mitchell" (Princeton University, 1961). Davis' *The Billy Mitchell Affair* (New York: Random House, 1967) affords excellent coverage of the court-martial, the author having full access to the records. Mitchell's biographers include Isaac D. Levine, *Mitchell, Pioneer of Air Power*, rev ed (New York: Duell, Sloan and Pearce, 1958); Roger Burlingame, *General Billy Mitchell: Champion of Air Defense* (New York: McGraw-Hill Book Co, 1952); and Ruth Mitchell, *My Brother Bill: The Life of General "Billy" Mitchell* (New York: Harcourt, Brace and Co, 1953).

James H. Doolittle has attracted several biographers, including Carroll V. Glines with two books, *Jimmy Doolittle: Daredevil Aviator and Scientist* (New York: The Macmillan Co, 1972), and *Jimmy Doolittle: Master of the Calculated Risk* (New York: Van Nostrand Reinhold Co, 1980); Carl Mann, *Lightning in the Sky: The Story of Jimmy Doolittle* (New York: McBride, 1944); and Quentin Reynolds, *The Amazing Mr. Doolittle: A Biography of Lieutenant General James H. Doolittle* (New York: Appleton-Century-Crofts, 1953). Infinitely superior to any of these is the biography by Lowell Thomas and Edward Jablonski, *Doolittle* (Garden City, N.Y.: Doubleday and Co, 1976).

Would-be biographers will find no end of subjects. John H. Scrivner's biography of Orvil A. Anderson, "Pioneer into Space" (Ph.D. dissertation, University of Oklahoma, 1971) still leaves room for a more detailed study of Anderson's work with airships and balloons.

In *A Few Great Captains: The Men and Events that Shaped the Development of U.S. Air Power* (Garden City, N.Y.: Doubleday and Co, 1980), DeWitt S. Copp followed Arnold, Andrews, Eaker, and Spaatz through the years preceding World War II. Those who would like to follow Copp to see where he had been will regret the almost complete absence of footnotes. Still, anyone interested in U.S. military aviation can read his story with great enjoyment and no little profit.

Biographies of key Army officials can be useful in connection with Army aviation in the 1920s and 1930s. These include Clarence H. Cramer, *Newton D. Baker* (Cleveland: World Publishing Co, 1961); Frank E. Vandiver, *Black Jack: The Life and Times of John J. Pershing*, 2 vols (College Station, Tex.: Texas A&M University Press, 1977); D. Clayton James, *Years of MacArthur: 1880–1941*, Vol I (Boston: Houghton Mifflin Co, 1970); Forrest C. Pogue, *George C. Marshall*, Vol I: *Education of a General, 1880–1939* (New York: Viking Press, 1963), Vol II: *Ordeal and Hope, 1939–1942* (New York: Viking Press, 1966), and Vol III: *Organizer of Victory, 1943–1945* (New York: Viking Press, 1973). The Chief of Staff's view of demobilization and Army reorganization in 1919–1920 is presented briefly by Peyton C. March, *The Nation at War* (Garden City, N.Y.: Doubleday, Doran, and Co, 1932).

Books and Monographs

Army airmen who wrote on aviation during the period between world wars include Mason M. Patrick, *The United States in the Air* (Garden City, N.Y.: Doubleday, Doran, and Co, 1928); Clayton Bissell, *Brief History of the Air Corps and Its Late Development* (Fort Monroe, Va.: Coast Artillery School Press, 1928); and Henry H. Arnold and Ira C. Eaker, *This Flying Game* (New York: Funk and Wagnalls Co, 1936). Better known and more influential are William Mitchell's *Our Air Force, the Keystone of National Defense* (New York: E. P. Dutton and Co, 1921); *Winged Defense: The Development and Possibilities of Modern Air Power— Economic and Military* (New York: G. P. Putnam's Sons, 1925); and *Skyways: A Book of Modern Aeronautics* (Philadelphia: J. P. Lippincott Co, 1930).

Air Corps officers and civilian employees contributed, or collaborated on, books in an aeronautical series published in New York by The Ronald Press Co: Henry H. Arnold, *Airmen and Aircraft: An Introduction to Aeronautics* (1926); Byron Q. Jones, *Practical Flying: A Training Manual for Airplane Pilots* (1928); Ralph H. Upson and Charles deF. Chandler, *Free and*

Captive Balloons (1926); Charles deF. Chandler and Walter S. Diehl, *Balloon and Airship Gases* (1926); Donald Duke, *Airports and Airways* (1927); and William C. Sherman, *Air Warfare* (1926). The last grew out of Major Sherman's earlier work, "Air Tactics," prepared in 1921 for the Air Service Field Officers School, and a classic Air Service text on air doctrine.

Bradley Jones wrote on *Avigation* (New York: John Wiley and Sons, 1931), Ashley C. McKinley on *Applied Aerial Photography* (New York: John Wiley and Sons, 1929), and William C. Ocker and Carl J. Crane on *Blind Flying in Theory and Practice* (San Antonio: The Naylor Co, 1932), all of which are books of great value in tracing developments in particular aspects of aviation. Harry E. Wimperis, the British naval officer who developed a bombsight used by the U.S. Army during and after World War I, included an interesting passage on that subject in his book, *Aviation* (London: Oxford University Press, 1945). Some other books on specialized topics pertinent to the present work include Louis Shores, *Highways in the Sky*: *The Story of the AACS* (New York: A. S. Barnes and Co (Cranbury, N.J.), 1947); Monte D. Wright, *Most Probable Position* (Lawrence, Kans.: The University Press of Kansas, 1972); Harry G. Armstrong, *Principles & Practice of Aviation Medicine* (Baltimore: Williams and Wilkins, 1952); Green Peyton [Wertenbaker], *50 Years of Aerospace Medicine* (Brooks Air Force Base, Tex.: Aerospace Medical Division, 1968); and Walter W. Ristow, *Aviation Cartography*: *A Historic-Bibliographic Study of Aeronautical Charts*, 2d rev ed (Washington: The Library of Congress, 1960).

Of the numerous works on airplanes, the multivolume series, *Aircraft in Profile*, Martin C. Windrow, ed (Garden City, N.Y.: Doubleday and Co, 1965–) is among the best. Edward Jablonski has an excellent book on the B–17s, *Flying Fortress*: *The Illustrated Biography of the B–17s and the Men Who Flew Them* (Garden City, N.Y.: Doubleday and Co, 1965), and Peter M. Bowers one on *Boeing Aircraft Since 1916* (London: Putnam, 1966). Kimbrough Brown and E. F. Heyn, *United States Army and Air Force Fighters, 1916–1961* (Fallbrook, Calif.: Aero Publishers, 1961), is useful as is Reed Kinert, *Racing Planes and Air Races*, 4 vols (Fallbrook, Calif.: Aero Publishers, 1969).

The development of Army air bases is well told by Jerold E. Brown, "Where Eagles Roost: A History of Army Airfields before World War II" (Ph.D. dissertation, Duke University, 1977). The Work Projects Administration's book on Randolph Field (New York: Devin-Adair Co, 1942) furnishes some good information about that base and the activities thereon, while "The History of Randolph Air Force Base" is the subject of an M.A. thesis (University of Texas, 1958) by Rossi L. Selvaggi. Charles R. Rowdybush wrote a thesis on the "History of Bolling Field, Anacostia, D.C., 1917–1948" (American University, 1957), and Carl W. Ryan a dissertation on "Modern Scott: A History of Scott Air Force Base" (St. Louis University, 1969). R.

Bruce Harley and associates in the history office at Headquarters Fifteenth Air Force published *The March Field Story, 1918–1978* (March Air Force Base, Calif., 1968), while Robert I. Curtis, *et al*, prepared a well-illustrated history of *Langley Field, The Early Years, 1916–1948* (Langley Air Force Base, Va.: 4500th Air Base Wing, 1977).

Turning to operations, one finds two excellent monographs about activities on the Mexican border by a person with first-hand experience: Stacy C. Hinkle, *Wings and Saddles: The Air and Cavalry Punitive Expedition of 1919* (Southwestern Studies, Monograph 19, El Paso: Texas Western Press, 1967), and *Wings Over the Border: The Army Air Service Armed Patrol of the United States-Mexican Border, 1919–1921* (Southwestern Studies, Monograph 26, El Paso: Texas Western Press, 1970). For the border patrol, Clarence C. Clendenen, *Blood on the Border: The United States Army and the Mexican Irregulars* (London: Collier-Macmillan, 1969) is also useful.

Paul Tillett has written an excellent monograph on the Air Corps airmail operation, *The Army Flies the Mail* (Tuscaloosa, Ala.: University of Alabama Press, 1955). Less valuable but still useful is Norman E. Borden, Jr., *Air Mail Emergency, 1934* (Freeport, Maine: Bond Wheelwright Co, 1938). Also useful is Francis A. Spencer, *Air Mail Payment and the Government* (Washington: Brookings Institution, 1941). Carroll V. Glines gives the Air Corps airmail operation his usual light treatment in *The Saga of the Air Mail* (Princeton: D. Van Nostrand Co, 1968).

Eldon W. Downs' Ph.D. dissertation (University of Wisconsin, 1959) on "Contributions of U.S. Army Aviation to Uses and Operation of Aircraft" deals with airmail, crop dusting, forest patrol, parachutes, night flying, airways, photography, radio, and navigation aids. Ray L. Bowers covered "The Transcontinental Reliability Test: American Aviation After World War I" (M.A. thesis, University of Wisconsin, 1960). Works concerning great flights include Clarence D. Chamberlin, *Record Flights* (Philadelphia: Dorrance and Co, 1928); Marian T. Place, *New York to Nome: The First International Cross- Country Flight* (New York: Macmillan Publishing Co, 1972); William J. Horvat, *Above the Pacific* (Fallbrook, Calif.: Aero Publishers, 1966), and Lowell Thomas, *The First World Flight* (Boston: Houghton Mifflin Co, 1925), in which the author, who interviewed the flyers, permits the participants to tell most of the story. Don Vorderman wrote on *The Great Air Races* (Garden City, N.Y.: Doubleday and Co, 1969). Douglas J. Ingells' work on test pilots, *They Tamed the Sky: The Triumph of American Aviation* (New York: D. Appleton-Century Co, 1947) contains but a few brief references to the period before World War II.

James H. Doolittle's experimental work on blind flying is covered in detail in two papers issued by the Daniel Guggenheim Fund for the Promotion of Aeronautics: *Solving the Problem of Fog Flying* (1929); and *Equipment Used in Experiments to Solve the Problem of Fog Flying* (1930).

The National Georgraphic Society published two technical papers in 1935 and 1936 on the stratosphere flights: *The National Geographic Society—U.S. Army Air Corps Stratosphere Flight of 1934 in the Balloon "Explorer,"* and *The National Geographic Society—U.S. Army Air Corps Stratosphere Flight of 1935 in the Balloon "Explorer II."*

Two studies of Army appropriations and one on government finance provide significant background concerning the funding of Army aviation: William F. Willoughby, *Financial Condition and Operations of the National Government, 1921–1930* (Washington: The Brookings Institution, 1931); John W. Killigrew, "The Impact of the Great Depression on the Army, 1929–1936" (Ph.D. dissertation, Indiana University, 1960); Elias Huzar, *The Purse and the Sword: Control of the Army by Congress through Military Appropriations, 1933–1950* (Ithaca, N.Y.: Cornell University Press, 1950). Irving B. Holley, Jr., *Buying Aircraft: Materiel Procurement for the Army Air Forces* [United States Army in World War II: Special Studies] (Washington: Office of the Chief of Military History, 1964) discusses Air Corps procurement policies and practices.

Legislative studies include Guido R. Perera, *A Legislative History of Aviation in the United States and Abroad for the Period 1907–1940* (Washington: Office of the Chief of the Air Corps, 1941), and Harry H. Ransom, "The Air Corps Act of 1926: A Study of the Legislative Process" (Ph.D. dissertation, Princeton University, 1954), the latter dealing with political aspects of the Mitchell affair. Edwin Rutkowski explored *The Politics of Military Aviation Procurement, 1926–1934: A Study in the Political Assertion of Consensual Values* (Columbus: Ohio State University Press, 1966). Army policy on aviation between world wars is the subject of James P. Tate's "The Army and Its Air Corps" (Ph.D. dissertation, Indiana University, 1976). Organization was treated by R. Earl McClendon, *Autonomy of the Air Arm*, rev ed (Maxwell Air Force Base, Ala.: Air University, 1954), and Curtis W. Tarr, "Unification of America's Armed Forces: A Century and a Half of Conflict, 1798–1947" (Ph.D. dissertation, Stanford University, 1962). Doctrinal studies include Lester H. Brune, "The Foundations of American Air Power—Aviation and National Defense, 1919–1933" (Ph.D. dissertation, University of Rochester, 1959); Raymond R. Flugel, "United States Air Power Doctrine: A Study of the Influence of William Mitchell and Giulio Douhet on the Air Corps Tactical School, 1921–1935" (Ph.D. dissertation, University of Oklahoma, 1965); and Thomas A. Fabyanic, *Strategic Air Attack in the United States Air Force: A Case Study* (Manhattan, Kans.: Military Affairs/Aerospace Historian, 1977).

The following are useful in connection with Reserve components: Eilene Galloway, *History of United States Military Policy on Reserve Forces, 1775–1957* (Washington: Library of Congress, 1957); Jim D. Hill, *The Minute Man in Peace and War: A History of the National Guard* (Harrisburg,

Pa.: Stackpole Books, 1964); Elbridge Colby, *The National Guard of the United States: A Half Century of Progress* (Manhattan, Kans.: Military Affairs/Aerospace Historian, 1977); Charles J. Gross, *Prelude to the Total Force: Air National Guard, 1943–1969* (Washington: Office of Air Force History, 1984); and Russell Stompler, "The Origin and Growth of the Reserve Officers' Training Corps, 1916–1950" (Masters thesis, University of Alabama, 1951).

The best work on demobilization after World War I is that authored by Benedict Crowell and Robert F. Wilson, *Demobilization: Our Industrial and Military Demobilization after the Armistice, 1918–1920* (New Haven: Yale University Press, 1921), this being the fourth volume in the series, *How America Went to War.* Preparations begun in the late 1930s for World War II are covered by Mark S. Watson, *Chief of Staff: Prewar Plans and Preparations* [U.S. Army in World War II: The War Department] (Washington: Historical Division, United States Army, 1950); Stetson Conn, Rose C. Engelman, and Byron Fairchild, *Guarding the United States and Its Outposts* [U.S. Army in World War II: The Western Hemisphere] (Washington: Office of the Chief of Military History, United States Army, 1964); and Volume I: *Plans and Early Operations, January 1939 to August 1942* of Craven and Cate, *The Army Air Forces in World War II* (cited under General Works).

Magazines, Journals, and Newspapers

Contemporary aviation magazines and journals are useful not only for news stories but for signed articles narrating and discussing events or presenting personal views and opinions. Such articles are often the work of persons intimately associated with the things they wrote about; some are first-hand accounts of historic events. The advertising often affords interesting and valuable information. Other contemporary magazines and journals occasionally carry articles about military aviation, the following being the more important: *Aerial Age Weekly*; *Aero Digest*; *Airports*; *The Annals of the American Academy of Political and Social Science*; *Atlantic Monthly*; *Aviation*; *Bulletin of the American Meteorological Society*; *The Coast Artillery Journal*; *Colliers*; *The Independent*; *Infantry Journal*; *Journal of the Franklin Institute*; *Literary Digest*; *The Nation*; *National Aeronautical Association Review*, later *Aeronautical Review*, *National Aeronautic Magazine*, *National Aeronautics*; *National Geographic Magazine*; *The New Republic*; *The Outlook*; *Popular Aviation*; *Review of Reviews*; *Saturday Evening Post*; *Scientific American*; *Slipstream*; *U.S. Air Service(s)*; *The World's Work.*

Most recent magazines and journals containing articles on specialized topics include: *The Air Power Historian*, later *Aerospace Historian*; *Air*

University Quarterly Review, later *Air University Review*; *American Aviation Historical Society Journal*; *Military Affairs*; *Smithsonian Annals* of *Flight*.

The *Army and Navy Journal*, *The New York Times*, and the *News Letter* of the Division of Military Aeronautics, Air Service, and Air Corps (title varies) are very useful for the study of military aviation in the years between the wars. The last of these is of special value. The best and most interesting issues of the *News Letter* are those of the period following the Armistice, when Horace M. Hickam, Air Service chief of information, took what he got from correspondents in the field and published it with little or no editing. The *News Letter* lost some of its color and something of its historical value after it became Arnold's responsibility in 1925. Nevertheless, a complete file furnishes an important record of events and constitutes a valuable source on U.S. Army aviation in the 1920s and 1930s. Publication on microfilm (Scholarly Resources, Incorporated, Wilmington, Delaware) makes the *News Letter* through 1935 readily available for research; lacking an index, it is almost useless for reference.

Government Documents

Published government documents constitute a major source for the study of Army aviation. Legislative items include the *Congressional Record*, hearings and reports of joint committees of Congress and of Senate and House committees on military affairs and appropriations. Related documents are the *United States Statutes at Large*, the *Opinions of the Attorney General*, and the compilation of *Military Laws of the United States* issued at intervals by the Office of the Judge Advocate General of the Army.

Published proceedings of investigative bodies include the *Hearings* and *Report* of the President's Aircraft Board (Morrow Board), 1925; *Final Report* of the War Department Special Committee on Army Air Corps (Baker Board), 1934; and *Report* of the Federal Aviation Commission (Howell Commission), 1935.

The official views of key War Department and Army officials are found in *War Department Annual Reports*, which contain reports by the Secretary and Assistant Secretary of War, the Chief of Staff, and the chiefs of the various arms and services, including the Air Service/Air Corps. The following publications of various departments and agencies concern Army aviation in one way or another:

War Department: Army Regulations; Field Service Regulations; Field Manuals; and General Orders.

Air Service/Air Corps: Circulars; *Aeronautical Bulletins, State Series*; *Aeronautical Bulletins, Route Information Series.*

Signal Corps: *Bulletins.*

Joint Army and Navy Board: *Joint Action.*

Post Office Department, Airmail Service: *Pilots' Directions, New York–San Francisco Route* (1921).

Department of Commerce: *Air Commerce Bulletins.*

National Advisory Committee for Aeronautics: *Annual Reports*; *Technical Reports.*

U.S. Air Force Historical Records

The United States Air Force Historical Research Center holds three large collections of documents of great value for the study of Army aviation between World Wars I and II. One is the Ernest L. Jones collection, mentioned above under Reference Works. Another is the Tactical School collection, containing materials of three broad classes (though not so identified and filed): administration; curriculum; and research and reference. The second, concerning the school's curriculum, consists of texts, manuals, lectures, faculty studies, and student reports, all being primary materials for the study of airpower ideas, concepts, and doctrines. The third class, made up of research and reference material, offers papers on nearly every aspect of U.S. Army aviation in the twenties and thirties. The last of the three major collections referred to above contains papers of the Army Air Forces' War Plans Division, which include project files, plans, studies, working papers, memorandums, correspondence, and other materials originally in the Air Service/Air Corps plans section.

Other documents scattered among the files of the AFHRC include War Department orders, regulations, and manuals, Air Service/Air Corps circulars, correspondence, staff studies, unit histories, base histories, tables of organization, organization charts, strength reports, biographical data, financial information, station lists, airplane inventories, aircraft record cards, airplane characteristics, engine characteristics, equipment manuals, test reports, accident investigation reports and (as indicated below) personal papers, oral histories, and historical studies, all of which contributed to this work.

Personal Papers

Many persons associated with military aviation have deposited personal papers in the AFHRC. In addition the Center holds microfilm of personal papers in other repositories. Collections containing information about the 1920s and 1930s include those of Orvil A. Anderson, Herbert A. Dargue, Muir S. Fairchild, Benjamin D. Foulois, Ernest L. Jones, William E. Kepner, Arno H. Luehman, Erik H. Nelson, Clifford C. Nutt, Henry C. Pratt, Ralph Royce, Oscar Westover, Thomas D. White, Ennis C. Whitehead, and John F. Whiteley.

Oral Histories

Though generally preferring contemporary accounts to reminiscences, the author found the AFHRC's oral history collection useful. Among the interviews with information about the 1920s and 1930s are those of Glenn O. Barcus, Orval R. Cook, Laurence C. Craigie, Howard C. Davidson, James H. Doolittle, Ira C. Eaker, Barney M. Giles, Albert F. Hegenberger, George C. Kenney, Laurence S. Kuter, Noel F. Parrish, Earle E. Partridge, Donald L. Putt, Frederick H. Smith, Jr., Joseph Smith, Carl Spaatz, Mrs. Carl Spaatz, Leigh Wade, and Donald Wilson.

AAF/USAF Historical Studies

The studies listed below are products of the historical programs of the U.S. Army Air Forces and the U.S. Air Force. Numbered studies issued by the AAF or USAF Historical Division include works produced by the Air Technical Service Command (later Air Materiel Command) as well as volumes prepared by the division's staff. The unnumbered studies in the list below are products of historical offices of various commands and subordinate units. Studies of both classes appeared in limited editions (some typewritten, some printed locally), principally for Air Force use. Other studies, printed in larger editions, either by the U.S. Government Printing Office or commercially, appear in this bibliographic note under other headings.

Numbered Studies

Claussen, Martin P. *Comparative History of Research and Development Policies Affecting Air Materiel, 1915–1944*. Study 20. 1945.

____________. *Materiel Research and Development in the Army Air Arm, 1914–1945*. Study 50. 1946.

Cohen, Victor H. *Classification and Assignment of Enlisted Men in the Army Air Arm, 1917–1945*. Study 76. 1953.

Dubuque, Jean H., and Gleckner, Robert F. *The Development of the Heavy Bomber, 1918 to 1944*. rev ed. Study 6. 1951.

Finney, Robert T. *History of the Air Corps Tactical School, 1920–1940*. Study 100. 1955.

Fogerty, Robert P. *Biographical Data on Air Force General Officers, 1917–1952*. 2 vols. Study 91. 1953.

Futrell, Robert F. *Command of Observation Aviation: A Study in Control of Tactical Air Power*. Study 24. 1952.

———. *Development of Aeromedical Evacuation in the USAF, 1909–1960*. Study 23. 1961.

———. *Development of AAF Base Facilities in the United States, 1939–1945*. Study 69. 1951.

———. *Ideas, Concepts, Doctrine: A History of Basic Thinking in the USAF, 1907–1964*. rev ed. Study 139. 1974.

Greer, Thomas H. *The Development of Air Doctrine in the Army Air Arm, 1917– 1941*. Study 89. 1953.

Holley, Irving B., Jr. *Development of Aircraft Gun Turrets in the AAF, 1917–1944* . Study 54. 1947.

———. *Evolution of the Liaison-Type Airplane, 1917–1944*. Study 44. 1946.

Layman, Martha E. *Legislation Relating to the Air Corps Personnel and Training Programs, 1907–1939*. Study 39. 1945.

Mooney, Chase C., and Layman, Martha E. *Organization of Military Aeronautics, 1907–1935 (Congressional and War Department Action)*. Study 25. 1944.

Mooney Chase C., and Williamson, Edward C. *Organization of the Army Air Arm, 1935–1945*. rev ed. Study 10. 1956.

Thompson, Robert L. *Barrage Balloon Development in the United States Army Air Corps, 1923 to 1942*. Study 3. 1943.

Walters, Raymond. *Weather Training in the AAF, 1937–1945*. Study 56. 1951.

Unnumbered Studies

Ackerman, Robert W. *The Maintenance of Army Aircraft in the United States, 1921–1939*. Patterson Field, Ohio: Air Technical Service Command, 1945.

Air Technical Service Command, Historical Office. *Case History of Quick Release Parachutes, March 1931–May 1945*. Patterson Field, Ohio: Air Technical Service Command, 1945.

Air University. *Fifty Years of Aviation History at Maxwell Air Force Base, 1910– 1960*. Maxwell Air Force Base, Ala.: Air University, 1960.

Bagwell, Margaret C. *Case History of Camouflage Paint, November 1917–October 1945*. 2 pts. Patterson Field, Ohio: Air Materiel Command, 1947. *N&———. *Case History of S–1 Bombsight and A–5 Automatic Pilot, January 1936– May 1945*. 2 vols. Patterson Field, Ohio: Air Technical Service Command, 1945.

Baker, Doris. *History of AMC Field Organization, 1917–1955*. Wright-Patterson Air Force Base, Ohio: Air Materiel Command, 1956.

Brown, Genevive. *Development of Transport Airplanes and Air Transport Equipment, 1917–1945.*Patterson Field, Ohio: Air Technical Service Command, 1946.

Bruce, Dorothy H. *The Evolution of the Storage System of the Air Technical Service Command, Part I, 1918–1940.*Patterson Field, Ohio: Air Technical Service Command, 1945.

Canham, Doris A. *Development and Production of Fighter Aircraft for the United States Air Force, 1911– 1949*. Wright-Patterson Air Force Base, Ohio: Air Materiel Command, 1949.

Frey, Royal D. *Evolution of Maintenance Engineering, 1907–1920*. 4 vols. Wright-Patterson Air Force Base, Ohio: Air Materiel Command, 1960.

Myers, Robert M. *History of the USAF Cataloging System, 1918–1951*. 3 vols. Wright-Patterson Air Force Base, Ohio: Air Materiel Command, 1953.

Parr, Esther, and Poage, Nancy. *Case History of Norden Bombsight and C–1 Automatic Pilot, March 1933– December 1944*. 2 pts. Patterson Field, Ohio: Air Technical Service Command, 1945.

Purtee, Edward O. *Development and Adaptation of Aircraft Instruments for Military Use, 1917–1945.* Patterson Field, Ohio: Air Technical Service Command, 1946.

———. *Development of AAF Clothing and Other Personal Equipment Peculiar to Air Operations, 1942–1944.* 3 vols. Patterson Field, Ohio: Air Technical Service Command, 1945.

———. *The Development of Light and Medium Bombers, 1918–1946.* Patterson Field, Ohio: Air Materiel Command, 1946.

———. *History of the Army Air Service, 1907– 1926.* Wright-Patterson Air Force Base, Ohio: Air Materiel Command, 1948.

———. *The Modification and Development of Training Aircraft for AAF Use, 1918–1945.* Patterson Field, Ohio: Air Materiel Command, 1946.

Self, Mary R. *History of the Development and Production of Heavy Bombardment Aircraft, 1917–1949.* Wright-Patterson Air Force Base, Ohio: Air Materiel Command, 1950.

———. *Reconnaissance Aircraft and Aerial Photographic Equipment, 1915–1945.* Patterson Field, Ohio: Air Materiel Command, 1946.

Toole, Virginia G. *Development of Bombing Equipment in the Army Air Forces, 1915– 1945.* Patterson Field, Ohio: Air Technical Service Command, 1945.

———. *The Development and Procurement of Parachutes for the Army Air Forces, 1920–1947.* 3 vols. Wright-Patterson Air Force Base, Ohio: Air Materiel Command, 1948.

Wampler, Ruth. *Thirty Year History of Hamilton AFB, California, 1934–1964.* Hamilton Air Force Base, Calif.: 28th Air Division, 1964.

Index

Index

INDEX

INDEX

INDEX

INDEX

INDEX

INDEX

INDEX

INDEX